Introduction to Fluorescence Sensing

Alexander P. Demchenko

Introduction to
Fluorescence Sensing

 Springer

Alexander P. Demchenko
Palladin Institute of Biochemistry,
National Academy of Sciences of Ukraine
9 Leontovich street
Kiev 01030
Ukraine

ISBN 978-1-4020-9002-8   e-ISBN 978-1-4020-9003-5

Library of Congress Control Number: 2008936100

Printed on acid-free paper

springer.com

Preface

The field of molecular sensing is immense. It is nearly the whole world of natural and synthetic compounds that have to be analyzed in a variety of conditions and for a variety of purposes. In the human body, we need to detect and quantify virtually all the genes (genomics) and the products of these genes (proteomics). In our surrounding there is a need to analyze a huge number of compounds including millions of newly synthesized products. Among them, we have to select potentially useful compounds (e.g., drugs) and discriminate those that are inefficient and harmful. No less important is to control agricultural production and food processing. There is also a practical necessity to provide control in industrial product technologies, especially in those that produce pollution. Permanent monitoring is needed to maintain the safety of our environment. Protection from harmful microbes, clinical diagnostics and control of patient treatment are the key issues of modern medicine. New problems and challenges may appear with the advancement of human society in the XXI century. We have to be ready to meet them.

Modern society needs the solution of these problems on the highest possible scientific and technological level. The science of intermolecular interactions is traditionally a part of physical chemistry and molecular physics. Now it becomes a strongly requested background for modern sensing technologies. The most specific and efficient sensors are found in the biological world and the sensors based on biomolecular recognition (biosensors) have acquired a strong impulse for development and application. A strong move is observed for improving them by endowing new features or even by making fully synthetic analogs of them. Modern electronics and optics make their own advance in providing the most efficient means for supplying the sensors with the input and output signals and now becomes oriented at satisfying the needs of not only researchers but a broad community of users.

This book is focused on one sensing technology, which is based on fluorescence. This is not only because of limited space or limited expertise of the present author. Indeed, fluorescence techniques are the most sensitive; their sensitivity reaches the absolute limit of single molecules. They offer very high spatial resolution; that with overcoming the light diffraction means the limit has reached molecular scale. They are also the fastest; their response develops on the scale of fluorescence lifetime and can be as short as 10^{-8}–10^{-10} s. However, their greatest advantage is versatility. Fluorescence sensing can be provided in solid, liquid and gaseous media and at all kinds of interfaces between these phases. It is because the fluorescence reporter and the detecting instrument are connected via light emission that fluorescence detection can be made non-invasive and equally well suited for remote industrial control and for sensing different targets within the living cells. All these features explain their high popularity.

The fascinating field of fluorescence sensing needs new brains. Therefore, this book is primarily addressed to students and young scientists. Together with a basic knowledge they will obtain an overview of different ideas in research and technology and will be guided in their own creative activity. Providing a link between the basic sciences needed to understand sensor performance and the frontiers of research, where new ideas are explored and new products developed, this book will make a strong link between research and education. For the active researcher it will also be a source of useful information in nearly all areas where fluorescence sensing is used.

Thus, this book is organized with the aim to satisfy both curious student and busy researchers. After a short introduction, a comparative analysis of basic principles used in fluorescence sensing will be made. We then provide a formal description of binding equilibrium and binding kinetics that are the background to sensing technologies. The focus will be on techniques of obtaining information from fluorescent reporters and on analysis of their structures and properties. The design of various types of recognition units will be reviewed, including those selected from large libraries. A deeper understanding of the basic mechanism of signal transduction in fluorescence sensing will be our focus, with special attention paid to the new possibilities provided by support structures, scaffolds and integrated units that expand the range of sensor applications. Non-conventional generation and transformation of response signals will also be described. Fluorescence sensing is realized with optical instrumentation, so these devices are overviewed, including microarrays, microfluidics and flow cytometry. Detection of different targets from physical, chemical and biological worlds is discussed, with many examples presented. We will also address the analytical means of detecting different targets inside the living cells based on modern microscopy. Finally, the frontiers of modern research are overviewed with the prospects for fluorescence sensing behind the horizon. Each chapter is terminated by the section "Sensing and thinking", in which, after a short summary, a series of questions and exercises is suggested for the reader.

Enjoy your reading.

Alexander P. Demchenko

Contents

Preface v

Introduction xxi

1 Basic Principles 1

   1.1 Overview of Strategies in Molecular Sensing 1
      1.1.1 Basic Definitions: Sensors and Assays, Homogeneous
             and Heterogeneous 2
      1.1.2 Principles of Sensor Operation 6
      1.1.3 Label-Free, General Approaches 7
      1.1.4 Label-Free, System-Specific Approaches................ 9
      1.1.5 Label-Based Approaches 10
   1.2 Labeled Targets in Fluorescence Assays 12
      1.2.1 Arrays for DNA Hybridization 13
      1.2.2 Labeling in Protein-Protein and Protein-Nucleic Acid
             Interactions 14
      1.2.3 Micro-Array Immunosensors................................. 14
      1.2.4 Advantages and Limitations of the Approach Based
             on Pool Labeling 14
   1.3 Competitor Displacement Assay 15
      1.3.1 Unlabeled Sensor and Labeled Competitor in
             Homogeneous Assays 16
      1.3.2 Labeling of Both Receptor and Competitor............. 18
      1.3.3 Competition Involving Two Binding Sites 19
      1.3.4 Advantages and Limitations of the Approach 19
   1.4 Sandwich Assays 20
      1.4.1 Sensing the Antigens and Antibodies 21
      1.4.2 Ultrasensitive DNA Detection Hybridization Assays............ 23
      1.4.3 Advantages and Limitations of the Approach 24
   1.5 Catalytic Biosensors 24
      1.5.1 Enzymes as Sensors... 24
      1.5.2 Ribozymes and Deoxyribozymes Sensors............... 25
      1.5.3 Labeling with Catalytic Amplification 26

1.5.4 Advantages and Limitations of the Approach   27
1.6 Direct Reagent-Independent Sensing   28
        1.6.1 The Principle of Direct 'Mix-and-Read' Sensing..................   28
        1.6.2 Contact and Remote Sensors   29
        1.6.3 Advantages and Limitations of the Approach   31
Sensing and Thinking 1: How to Make the Sensor?
Comparison of Basic Principles   32
Questions and Problems   32
References   34

2 Theoretical Aspects   37
    2.1 Parameters that Need to Be Optimized in Every Sensor   38
            2.1.1 The Limit of Detection and Sensitivity.......................   38
            2.1.2 Dynamic Range of Detectable Target Concentrations...........   39
            2.1.3 Selectivity   40
    2.2 Determination of Binding Constants   42
            2.2.1 Dynamic Association-Dissociation Equilibrium   42
            2.2.2 Determination of K_b by Titration   44
            2.2.3 Determination of K_b by Serial Dilutions...................   47
    2.3 Modeling the Ligand Binding Isotherm   48
            2.3.1 Receptors Free in Solution or Immobilized to a Surface........   48
            2.3.2 Bivalent and Polyvalent Reversible Target Binding   49
            2.3.3 Reversible Binding of Ligand and Competitor..................   51
            2.3.4 Interactions in a Small Volume...............................   54
    2.4 Kinetics of Target Binding..   55
    2.5 Formats for Fluorescence Detection.....................................   57
            2.5.1 Linear Format   57
            2.5.2 Intensity-Weighted Format   59
Sensing and Thinking 2: How to Provide the Optimal
Quantitative Measure of Target Binding   61
Questions and Problems   62
References   63

3 Fluorescence Detection Techniques   65
    3.1 Intensity-Based Sensing   66
            3.1.1 Peculiarities of Fluorescence Intensity Measurements..........   67
            3.1.2 How to Make Use of Quenching Effects   68
            3.1.3 Quenching: Static and Dynamic   69
            3.1.4 Non-linearity Effects...   71
            3.1.5 Internal Calibration in Intensity Sensing   71
            3.1.6 Intensity Response as a Choice for Fluorescence Sensing   75
    3.2 Anisotropy-Based Sensing and Polarization Assays......................   76
            3.2.1 Background of the Method   77

	3.2.2	Practical Considerations	78
	3.2.3	Applications	80
	3.2.4	Comparisons with Other Methods of Fluorescence Detection	83
3.3	Lifetime-Based Fluorescence Response	83
	3.3.1	Physical Background	84
	3.3.2	Technique	85
	3.3.3	Time-Resolved Anisotropy	85
	3.3.4	Applications	86
	3.3.5	Extension to Reporter Response Based on Phosphorescence	87
	3.3.6	Comparison with Other Fluorescence Detection Methods	88
3.4	Excimer and Exciplex Formation	88
	3.4.1	Application in Sensing Technologies	89
	3.4.2	Comparison with Other Fluorescence Reporter Techniques	90
3.5	Förster Resonance Energy Transfer (FRET)	91
	3.5.1	Physical Background of the Method	91
	3.5.2	FRET Modulated by Light	94
	3.5.3	Applications of FRET Technology	96
	3.5.4	FRET to Non-fluorescent Acceptor	98
	3.5.5	Comparison with Other Detection Methods	99
3.6	Wavelength-Shift Sensing	99
	3.6.1	The Physical Background Behind the Wavelength Shifts	100
	3.6.2	The Measurements of Wavelength Shifts in Excitation and Emission	102
	3.6.3	Wavelength-Ratiometric Measurements	103
	3.6.4	Application in Sensing	104
	3.6.5	Comparison with Other Fluorescence Reporting Methods	105
3.7	Two-Band Wavelength-Ratiometric Sensing with a Single Dye	106
	3.7.1	Generation of a Two-Band Ratiometric Response by Ground-State Isoforms	106
	3.7.2	Excited-State Reactions Generating a Two-Band Response in Emission	107
	3.7.3	Excited-State Intramolecular Proton Transfer (ESIPT)	109
	3.7.4	Prospects for Two-Band Ratiometric Recording	111
Sensing and Thinking 3: The Choice of Fluorescence Detection Technique and Optimization of Response	112
Questions and Problems	112
References	114

4   Design and Properties of Fluorescence Reporters 119

4.1   Organic Dyes 119
      4.1.1   General Properties of Fluorescence Reporter Dyes 120
      4.1.2   Dyes for Labeling 123
      4.1.3   The Dyes Providing Fluorescence Response 129
      4.1.4   The Environment-Sensitive (Solvatochromic) Dyes 130
      4.1.5   Hydrogen Bond Responsive Dyes 132
      4.1.6   Electric Field Sensitive (Electrochromic) Dyes 134
      4.1.7   Supersensitive Multicolor Ratiometric Dyes 136
      4.1.8   The Optimal FRET Pairs 141
      4.1.9   Phosphorescent Dyes and the Dyes with
              Delayed Fluorescence 141
      4.1.10 Combinatorial Discovery and Improvement of
              Fluorescent Dyes 143
      4.1.11 Prospects 143
4.2   Luminescent Metal Complexes 144
      4.2.1   Structure and Spectroscopy of Complexes
              of Lanthanide Ions 145
      4.2.2   Lanthanide Chelates as Labels and Reference Emitters 148
      4.2.3   Dissociation-Enhanced Lanthanide Fluoroimmunoassay
              (DELFIA) 149
      4.2.4   Switchable Lanthanide Chelates 149
      4.2.5   Transition Metal Complexes that Exhibit
              Phosphorescence 151
      4.2.6   Metal-Chelating Porphyrins 152
      4.2.7   Prospects 153
4.3   Dye-Doped Nanoparticles and Dendrimers 154
      4.3.1   The Dye Concentration and Confinement Effects 154
      4.3.2   Nanoparticles Made of Organic Polymer 156
      4.3.3   Silica-Based Nanoparticles 158
      4.3.4   Dendrimers 159
      4.3.5   Applications of Dye-Doped Nanoparticles in Sensing 160
      4.3.6   Summary and Prospects 162
4.4   Semiconductor Quantum Dots and Other Nanocrystals 162
      4.4.1   The Properties of Quantum Dots 163
      4.4.2   Stabilization and Functionalization of Quantum Dots 165
      4.4.3   Applications of Quantum Dots in Sensing 166
      4.4.4   Nanobeads with Quantum Dot Cores 169
      4.4.5   Porous Silicon and Silicon Nanoparticles 169
      4.4.6   Other Fluorescent Nanocrystal Structures 170
      4.4.7   Prospects 171
4.5   Noble Metal Nanoparticles and Molecular Clusters 171
      4.5.1   Light Absorption and Emission by Noble
              Metal Nanoparticles 172

4.5.2   Preparation and Stabilization 173
4.5.3   Gold and Silver Nanoparticles as Fluorescence
        Quenchers 173
4.5.4   Nanoparticles and Molecular Clusters as Emitters 174
4.5.5   Metal Nanoclusters 174
4.5.6   Prospects 176
4.6   Fluorescent Conjugated Polymers 176
4.6.1   Structure and Spectroscopic Properties 177
4.6.2   Possibilities for Fluorescence Reporting in
        Sensor Design 179
4.6.3   Nanocomposites Based on Conjugated Polymers 181
4.6.4   Prospects 181
4.7   Visible Fluorescent Proteins 182
4.7.1   Green Fluorescent Protein (GFP) and Its
        Colored Variants 182
4.7.2   Labeling and Sensing Applications of
        Fluorescent Proteins 184
4.7.3   Other Fluorescent Proteins 184
4.7.4   Finding Simple Analogs of Fluorescent Proteins 185
4.7.5   Prospects 185
Sensing and Thinking 4: Which Reporter to Choose for
Particular Needs? 185
Questions and Problems 186
References 188

5   Recognition Units 197
5.1   Recognition Units Built of Small Molecules 197
5.1.1   Crown Ethers, Cryptands, Polyhydroxilic and
        Boronic Acid Derivatives 198
5.1.2   Cyclodextrins 200
5.1.3   Calixarenes 203
5.1.4   Porphyrins 206
5.1.5   Dendrimers 207
5.1.6   Prospects 208
5.2   Antibodies and Their Recombinant Fragments 209
5.2.1   The Types of Antibodies Used in Sensing 209
5.2.2   The Assay Formats Used for Immunoassays 211
5.2.3   Prospects for Antibody Technologies 212
5.3   Ligand-Binding Proteins and Protein-Based Display Scaffolds 213
5.3.1   Engineering the Binding Sites by Mutations 213
5.3.2   Bacterial Periplasmic Binding Protein (PBP) Scaffolds 215
5.3.3   Engineering PBPs Binding Sites and the Response
        of Environment-Sensitive Dyes 216
5.3.4   Scaffolds Based on Proteins of the Lipocalin Family 217

          5.3.5   Other Protein Scaffolds 218
          5.3.6   Prospects 219
   5.4   Designed and Randomly Synthesized Peptides 219
          5.4.1   Randomly Synthesized Peptides,
                   Why They Do Not Fold? 220
          5.4.2   Template-Based Approach 221
          5.4.3   The Exploration of the 'Mini-Protein' Concept 221
          5.4.4   Molecular Display Including Phage Display 223
          5.4.5   Antimicrobial Peptides and Their Analoges 224
          5.4.6   Advantages of Peptide Technologies and Prospects
                   for Their Development 225
   5.5   Nucleic Acid Aptamers 225
          5.5.1   Selection and Production of Aptamers 226
          5.5.2   Attachment of Fluorescence Reporter, Before or
                   After Aptamer Selection? 226
          5.5.3   Obtaining a Fluorescence Response and Integration
                   into Sensor Devices 228
          5.5.4   Aptamer Applications 231
          5.5.5   Comparison with Other Binders: Prospects 232
   5.6   Peptide Nucleic Acids 233
          5.6.1   Structure and Properties 233
          5.6.2   DNA Recognition with Peptide Nucleic Acids 234
   5.7   Molecularly Imprinted Polymers 236
          5.7.1   The Principle of the Formation of an
                   Imprinted Polymer 236
          5.7.2   The Coupling with Reporting Functionality 237
          5.7.3   Applications 238
   Sensing and Thinking 5: Selecting the Tool for Optimal
   Target Recognition 238
   Questions and Problems 239
   References 240

6 Mechanisms of Signal Transduction 249

   6.1   Basic Photophysical Signal Transduction Mechanisms 250
          6.1.1   Photoinduced Electron Transfer (PET) 250
          6.1.2   Intramolecular Charge Transfer (ICT) 254
          6.1.3   Excited-State Proton Transfer 259
          6.1.4   Prospects 260
   6.2   Signal Transduction via Excited-State Energy Transfer 261
          6.2.1   Directed Excited-State Energy Transfer in
                   Multi-fluorophore Systems 262
          6.2.2   Light-Harvesting (Antenna) Effect 264
          6.2.3   Peculiarities of FRET with and Between
                   Nanoparticles 266

6.2.4 The Optimal Choice of FRET Donors 267
6.2.5 Lanthanides as FRET Donors 267
6.2.6 Quantum Dots as FRET Donors 269
6.2.7 The Optimal Choice of FRET Acceptors 269
6.2.8 Prospects 270
6.3 Signal Transduction via Conformational Changes 271
6.3.1 Excited-State Isomerism in the Reporter Dyes and
       Small Molecules 272
6.3.2 Conformational Changes in Conjugated Polymers 273
6.3.3 Conformational Changes in Peptide Sensors
       and Aptamers 273
6.3.4 Molecular Beacons 276
6.3.5 Proteins Exhibiting Conformational Changes 278
6.3.6 Prospects 281
6.4 Signal Transduction via Association and Aggregation
     Phenomena 282
6.4.1 Association of Nanoparticles on Binding a
       Polyvalent Target 282
6.4.2 Association-Induced FRET and Quenching 283
6.5 Integration of Molecular and Digital Worlds 284
6.5.1 The Direct Recording of Digital Information
       from Molecular Sensors 284
6.5.2 Hybrid Molecular-Digital Systems 285
6.5.3 Logical Operations with Fluorescent Dyes 286
6.5.4 Prospects 289
Sensing and Thinking 6: Coupling Recognition and
Reporting Functionalities 289
Questions and Problems 290
References 291

7 Supramolecular Structures and Interfaces for Sensing 299
7.1 Building Blocks for Supramolecular Sensors 299
7.1.1 Carbon Nanotubes 299
7.1.2 Core-Shell Compositions 300
7.1.3 Polynucleotide Scaffolds 301
7.1.4 Peptide Scaffolds 301
7.2 Self-Assembled Supramolecular Systems 302
7.2.1 Affinity Coupling 303
7.2.2 Self-Assembly 304
7.2.3 Two-Dimensional Self-Assembly of S-Layer Proteins 306
7.2.4 Template-Assisted Assembly 307
7.2.5 Micelles: The Simplest Self-Assembled Sensors 308
7.2.6 Prospects 311
7.3 Conjugation, Labeling and Cross-linking 312

7.3.1   Conjugation and Labeling.. 312
7.3.2   Co-synthetic Modifications... 313
7.3.3   Chemical and Photochemical Cross-linking........................ 314
7.4   Supporting and Transducing Surfaces 314
7.4.1   Surfaces with a Passive Role: Covalent Attachments............ 314
7.4.2   Self-Assembled Monolayers.. 316
7.4.3   Langmuir-Blodgett Films 318
7.4.4   Layer-by-Layer Approach 320
7.4.5   Prospects 321
7.5   Functional Lipid Bilayers 322
7.5.1   Liposomes as Integrated Sensors....................................... 323
7.5.2   Stabilized Phospholipid Bilayers.. 325
7.5.3   Polymersomes.. 326
7.5.4   Formation of Protein Layers over Lipid Bilayers 326
7.5.5   Prospects 327
Sensing and Thinking 7: Extended Sensing Possibilities
with Smart Nano-ensembles.. 327
Questions and Problems 328
References 329

8   Non-Conventional Generation and Transformation of Response 335

8.1   Chemiluminescence and Electrochemiluminescence........................ 335
8.1.1   Chemiluminescence... 336
8.1.2   Enhanced Chemiluminescence 337
8.1.3   Electrochemiluminescence 338
8.1.4   Cathodic Luminescence... 339
8.1.5   Solid-State Electroluminescence 341
8.1.6   Essentials of the Techniques and Their Prospects 342
8.2   Bioluminescence.. 342
8.2.1   The Origin of Bioluminescence.. 343
8.2.2   Genetic Manipulations with Luciferase............................... 343
8.2.3   Bioluminescence Resonance Energy Transfer....................... 344
8.2.4   Prospects 345
8.3   Two-Photon Excitation, Up-Conversion and Stimulated
Emission 345
8.3.1   Two-Photon and Multi-Photon Fluorescence 345
8.3.2   Up-Conversion Technique with Nanocrystals
Possessing Lanthanine Guests.. 348
8.3.3   Sensors as Lasers and Lasers as Sensors 350
8.4   Direct Generation of the Electrical Response Signal 352
8.4.1   Light-Addressable Potentiometric Sensors (LAPS) 352
8.4.2   Photocells as Sensors... 353
8.5   Evanescent-Wave Fluorescence Sensors 354

8.5.1   Excitation by the Evanescent Field.. 354
8.5.2   Applications in Sensing 356
8.6   Plasmonic Enhancement of Emission Response 357
8.6.1   Surface Plasmon-Field Enhanced Fluorescence.................... 358
8.6.2   Enhancement of Dye Fluorescence Near
        Metal Nanoparticles 359
8.6.3   Application of Metal-Nanoparticle Enhancement................. 363
8.6.4   Microwave Acceleration of Metal-Enhanced Emission 364
8.6.5   Prospects.. 365
Sensing and Thinking 8: Eliminating Light Sources and
Photodetectors: What Remains?.. 366
Questions and Problems 366
References 367

9   The Sensing Devices... 371

9.1   Instrumentation for Fluorescence Spectroscopy............................... 372
9.1.1   Standard Spectrofluorimeter... 373
9.1.2   Light Sources 374
9.1.3   Light Detectors 376
9.1.4   Passive Optical Elements.. 377
9.1.5   Integrated Systems.. 378
9.1.6   Prospects 379
9.2   Optical Waveguides, Optodes and Surface-Sensitive Detection 380
9.2.1   Optical Fiber Sensors with Optode Tips................................ 381
9.2.2   Evanescent-Field Waveguides 382
9.3   Multi-Analyte Sensor Chips and Microarrays................................... 383
9.3.1   Fabrication 384
9.3.2   Problems with Microarray Performance................................ 385
9.3.3   Read-Out and Data Analysis 386
9.3.4   Applications of Microarrays 387
9.3.5   Prospects 387
9.4   Microsphere-Based Arrays... 388
9.4.1   Barcodes for Microsphere Suspension Arrays 389
9.4.2   Reading the Information from Microparticles........................ 390
9.4.3   Prospects 390
9.5   Microfluidic Devices 391
9.5.1   Fabrication and Operation of a Lab-on-a-Chip...................... 391
9.5.2   Microfluidic Devices as Microscale Reactors
        and Analytical Tools... 393
9.5.3   Fluorescence Detection in Microfluidic Devices................... 394
9.5.4   Prospects 395
9.6   Devices Incorporating Whole Living Cells 396
9.6.1   Cellular Microorganisms or Human Cultured Cell Lines?..... 396

9.6.2   Living and Fixed Cells 397
9.6.3   Single Cells in Microfluidic Devices 398
9.6.4   Bacterial Cells with Genetically Incorporated Sensors 399
9.6.5   The Cultured Human Cells 400
9.6.6   Whole Cell Arrays 400
9.6.7   Prospects 400
Sensing and Thinking 9: Optimizing Convenience,
Sensitivity and Precision to Obtain the Proper Sensor Response 401
Questions and Problems 401
References 402

10   Focusing on Targets 407

10.1   Temperature, Pressure and Gas Sensing 407
          10.1.1   Molecular Thermometry 408
          10.1.2   Molecular Barometry 410
          10.1.3   Sensors for Gas Phase Composition 410
10.2   Probing the Properties of Condensed Matter 411
          10.2.1   Polarity Probing in Liquids and Liquid Mixtures 412
          10.2.2   Viscosity and Molecular Mobility Sensing 413
          10.2.3   Probing Ionic Liquids 417
          10.2.4   The Properties of Supercritical Fluids 418
          10.2.5   The Structure and Dynamics in Polymers 420
          10.2.6   Fluorescence Probing the Interfaces 421
10.3   Detection of Small Molecules and Ions 422
          10.3.1   pH Sensing 422
          10.3.2   Oxygen 423
          10.3.3   Heavy Metals 425
          10.3.4   Glucose 425
          10.3.5   Cholesterol 427
10.4   Nucleic Acid Detection and Sequence Identification 427
          10.4.1   Detection of Total Double-Stranded DNA 427
          10.4.2   Detection of Single-Stranded DNA and RNA 429
          10.4.3   Sequence-Specific DNA Recognition 429
          10.4.4   'DNA Chip' Hybridization Techniques 430
          10.4.5   Sandwich Assays in DNA Hybridization 434
          10.4.6   Molecular Beacon Technique 435
          10.4.7   DNA Sensing Based on Conjugated Polymers 436
          10.4.8   Concluding Remarks and Prospects 437
10.5   Recognition of Protein Targets 438
          10.5.1   Total Protein Content 438
          10.5.2   Specific Protein Recognition 438
          10.5.3   Protein Arrays 439
10.6   Polysaccharides, Glycolipids and Glycoproteins 441
10.7   Detection of Harmful Microbes 442

10.7.1 Detection and Identification of Bacteria............................ 442

10.7.2 Bacterial Spores.. 444

10.7.3 Detection of Toxins 444

10.7.4 Sensors for Viruses 444

10.7.5 Conclusions and Prospects 445

Sensing and Thinking 10: Adaptation of Sensor Units for a
Multi-scale and Hierarchical Range of Targets 445

Questions and Problems 446

References 447

11 Sensing Inside Living Cells and Tissues.. 455

11.1 Modern Fluorescence Microscopy 455

11.1.1 Epi-fluorescence Microscopy 457

11.1.2 Total Internal Reflection Microscopy 458

11.1.3 Confocal Microscopy 460

11.1.4 Two-Photon and Three-Photon Microscopy...................... 461

11.1.5 Time-Gated and Time-Resolved Imaging 463

11.1.6 Breaking the Diffraction Limit: Near-Field
Microscopy 464

11.1.7 Stimulated Emission Depletion Microscopy
in Breaking the Diffraction Limit...................................... 465

11.1.8 Considerations on the Problem of Photobleaching 466

11.1.9 Critical Comparison of the Techniques 468

11.2 Sensing on a Single Molecule Level 468

11.2.1 Single Molecules in Sensing 469

11.2.2 Detection of Single Molecules Inside the Living Cells..... 472

11.2.3 Fluorescence Correlation Spectroscopy and
Microscopy 474

11.2.4 Additional Comments.. 475

11.3 Site-Specific Intracellular Labeling and Genetic Encoding 476

11.3.1 Attachment of a Fluorescent Reporter to
Any Cellular Protein.. 476

11.3.2 Genetically Engineered Protein Labels 478

11.3.3 The Co-synthetic Incorporation of
Fluorescence Dyes... 481

11.3.4 Concluding Remarks 482

11.4 Advanced Nanosensors Inside the Cells 482

11.4.1 Fluorescent Dye-Doped Nanoparticles in
Cell Imaging 482

11.4.2 Quantum Dots Applications in Imaging............................ 483

11.4.3 Self-Illuminating Quantum Dots 485

11.4.4 Extending the Range of Detection Methods...................... 485

11.5 Sensing the Cell Membrane 486

11.5.1 Lipid Asymmetry and Apoptosis....................................... 486

11.5.2   Sensing the Membrane Potential 487
11.5.3   Membrane Receptors 489
11.5.4   Future Directions 489
11.6   Molecular Recognitions in the Cell's Interior 490
11.6.1   Ion Sensing 490
11.6.2   Tracking Cellular Signaling.. 492
11.6.3   Location of Metabolites and Tracking
          Metabolic Events 494
11.6.4   In Situ Hybridization 494
11.6.5   Looking Forward 494
11.7   Sensing the Whole Body 495
11.7.1   Optimal Emitters for the Human Body 495
11.7.2   Contrasting the Blood Vessels 496
11.7.3   Imaging Cancer Tissues.. 497
11.7.4   Surgical Operations Under the Control of
          Fluorescence Image: Fantasy or Close Reality?............... 498
Sensing and Thinking 11: Intellectual and Technical Means
to Address Systems of Great Complexity.. 499
Questions and Problems 499
References 500

12   Opening New Horizons... 507
12.1   Genomics, Proteomics and Other 'Omics' 507
12.1.1   Gene Expression Analysis... 508
12.1.2   The Analysis of Proteomes... 510
12.1.3   Addressing Interactome.. 513
12.1.4   Outlook 515
12.2   Sensors to Any Target and to an Immense Number of Targets 516
12.2.1   The Combinatorial Approach on a New Level.................. 516
12.2.2   Toxic Agents and Pollutants Inconvenient
          for Detection 518
12.2.3   The Problem of Coding and Two Strategies
          for Its Solution 519
12.2.4   Prospects.. 521
12.3   New Level of Clinical Diagnostics.. 522
12.3.1   The Need for Speed 522
12.3.2   Whole-Blood Sensing.. 523
12.3.3   Testing Non-invasive Biological Fluids............................ 523
12.3.4   Gene-Based Diagnostics.. 524
12.3.5   Protein Disease Biomarkers... 525
12.3.6   Prospects.. 526
12.4   Advanced Sensors in Drug Discovery 526
12.4.1   High-Throughput Screening... 526
12.4.2   Screening for Anti-cancer Drugs..................................... 528
12.4.3   Future Directions 528

12.5   Towards a Sensor that Reproduces Human Senses 529
        12.5.1   Electronic Nose 529
        12.5.2   Electronic Tongue.. 530
        12.5.3   Olfactory and Taste Cells on Chips and
                  Whole-Animal Sensing 531
        12.5.4   Lessons Obtained for Sensing 532
12.6   Sensors Promising to Change Society.. 532
        12.6.1   Industrial Challenges and Safe Workplaces 533
        12.6.2   Biosensor-Based Lifestyle Management........................... 534
        12.6.3   Living in a Safe Environment and Eating
                  Safe Products 536
        12.6.4   Implantable and Digestible Miniature Sensors
                  Are a Reality.. 537
        12.6.5   Prospects... 538
        Sensing and Thinking 12: Where Do We Stand
        and Where Should We Go?... 539
        Questions and Problems 540
        References 541

Epilogue 545

Appendix   Glossary of Terms Used
in Fluorescence/Luminescence Sensing.. 549

Index.. 561

Color Plates.. 571

Introduction

The simplest common definition of a sensor is that "a sensor is something that senses", i.e., receives information and transforms it into a form compatible with our perception, knowledge and understanding. Our body is full of sensors that respond to light, heat, taste, etc. With the development of civilization they became insufficient for the orientation of personality, community or the whole society in new conditions. More and more we need objective knowledge on what is happening inside and outside of our body and what are benefits and threats to the whole society. There is a necessity to know what compounds are useful and what are harmful, how safe and healthful is our environment and to monitor them continuously. Different industrial processes, including that of production of agricultural goods, food processing and storage need to be controlled as well. The human genome is a very useful piece of knowledge only when we can analyze gene expression and find its relation with individuality, age and disease. We need to know the distribution inside living cells of many compounds, including enzymes and their regulators and also substrates and products of these reactions. This information may need to be obtained throughout the whole cell life cycle including its division, differentiation, aging and apoptosis. All that can be accomplished only with the help of man-made sensors. They will be the subject of the present book.

The man-made sensors are often called 'chemical sensors' and those of them that involve biology-related compounds and/or biospecific target binding – 'biosensors'. According to the definition approved by IUPAC, 'a chemical sensor is a device that transforms chemical information, ranging from the concentration of a specific sample component to total composition analysis, into an analytically useful signal'. Thus the sensor can be regarded as both a designed molecule and a miniaturized analytical device that delivers real-time and online information on the presence of specific compounds in complex samples (Thevenot et al. 1999). In a narrow sense, the sensor is a molecule or assembled supra-molecular unit (or nanoparticle), which is able to selectively bind the target molecule (or supra-molecular structure, living cell) and provide information about this binding. In a broader sense it should include control and processing electronics, interconnecting networks, software and other elements needed to make the signal not only recordable but understandable.

Molecular, supra-molecular and cellular mechanisms of acquisition of primary information on the presence and amount of target compounds, particles and cells and of reporting about that in the form of fluorescence signal will be of primary concern in this book. The relevant analytic devices will also be discussed in due course but in a much lesser extent. Our view is that the immense world of potential target compounds is incomparably larger than the variations of instrumental design based on modern electronics and optics. In principle, each member of this world needs its own sensor. It is a great challenge to create them.

Because of this broad range of potential (Cooper 2003a) applications, sensing techniques are attracting an increasing interest of researchers. A number of excellent reviews have been published in the field of chemical sensors, biosensors and nanosensors. By addressing a number of publications one can make a comparative analysis of different sensing strategies: electrochemical (Palecek et al. 2002; Warsinke et al. 2000); microcantilever (Carrascosa et al. 2006); optical (Baird et al. 2002; Baird and Myszka 2001) including surface plasmon resonance (SPR) (Baird et al. 2002; Homola 2003) and microrefractometric-microreflectometric techniques (Gauglitz 2005). Regarding fluorescence sensing techniques one can find important information in the books of Lakowicz and Valeur (Lakowicz 1999, 2007; Valeur 2002) and reviews (de Silva et al. 1997, 2001; Geddes and Lakowicz 2005). In some reviews the applications in particular areas are outlined: food safety (Patel 2002), clinical applications and environment monitoring (Andreescu and Sadik 2004, 2005; Nakamura and Karube 2003), detection of biological warfare agents (Gooding 2006), pharmacology and toxicology (Cooper 2003a). Particular recognition units were highlighted from antibodies (Luppa et al. 2001) and aptamers (O'Sullivan 2002; Tombelli et al. 2005) to functional nanoparticles (Apostolidis et al. 2004) and to whole living cells (Pancrazio et al. 1999). Sensing technologies have started to be used not only in cells but also on the level of whole human bodies (Wilson and Ammam 2007). It is difficult to become oriented in this broad and permanently increasing mass of information. Therefore, a systematization of obtained results and their critical evaluation are badly needed.

The general problem in any sensor technology can be formulated as follows. We have the target molecule, particle or cell dispersed in a medium that may contain many similar molecules, particles or cells. We have to provide a sensor that has to be incorporated into this medium or exposed to contact with it. The presence of a target should be revealed by its selective binding to the sensor. This binding should be detected and, if necessary, quantified in target concentration. This requires some transduction mechanism that connects the binding (molecular event) and its detection by the instrument, on the scale of our vision and understanding.

The sensor and biosensor technologies used for performing this task are based on different physical principles. They develop in parallel, competing with and enriching each other. Some transduction principles, as those used in surface plasmon resonance (SPR) sensors, acoustic sensors, microcantilevers and microcalorimeters, can be applied to any molecular interaction because they are based on the changes in mass or in heat, which are general features of complex formation (Cooper 2003b). However, these approaches generally require sophisticated instrumentation, restricting their use to research purposes. In contrast, electrochemical sensors that are based on redox reactions at electrodes (Palecek et al. 2002) are very simple since they allow a direct conversion of a signal on target binding into an electrical signal. But they are still not always applicable because the sensing mechanism is not general enough. For instance, in biosensing they are mostly based on enzymatic activity generating a detectable product and are therefore restricted to the monitoring of the substrate(s) or effector(s) of a particular enzyme. Therefore, there is a need for generic sensing strategies that can be applied to the detection of any target, rely on low cost and easy-to-use instrumentation and are suitable for on-the-spot or field analysis. In addition, as required in some applications, the response should be very fast and the spatial resolution high enough to allow obtaining microscope images of target distribution and reading from sensors assembled in microarrays containing thousands of spots. This method exists, it is fluorescence.

Thus, what distinguishes fluorescence from all other methods suggested for reporting about sensor-target interaction? Primarily it is its *ultra-high sensitivity* (Lakowicz 1999, 2007; Valeur 2002). This feature is especially needed if the analyte exists in trace amounts. High sensitivity may allow avoiding time-consuming and costly enrichment steps. Meanwhile, one has to distinguish the absolute sensitivity, which is the sensitivity of detecting the fluorescent dye (or particle) from sensitivity in response to target binding. With proper dye selection and proper experimental conditions, the absolute sensitivity may reach the limit of single molecules. This is sufficient and very attractive for many applications, particularly for those in which the dyes are used as labels and the primary response from them is not required. High sensitivity is necessary to achieve to provide the necessary dynamic range of variation of the recorded fluorescence parameters in detecting the sensor-target interaction. This is a much harder task, which we will discuss in detail.

The second distinguishing feature of fluorescence is the *high speed* of response. This response can be as fast as $10^{-8}-10^{-10}$ s and is limited by fluorescence lifetime and the speed of the photophysical or photochemical event that provides the response. Usually, such high speed is not needed but sometimes it is essential. For instance, probing the rate of action potential propagation in excitable cells needs submicrosecond time resolution. The speed of sensor response is not commonly limited by fluorescence reporting. It is limited by other factors, such as the rate of target – sensor mutual diffusion and the establishment of the dynamic equilibrium between bound and unbound target.

The *very high spatial resolution* that can be achieved with fluorescence is important. It allows detecting cellular images and operating with dense multi-analyte sensor arrays. This resolution in common microscopy is limited to about 500 nm (in visible light). The limit is due to the effect of the diffraction of light when the dimensions become similar or shorter than the wavelength. Even this limit can be overcome in special conditions.

The *non-destructive* and *non-invasive* character of fluorescence sensing may be beneficial primarily for many biological and medical applications. In fluorescence sensing the reporter dye and the detecting instrument are located at a certain distance and connected via the propagation of light. This is why fluorescence detection is equally well suited to the remote control of chemical reactions in industry and to sensing different targets within living cells.

The greatest advantage of fluorescence reporting is its *versatility*, coming from the basic event of the fluorescence response. It is essentially a photophysical event coupled to a molecular event of sensor-target interaction. That is why it can be achieved in any environment: in solid, liquid and gaseous media and at all kinds of interfaces between them. The basic mechanism of response remains always the same. It does not impose any limit on the formation of any supra-molecular structures, incorporation of reporter into any nano-composite, attachment to solid support, etc. This allows not only creating smart molecular sensing devices: their attachment to the surfaces in heterogeneous assays or integration into nanoparticles endows new functional possibilities. Due to these facts, homogeneous assays in liquid media develop into nanosensor technologies in which, in addition to fluorescence, different self-assembling, magnetic and optical properties can be explored. In microfluidic devices the detection volume can be reduced to nanoliters. Heterogeneous assays develop into multi-analyte microarrays (sensor chips), which allow the simultaneous detection of several hundreds and thousands of analytes. Two-photon and confocal microscopies allow one to obtain 3-dimensional images, which allows localizing target compounds in space.

In any sensing technology the sensor should switch between two distinguishable states – free and with bound target. There are two possibilities for reporting about a binding event and for providing a signal to distinguish these states and both of them can be realized in fluorescence sensing. First is *indirect*: to label one of these states and then to provide a quantitative measure for labeling that will be connected with the quantity of bound target. This needs additional reagents and special treatments to separate the bound and unbound label and that is why we call this approach indirect. The reporter in this case should provide a stable and bright fluorescence response and additional manipulation with the sample makes this response informative. The other possibility is *direct*: the sensor reports immediately and without any treatments on the primary act of sensor-target interaction. This requires a different property from the fluorescence reporter: to change the parameters of its emission to the very act of target binding. It is the variation of this parameter that can be calibrated in target concentration. Both possibilities show their merits and weaknesses and both of them allow broad possibilities for technical solutions.

The last decade has seen tremendous progress in the development of molecular binders – recognition units of molecular sensors, nanosensors and sensing devices. Each of them should exhibit a high affinity to a target analyte and a high level of discrimination against the species of a similar structure. This can be achieved in many ways: by using complementarity in DNA and RNA sequences, by applying monoclonal antibodies and their recombinant fragments, natural protein receptors and their analogs, combinatorial peptide and polynucleotide libraries, compounds forming inclusion complexes, imprinted polymers, etc. Imagine that one succeeded to select or to design the whole range of necessary binders. In order to benefit from that and to make efficient sensors an efficient mechanism of transduction of this effect of binding into a detectable signal should be applied. The response has to be developed based on available fluorescent dyes and a preferable fluorescence parameter to be recorded. Synthetic chemistry offers tremendous numbers of fluorescent organic dyes (fluorophores) plus many types of nanoparticles and nanocomposites, whereas optical detection methods are much more limited. They involve measuring only several parameters, such as fluorescence excitation and emission spectra and also relative intensities, anisotropies (or polarizations) and lifetimes at particular wavelengths. The optimal choice among these possibilities means the optimal strategy in fluorescence sensing technology.

Particular attention should be given to the coupling of sensing elements with fluorescent reporter dyes and to the methods for producing efficient fluorescence response. Fluorescence reporter units are commonly referred to as "dyes". Indeed, in most cases they are organic dyes that contain extended π-electronic systems with excitation and emission in the convenient visible range of spectrum. In addition, coordinated transition metal ions can be used since they produce a luminescence emission with extended lifetimes. (In this and similar cases it should be more correct to use the term 'luminescence sensing' but it is still not in common use.) Some semiconductor nanoparticles known as quantum dots can generate a narrow-band emission and this property can also be used in sensing. The other possibility that is very attractive for intracellular studies is related to green fluorescent protein (GFP) and its analogs. In this case, the fluorescent moiety appears as a result of a folding of the polypeptide chain and a reaction between proximate amino acid side groups.

One cannot predict the long-run future developments of sensor technologies. But what is sure, they are rapidly becoming a part of everyday life. Thus, for helping diabetes patients the color-changing glucose sensor molecules are already incorporated into plastic eye contact lenses (Badugu et al. 2003) and there has been a reported development of 'an ingestible one-use nanotechnology biosensor' that can be swallowed like a vitamin to report in fluorescent light about the pathological changes in human tissues (Kfouri et al. 2008). So, what will happen next?

References

Andreescu S, Sadik OA (2004) Trends and challenges in biochemical sensors for clinical and environmental monitoring. Pure and Applied Chemistry 76:861–878

Andreescu S, Sadik OA (2005) Advanced electrochemical sensors for cell cancer monitoring. Methods 37:84–93

Apostolidis A, Klimant I, Andrzejewski D, Wolfbeis OS (2004) A combinatorial approach for development of materials for optical sensing of gases. Journal of Combinatorial Chemistry 6:325–331

Badugu R, Lakowicz JR, Geddes CD (2003) A glucose sensing contact lens: a non-invasive technique for continuous physiological glucose monitoring. Journal of Fluorescence 13:371–374

Baird CL, Myszka DG (2001) Current and emerging commercial optical biosensors. Journal of Molecular Recognition 14:261–268

Baird CL, Courtenay ES, Myszka DG (2002) Surface plasmon resonance characterization of drug/liposome interactions. Analytical Biochemistry 310:93–99

Carrascosa LG, Moreno M, Alvarez M, Lechuga LM (2006) Nanomechanical biosensors: a new sensing tool. Trac-Trends in Analytical Chemistry 25:196–206

Cooper MA (2003a) Biosensor profiling of molecular interactions in pharmacology. Current Opinion in Pharmacology 3:557–562

Cooper MA (2003b) Label-free screening of bio-molecular interactions. Analytical and Bioanalytical Chemistry 377:834–842

de Silva AP, Gunaratne HQN, Gunnaugsson T, Huxley AJM, McRoy CP, Rademacher JT, Rice TE (1997) Signaling recognition events with fluorescent sensors and switches. Chemical Reviews 97:1515–1566

de Silva AP, Fox DB, Moody TS, Weir SM (2001) The development of molecular fluorescent switches. Trends in Biotechnology 19:29–34

Gauglitz G (2005) Direct optical sensors: principles and selected applications. Analytical and Bioanalytical Chemistry 381:141–155

Geddes CD, Lakowicz JR, eds. (2005) Advanced concepts in fluorescence sensing. Topics in fluorescence spectroscopy, v. 10. Springer, New York

Gooding JJ (2006) Biosensor technology for detecting biological warfare agents: recent progress and future trends. Analytica Chimica Acta 559:137–151

Homola J (2003) Present and future of surface plasmon resonance biosensors. Analytical and Bioanalytical Chemistry 377:528–539

Kfouri M, Marinov O, Quevedo P, Faramarzpour N, Shirani S, Liu LWC, Fang Q, Deen MJ (2008) Toward a miniaturized wireless fluorescence-based diagnostic Imaging system. IEEE Journal of Selected Topics in Quantum Electronics 14:226–234

Lakowicz JR (1999) Principles of fluorescence spectroscopy. Kluwer, New York

Lakowicz JR (2007) Principles of fluorescence spectroscopy. Springer, New York

Luppa PB, Sokoll LJ, Chan DW (2001) Immunosensors - principles and applications to clinical chemistry. Clinica Chimica Acta 314:1–26

Nakamura H, Karube I (2003) Current research activity in biosensors. Analytical and Bioanalytical Chemistry 377:446–468

O'Sullivan CK (2002) Aptasensors–the future of biosensing? Analytical and Bioanalytical Chemistry 372:44–48

Palecek E, Fojta M, Jelen F (2002) New approaches in the development of DNA sensors: hybridization and electrochemical detection of DNA and RNA at two different surfaces. Bioelectrochemistry 56:85–90

Pancrazio JJ, Whelan JP, Borkholder DA, Ma W, Stenger DA (1999) Development and application of cell-based biosensors. Annals of Biomedical Engineering 27:697–711

Patel PD (2002) (Bio)sensors for measurement of analytes implicated in food safety: a review. Trac-Trends in Analytical Chemistry 21:96–115

Thevenot DR, Toth K, Durst RA, Wilson GS (1999) Electrochemical biosensors: recommended definitions and classification - (Technical Report). Pure and Applied Chemistry 71:2333–2348

Tombelli S, Minunni M, Mascini M (2005) Analytical applications of aptamers. Biosensors & Bioelectronics 20:2424–2434

Valeur B (2002) Molecular fluorescence. Wiley-VCH, Weinheim

Warsinke A, Benkert A, Scheller FW (2000) Electrochemical immunoassays. Fresenius Journal of Analytical Chemistry 366:622–634

Wilson GS, Ammam M (2007) In vivo biosensors. FEBS Journal 274:5452–5461

Chapter 1
Basic Principles

The boom in sensor technologies is a response to a strong demand in society. As a result, almost every physical principle and technique that can detect interactions between molecules, particles and interfaces has been suggested and tested for application in sensing. In this Chapter we will provide a short survey of these techniques and try to determine the role of those of them that are based on fluorescence detection. In recent years, intensive research and development led to the establishment of several important strategies for sensor operation. Some of them are of a rather general nature and some are specific for fluorescence techniques. An overview is given below.

1.1 Overview of Strategies in Molecular Sensing

No interaction – no information. This principle is clearly seen in the background of all sensing technologies. In every interaction, we have at least two partners. One is, of course, the object that has to be detected. It is commonly called the *target* or *analyte*. It can be an object of any size and complexity, starting from protons, small molecules and ions up to large particles and living cells. The other partner, designed or selected for target detection, is the *sensor*.

The sensor has two functions. The first is to provide interaction with the target in a highly selective way, recognizing it from other objects of similar structure and properties that can be present in the probed system. Addressing the demand of target detection, it can also be the object of any size and complexity. The structure responsible for that is called *recognition unit* or *receptor*.

The other function is to 'visualize' this interaction, to report about it by providing a signal that can be analyzed and counted. The structure responsible for the generation of this signal is called a *reporter*. The transformation of the signal about a binding event into a response of a reporter is called transduction and if additional elements of the structure are needed for that, they are named *transducers*. Sensors in a broad sense also involve instrumentation, which is important as an interface between the micro-world of molecules and the macro-world in which we live.

A.P. Demchenko, *Introduction to Fluorescence Sensing,*
© Springer Science + Business Media B.V. 2009

In the micro-world the elementary events in sensing occur and we have to analyze the results and make our decisions in the macro-world.

1.1.1 Basic Definitions: Sensors and Assays, Homogeneous and Heterogeneous

Assay is a broader term than sensing, since it may involve different manipulations with the tested system that may include different chemical and biochemical reactions. *Sensing* is always a more direct procedure, which is based on an interaction with the target of a particular sensor unit and on a detectable response to this interaction. The latter can be a molecule, a particle or a solid surface in which additional reactions are not needed or can be used only for the amplification of a primary effect of this interaction. Both assays and sensors can operate in heterogeneous or homogeneous formats.

Heterogeneous formats for assays and sensors are those that require the separation of the sensor-analyte complex for subsequent detection by any analytical method, including fluorescence. Most conveniently, this can be done in a *heterogeneous system*, in which the sensor elements are immobilized on a solid surface. In this case, after incubation in the tested medium, the unreacted species that are present in the system can be removed simply by washing and, if necessary, supplied with additional reagents for the visualization or generation of a response signal (Fig. 1.1).

In a solution, this type of assay is also possible but it requires separation by chromatography, electrophoresis, etc. As we will see below, the employment of unbound and specifically bound species of this assay principle in sensor technologies allows us to achieve the broadest dynamic range of quantitative determination of the target. Because of the implied washing step, the result is less sensitive to interference from non-specifically bound components of the test system. However, the assay is limited to high-affinity binding, since only in this case will the target-sensor complexes not be destroyed during manipulations of the sample. All additional operations, such as separation and washing, are time-consuming, which does not allow one to obtain immediate results.

Homogeneous assays are those that provide the necessary signal upon target binding in the test medium without any separation or washing (Fig. 1.2). Therefore, they are often called 'mix-and-read' assays. Many sensor technologies operate according to this principle. They use different physical mechanisms of response to primary sensor-target binding. Such sensors are often called *direct* sensors (Altschuh et al. 2006). If such a response is provided, then there is no need for separation, reagent addition or washing. Therefore, such sensors can also be called *reagent-independent* sensors. We will see below that there is no general simple and straightforward way to provide such direct responses. Nevertheless, there are many possibilities, especially in the cases of high target binding affinity.

A strong advantage of homogeneous assay formats is the possibility of the quantitative determination of analytes of relatively low affinities, in which a dynamic concentration-dependent equilibrium is established between a free and bound target.

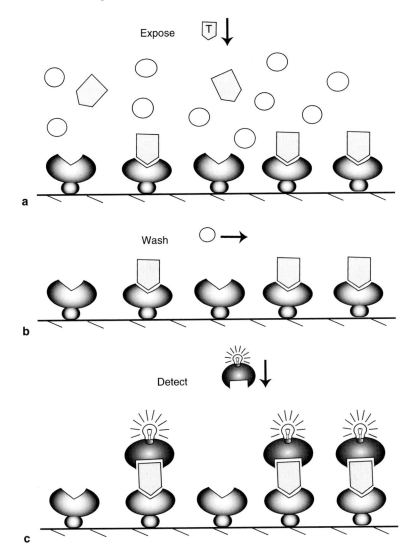

Fig. 1.1 An example of an heterogeneous assay, commonly consisting in three steps. **(a)** The plate with immobilized receptor molecules is exposed to the probed sample. The targets are strongly and specifically bound, whereas contaminating compounds remain in solution. **(b)** The plate is washed. It contains bound targets and all unbound components of the mixture are removed. **(c)** The plate is exposed to a solution of molecules or particles that are able to recognize the target and bind to it at a different site (indicator). The indicator contains a fluorescent label. Here and in other illustrations below, geometrical fitting indicates specific binding (recognition)

In this case, the *dynamic range* of the assay (the target concentration range, in which the variation of reporter signal is detected) is much narrower but the range of possible applications is dramatically increased. Direct sensors are especially desirable for different practical applications, due to the possibility of obtaining the results on-line.

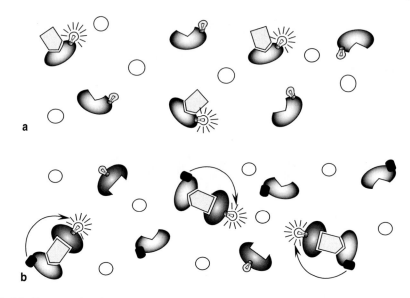

Fig. 1.2 Two examples of an homogeneous assay. (**a**) Sensor molecules comprises receptor and reporter groups that change their fluorescence on target binding. (**b**) Two different sensor molecules possessing different reporters bind the target at two different binding sites. The transduction of the reporter signal is generated on interaction between the reporter units. This is possible only in the case when they bind the same target and thus appear in close proximity. In both cases, the assay may occur without the sensor immobilization and without the separation of the complex from unreacted components

However, the non-separation nature of these assays sets some major limitations to their performance. They can be very prone to interference from non-specific (non-target) binding. Moreover, the researcher is always restricted in finding a mechanism of response that is far from being universal.

It is a matter of terminology but some authors prefer to call heterogeneous assays the 'assays' while homogeneous assays are the 'sensors'. This is because in homogeneous assays it is much easier to develop sensors that will not need any manipulation in the course of or after measurement and that are applicable for the continuous monitoring of target concentration. According to the International Union of Pure and Applied Chemistry (IUPAC) nomenclature recommendations, a *biosensor* (that can be extended to any type of sensor) is defined as a self-contained analytical device, which is capable of providing quantitative or semi-quantitative analytical information using a biological recognition element either integrated within or intimately associated with a physicochemical transducer (Thevenot et al. 2001). It is clearly affirmed that "*a biosensor should be clearly distinguished from a bioanalytical system, which requires additional processing steps, such as the addition of reagents*". Meantime, this definition is not supported by many researchers and other definitions exist (Kellner et al. 2004). For instance, some authors suggest that sensors are the devices in which the response is produced in a chemical reaction with the analyte.

The distinction between 'sensors' and 'assays' is still not very clear. Moreover, the same molecules with a selective target-binding function can serve in both assay and sensor techniques. Therefore, it is important how the reporter signal is provided.

An example is the antibody, which is the most frequently used molecule with the function of biological recognition (Fig. 1.3). For participating in bioassays, this molecule can be labeled at the periphery, at any site outside its target-recognition site. This is enough to provide a response in heterogeneous format to the immobilized

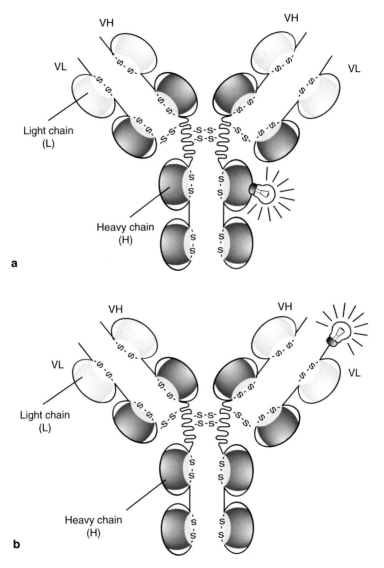

Fig. 1.3 The labeling of antibodies with fluorescence dyes for use in bioassays and biosensors. The IgG antibody is shown. It is a protein molecule composed of two light (L) and two heavy (H) chains forming domains, cross-linked by disulfide (-S-S-) bonds. The VL and VH domains form the target antigen binding site. The site of labeling determines the range of the applications of the antibodies. (**a**) The fluorescent dye is bound at the periphery of the antibody molecule and serves only for the purpose of labeling. Its response is insensitive to antigen binding. (**b**) The reporter dye is located close to the antigen binding site so that the interaction with antigen changes its fluorescence parameters providing the reporting signal

target. In contrast, for its participation in sensing in homogeneous format, stringent additional requirements should be satisfied. The label should directly or indirectly sense the interaction with the target and provide a detectable reporting signal. One of the solutions to achieving this is to locate the label at the target-binding site.

1.1.2   Principles of Sensor Operation

Different physical principles can be applied in sensor technologies to generate a measurable signal in response to sensor-target interaction (Fig. 1.4). The primary event of target binding by *receptor* (recognition element of the sensor) should generate some response, which is provided by a *reporter element*. These functions are different but they must be strongly coupled. This response (optical, heat, mass, etc.) is transformed into a measurable electrical signal.

The sensors can be classified according to three basic elements of their operation.

1. *The recognition element.* It can be as small as a group of atoms chelating a metal cation and as big as large protein molecule, DNA, membrane and even the living cell.
2. *The reporter element and the principle of reporting.* Optical (light absorption, reflection, fluorescence), electrochemical (amperometric, conductometric), mass, heat and acoustic effects can be used for this purpose.

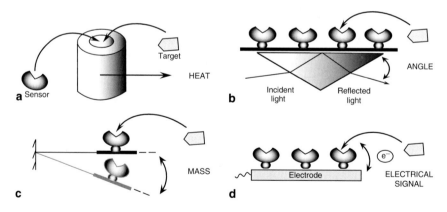

Fig. 1.4 Schematics illustrating the operation principles of sensing based on several alternative technologies. (**a**) Calorimetry. When target and receptor are mixed in a calorimeter, they produce heat effects. (**b**) Surface plasmon resonance. The layer of receptors is formed on the surface of a thin layer of gold or silver. The target binding causes the change of the refractive index close to this surface that modulates the angular dependence of the reflected light beam. (**c**) Micro-cantilever technique. The target binding results in the mechanical signal that is detected as the change of mass. (d) Electrochemistry. The target binding usually results in an electron transfer between the electroactive compound and the transducer (electrode) or in a change of the existing electrochemical signal

3. *The mechanism of coupling.* Electronic, conformational, through the medium (e.g., due to pH change), or by coupled chemical or electrochemical reaction.

For efficient sensor operation, some technologies need the immobilization of sensor molecules on planar surfaces or nanoparticles and some need the incorporation of labels (e.g., absorbing light or producing fluorescence) serving as molecular reporters. This creates a versatility of approaches. Of great importance are the medium conditions, in which the sensor-target binding occurs. This can be the liquid phase or liquid-solid interface. In the former case, in order to obtain the response, the labeling and/or separation of target-sensor complexes may be needed. In the latter case, the washing-out of unbound material may be necessary but some methods avoid this step. Using several examples, we will illustrate below the advantages and disadvantages of sensor immobilization and labeling.

1.1.3 Label-Free, General Approaches

Label-free techniques that are developed to the level of research and practical application involve those with optical detection, such as surface plasmon resonance (SPR) and also acoustic, cantilever and calorimetric biosensors.

Calorimetry (Fig. 1.4a) is an example of a general label-free approach that allows measurements in solutions without the necessity of immobilizing the sensor molecules or labeling them (Ladbury and Chowdhry 1996). This method is based on the principle of the generation of heat effects that appear on sensor-target binding. Since the detection is based on a general property of obtaining thermodynamic balance in the system after the new target-receptor bonds are formed, this method seems to be broadly applicable. The process of binding can be detected as a *heat effect in solution*, therefore the sensor immobilization is not needed. It should be noted, however, that this method does not possess any structural resolution. In complex mixtures, a strong binding by specific targets existing in small concentrations is not distinguishable from that of nonspecific binding at high concentrations. Thus, the advantage of measurement in label-free conditions in solutions is compromised by a very restricted range of applications. However, such applications exist, for instance, in the detection of explosives and other highly energetic compounds (Liu et al. 2005).

Let us now take a closer look at *surface plasmon resonance* (SPR) sensing, which is a typical example of an immobilization-based but label-free technique (Fig. 1.4b). A description of SPR principles and a review of its applications can be found in many articles (Homola 2003; Homola et al. 2005; Rich and Myszka 2006). The method is based on probing the change in the refractive index close to the surface of a thin (~50 nm) metal (usually gold) film, illumination of which creates the so-called *evanescent wave* that generates an optical signal. On this surface the receptor molecules or particles are immobilized. The target, when bound from a contacting solution phase, produces a detectable local increase in the refractive index. This signal is observed as the change of the angle at which a strong decrease of intensity of reflected light occurs. The evanescent wave decreases exponentially as a function of distance from the surface of the sensor film on a scale up to 200 nm, which is convenient for the detection of large molecules. Unbound molecules and particles remain in the sample solution phase outside the detection layer and do not interfere with the measurement. Meanwhile, as every label-free method, SPR cannot distinguish specific and non-specific binding and therefore is not secured from false positive results.

The SPR method can in principle be applied to any molecular interaction because the transduction is based on a general property of complex formation (the local values of the refractive index of many molecules are much larger than that of water). This method has many attractive features. It records the target binding in real time without the necessity of pre-treatments or after-treatments to visualize the result. It has become a valuable technique not only for determining the receptor-target affinities but also for target sorption-desorption kinetics. Its application for simultaneous determination of multiple targets is difficult but this problem can be overcome as observed in recent reports (Homola et al. 2005).

It should be noted however, that for the application of SPR it is obligatory that one of the partners in intermolecular interactions (usually the receptor) should be immobilized on a specially prepared surface. Attachment to the surface can modify its binding affinity and this may impose a significant limitation in the application of this method. However, a very important effect is achieved: *only the bound target molecules produce the response*. Therefore, if the sample contains a complex mixture of different compounds, only the target will produce the signal and no separation or washing steps will be needed. That is why this method allows direct kinetic measurements.

It is interesting to note that a hybrid between fluorescence and SPR can be created. The method is called *evanescent wave fluorescence* and will be discussed in more detail in Section 8.5. Like SPR, it can measure surface-specific binding events in real time. In this method, the excitation is by evanescent wave and the response is provided by the detection of fluorescence. Thus, only the species bound to the surface will be excited.

The developments in sensors based on the change of mass on target binding (Fig. 1.4c) are very interesting. Two techniques have been developed using this principle. One is *acoustic* sensing. It is based on a highly sensitive detection of mass changes measured via surface acoustic wave (Lange et al. 2008). The sensor needs to be coupled to the surface of a piezoelectric material, such as quartz. Its applicability for the accurate detection of protein targets has been demonstrated. The second one uses a *microcantilever* (vibrating microbalance). The spring constant of this vibrating nanomechanical device is directly related to the increase of mass on target binding (Battiston et al. 2001).

Electrochemical sensors are very popular because they use simple instrumentation and allow direct recording of target binding to immobilized receptors as an electrical signal (Fig. 1.4d). An electrical signal can be obtained in oxidation-reduction reaction at the electrode. This can be either a direct reaction with the participation of a target or a coupled reaction that may involve a catalyst (Palecek 2005; Palecek and Jelen 2005).

Thus, the obvious positive features of SPR are the label-free, direct and rapid detection that allows kinetic measurements. Of some prospect are the sensors detecting a change of mass. These technologies need a *sensor immobilization to a specially prepared surface with special properties.* Ideally, only the targets, interacting with the sensor, become captured on the surface and are detected. An advantage is a typically small sample size. The disadvantages are the complexity and cost of instrumentation and not only the complexity and cost of plates for sensor immobilization but also the absolute necessity of their use. Electrochemical sensors are attractive for their simplicity but sensor immobilization is always needed and in most cases the protocol involving the coupled electrochemical reaction needs to be developed.

There are many possibilities to apply the *label-free approach in solution.* For instance, one can provide the sensor-target binding with a subsequent chromatographic separation of the formed complexes and their analysis by different analytical techniques, such as mass spectrometry, spectroscopy, enzyme inhibition or antibody binding. However, such procedures are always difficult and time-consuming. They are limited to very strong sensor-target interactions so that these interactions do not have to be disrupted by separation procedures. Finally, it should be noted that the high cost of instrumentation is a characteristic feature of all general label-free and immobilization-free approaches.

1.1.4 Label-Free, System-Specific Approaches

A second group of label-free approaches uses a specific reaction. Commonly it is limited to biospecific interactions and is mainly based on the detection of *enzyme activity* (D'Orazio 2003; de Castro and Herrera 2003). The measurement of enzyme activity requires the transformation of an additional reagent (substrate) into a reaction product, which generates the reporter signal. According to the definition presented above, these are assays not sensors. In the meantime, they are often recognized as '*catalytic biosensors*' in contrast to '*affinity-based sensors*', which do not use any coupled reactions.

In the case of catalytic biosensors, the range of possible targets usually does not go beyond substrates or effectors of enzyme reactions, though attempts to induce artificial enzyme activity are also known. The enzyme here serves as the receptor element and the chosen transduction principle is determined by the properties of the reaction substrate or product. In this group of sensors the generated transduction principle is often *electrochemical*, since it allows the easy production of a reporter signal when an enzyme is immobilized at the electrode. The first known example of such a biosensor is an amperometric glucose sensor, in which the enzyme glucose oxidase was coupled with an amperometric electrode sensing P_{O2} (Palecek 2005). Fluorescence-based enzymatic sensors also exist. They will be discussed in Section 1.5.

Enzyme reactions can easily be observed in solutions. Nevertheless, most of the methods employing enzyme biosensors are based on *immobilized enzymes.* Immobilization solves several important problems, allowing one to decrease the amount of enzyme needed for the assay to maintain its stability and to reduce the time of enzymatic response (Amine et al. 2006). The choice of surface for enzyme immobilization is determined by the detection method, so that the surface may serve for transducing the response signal. Not only flat but also porous surfaces, polymer gels and nanoparticles can be used for this purpose. It should be stressed that the advantages obtained from immobilization could be so significant that the development of immobilization protocols is justified for almost every particular case.

Finally, it should be noted that the label-free *electrochemical immunoassays* (Warsinke et al. 2000) have been described for a particular antigen-antibody interaction, when the complex formation results in variations in charge density or conductivity. Since the mode of transduction is generally based on the properties of a reaction product, these sensing principles apply to the detection of particular targets (substrates or effectors) only. This severely restricts their use.

1.1.5 Label-Based Approaches

The *label-based* techniques are in principle applicable to any molecular interaction, provided that at least one of the interacting partners can be modified by attaching the label without loss of binding affinity. The disadvantage in this approach is that the labeling may affect the target binding. The label-based techniques can be used in conjunction with analytical devices such as fiber-optic or gravimetric sensors (Luppa et al. 2001). But the most important advantage is that they can be used for constructing sensors, in which the recognition element is labeled in such a way that a measurable physico-chemical property of the construct changes upon the complex formation in solution. This allows homogeneous assays that avoid any separation steps.

Molecular constructs can be made that integrate both the recognition and reporter elements of a sensor, allowing target detection in a direct way. Such sensors are essentially '*direct sensors*' since the transduction of a signal produced on a complex formation does not need any additional steps or third interacting partners. Such molecular sensors are '*affinity-based*' because the signal is produced by the analyte-sensor interaction itself, in contrast to '*catalytic sensors*', in which the signal is caused by a catalytic substrate transformation. The label in most cases is a fluorescent dye or nanoparticle (*a fluorophore*), leading to a signal of fluorescence emission that is relatively easy to record. The labeling of the interacting molecules with redox compounds or enzymes has also been described (Warsinke et al. 2000). Fluorescence is often very sensitive to the changes of molecular environment and not only to binding. This property can be also applied in sensing.

A technology based on fluorescent labeling is, in principle, of *general applicability*; it is easy to implement and inexpensive. In sensing applications, fluorescence detection is based on a variation in fluorescence properties upon complex formation and the sensor operation can be made 'direct', without additional manipulations or reagent additions (Gauglitz 2005).

The *versatility* of this approach is also very important. Indeed, in contrast to many other methods used for the studies of macromolecular interactions and employed for the construction of biosensors (electrochemical and piezoelectric sensing, SPR, etc.), fluorescence sensors can operate as separate molecules in solutions to allow homogeneous assays without any immobilization. The absence of a solid support allows the sensor molecules to move easily in solutions and to integrate spontaneously into nanoscale particles, porous materials and polymer gels and penetrate into living cells, providing a measurable response. They can also be immobilized on the surfaces of optical elements forming key parts of sensing devices. Also, they can be deposited in nanogram quantities in a mosaic way forming arrays for the simultaneous detection of many analytes. Their incorporation into living cells allows the non-invasive investigation of diagnostically valuable compounds, using all the capabilities of fluorescence microscopy and flow cytometry. In a broader sense, fluorescence allows the visualization of sensor-target interactions with a high spatial resolution (Fig. 1.5).

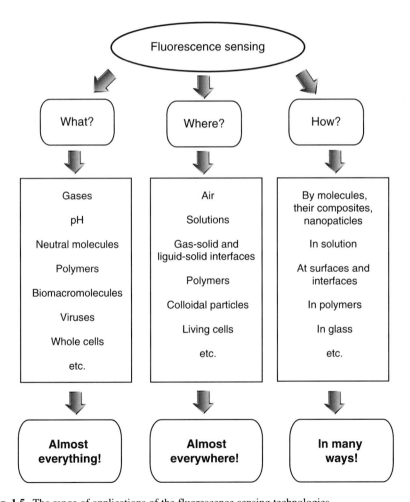

Fig. 1.5 The range of applications of the fluorescence sensing technologies

Thus, fluorescence sensing is essentially a label-based approach (intrinsic fluorophores, such as tryptophan in proteins, can be considered as naturally imposed labels).
When comparing this approach with label-free methods based on the detection of
changes of refractive index, charge or mass at the sensor surface or heat effects, in solutions, one has to note the following important point. Testing in real systems always
involves side effects produced by contaminants. These contaminants also have mass and
charge; they change refractive index and produce heat effects on binding. This results
in a background signal that may induce a serious complication when these techniques
are used for testing realistically complex samples, such as blood serum. In contrast,
specific labeling in sensing technologies always makes the response more specific leading
to the reduction of false positive complications in results.

1.2   Labeled Targets in Fluorescence Assays

Historically, fluorescence sensing came into play to substitute the *radioactive isotope*
detection technique and to a large extent borrowed its methodology. Radioisotopes
are insensitive to intermolecular interactions and all the methods based on their
usage allow only their quantitative detection in the desired locations. Their substitution by fluorescent dyes made the analysis cheaper, safer and much more sensitive.
Many fluorescence technologies were developed following this line. Here the
fluorescence of the dyes indicates the presence of a given target compound in
the analyzed medium and provides a quantitative measure of this compound. Such
dyes are in fact a part of the target system but not of the sensor system.

In this technology, the whole *pool of potential targets* has to be labeled
(Fig. 1.6). Then, after reacting with the receptor, the receptor-target complexes
should be isolated from the pool. If the receptor is immobilized on the surface, this

Fig. 1.6 The illustration of principle behind the initial labeling of the whole pool of potential
targets present in the sample. The sensors are represented only by the receptors that remain unlabeled. They bind the targets and immobilize them on the surface leaving all *labeled* contaminating
compounds in solution

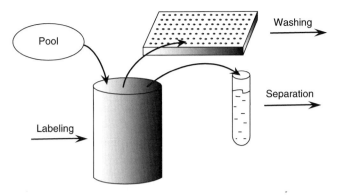

Fig. 1.7 Schematic illustration of procedures involved in sensing requiring pool labeling. In the case of immobilized receptors, the unbound labeled species are removed by washing and in the case of an assay in solution, a separation of complexes and unbound components is required

can be easily done by the procedure of washing-out the unreacted target analogs. In the case when the reaction occurs in solutions, there should be a chromatographic or electrophoretic separation of a receptor-target complex (Fig. 1.7).

1.2.1 Arrays for DNA Hybridization

The labeling of the pool of all potential target molecules is a procedure that is commonly used in very popular DNA or RNA *hybridization* assays (Freeman et al. 2000), as described in more detail in Section 10.4. Commonly, solid-state hybridization is used, in which the sensing nucleic acid strands tethered to a solid support ('probe' strands) bind strands of single-chain DNA or RNA molecules from a solution ('target' strands).

To identify specific recognition between complementary sequences, the whole pool of poly- or oligonucleotides that may contain the 'target' strands is labeled with a fluorescent dye before application to the sensor. During the next step, the supportive plate is washed to remove unbound material and is applied to a fluorescence reader. If comparative studies are needed, the two nucleic acid pools can be labeled using different fluorescent dyes (usually a red-emitting cyanine dye Cy5 and a green-emitting dye Cy3) and applied to the same plate after mixing.

DNA or RNA microarrays (sometimes called microchips) are the results of the miniaturization of this technology (Shoemaker and Linsley 2002; Venkatasubbarao 2004). Using hybridization on microarrays, thousands of solid-phase hybridization reactions can be performed simultaneously to determine gene expression patterns or to identify genotypes. This technique features a variety of important applications, from finding peculiarities of gene expression in disease, such as cancer, to analysis of its changes in ontogenesis and aging.

1.2.2 Labeling in Protein-Protein and Protein-Nucleic Acid Interactions

Detection of protein targets based on microarrays or any other high-throughput technology is highly necessary for research and various applications. The development of 'protein function' arrays that should consist of thousands of native proteins immobilized in a defined pattern and used for massive parallel testing of protein functions, is very important (Kodadek 2001, 2002). Such techniques are needed in proteomics research and also to satisfy the practical needs of pharmacology. The other type of arrays, developed as 'protein detection' devices with arrayed protein-binding agents, are needed for proteomics. Examples of their application could include cytokines, growth factors, apoptosis-related and other biological systems-related arrays. They must allow detection from dozens up to several thousands of analytes.

In the meantime, compared to labeling the whole pool of DNA or RNA molecules, the fluorescent labeling of a protein pool is much more difficult and much less efficient. One may claim that covalent labeling at amino groups (N-terminal and belonging to Lys residues) is a very common procedure in protein chemistry. However, the fact is that every protein in the pool will bind the dye at different sites and in different quantities and this property is determined by its three-dimensional structure. The charged Lys residues are commonly exposed to the surface of protein molecules and their modification may hamper the binding to the receptor. This does not allow a proper calibration for quantitative measurements.

Thus, the labeling of protein pools has to be avoided; other methods described in this book that allow omitting this procedure could be more appropriate for protein assays.

1.2.3 Micro-Array Immunosensors

A similar approach to that described above is used in antibody micro-array techniques (micro-array immunosensors). In a number of currently used assays, immobilized, unlabeled antibodies assembled into micro-arrays are used for testing protein antigens. In order to record selective binding, all potential antigens in the sample have to be preliminarily labeled with fluorescent dye (MacBeath and Schreiber 2000). Thus, this approach shows the same disadvantages as the DNA and RNA hybridization assays, with possibly greater errors. They can arise from nonspecific binding of analyte antigens and non-analytes present in the sample.

1.2.4 Advantages and Limitations of the Approach Based on Pool Labeling

Pool labeling is a simple, usually one-step, procedure. Nevertheless, this approach has a number of limitations. Some of them are technical and some derive from the fact that in such systems it is hard to achieve a true thermodynamic equilibrium between bound and unbound species. This is because of the following.

1. The correct quantitative measurement of target concentration is possible only when all target molecules (or particles) present in the system are bound to the sensor (unless separation can be made based on the spectral change induced by the binding). Therefore, measurements cannot be made in the studied system directly but only in aliquots taken from it. In order to achieve complete target binding, the concentration of receptors should be higher than that of the targets.
2. The procedure requires a step of labeling with the dye, an incubation step for target binding and the washing step to remove unbound target analogs that contain the fluorescent label. These steps are time-consuming. Therefore, this method is not applicable for continuous monitoring of the target concentration.
3. The method can be applied only to strong binders, i.e., to the cases in which the sensor-target complex can remain intact under the washing procedure. Therefore, targets with relatively low affinities (dissociation constant K_d 10^{-5}–10^{-7} M) which are of the order of magnitude in enzyme-substrate and competitive enzyme-inhibitor complexes, cannot be analyzed reliably using this approach.
4. The label itself may influence the target recognition. There should be independent proof that there is no modification of target affinity by the attached fluorescent label.

It is not surprising therefore, that this approach is attractive only when the tested compounds are of the same chemical nature and their labeling can be achieved quantitatively using the same general procedure. Fortunately, this is the case for many DNA and RNA hybridization assays.

The major advantages of this approach are its simplicity and the absence of limitation imposed on reporter dyes. This approach is not limited to sensing molecules. It can be extended to sensing large particles and even living cells. The label can be introduced to their pool and specific targets can be analyzed after a separation step using electrophoresis, chromatography or flow cytometry.

1.3 Competitor Displacement Assay

Many scientists are trying to develop technologies that could avoid the introduction of a fluorescence label into analyzed samples and competitor displacement assay is one of these possibilities. In this technology, neither receptor nor target may contain a fluorescence label. Instead, an additional player in the target recognition mechanism is introduced, the fluorescent *competitor* (see Fig. 1.8). Usually, a competitor is the labeled target analog that binds to the same site on the receptor as the target. After this binding, a mixture of different compounds that contains the target is applied. Due to its specific interaction (molecular recognition) with the receptor, the target displaces the competitor from the binding site. The target can be analyzed quantitatively because it is the only species that

Fig. 1.8 The illustration of principle behind the competitor displacement assay. The response to target binding is obtained when the labeled competitor, competing for the binding site on the receptor, is displaced to the solution with a change in parameters of its fluorescence emission

can displace the competitor. The competitor reports about its displacement by changing the parameters of its fluorescence. These changes can be calibrated as a function of target concentration.

There can be other configurations in competitor displacement assays. For instance, the competitor can be unlabeled and immobilized on the surface. It binds the sensor molecules possessing the reporting units added to solution. When the sample that contains the target is applied, the target interacts with the sensor. The latter dissociates from the surface with the generation of a reporter signal.

1.3.1 Unlabeled Sensor and Labeled Competitor in Homogeneous Assays

Although competitor displacement assays can be performed with *immobilized receptors*, this method allows *homogeneous assays* in solutions. In this case, there is no need to separate the receptor-target complexes from the target-free receptor molecules. It is essential that in this configuration the competitor does not interact directly with the target but only with the receptor via reversible non-covalent binding. Upon release from the complex, the competitor experiences a change of interactions from those existing within its complex with the receptor to those with solvent molecules (Wiskur et al. 2001). The fluorescence reporter signal in this case can be obtained by the change of different fluorescence parameters, such as intensity, polarization and lifetime (see Chapter 3). Therefore, the target is not involved in the transduction mechanism, it just substitutes the reporting competitor.

The competition assay is most convenient when the target is of a relatively low molecular mass and if it can be coupled with a fluorescent dye to produce a labeled competitor. The competition can then be exploited between labeled and unlabeled target molecules for a limited number of receptors. In the presence of a fixed amount of labeled competitor, the mole fraction bound to the receptor varies inversely with the target concentration. In this competitive format, it is necessary that some fluorescence signal should change between the two states of the competitor, its free and bound forms. Alternatively, there should be a need for a separation of the bound and free competitor. As a general feature, this method requires keeping the competitor concentration on a strictly defined level; it lacks adaptability to continuous monitoring of target concentrations (Piatek et al. 2004).

Many proteins can bind fluorescent dyes at their ligand-binding sites. When the ligand replaces the dye, large changes in the dye fluorescence may be observed. Based on these observations, different rather simple procedures for the determination of particular ligands can be developed. It is known, for example, that avidin and biotin form a very strong and highly specific complex that can be easily quantified using the fluorescence probe 2,6-ANS (2-anilinonaphthalene-6-sulfonic acid) (Mock et al. 1988). The binding of 2,6-ANS produces a dramatic increase in avidin fluorescence and biotin, when present in the system, completely displaces it with an easily recordable fluorescence quenching.

Serum albumin has a function of binding and transporting many ligands in the blood, including different drugs. It can also bind some fluoresce dyes with high affinity at the same sites where it binds the drugs (Ercelen et al. 2005). Therefore, displacement of fluorescent ligands by the target ligands may provide a simple and efficient estimate of their concentrations in unknown media. Moreover, the tests for screening different ligands for their albumin-binding affinities can be established. Such tests, being one of the determinants of the lifetimes of toxic compounds and the drug in the blood circulation, may be used in toxicology and drug discovery.

A displacement assay using fluorescent *intercalator dyes* may serve as another example (Tse and Boger 2004). It can be applied for establishing DNA binding affinity and sequence selectivity for different compounds. Intercalator dyes are the dyes such as ethidium bromide or thiasole orange, which can insert spontaneously between nucleic acid bases. These dyes are almost non-fluorescent in their free form in an aqueous buffer (see Section 4.1). They attain strong fluorescence only in the DNA bound form and their binding occurs only when the DNA has a double-strand helical structure. Intercalation between nucleic acid bases screens them from the contact with water that produces high-intensity emissions. The application of target DNA binding compounds that disrupt the double helix destroys the binding sites for intercalator dyes with a dramatic decrease of fluorescence intensity (quenching).

Such a possibility for generating the reporter signal indicates the fact that there is no absolute necessity for a competitor to be replaced by the target from the same binding site. The only requirement is that the sensor-target interaction should destroy the binding site for the competitor, which can be done in different ways.

1.3.2   Labeling of Both Receptor and Competitor

There is an alternative possibility for implementing displacement assays, which finds a broad range of applications. The receptor can be additionally labeled to provide long-range interaction with the competitor, which will allow FRET (Förster resonance energy transfer) measurements. In this case (see Section 3.5), the FRET effect can be observed only in a competitor complex with a labeled receptor. When the unlabeled target replaces the competitor, the latter diffuses into the medium and the optical effect generated by the energy transfer is lost. This event can report on the act of target binding.

An interesting realization of this idea is a fluorescent immunoassay in which the competitor is designed by the conjugation of a target compound to a fluorescently labeled protein. Bovine serum albumin can be used and the whole complex can serve as a competitor in the determination of target binding to antibodies (Schobel et al. 1999). If the antibody is labeled with a FRET donor and albumin by a FRET acceptor, an enhancement of fluorescence emission of the donor can be observed at competitor displacement. This response can be analyzed in terms of target concentration. In such assays, the labeled albumin can be successfully substituted by Quantum Dots (see Section 4.4) or another nanoparticle.

This approach can be illustrated by the operation of the competitive glucose biosensor, the design of which has been reported by many authors. Glucose-binding proteins concanavalin A (Ballerstadt et al. 2004) or apo-glucose oxidase (Chinnayelka and McShane 2004) form stable complexes with labeled dextran, so that FRET occurs between the two labels. Glucose displaces dextran in the complex and this effect, dependent on glucose concentration, can be used for its determination.

In DNA hybridization assays, the following procedure can be applied. The sensor sequence labeled with a fluorescent dye is first hybridized with a competitor sequence labeled with a FRET acceptor (quencher). Hybridization with target DNA displaces the quencher and induces enhancement of fluorescence (Shoemaker and Linsley 2002).

Another example comes from antibody sensing. In the sensor targeted at determining the concentration of the explosive 2,4,6-trinitrotoluene (TNT) in aqueous environments, an antibody fragment was bound to a Quantum Dot, which served as a FRET donor. Trinitrotoluene analog was labeled with a dye that served as an acceptor (Goldman et al. 2003). In this case, the binding of the target TNT displaces its analog and induces a fluorescence enhancement.

Interesting examples of the application of competitor displacement assays can be found with aptamer sensors, which are the oligonucleotide binders selected from large libraries (Section 5.5). A fluorescently labeled DNA aptamer that detects oligonucleotide sequences can form a complex with short oligonucleotide, which contains a quencher. Dissociation of this complex, which occurs by substitution in target binding, brings about the fluorescence enhancement (Nutiu and Li 2004).

1.3.3 Competition Involving Two Binding Sites

An interesting approach has been applied recently in chemical sensing of ions (Royzen et al. 2005). Two ions are involved in this assay, Cu^{2+} as the target and Cd^{2+} as the competitor, together with two fluorescent metal chelating ligands (sensors). The latter were the dyes Calcein Blue and FluoZin-1, both of which bind the two types of ions with different affinities (Ligands 1 and 2, see Fig. 1.9). Both free ligands and their complexes with Cu^{2+} are almost non-fluorescent, whereas complexes with Cd^{2+} are strongly fluorescent with significant shifts of band maxima. In the absence of Cu^{2+} ions, the Cd^{2+} ions interact primarily with the ligand possessing a higher affinity. This gives a spectrum located at shorter wavelengths. When Cu^{2+} ions are present in the system, they replace Cd^{2+} ions and make Ligand 1 non-fluorescent. Being replaced, these ions become bound to the weaker Ligand 2. This results in the appearance of a strong fluorescence emission located at longer wavelengths. Thus, a convenient self-referencing two-band ratiometric recording of Cu^{2+} binding can be achieved.

1.3.4 Advantages and Limitations of the Approach

In principle, the approach based on competitor displacement is advantageous over techniques requiring the labeling of a whole pool containing the required target. However, in practice it is difficult to implement it in many particular cases and it is

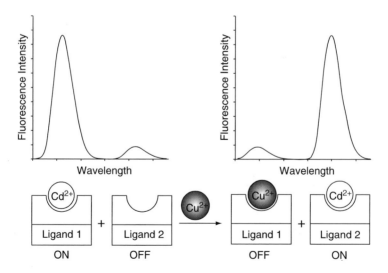

Fig. 1.9 Schematic representation of a ratiometric Cu^{2+} sensor system. In the absence of Cu^{2+}, Ligand 1 binds Cd^{2+}, which causes an enhancement of fluorescent ('ON' state). At the same time, Ligand 2 is in the uncomplexed form and is thus relatively poorly fluorescent ('OFF' state). In the presence of Cu^{2+}, Cd^{2+} is displaced from Ligand 1. Formation of [Cu(1)] complex quenches the fluorescence, therefore switching Ligand 1 'OFF'. Reproduced with permission from (Royzen et al. 2005)

even more difficult to make it generic. The affinity of receptor to target should be high enough to replace the competitor but its affinity to competitor should also be high in order to withstand chromatographic separation (if the assay is performed in solution) or washing the plates (if the sensor is immobilized). Therefore, it is often not reasonable to use the attached dye only as a label; this methodology develops in a different direction, with the application of responsive dyes (see Chapter 3). If the dye serving as a competitor responds to dissociation from the sensor by changing its fluorescence parameters, there may be no need to apply the separation/washing steps.

Some authors distinguish between the competition and displacement assays. In displacement assays the analyte should possess a high affinity to the receptor and be totally bound, substituting the labeled competitor that may be of lower affinity but present in saturating concentrations. This condition allows operating with multi-analyte sensor arrays and reading fluorescence after washing steps. In contrast, in competition assays both the target analyte and competitor may be of a lower affinity though an equilibrium condition should be established in their binding.

1.4   Sandwich Assays

The sandwich as a component of everyday food is known to everyone. It has three layers: bread at the bottom, then ham, cheese or butter in the middle and, again, bread on the top. The same is the composition of the *sandwich-type fluorescence assay unit*, in which three layers are formed on target binding (Fig. 1.10a).

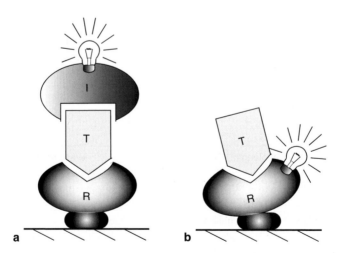

Fig. 1.10   The illustration of principles of operation of 'sandwich assay' with fluorescence detection and of the sensors based on the 'direct sensor' concept. **(a)** Sandwich assay. After the target (T) binding to immobilized receptor (R) and forming their complex on the surface, an indicator (I) that interacts with the target is introduced. The indicator is labeled, which provides labeling of the complex. **(b)** Direct sensor. Here the fluorophore labeling of the receptor is the integral part of the sensor. Its response is coupled directly with the target recognition event

In a sandwich configuration the receptor molecules are at the bottom. Being attached to a solid support they form a layer. In immunosensor techniques the receptors are antibodies that recognize large-size antigens or the antigens that recognize antibodies. The target molecules, when bound to receptor molecules, are in the middle of such a sandwich. Both sensors and targets are not labeled, so their complex formation does not produce any response signal. Therefore, to detect the sensor-target interaction, a third component is needed, which we will call the *indicator*. Its function is to visualize the formed complex with the aid of an attached label, so it should be applied on the top. Usually, it is the antibody that is specially prepared to recognize the immobilized target at a different binding site than the receptor. In the case of the receptor-target complex being formed, the indicator has to label this complex and if not – it has to be removed.

1.4.1 Sensing the Antigens and Antibodies

This type of analysis is important primarily in clinical practice. Antigens are the viruses and bacteria as well as different foreign materials that can penetrate the human body or with which it can come into contact. Antibodies are the molecules forming a response to antigens. Their detection is also very important. Clinical diagnostics, responses of individuals to treatment, prognosis and other issues can be resolved with this analysis. In both the tasks of sensing the antigens or the antibodies, the sandwich approach is commonly used. This is because of strong binding between antigens and antibodies (with K_d on the level of picomoles or nanomoles). There is often the possibility to substitute the real antigen (e.g., microbe) by a molecule of a rather small size (e.g., peptide or glycolipide), which retains the antibody-binding properties. The structural elements of an even smaller size, called antigenic determinants (epitops), are recognized by antibodies. The assay configurations for the detection of antigens and antibodies are schematically represented in Fig. 1.11.

The secondary antibody serving as the indicator (the capture antibody) can be modified in different ways to provide the most sensitive response. Because of this, the mechanism of indicator operation is rather general. In immunoassays detecting the antibodies, it is simply a different antibody reacting with the target antibody. This prevents many problems. The function of the indicator is universal, since it does not need to display any site-sensitivity. There is also no need for the reporting function of the label. The label serves only to measure the amount of bound indicator (Marquette and Blum 2006).

With the application of fluorescence dyes, the sandwich assays can demonstrate a high sensitivity (Swartzman et al. 1999). These assays can be made even more sensitive, since they allow catalytic amplification of the sensing signal. In a popular version of this assay, called *enzyme-linked immunosorbent assay* (ELISA), the indicator antibody can be fused to an enzyme, which, on the addition of substrate, produces a colored (light-absorbing) reaction product (Mendoza et al. 1999). Upon incubation, each such molecular complex can generate many colored molecules. The concentration producing the response can be analyzed as a function of target concentration. Hence, the substantial increase in sensitivity.

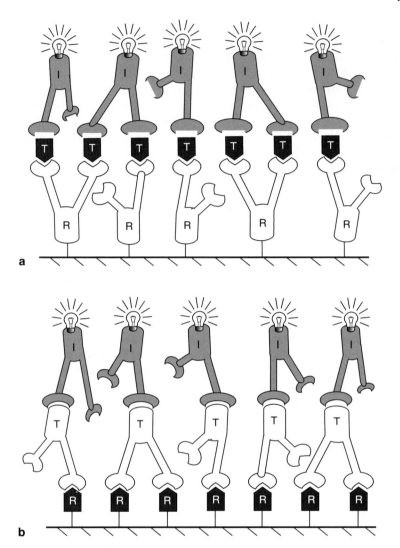

Fig. 1.11 Schematic representation of assay formats for sensing the antigens and the antibodies. (a) Detection of antigens. Antibodies serving as receptors (R) are immobilized on a solid support. Each antibody (usually, bivalent) captures its respective target (T) (called antigen) from a complex mixture. Detection is by recognizing the bound antigen by a second antibody serving as an indicator (I) that contains an optical label or enzyme producing optically detectable product (ELISA). (b) Detection of antibodies. Antigen is the receptor (R) that is immobilized on a solid support. Antibodies specific to this antigen are the targets (T) captured from a complex mixture (e.g., blood serum). Detection occurs by recognizing the bound antibodies by other antibodies serving as indicators (I). They react with the target antibodies as with their own antigens. The latter antibodies can be labeled for providing an optical signal as in case (a)

The application of *fluorogenic* (producing fluorescence) substrates allows a substantial increase in precision and sensitivity (Meng et al. 2005). Some enzyme-labeled fluorescence kits use phosphatase substrate, which has weak blue fluorescence in solution but upon enzymatic cleavage forms a bright and highly photostable yellow-green fluorescent precipitate. The sites of the binding of the antibody-phosphatase complex can be detected by fluorescent precipitate deposited at the site of enzymatic activity. A very high sensitivity of these assays can be achieved, which ranges from picomoles to nanomoles in target concentrations. This is quite sufficient to satisfy many demands in clinical diagnostics. In many cases however, the labeling without catalytic amplification can be quite sufficient for achieving this level of sensitivity, especially when ultra-bright particles, such as dye-doped nanoparticles or Quantum Dots (see Chapter 4), are used for labeling.

1.4.2 Ultrasensitive DNA Detection Hybridization Assays

The sandwich assay format can be applied for the detection of all types of analytes that contain two distinct molecular recognition sites, one for binding to the receptor and the other for attaching the indicator. DNA hybridization assays are remarkable in this respect. The recognition and detection of the single-chain DNA molecule can be provided in such a way that one part of it interacts with the receptor DNA, forming a double helix. The remaining free part can interact with the indicator DNA, which can carry a fluorescent label. Here, the most specific interaction with the formation of double helix can be realized as well.

The brightest (in all aspects) application of this principle can be found in the work with brightly fluorescent dye-doped silica nanoparticles (Fig. 1.12). Their application allows detection of the DNA target molecules down to subpicomolar levels (Zhao et al. 2003).

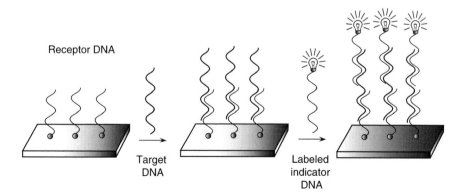

Fig. 1.12 Schematic representation of a sandwich DNA assay based on an indicator DNA conjugated with dyes or nanoparticles

1.4.3 *Advantages and Limitations of the Approach*

The successful application of sandwich assays mostly depends on the possibility of recognizing *two independent binding sites in the target*. This is because with this technology the target has to bind the receptor and the indicator independently. This limits the assay to rather large molecules (e.g., proteins) or to supramolecular structures, in which the binding to two different sites can be independent. Many small target molecules do not possess multiple structurally separate sites for binding and therefore cannot be detected by sandwich assays.

Since the indicator can be coupled with an enzyme generating a light-absorbing or fluorescent product, signal amplification can be achieved, leading to a very high sensitivity of the assay. However, in all cases the response *is not immediate and not direct,* so that a separation of the bound and unbound indicators is always needed.

The sandwich assay technique is well adapted to antibodies serving either as the receptors or the targets. Regarding applications on a more general scale, their range is very limited. When used for the detection of single-strand DNA, two sites of hybridization have to be used. This technique is costly and time-consuming. In addition, it is subject to non-specific absorption effects that increase with the number of used components, especially of proteins that can denature and provide nonspecific binding.

1.5 Catalytic Biosensors

Here we concentrate on fluorescence-based *catalytic biosensors*. These are sensors, the mechanism of operation of which involves a catalytic event with generation (or consumption) of fluorescent species and in a strict sense they cannot be considered as 'sensors'. The target can be detected as the modulator of this biocatalytic activity. This modulator effect is specific for a particular catalyst (enzyme, catalytic antibody, ribozyme) and can be either competitive with a fluorogenic substrate or can be bound at a regulatory site and provide the allosteric modulation of enzyme activity.

There is also a different possibility for using species with a catalytic function in sensing. Catalysis can be used for the *amplification of the reporting signal* produced by other sensors. Techniques based on this principle will also be considered.

Enzyme activity itself is the focus of analytical methods. Many of them are actively used for the determination of enzyme activity, being based on the detection of fluorescent substrates or products in direct or coupled reactions. These cases will not be discussed here and the reader is referred to the vast biochemical literature.

1.5.1 *Enzymes as Sensors*

It is well-known that the activity of many enzymes can be modulated by binding specified molecules to the sites different from a catalytic site. Such enzymes are called *allosteric enzymes*, the sites of binding are called *allosteric sites* and the

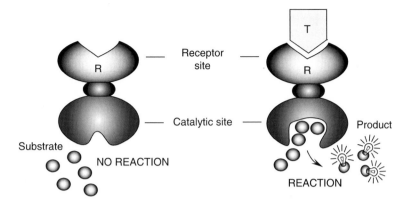

Fig. 1.13 The principle of allosteric enzyme sensors. Binding of a target at the allosteric binding site influences enzyme catalytic activity with the production of a chromogenic or fluorescent reaction product. Fluorescence of this product can serve as a measure of target binding

molecules producing the modulation of activity are the *allosteric effectors*. These allosteric effectors can be determined as the targets. In this case, the rate of catalytic transformation of substrate with the appearance of fluorescent products can be used as a quantitative measure of target concentration.

The binding of the effector modulates catalytic activity by coupling with a conformational change transmitted to the substrate binding site. In *allosteric sensors*, the catalytic activity of an enzyme is used not only to provide but also to amplify the response resulting from the binding of an effector. This is simply because a single binding event triggers the catalytic transformation of many substrate molecules and in this way the reporting signal can be accumulated. This idea is illustrated in Fig. 1.13.

Thus, in such catalytic sensors the targets are analyzed as the effectors that bind to a site distinct from the active rather than the catalytic site and the binding signal is transmitted to a catalytic site through a conformational change, resulting in a modulation of catalytic activity (Villaverde 2003). This principle of catalytic amplification is not limited to enzyme reactions. It can be applied to organometallic and supramolecular catalysis (Zhu and Anslyn 2006).

1.5.2 Ribozymes and Deoxyribozymes Sensors

Nucleic acids that behave as enzymes are called *ribozymes* and *deoxyribozymes*. Many of their mutated forms can be identified and some of them exhibit allosteric properties. They have been used for the purpose of the design of catalytic biosensors (Stojanovic and Kolpashchikov 2004) based on switching between two or more catalytic regimes. Engineered molecular switches can be used to report on the presence of specific analytes in complex mixtures, making possible the creation of new types of biosensor devices and genetic control elements (Penchovsky and Breaker 2005).

One of the first impressive applications of ribozymes was a sensor system developed for the determination of theophylline concentrations based on a theophylline-dependent allosteric ribozyme in combination with an RNA substrate (Soukup and Breaker 1999). This substrate is double-labeled with a fluorophore and a quencher dye. In the presence of theophylline, a hammerhead ribozyme domain is switched into an active conformation by the action of a theophylline-binding aptamer domain. Upon substrate cleavage, the quencher is removed from the vicinity of the fluorophore, causing an increased fluorescence signal. Real-time analysis of the cleavage reactions, both under single and multiple turnover conditions, revealed a dependence of theophylline concentration on the cleavage rate within a range from 0.01 to 2 mM. The structurally similar molecule caffeine, however, had no detectable influence on the fluorescence signal (Frauendorf and Jaschke 2001).

Ribozymes and deoxyribozymes can also be applied to construct catalytic molecular beacons (Section 6.3) and bring to this technology the ability of signal amplification. For example, the specific oligonucleotide recognition property of the stem-loop structure can be coupled with a catalytic activity, such as the deoxyribozyme RNase activity. These catalytic molecules can monitor the concentrations of not only oligonucleotides but also proteins or a variety of other molecules (Navani and Li 2006).

1.5.3   Labeling with Catalytic Amplification

High sensitivity in a detection system is a very desirable factor, which in many cases determines the strategy of sensor development. Coupling the response function with the catalytic or biocatalytic transformation of a reporter signal is one of the earliest ideas in the field of sensing and it is still very actively explored. Most of these applications are related to sandwich (e.g., ELISA) techniques, where a second binder (an indicator) carries the label and it is the brightness of this label that determines the overall sensitivity (Section 1.4). Here the idea of connecting sensing and catalysis is simple. An indicator is chemically coupled to an enzyme which, upon addition of substrate, generates a colored (measured by light absorption) or fluorescent product at the site of binding. Then, if an excess of reaction substrate is provided, the *sensitivity will depend on the incubation time* needed to generate the detectable product. This situation is depicted schematically in Fig. 1.14.

The effect of amplification appears due to *multiple substrate transformations at a single site*. In common affinity-based sensors the sensitivity is limited by the 1:1 binding stoichiometry between the target and the sensor, which allows the generation of a response signal only in a single reporter molecule. As a consequence, most biosensors, whatever the transduction principle, are less sensitive than techniques that use catalytic amplification, such as ELISA. This methodology is not focused on the determination of catalytic activity or substrate concentration, although such information can be obtained. The allosteric modulation of activity in this application is also not needed. Catalyst and substrate are selected here to provide the most efficient response. In the meantime, it is essential to note that the response relies on the correct assembly of the whole complex (R) – (T) – (I) – (E), which limits this methodology to heterogeneous sensor formats.

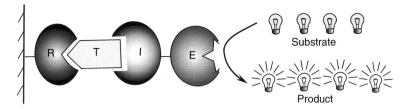

Fig. 1.14 The scheme illustrating operation of an assay using the labeled indicator with catalytic amplification. Primarily, the sensor exposing the recognition unit (R) captures the target (T) that on the next step is recognized by an indicator (I). The indicator is fused to the enzyme (E). When the R – T – I complex is formed, it can then be detected by the transformation of a nonfluorescent substrate into a fluorescent product catalyzed by the enzyme

As shown above, such a limit does not exist if the catalytic transformation is provided by an allosteric catalyst and the modulation of its activity is directly coupled with the target binding. Analyzing this difference, we can find an analogy with the difference between sensing in sandwich assay format and direct sensing (which will be discussed in the next section). The determination of the *allosteric effector* as the target resembles direct fluorescence sensing because in both cases a very efficient and specific transduction mechanism is needed. In direct sensing, the proximity effects or conformational changes in the sensor are important for providing the response. Likewise, these changes are needed for the operation of an allosteric enzyme as a sensor. They are also required here for communication between the target-binding site and the fluorescence reporter site. This restricts the area of application of such sensors in three respects. There should be (a) a proper allosteric binding site specific for the target, (b) a proper catalytic site and (c) a proper allosteric coupling between the two.

It needs to be stressed again that these conditions are not required if the enzyme (or catalyst, in general) functions only as the catalytic amplifier. Here, the enzyme is used for generating the reporter signal and all recognition roles are played by other partners in a complex formed on target recognition. This is essentially the sandwich assay.

1.5.4 Advantages and Limitations of the Approach

The sensors with catalytic amplification have a great capacity to increase sensitivity. In principle, when the enzyme reaction is triggered by target binding, the accumulation of a light-absorbing or fluorescent product will continue until the reaction is stopped or the substrate is exhausted. However, introduction into the sensor system of an additional player, such as an enzyme, requires keeping the medium conditions (such as temperature and pH) within very narrow ranges. In addition, the catalyst and fluorogenic substrate concentrations should be set constant.

Thus, in the application of catalytic biosensors there is a great deal of compromise (de Castro and Herrera 2003). On the one hand is the simplicity of this approach and on the other hand the necessity in many real cases to use an additional chromatographic separation. On one side is a high sensitivity that is achieved due to the accumulation of an analyzed reaction product and on the other – a lengthy incubation step needed for this accumulation. In all cases, manipulation of the sample, such as washing and the application of additional reagents, is needed.

1.6   Direct Reagent-Independent Sensing

It is known that different parameters of fluorescence emissions can be modulated in broad ranges by weak intermolecular interactions of the dyes (see Chapter 3). It must be a good idea to use this property in a straightforward and most efficient way, by making the sensor with a *direct response* to sensor-target interactions. If we are successful, we may obtain a simple and universal means to provide fluorescence sensing without any limitations on the assay format.

1.6.1   The Principle of Direct 'Mix-and-Read' Sensing

On the pathway to the realization of the idea of *direct sensing*, special requirements should be imposed on the properties of fluorescent dyes. Instead of being simple labels or tags they have to attain all the features of molecular instruments reporting on particular changes in van der Waals interactions, hydrogen bonds and electrostatic forces. These changes should provide the reporting signal. In many cases, sophisticated transduction mechanisms have to be developed to obtain the signal on an *elementary act of sensor-target binding* and transform it into the signal from the fluorescence reporter.

Let us specify what we wish to achieve. First, these new instruments should be *molecular* sensors. This means that their operation (both sensing and response) has to take place at a molecular level. If necessary, they can easily be integrated into supramolecular and macroscopic devices, although their basic operation should not require this. Therefore, they can be used for assays either in the homogeneous phase (in solutions) or on their attachment to solid support. Secondly, they should be *direct* sensors. This means that they should avoid any intermediate signal-transduction factors or steps in recording the sensing event. In addition, they should be *reagent-independent* (sometimes called reagentless) sensors, so that their operation does not require the supply and consumption of additional reagents. The incorporation into sensor molecules of a fluorescent dye that changes its emission properties during the recognition event is the most promising approach for detecting the sensor – target binding. There may be many possibilities for the operation with different kind of nanoparticles. They may serve as nano-size support, as reporters or as modulators and amplifiers of the report.

The fact that no additional factors are required for the operation of direct sensors is very important. A device made from such sensors can operate in a broad range of external conditions and in any heterogeneous environment, be it a waveguide surface, a microplate or the interior of a living cell. Moreover, since this technique does not require any separation or washing steps and allows a distinction between bound and unbound analyte in *thermodynamic equilibrium*, it may serve in a very broad range of target affinities (Chapter 2).

1.6.2   Contact and Remote Sensors

In the case of direct sensing, the fluorescent reporter is an integral part of the sensor (Fig. 1.10b) and its function is to provide the change of fluorescence parameters on analyte binding. This function can be introduced by covalent labeling of the recognition unit by the dye, to form a molecular sensing device. In the meantime, this straightforward and simple construction with an efficient response function is often difficult to achieve and there is no general methodology for this. For the transduction of the sensing event directly into a fluorescence response, one has to choose between several possibilities (Fig. 1.15).

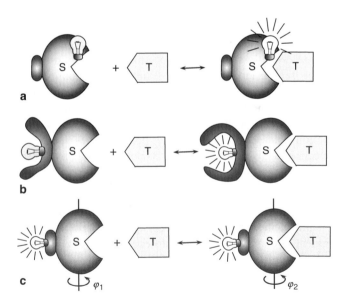

Fig. 1.15 Different possibilities for the direct sensing of target (T) by the sensor (S) labeled with fluorescence reporter. (a) In contact sensing, the reporter is involved in direct contact with the bound target. (b) In remote (allosteric) sensing, the reporter responds to conformational change in the sensor molecule occurring on target binding and its direct contact with the target is not needed. (c) In anisotropy sensing, the rotational mobility of the sensor decreases on the target binding (due to an increased mass of a complex) and this change is reflected as an increase of fluorescence anisotropy

1. *Contact sensing.* Here the fluorescent reporter is involved in direct contact with the target and is a part of the recognition process. As a result of reporter-target interactions, one or more of its fluorescence parameters change. They can be scaled in target concentration. Since the reporter has to be present in the sensor-target contact area, there is always a danger that it may influence the recognition of the target by producing steric and energetic barriers. Therefore, a compromise should be found between providing its strong involvement in analyte binding with the production of a strong fluorescence response (that may also hamper recognition) and a weaker involvement of the dye in binding that could sometimes be insufficient to provide a detectable response in fluorescence. Direct contact between the target and fluorescence reporter can also be beneficial when it allows modulating the excited-state reaction occurring in the dye: electron, charge or energy transfers (see Chapter 6).

The principle of contact sensing has found optimal realization in the design of small chemical sensors detecting ions (de Silva et al. 1997; Valeur 2002). It has also found application in sensing with peptides and proteins. In particular, antibody sensors were designed in which the environment-sensitive dye is attached close to the antigen binding site (Renard et al. 2002); the knowledge of a three-dimensional structure of antibody allowed suggesting this site. This is an example of a structure-based '*rational design*' approach. Meantime, the approaches using *peptide or nucleic acid libraries* in which the dye is included into their covalent structure may appear extremely efficient due to the possibility of selection based on the optimization of both sensing and reporting functions.

2. '*Remote*' sensing. In the case of remote (e.g., allosteric) sensing, the reporting dye is located distantly from the site of analyte recognition. To provide the response, the sensor molecule should exhibit some conformational change reflected by reporter dye located at the periphery, away from the binding site. The possibilities of the realization of this principle are limited to those sensor molecules, in which relatively strong global conformational changes accompany the recognition event. But if these changes do occur, they allow many possibilities for selecting the dye and optimizing its positioning in the sensor structure.

The changes of protein conformation on binding various ligands, including protein-protein and protein – nucleic acid interactions, are very common (Gerstein et al. 1994), up to the complete unfolding-folding (Demchenko 2001). This allows realizing many possibilities in sensor design. Recent applications of this principle include the rearrangement of domains in sensing proteins and disruption of beacons, as in oligonucleotide sensors. Sensing with ligand-binding proteins (Dwyer and Hellinga 2004; Sapsford et al. 2006) often follows the principle of remote fluorescence detection. The sensing with molecular beacons and other polymers and biomolecules exhibiting conformational change deserves discussion in more detail (Section 6.3).

3. A general property of the sensor should change on analyte binding, such as *rotational mobility* that can be observed as the change of *emission anisotropy* (Section 3.2).

The mechanisms coupling the binding event with a signal transduction for the generation of a response in emission will be discussed in different sections of this book. We will observe that the level of structural integration between the receptor and the reporter functional elements can be very different. This produces the basis for the development of many simple direct 'mix-and-read' methods.

1.6.3 Advantages and Limitations of the Approach

A very attractive feature of direct sensors is the possibility of using nearly all possible biopolymer and mimetic systems, in which molecular recognition can be realized or simulated. This makes the background for the development of a wide range of molecular biosensor technologies, which include the use of antibodies and their fragments, ligand-binding proteins (natural and raised by selection using several scaffolds), enzymes and their chemically modified products, synthetic peptides and natural or synthetic polynucleotides. A wide variety of targets can also be analyzed, from small ions to supramolecular structures and living cells.

Still, the sensitivity of direct sensors is not as high as of those indirect sensors and immunosensors that operate involving the multi-step coupled reactions and that use signal amplification, e.g., ELISA (McHugh 1994; Mendoza et al. 1999; Swartzman et al. 1999). Different methods for the dramatic sensitivity increase have been suggested recently and now we observe that the concept of a direct regent-free sensing offers new fascinating possibilities for the development of sensing systems.

The strong advantage of direct sensors is due to the fact that they explore a potentially simple *one-step binding reaction with the response on a time scale of measurement*. Since this reaction allows the immobilization and integration of sensor molecules into arrays, all the range of molecular, nanoscale, microscopic and macroscopic devices can be adapted to this concept. Multiple advantages make direct sensors progressive and prospective for future applications.

Another major advantage of direct sensors must be their *versatility*, which is out of competition. Direct sensors do not need the plates or any other solid supports but can use them to achieve the spatial resolution for providing a simultaneous response to many targets. In direct sensing the reaction of target binding can be performed with the sensor molecules already in dilute solutions and they can also be incorporated into nanoparticles, membrane vesicles, living cells, etc.

No less important is the advantage of the *high speed of analysis* that is limited only by target diffusion in the medium (Brune et al. 1994) and by the rate of conformational changes in the sensor molecule if they occur on target binding to provide fluorescence response.

With direct sensors, the *range of quantitatively determined target concentrations* can be extremely broad, which can be achieved by a variation of the binding constants as described in Chapter 2. This is the result of the rapid establishment of dynamic equilibrium between free and bound targets in accordance to *mass action law*. Only the bound analyte generates the fluorescence response, which allows easy determination of *total concentration* in the volume of any size. Such a sensor can be adapted for a particular range of target concentrations by a variation of the binding constant. Moreover, since direct sensors allow to measure target binding at equilibrium, they, in principle, can be used *many times* without regeneration and may allow *continuous monitoring*.

Thus, as an idea, the direct sensor format is always preferable. But in practice it has to compete with other formats in terms of sensitivity and adaptability to special instrumental conditions.

Sensing and Thinking 1: How to Make the Sensor? Comparison of Basic Principles

The reader must be very much surprised by the fact that in such an important field as chemical sensing and biosensing there is no single leading idea and dominant technology. A variety of technologies develop in parallel and every one of them has its advocates and critics. For instance, for glucose sensing, enzymatic electrochemical techniques, light-addressable potentiometric sensors, enzyme-containing photosensitive membranes based on the microcantilever and other techniques were suggested in addition to sensing based on fluorescence detection. In this respect, fluorescence sensing develops in a very stimulating and productive environment. Its positions are very strong but not dominant and in many application areas, the competition with other technologies is very tough.

In addition, a strong competition exists between different approaches based on fluorescence, in which fluorescence sensing of glucose can be a characteristic example. Thus, lectins, glucose-binding proteins, enzymes of glucose metabolism and boronic acid derivatives are used for glucose recognition together with a response based on intensity, lifetime, spectral shifts, the resonance energy transfer and their combinations (Moschou et al. 2006; Pickup et al. 2005a). In the meantime, the diabetes patients need simple and convenient glucose monitoring *in vivo*. The result that could fully satisfy them is still not achieved (Pickup et al. 2005b). Therefore, the competition continues.

Answering the questions listed below will help the reader to move from passive reading to thinking, analysis and creativity.

Questions and Problems

1. List all possible areas of applications of molecular sensors that you know. Explain what advantages appear on sensor immobilization to the surface compared to an assay in solution and how this expands or limits their application areas.

2. Explain why the potentially most general approach based on a label-free and immobilization-free principle has a rather narrow range of applications in practice.
3. What is the range of the applicability of the sensing coupled with enzyme reactions? What is the difference in application between sensing based on an enzyme reaction and a ligand-binding (inhibitor-binding) allosteric sensing with the same or similar enzyme?
4. Evaluate comparatively the label-based methods using fluorescence detection. What is better to label, the receptor, the target together with the whole pool of potential targets, the capture antibody or the competitor? What range of applications is expected on the application of any of these possibilities?
5. What are the areas of application of indirect and direct sensors? Which are broader? In the case that both techniques are applicable, what is preferable based on such criteria as complexity, time consumption and cost?
6. In what sense and under which conditions are the techniques based on radioactive labeling (isotope tracing) and fluorescent labeling analogous? What are the requirements for the fluorescence reporters in these applications?
7. How does DNA hybridization occur? How to use this process for detecting particular DNA sequence? What is the role of fluorescent labels in these assays and how in this case can the fluorescence response be technically realized?
8. Describe the problems that appear in the labeling of the protein targets. Why is it better to avoid labeling of a protein pool? Why is this approach not applicable to low-affinity targets?
9. What is the role of a fluorescence reporter in competitor displacement assay? Suggest examples for proper applications of these assays. Can the sensor operate with an immobilized competitor? How do such sensors work in the case of DNA intercalating dyes? What can be additionally achieved by labeling both sensor and competitor?
10. Think about the application of competitor displacement assay in complex situations, such as the case of two or more types of binding sites with different affinities.
11. Why in sandwich assays is the third partner in target recognition, the indicator (e.g., capture antibody) needed? Why is such a complex composition devised? What is gained and what is lost in its application?
12. Explain in detail how ELISA works. How is the reporting signal generated and how is the effect of amplification achieved? Why is this assay so good with antibodies but is less practical with other binders?
13. The dyes can respond to a target binding directly, by the change of parameters of their fluorescence. Where should they be located in this case? What properties of the sensor should correspond to different locations of the reporting dye?
14. Select one protein target (let it be insulin) and explain how you will apply all sensor strategies described above. Provide their comparative analysis in terms of speed, complexity (number of needed manipulations), skills needed by the performers, etc.

References

Altschuh D, Oncul S, Demchenko AP (2006) Fluorescence sensing of intermolecular interactions and development of direct molecular biosensors. Journal of Molecular Recognition 19:459–477

Amine A, Mohammadi H, Bourais I, Palleschi G (2006) Enzyme inhibition-based biosensors for food safety and environmental monitoring. Biosensors & Bioelectronics 21:1405–1423

Ballerstadt R, Polak A, Beuhler A, Frye J (2004) In vitro long-term performance study of a near-infrared fluorescence affinity sensor for glucose monitoring. Biosensors & Bioelectronics 19:905–914

Battiston FM, Ramseyer JP, Lang HP, Baller MK, Gerber C, Gimzewski JK, Meyer E, Guntherodt HJ (2001) A chemical sensor based on a microfabricated cantilever array with simultaneous resonance-frequency and bending readout. Sensors and Actuators B-Chemical 77:122–131

Brune M, Hunter JL, Corrie JET, Webb MR (1994) Direct, real-time measurement of rapid inorganic phosphate release using a novel fluorescent probe and its application to actomyosin subfragment 1 ATPase. Biochemistry 33:8262–8271

Chinnayelka S, McShane MJ (2004) Resonance energy transfer nanobiosensors based on affinity binding between apo-enzyme and its substrate. Biomacromolecules 5:1657–1661

de Castro MDL, Herrera MC (2003) Enzyme inhibition-based biosensors and biosensing systems: questionable analytical devices. Biosensors & Bioelectronics 18:279–294

de Silva AP, Gunaratne HQN, Gunnaugsson T, Huxley AJM, McRoy CP, Rademacher JT, Rice TE (1997) Signaling recognition events with fluorescent sensors and switches. Chemical Reviews 97:1515–1566

Demchenko AP (2001) Recognition between flexible protein molecules: induced and assisted folding. Journal of Molecular Recognition 14:42–61

D'Orazio P (2003) Biosensors in clinical chemistry. Clinica Chimica Acta 334:41–69

Dwyer MA, Hellinga HW (2004) Periplasmic binding proteins: a versatile superfamily for protein engineering. Current Opinion in Structural Biology 14:495–504

Ercelen S, Klymchenko AS, Mely Y, Demchenko AP (2005) The binding of novel two-color fluorescence probe FA to serum albumins of different species. International Journal of Biological Macromolecules 35:231–242

Frauendorf C, Jaschke A (2001) Detection of small organic analytes by fluorescing molecular switches. Bioorganic & Medicinal Chemistry 9:2521–2524

Freeman WM, Robertson DJ, Vrana KE (2000) Fundamentals of DNA hybridization arrays for gene expression analysis. Biotechniques 29:1042–1055

Gauglitz G (2005) Direct optical sensors: principles and selected applications. Analytical and Bioanalytical Chemistry 381:141–155

Gerstein M, Lesk AM, Chothia C (1994) Structural mechanisms for domain movements in proteins. Biochemistry 33:6739–6749

Goldman ER, Anderson GP, Lebedev N, Lingerfelt BM, Winter PT, Patterson CH, Mauro JM (2003) Analysis of aqueous 2,4,6-trinitrotoluene (TNT) using a fluorescent displacement immunoassay. Analytical and Bioanalytical Chemistry 375:471–475

Homola J (2003) Present and future of surface plasmon resonance biosensors. Analytical and Bioanalytical Chemistry 377:528–539

Homola J, Vaisocherova H, Dostalek J, Piliarik M (2005) Multi-analyte surface plasmon resonance biosensing. Methods 37:26–36

Kellner R, Mermet J-M, Otto M, Valcarcei M, Widmer HM (2004) Analytical chemistry. Wiley-VCH, New York

Kodadek T (2001) Protein microarrays: prospects and problems. Chemistry & Biology 8:105–115

Kodadek T (2002) Development of protein-detecting microarrays and related devices. Trends in Biochemical Sciences 27:295–300

Ladbury JE, Chowdhry BZ (1996) Sensing the heat: the application of isothermal titration calorimetry to thermodynamic studies of biomolecular interactions. Chemistry & Biology 3:791–801

Lange K, Rapp BE, Rapp M (2008) Surface acoustic wave biosensors: a review. Analytical and Bioanalytical Chemistry 391:1509–1519

Liu YS, Ugaz VM, Rogers WJ, Mannan MS, Saraf SR (2005) Development of an advanced nano-calorimetry system for material characterization. Journal of Loss Prevention in the Process Industries 18:139–144

Luppa PB, Sokoll LJ, Chan DW (2001) Immunosensors - principles and applications to clinical chemistry. Clinica Chimica Acta 314:1–26

MacBeath G, Schreiber SL (2000) Printing proteins as microarrays for high-throughput function determination. Science 289:1760–1763

Marquette CA, Blum LJ (2006) State of the art and recent advances in immunoanalytical systems. Biosensors & Bioelectronics 21:1424–1433

McHugh TM (1994) Flow microsphere immunoassay for the quantitative and simultaneous detection of multiple soluble analytes. Methods in Cell Biology 42 Pt B:575–595

Mendoza LG, McQuary P, Mongan A, Gangadharan R, Brignac S, Eggers M (1999) High-throughput microarray-based enzyme-linked immunosorbent assay (ELISA). Biotechniques 27:778–788

Meng Y, High K, Antonello J, Washabaugh MW, Zhao QJ (2005) Enhanced sensitivity and precision in an enzyme-linked immunosorbent assay with fluorogenic substrates compared with commonly used chromogenic substrates. Analytical Biochemistry 345:227–236

Mock DM, Lankford G, Horowitz P (1988) A study of the interaction of avidin with 2-Anilinonaphthalene-6-Sulfonic Acid (2,6 Ans) as a probe of the biotin binding-site. Clinical Research 36:A895

Moschou EA, Bachas LG, Daunert S, Deo SK (2006) Hinge-motion binding proteins: unraveling their analytical potential. Analytical Chemistry 78:6692–6700

Navani NK, Li YF (2006) Nucleic acid aptamers and enzymes as sensors. Current Opinion in Chemical Biology 10:272–281

Nutiu R, Li YF (2004) Structure-switching signaling aptamers: Transducing molecular recognition into fluorescence signaling. Chemistry – A European Journal 10:1868–1876

Palecek E (2005) Electroactivity of proteins and its possibilities in biomedicine and proteomics (Ch. 19). In: Palecek E, Scheller F, Wang J (eds) Electrochemistry of nucleic acids and proteins. Towards electrochemical sensors for genomics and proteomics. Elsevier, Amsterdam, pp. 690–750

Palecek E, Jelen F (2005) Electrochemistry of nucleic acids. In: Palecek E, Scheller F, Wang J (eds) Electrochemistry of nucleic acids and proteins. Towards electrochemical sensors for genomics and proteomics. Elsevier, Amsterdam, pp. 74–174

Penchovsky R, Breaker RR (2005) Computational design and experimental validation of oligonucleotide-sensing allosteric ribozymes. Nature Biotechnology 23:1424–1433

Piatek AM, Bomble YJ, Wiskur SL, Anslyn EV (2004) Threshold detection using indicator-displacement assays: An application in the analysis of malate in Pinot Noir grapes. Journal of the American Chemical Society 126:6072–6077

Pickup JC, Hussain F, Evans ND, Rolinski OJ, Birch DJS (2005a) Fluorescence-based glucose sensors. Biosensors & Bioelectronics 20:2555–2565

Pickup JC, Hussain F, Evans ND, Sachedina N (2005b) In vivo glucose monitoring: the clinical reality and the promise. Biosensors & Bioelectronics 20:1897–1902

Renard M, Belkadi L, Hugo N, England P, Altschuh D, Bedouelle H (2002) Knowledge-based design of reagentless fluorescent biosensors from recombinant antibodies. Journal of Molecular Biology 318:429–442

Rich RL, Myszka DG (2006) Survey of the year 2005 commercial optical biosensor literature. Journal of Molecular Recognition 19:478–534

Royzen M, Dai ZH, Canary JW (2005) Ratiometric displacement approach to Cu(II) sensing by fluorescence. Journal of the American Chemical Society 127:1612–1613

Sapsford KE, Pons T, Medintz IL, Mattoussi H (2006) Biosensing with luminescent semiconductor quantum dots. Sensors 6:925–953

Schobel U, Egelhaaf HJ, Brecht A, Oelkrug D, Gauglitz G (1999) New-donor-acceptor pair for fluorescent immunoassays by energy transfer. Bioconjugate Chemistry 10:1107–1114

Shoemaker DD, Linsley PS (2002) Recent developments in DNA microarrays. Current Opinion in Microbiology 5:334–337

Soukup GA, Breaker RR (1999) Engineering precision RNA molecular switches. Proceedings of the National Academy of Sciences of the United States of America 96:3584–3589

Stojanovic MN, Kolpashchikov DM (2004) Modular aptameric sensors. Journal of the American Chemical Society 126:9266–9270

Swartzman EE, Miraglia SJ, Mellentin-Michelotti J, Evangelista L, Yuan PM (1999) A homogeneous and multiplexed immunoassay for high-throughput screening using fluorometric microvolume assay technology. Analytical Biochemistry 271:143–151

Thevenot DR, Toth K, Durst RA, Wilson GS (2001) Electrochemical biosensors: recommended definitions and classification. Biosensors & Bioelectronics 16:121–131

Tse WC, Boger DL (2004) A fluorescent intercalator displacement assay for establishing DNA binding selectivity and affinity. Accounts of Chemical Research 37:61–69

Valeur B (2002) Molecular fluorescence. Wiley-VCH, Weinheim

Venkatasubbarao S (2004) Microarrays - status and prospects. Trends in Biotechnology 22:630–637

Villaverde A (2003) Allosteric enzymes as biosensors for molecular diagnosis. FEBS Letters 554:169–172

Warsinke A, Benkert A, Scheller FW (2000) Electrochemical immunoassays. Fresenius Journal of Analytical Chemistry 366:622–634

Wiskur SL, Ait-Haddou H, Lavigne JJ, Anslyn EV (2001) Teaching old indicators new tricks. Accounts of Chemical Research 34:963–972

Zhao X, Tapec-Dytioco R, Tan W (2003) Ultrasensitive DNA detection using highly fluorescent bioconjugated nanoparticles. Journal of American Chemical Society 125:11474–11475

Zhu L, Anslyn EV (2006) Signal amplification by allosteric catalysis. Angewandte Chemie-International Edition 45:1190–1196

Chapter 2
Theoretical Aspects

Sensing always involves an interaction between the target and the system that is able to detect it (the sensor). The essence of sensing methodologies is the ability to tell about the presence and quantity of target species in a tested sample from a signal produced by their small fraction bound to the sensor.

The amount of target binding sites (*receptors*) composing the sensors is commonly smaller by many orders of magnitude than the number of targets (*analytes*) in the tested system. For the correct detection of this number, different strategies can be applied.

One of these strategies is to take out a small aliquot of sample from a tested system, expose it to strong or even irreversible binding by the sensor *receptor* and then, after calculating the number of bound targets, extrapolate this result to a whole tested system. An example of such an approach is the analysis of blood or urine samples in clinical laboratories.

The other strategy allows avoiding taking the aliquots. The sensor is designed in such a way that the target-receptor binding is *reversible*. Upon exposure to the tested sample, only a part (often, a small number) of the target species bind to a sensor and this binding can be sufficient to tell us about the true target concentration. In this case, the testing is not limited to the size of the sample. It can even be the whole sea in which the sensor can be immersed for continuous monitoring of its pollution. In this case, we must describe and use the regularities existing between the bound and unbound species to determine their whole number, keeping in mind that only the bound species produce the sensor response signal.

The interactions on every structural level involve the *dimension of time*. The target and receptor have to approach and 'recognize' each other by many translational and rotational diffusion steps. A 'conformational adaptation' can be involved in these steps with time-dependent strengthening of these interactions with the true target and discrimination of target analogs if they are present in the system.

Thus, the target-receptor binding involves two important aspects. (a) *Thermodynamic*, describing the equilibrium between the bound and unbound analyte. (b) *Kinetic*. The target-receptor binding needs time for their mutual diffusion and optimizing their interactions. The true equilibrium may not be reached in the course of testing, which means that the kinetic variables may significantly influence the readout signal. In this chapter, after introducing the general parameters that characterize the sensor, we concentrate on these issues.

2.1 Parameters that Need to Be Optimized in Every Sensor

The target concentration ranges that are needed to be analyzed in solutions are tremendous – starting from molar ones down to single molecules (Table 2.1). This puts stringent demands on many sensor parameters. Primarily, it is the *absolute sensitivity*, which is the ability to detect the smallest amounts of target in the tested system. Next is the *dynamic range* of target concentrations that can be detected. The sensor developed for detecting picomolar target concentrations may not allow the detection of variations of these concentrations on millimolar levels. Very important is also the *selectivity*, which is the ability to discriminate in detection the target from its close analogs that can be also present in the tested system. These parameters are not independent, they are strongly interconnected.

2.1.1 The Limit of Detection and Sensitivity

The sensor can be characterized by the *limit of detection* (LOD). It is the lowest concentration of an analyte that the analytical process can reliably detect (Kellner et al. 2004). This concentration, C_{LOD}, has to produce a detectable signal, X_{LOD}, that is statistically distinguishable from the blank or background signal. The *sensitivity*, S, can be defined as the slope of the calibration curve (the dependence of the analytical signal, X, on analyte concentration, C):

Table 2.1 The scale of concentrations used in molecular sensing

Notation	Concentration (mol/l)	Molecules per 1 µl of solution	Examples
Molar (M)	1	~6×10^{17}	Saturated salt (NaCl, KCl) solutions
Millimolar (mM)	10^{-3}	~6×10^{14}	Normal concentration of glucose in blood (~10 mM)
Micromolar (µM)	10^{-6}	~6×10^{11}	Intracellular concentration of NADH
Nanomolar (nM)	10^{-9}	~6×10^{8}	Intracellular concentration of cyclic AMP
Picomolar (pM)	10^{-12}	~6×10^{5}	Single molecule in a volume ~1.7×10^{-12} l
Femtomolar (fM)	10^{-15}	~6×10^{2}	Single molecule in a volume ~1.7×10^{-9} l
Attomolar (aM)	10^{-18}	~0.6	Single molecule in a volume ~1.7×10^{-6} l (~1.7 µl)
Zeptomolar (zM)	10^{-21}	~0.0006	Single molecule in a volume ~1.7×10^{-3} l (~1.7 ml)

$$S = dX / dC \approx \Delta X / \Delta C. \tag{2.1}$$

The sensitivity S and the limit of detection are connected. The higher the sensitivity, the lower the limit of detection. It is usually accepted that C_{LOD} can be expressed as a function of S and S_B, which is a standard deviation of a set of blank signals obtained in serial measurements without an analyte:

$$C_{LOD} = 3S_B / S \tag{2.2}$$

The *limit of quantitation* (LOC) determines the analytical significance of the apparent analyte concentration. It is definitely above the limit of detection and it is recommended to estimate it on the level of $10S_B/S$ (Anonymous 1980).

The limit of detection is determined by two very different factors, the response of the detection system and the target-receptor affinity. Let us consider these factors in more detail.

(a) The limit imposed by the *mechanism of detection* and the *response of the detection system*. Each detection system records the meaningful signal on the background of some *noise*, which is the statistical fluctuation of the measured parameter. The noise cannot be eliminated totally but it can be suppressed by increasing the intensity of the useful signal. Thus, the limit appears as an inability to resolve the useful signal on the background of the noise level. The *signal-to-noise ratio* is an important characteristic of the sensor system. In addition, the interfering signal can appear as the background. In fluorescence sensing, this can be the signal from different fluorescent or light-scattering species, which cannot be eliminated by instrumental means. If this background is variable, this may be an additional source of error.

(b) The limit imposed by *target-receptor affinity*. Imagine the case when the sensor is able to detect picomolar amounts of target, the target is present in the tested system in picomolar concentrations but the sensor response is zero. This happens because the sensor responds to the amount of *bound target* and the target-receptor affinity can be so low that no target is bound. In these cases, to shift the target-receptor equilibrium towards binding, the affinity should be dramatically increased. This cannot always be done, therefore this limit of detection commonly exists.

2.1.2   Dynamic Range of Detectable Target Concentrations

No sensor can be operative throughout the whole concentration range from moles to single molecules. The lower limit is determined by the sensor sensitivity and the higher limit appears due to the effects of saturation. This higher limit can also be affinity-based but it can also be related to the mechanism of detection.

(a) *Affinity-based limit.* Imagine that we start to apply the sensor that detects picomoles to the samples that detect nanomolar and then micromolar concentrations. The sensor will respond to increasing target concentrations up to the point where all the sensor *receptor sites* become *saturated.* Since further increase will not cause additional binding, there will be no response of the sensor. This means that the sensor affinity must be strongly adapted to the required concentration range. If an extended range (covering several orders of magnitude) of target concentrations needs to be detected, a series of sensors with different affinities have to be applied to the same sample.

(b) *Saturation effects in the detection system.* These effects may exist but in fluorescence sensing they are usually not important. Particularly, they can appear at high concentrations of light-absorbing and fluorescent species and if the light passes through a large volume (Lakowicz 2007). They can be completely ignored in miniaturized sensor systems.

For correct measurements based on reversible binding, the analyte concentration has to be much larger than the concentration of receptors. This is because the binding to a sensor should not substantially change its free concentration.

2.1.3 Selectivity

Selectivity is the basic characteristic of any analytical method that determines the accuracy of results. It is defined as the ability of the method to produce signals that are exclusively dependent on the analyte in the sample (Kellner et al. 2004).

In some rare cases, the target is so different from the non-target species present in the analyzed system that there is no competition from these species to the target binding to the sensor. In a more general case, such interference can exist, especially if the interfering species are present at much higher concentrations than the analyte. Thus, the *selectivity* can be viewed as an interplay of affinities weighted by the concentrations of the target and interfering species. In many cases, the ability to achieve the highest selectivity is vital for sensor applications. For instance, inside the living cells, the calcium ions have to be determined in the presence of much higher concentrations of magnesium ions and the concentrations of sodium ions are much lower from that of potassium. These ion pairs have the same charge and the difference in size is not very significant. Therefore, the usefulness of these sensors depends strongly on the selectivity factor.

The term *specificity* is frequently used in biochemical literature. The binding is often called nonspecific when the ligand-receptor binding isotherm (see Fig. 2.1 below) does not show the effect of saturation in the studied concentration range. Mechanistically, this means that the interaction is of low affinity, so that such saturation must exist but is shifted to much higher concentrations. To describe the ability of the sensor to discriminate between the true target analyte and other compounds that are close in properties, many authors use the term '*cross-reactivity*'.

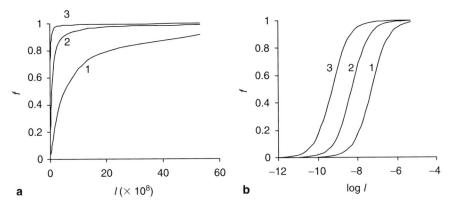

Fig. 2.1 Isotherms of ligand-receptor binding obtained by sequential addition of ligand concentrations (titration curve) in linear (a) and semi-logarithmic (b) coordinates. Plots of fractional saturation f as a function of total ligand concentration l were obtained with variation of l from 1.0×10^{-11} to 6.55×10^{-7} M for the receptor concentration 1×10^{-10} M and the following K_b values: 2×10^7 M^{-1} (1); 2×10^8 M^{-1} (2); 2×10^9 M^{-1} (3)

The estimation of cross-reactivity is commonly reduced to a comparison of dissociation constants between two or more species.

Specificity usually strongly correlates with the *affinity*. The stronger the ligand-receptor binding, the more significant can be the discrimination between the binding of specific and non-specific ligands (Eaton et al. 1995). This correlation is easy to understand. Specific binding involves multi-point contacts between ligand and receptor. The noncovalent interactions, providing these contacts, being individually weak, produce a collective effect increasing both affinity and specificity. The reader must not be confused by the fact that sometimes the weaker interaction allows revealing more specific changes in the target structure. Thus, the application of high-affinity probes for the detection of mismatched DNA and RNA sites shows that the increase of affinity in the hybridization of complementary chains may lead to a decrease in the ability to detect single mismatches in the sequence (Demidov and Frank-Kamenetskii 2004). This is because on the background of the lower affinity it is easier to detect its variations associated with the changes of a target structure.

Affinity is directly related to thermodynamic parameters characterizing the binding at equilibrium: the binding free energy ΔG and the changes of enthalpy ΔH and entropy ΔS on binding:

$$\Delta G = -RT \ln(K_b) = \Delta H - T\Delta S \qquad (2.3)$$

Here R is the gas constant and T is the temperature. K_b is the binding constant, which is the quantitative measure of affinity that will be defined below. Both enthalpy and entropy contributions are important for target binding. ΔH reflects the bond formations and is responsible for heat effects. The entropy term determines the change of molecular order and therefore depends strongly on the temperature. Though the thermodynamic approach cannot allow an accurate determination of affinities, it allows some estimates. For instance, in the case of the involvement of electrostatic interactions, the affinity will be decreased at higher electrolyte concentrations (ionic screening) and the introduction into the sensor of mobile groups that lose their mobility on binding decrease the affinity by influencing the entropy term.

We observe that there are basic limitations to the creation of a sensor that will be useful for target determination in all possible ranges of target concentrations and in all possible ranges of concentrations of low-specific target analogs that may compete with the assay. The regularities describing these systems will be analyzed below.

Since these regularities are valid for any intermolecular interactions not always related to sensing, here we will use rather general terms – the *receptor* that stands for the sensor element responsible for the target binding function and the *ligand*, which can be the target analyte. Not only can molecules be the receptors or ligands but also the particles of different size and even the whole living cells.

2.2   Determination of Binding Constants

In the case of reversible binding there is always a limit on the lowest and highest ligand concentrations that can be detected by a sensor. For a monovalent reversible ligand binding, the analytically useful concentration range is typically restricted to one order of magnitude below and one order above the ligand-receptor dissociation constant K_d, which is approximately the range between 20% and 80% of a receptor fractional saturation. As explained below, this statement follows directly from the mass action law when it is applied to interaction between two species at equilibrium. The heterogeneity of binding usually extends this range but this may happen only to a small extent. If the dynamic range of measured target concentrations needs to be broader, one has to use a combination of sensors with a similar specificity but of different affinities towards the analyte.

2.2.1   Dynamic Association-Dissociation Equilibrium

The simplest example of receptor-ligand interactions is when a single ligand (L) interacts with a single receptor (R) to form a single complex (LR) in the conditions of dynamic equilibrium.

$$L + R \Leftrightarrow LR \qquad (2.4)$$

In the conditions of equilibrium, a rapid binding-dissociation occurs all the time but the concentrations $[R]$, $[L]$ and $[LR]$ do not change over time. In this state the forward reaction of ligand binding proceeds with the same rate as the reverse reaction of its dissociation. This is why such equilibrium is called *dynamic*. It allows a simple description based on *mass action law*.

Let the kinetics of ligand binding be expressed by a rate constant k_1 and that of dissociation of the complex by rate constant k_2. Then the equilibrium can be characterized by a ratio of these constants. This can be either the *binding constant* $K_b = k_1/k_2$ or its reverse function, the *dissociation constant* $K_b = k_2/k_1$. K_b is often also called the stability or affinity constant, it is expressed in reverse molarity units. (M^{-1}), whereas $K_d = 1/K_b$ is expressed in molar (M) units. Table 2.1 presents the whole scale of concentrations used in molecular sensing.

It should be recalled that *molarity* (M) denotes the number of moles of a given substance per liter volume. The *mole* (mol), in contrast, is the unit that measures an amount of substance. One mole contains Avogadro's number (approximately 6.022 $\times 10^{23}$) of entities. The molarity dimension of K_d corresponds to the ligand molar concentration [L] at which one half of the receptor binding sites are occupied. Half of the ligands in this case exist in the bound and the other half in the free form. Table 2.2 provides an estimate of the typical ranges of the binding constants that can be useful for sensing.

According to mass action law, the equilibrium concentrations of the ligand, [L], of receptor, [R] and of their complex [LR] are related by the binding constant K_b:

$$K_b = \frac{k_1}{k_2} = \frac{[LR]}{[L] \times [R]} \tag{2.5}$$

If the total concentrations of receptor and ligand in the system are constant, then only one concentration of the three species ([L], [R] and [LR]) is independent. Therefore, if [R] is well determined (as usually happens in sensing), Eq. 2.5 connects [L] and [LR].

Table 2.2 Examples of intermolecular interactions with different binding constants K_b

Type of interaction	K_b (M^{-1})	Examples
Weak	Less than 10^5	Interactions of detergents in aqueous micelles
Medium	10^5–10^7	Common enzyme-substrate and enzyme-inhibitor interactions
Strong	10^7–10^{10}	Interactions of phospholipids in biomembranes; of steroids with their cell receptors; of pharmaceutically useful drugs with their targets
Very strong	More than 10^{10}	Interactions of antibodies with highly specific antigens; of some 'suicide' enzyme inhibitors with their targets; of avidin with biotin (10^{15} M^{-1})

2.2.2 *Determination of K_b by Titration*

The solution of many problems in sensor technologies needs the correct determination of ligand-receptor affinities. This can be done on the conditions of the application of mass action law. In this case, a titration experiment allows gradually changing the concentration ratio between ligand and receptor, usually by variation of the ligand concentration.

When, in the system that contains a fixed amount of receptors, the ligand concentration is gradually increased, the dependence of the *concentration of bound ligand* on *total ligand concentration* (titration curve) is a curve of hyperbolic shape. With the increase of ligand concentration and of the occupancy of the binding sites, the equilibrium becomes more and more shifted towards the free ligand until all binding sites become occupied and all added ligand remains unbound. It can be seen that this curve tends to saturation (Fig. 2.1a).

The half-saturation point on this curve corresponds to K_d. Thus, K_d expressed in concentration units is a convenient characteristic of a sensor characterizing the dynamic range of sensed concentrations. Meantime, since complete saturation is commonly not reached, such determination of K_d is not precise. The precision can be increased by taking into account all the points in the *titration experiment*. This can be done by presenting the titration curve in semi-logarithmic coordinates (Fig. 2.1b), in which it attains sigmoid shape, in which the half-saturation point is more clearly seen. For quantitative determinations there is a need for using non-linear regression analysis or, alternatively, by transformation of the curve to a linear form, in which all the data points could be efficiently used for building a straight line. The least square method allows fitting the experimental data to a straight line in an optimal way, which allows the calculation of precise values of the slope and axis intercepts that are connected with the values of K_b or K_d.

Two types of linearized binding curves, associated with the names of Klotz and Scatchard, are presently popular and they will be considered below. For convenience, Eq. (2.5) can be re-written using the following notations. Let l be the total concentration of the ligand and r_0 – the total concentration of receptors. $c = [LR]$ is the concentration of the ligand-receptor complex. If every acceptor is of valence m (has m binding sites for the ligand), then mr_0 is the total concentration of receptor binding sites. The concentration of the unbound ligand is then l-c and the concentration of free binding sites is mr_0-c. Thus, we have:

$$K_b = \frac{c}{(l-c)\times(mr_0 - c)} \tag{2.6}$$

It is convenient to introduce the *fractional saturation, f*, which is the fraction of the bound ligand (or of sensor binding sites that are occupied by the ligand). If l-$c = f$, then:

$$K_b = \frac{c}{f(m\times r_0 - c)} \tag{2.7}$$

The Klotz approaches are based on the simple algebraic transformation of Eq. (2.7) into a linear relationship either between the reverse concentration of the bound ligand and reverse concentration of the unbound ligand, or between the ratio of the unbound ligand to bound and the concentration of the unbound ligand, respectively. The linear dependence of $1/c$ on $1/f$ is presented in Eq. (2.8):

$$\frac{1}{c} = \frac{1}{mr_0 K_b} \times \frac{1}{f} + \frac{1}{mr_0} \tag{2.8}$$

The graph of this dependence (the Klotz graph) is the straight line with the slope $1/mr_0 K_b$ that crosses the ordinate axis at the point $1/mr_0$. Thus, with the knowledge of r_0 and using the values of $1/c$ and $1/f$ obtained in experiments, one can construct the linear dependence similar to that in Fig. 2.2a and determine from its slope and from its intercept with the ordinate the affinity of interaction (expressed by K_b) and the valence of receptor, m.

According to Scatchard, the linear relationship is observed between the ratio of the bound to unbound ligand, c/f and the concentration of bound ligand, c:

$$\frac{c}{f} = K_b(mr_0 - c) \tag{2.8a}$$

Its graph is a declining linear function (Fig. 2.2b) with the slope equal to K_b. It crosses the ordinate axis at the point $K_b mr_0$ and the abscissa axis at the point mr_0.

It may be noted that in many cases the experiments on the determination of K_b are made with the excess of ligand (or receptor). In this case, the total concentration of the component that is in excess substantially exceeds the concentration of the formed complex, $l \gg c$. This allows the approximation of the free concentration of

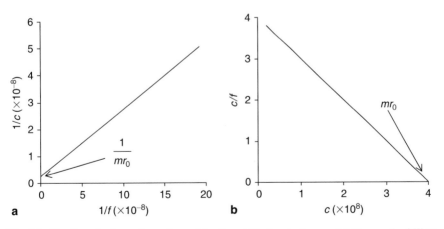

Fig. 2.2 The linearized transformations of the ligand binding isotherms. (a) The graph of Klotz, the dependence of $1/c$ on $1/f$. The tangent of the slope of straight line is equal to $1/mr_0 k_b$. (b) The Scatchard graph, the dependence of c/f on c. The tangent of the slope of straight line equals K_b. The graphs are calculated for: $m = 4$, $r_0 = 1.0 \times 10^{-8}$ M, $K_b = 1.0 \times 10^8$ M^{-1}. Ligand concentration changes from $l_1 = 2.5 \times 10^{-9}$ to $l_i = 2.05 \times 10^{-5}$ M

this component equal to its total concentration ($f \approx 1$), which avoids the necessity for its measurement.

In the case of two types of binding sites with different affinities, the Klotz graph may be represented by two segments, as depicted in Fig. 2.3. The Scatchard graph is more suitable for obtaining K_b data in the case of two types of ligand binding sites of different affinities. In this case the system of four equations should be solved with four unknowns, K_{b1}, K_{b2}, m_1 and m_2. Their solution is described in the literature (Klotz and Hunston 1971). When two linear segments are resolved (Fig. 2.3), an approximate graphical determination of these parameters using a Scatchard graph is possible.

It should be stressed that linear graphs in Klotz or Scatchard coordinates are observed only if the following conditions are satisfied:

(a) The receptor has identical binding sites for the ligand.
(b) The binding of one (or several) ligand(s) by one (or several) site(s) of the receptor does not change the affinity of its other binding sites, i.e., binding occurs without cooperative (positive or negative) effect.
(c) The ligand-receptor interaction obeys the law of mass action, which implies that all measurements have been taken at the state of equilibrium.

Several other methods of linearization of the same binding curves can be found in the literature. They are based on the same Eq. (2.4) and contain the same information. Meanwhile, it has to be mentioned that as in any linear transformations, such operations provide different statistical values to different data points along the binding

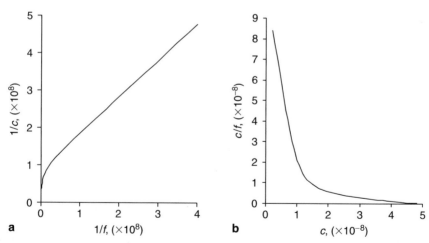

Fig. 2.3 The graphs of Klotz (a) and Scatchard (b) calculated for the binding ligand by two receptors of different affinities: $m_1 = 1$, $r_{01} = 1.0 \times 10^{-8}$ M, $K_{b1} = 1.0 \times 10^{8}$ M^{-1}, $m_2 = 1$, $r_{02} = 4.0 \times 10^{-8}$ M, $K_{b1} = 1.0 \times 10^{6}$ M^{-1}. The ligand concentration l changes from 2.5×10^{-9} to 2.05×10^{-5} M. Note that the contributions of two types of binding (stronger and weaker) in these curves are different and the Scatchard plots even allow approximate graphical deconvolution of the curve into two linear segments

curve, which may cause significant errors in extrapolated values. This issue was discussed in the literature (Klotz 1983). Non-linear approximations and direct fitting of data into Eq. (2.7) may lead to values that are more precise (Munson 1983).

Considerations regarding the estimation of error in determined f values on precision in determining K_b can be found in Tetin and Hazlett (2000). The lowest influence of statistical error in f is introduced with the titration points when $f \approx 0.5 \, K_d$.

2.2.3 Determination of K_b by Serial Dilutions

In some practical cases, the methodology based on ligand titration is inconvenient or even inapplicable, since the ligand and the receptor cannot always be obtained in purified form but their interaction can be studied in homogeneous conditions, in solutions. In these cases, an approach based on serial dilutions with the solvent of the sample that contains both receptor and ligand can be applied (Bobrovnik 2005a, b). With the dilution, the dynamic equilibrium will be shifted to dissociated forms. This shift depends on K_b and the determination of concentrations of the bound ligand in the course of serial dilutions allows obtaining this value. After dilution of the system by d_i times, a new equilibrium is established and we obtain:

$$K_b = \frac{c_i}{\left(\dfrac{l}{d_i} - c_i\right)\left(\dfrac{mr_0}{d_i} - c_i\right)} \tag{2.9}$$

Thus, in principle, the equilibrium constant can be obtained from re-distribution of the ligand between free and bound states as a function of dilution factor d_i. In a practically useful case when $l/d_i >> c_i$ Eq. (2.9) can be reduced to:

$$K_b = \frac{c_i}{\dfrac{l}{d_i}\left(\dfrac{mr_0}{d_i} - c_i\right)} \tag{2.10}$$

Transforming Eq. (2.10) one can obtain the dependence of c_i on d_i.

$$c_i = \frac{mr_0 l K_b}{d_i(d_i + l K_b)} \tag{2.11}$$

Several possibilities exist for the linearization of $1/c_i d_i$ and $l/c_i d_i$ functions of dilution factor d_i (Bobrovnik 2005a, b).

This method can be successfully used when a ligand-receptor mixture already exists (for example, in some experimental system *in vivo*) and obtaining a series of samples with a constant concentration of receptors and various concentrations of ligand cannot be provided.

The methods discussed above are applicable with the use of a common fluorescence technique. Meantime, in the case of very strong binding, very high dilutions are needed to achieve the presence of a ligand-dissociated form of receptor together with the bound form. In this case, the *photon-correlation spectroscopy* that allows increasing the dilution manifold is a good choice. Based on the detection of single molecules entering and leaving the illuminated volume, this method allows determining K_b at very low concentrations (Sanchez and Gratton 2005; Tetin and Hazlett 2000).

2.3   Modeling the Ligand Binding Isotherm

The *sensor response function* (often called *sensogram*) is the function describing the correlation between the target analyte concentration and the response of the analytical system. Since different parameters used for fluorescence response (Chapter 3) are commonly proportional (or proportional with weighting, Section 2.5) to the amount of bound ligand (analyte), the major task here is to establish a correlation between the amount (concentration) of the analyte in the tested system l and its fraction bound by the sensor (fractional saturation) f. This function is called the *ligand binding isotherm*. It has to be determined for particular sensor operation conditions. The receptor functional elements of sensors can be composed of molecules or particles distributed in a defined volume or they can be attached to a surface.

2.3.1   Receptors Free in Solution or Immobilized to a Surface

Let us consider the case when the sensors are attached to a surface S that accommodates their receptor units with the density σ. The total amount of receptors on this sensor surface will then be $N = \sigma \times S$. The sensors interact with the ligand (analyte) solution with concentration l in volume V and their affinity is characterized by binding constant K_b. Let l and K_b be known and our aim is to determine the part of the receptors that forms the complexes with the ligand and the part that remains unoccupied.

When the solution volume is large enough ($l \times V \gg N$) so that the ligand-receptor binding does not influence the ligand concentration in solution, the approximation l = const is acceptable. Let the number of receptors forming a complex with ligand be n, then the number of free receptors is $N - n$. Then, the apparent concentration of the binding sites will be N/V. Let the equilibrium condition of the binding be reached and let n receptors (out of N) bind the ligand and $N\text{-}n$ remain free. Then their apparent concentrations will be n/V and $(N - n)/V$. Applying the mass action law and providing simple transformations we obtain:

$$f = \frac{n}{N} = \frac{lK_b}{1+lK_b} \tag{2.12}$$

Equation (2.12) represents the *ligand binding isotherm* by the sensor, which with the account of the response of the sensor system can be transformed into a sensogram. The graphical forms of the function $f(l)$ in linear and logarithmic coordinates are presented in Fig. 2.1. We observe that the shape of this function does not change by any form of immobilization of receptors. It is solely determined by the K_d value.

A very important consequence follows from this simple analysis. There is a *fundamental restriction on the concentration range of sensor response*. Irrespective of sensor and sample geometry, this range is limited to roughly two orders of magnitude, one below and one above K_d (~0.1--~10 K_d). Outside this range the sensor becomes insensitive and the determination of target concentration not possible.

It also follows that a non-specific ligand, if it is present in the same concentration as the specific target ligand, cannot significantly influence the determination of the specific ligand if its binding is weaker by two orders of magnitude or more. Thus, the sensors operating in the conditions of equilibrium binding *can be very selective if they display higher affinity towards a specific target*. Accordingly, for sensing specific targets in a broad range of concentrations, one has to use a series of sensors with different affinities. For instance, this was done with mutants of maltose binding protein as saccharide receptors (Marvin et al. 1997).

2.3.2 Bivalent and Polyvalent Reversible Target Binding

A dramatically increased affinity of receptors present as dimers compared to monomers is explored in the design of fluorescent dyes. Thus, the best fluorescent nucleic acid binders are the dimers of intercalating dyes. The intrinsic DNA binding affinity constants of typical intercalator dye ethidium bromide and its homo-dimer are reported to be 1.5×10^5 and 2×10^8 M^{-1} correspondingly (Gaugain et al. 1978). Thus, dimerization increased the affinity by three orders of magnitude. The other example is the dramatic (2,500- to 170,000-fold) increase of binding constants between the nanodevices and folate binding protein through multivalency (Hong et al. 2007). The affinity enhancement effect produced by multivalency can be explained and to some extent predicted on a thermodynamic basis (Kitov and Bundle 2003). The effect of bivalent binding can be analyzed based on a formally applied concept that the local target (or receptor) concentration is increased on the formation of primary complex. In this case there is a difference, if the linker between binding sites is rigid or flexible (Bobrovnik 2007).

Polyvalent binding at equilibrium means establishing the dynamic equilibrium between the unbound form of the receptor and its bound forms with saturation of different valences m:

$$L + R_m \Leftrightarrow LR_1 + 2LR_2 + \ldots + mLR_m \tag{2.13}$$

Instead of a single K_b value, this system will be characterized by a matrix of K_{ij} values describing the equilibrium in interaction of each type of the complex with all the rest. In view of the experimental data cited above, we can consider the strength of the complex expressed by K_{ij} values to increase with the ligand saturation, reaching the maximum value for the mLR_m complex. This cannot be true in a general case but we can always find the complex xLR_x, for which affinity is the highest. In order to understand better the receptor response in such a complex system, we can consider the following limiting cases:

(a) The sensor responds to a binding of the first target with the formation of a LR_l complex. Since this complex is commonly weak, it can be formed only at very high concentrations. However, at these concentrations the complex xLR_x should be formed already with all population of receptor molecules due to high affinity. Therefore, the sensor response will be seen around some efficient K_b value describing equilibrium in interactions of the xLR_x complex with free ligand and all other complexes.

(b) The sensor responds only to the formation of the complex mLR_m, i.e., when all its valences are saturated by the ligand binding. When it is the case of a strongest complex, then we will have the situation similar to case (a) for xLR_x with the exception that intermediate complexes are not observed in response.

(c) The sensor responds in a well-resolvable manner to the formation of each of these complexes. For instance, ketocyanine dye responds to the formation of high-affinity hydrogen bond complexes with protic co-solvent molecules at their low concentrations and then a different change in spectra occurs on the formation of additional complexes of lower affinity observed at higher co-solvent concentrations (see Section 4.1). In this case two K_b values can be obtained in a titration experiment (Pivovarenko et al. 2000). The range for determining the protic co-solvent concentrations is extended to two areas around these K_b values. The other example is the design of a sensor for an extended range of pH. In fact, it is the sensor for proton binding-release to the sensor titrating groups. In the case of distinguishable or additive response of sensor groups titrating at a different pH, we obtain the sensor for an extended pH range (Li et al. 2006).

Consider now in more detail the simplest but practically important case when there are two ligands in the system that bind to the same receptors with different affinities. Let one (more specific) ligand, being in concentration l_1 bind the receptor with the higher binding constant K_{b1} and the other (less specific) in concentration l_2 possesses the binding constant K_{b2}. Then, in the conditions of equilibrium, the number of sensor receptors with the bound first ligand will be n_1 and their number with the bound second ligand – n_2. Based on the mass action law, we obtain two equations representing this case:

$$K_{b1} = \frac{\dfrac{n_1}{V}}{\left(\dfrac{N - n_1 - n_2}{V}\right)\left(l_1 - \dfrac{n_1}{V}\right)} = \frac{n_1 V}{\left(N - n_1 - n_2\right)\left(l_1 V - n_1\right)} \tag{2.14}$$

$$K_{b2} = \frac{\dfrac{n_2}{V}}{\left(\dfrac{N - n_1 - n_2}{V}\right)\left(l_2 - \dfrac{n_2}{V}\right)} = \frac{n_2 V}{\left(N - n_1 - n_2\right)\left(l_2 V - n_2\right)} \qquad (2.15)$$

Attempts to find analytical solutions for two unknowns n_1 and n_2 lead to the necessity of solving complex cubic equations. Meantime, numerical solution of Eqs. (2.13)–(2.14) allows easy determination of n_1 and n_2 for known k_{b1}, k_{b2}, l_1 and l_2. Alternatively, one may find l_1 and l_2 by assigning the values of k_{b1}, k_{b2}, n_1 and n_2. Consequently, one can calculate the *ligand binding isotherm* for this particular case.

2.3.3 Reversible Binding of Ligand and Competitor

Now we consider the case of two different ligands in which the binding of the *testing ligand* (e.g., fluorescent competitor) can provide an informative signal for the binding of the *target ligand*. For testing the ligand, the concentration and affinity of binding are known; they can be measured in a preliminary test. This ligand can be a key player as a competitor in competition assays (Section 1.3). Is there any possibility of finding in such an assay the target concentration in the case when its affinity is unknown? We will try to resolve this issue.

Let the unknown concentration of the target ligand be l_x and its unknown binding constant K_{bx}. The binding constant of the testing ligand is known, it is K_{b1}. In our experiment we can add to the system any desired concentration of this ligand, l_1. These two ligands can bind with the sensor, competing with each other so that the quantity of the bound testing ligand n_1 can be measured (for instance, by the response of the fluorescence reporter). Thus, all three of its important parameters, K_{b1}, l_1 and n_1 are known. In contrast, the amount of target ligand, n_{x1}, cannot be measured directly, since it does not provide any response.

If we measure the bound testing ligand at a different concentration, l_2, then the sensor will bind a different amount of this ligand, n_2, which can also be measured. The amount of analyte ligand, n_{x2}, remains unknown but now we are able to compose four linearly independent equations with four unknowns, l_x, K_{bx}, n_{x1} and n_{x2}. Their solutions will provide numerical values for these unknowns.

Based on the mass action law, we obtain the equations describing correlations between equilibrium constants K_{b1} or K_{bx} and the concentrations of free ligands or of their complexes.

Numerical solution of Eq. (2.15) can yield l_x, K_{bx}, n_{x1} and n_{x2}. Practical examples of these calculations will be presented at the end of this chapter.

$$K_{b1} = \frac{n_1 V}{(N - n_1 - n_{x1})(l_1 V - n_1)}$$

$$K_{b1} = \frac{n_2 V}{(N - n_2 - n_{x2})(l_2 V - n_2)}$$

$$K_{bx} = \frac{n_{x1} V}{(N - n_1 - n_{x1})(l_1 V - n_{x1})} \qquad (2.16)$$

$$K_{bx} = \frac{n_{x2} V}{(N - n_2 - n_{x2})(l_2 V - n_{x2})}$$

When the concentrations of both these ligands are relatively high and their binding constants low, the fractional concentrations of bound ligands will be small compared to their total concentrations. Thus the changes of free ligand concentrations on binding can be ignored, which simplifies the analysis. We can apply two linearly independent equations based on the mass action law:

$$K_{b1} = \frac{c_1}{(C - c_1 - c_x)l_1} \qquad (2.17)$$

$$K_{bx} = \frac{c_x}{(C - c_1 - c_x)l_x} \qquad (2.18)$$

Here l_1 and l_x are the concentrations of two ligands in solution. They bind to the sensor with affinity constants K_{b1} and K_{bx}. C is the efficient concentration of receptors, which can be calculated knowing their amount, N and solution volume V by formulae: $C = N/V \times N_a$. Here N_a is Avogadro's number and c_1 and c_x are the efficient concentrations of the *bound* testing and target ligands correspondingly.

After simple algebraic transformations, Eqs. (2.16) and (2.17) can be presented as:

$$c_1 = \frac{l_1 K_{b1} C}{1 + l_1 K_{b1} + l_x K_{bx}} \qquad (2.19)$$

$$c_x = \frac{l_x K_{bx} C}{1 + l_1 K_{b1} + l_x K_{bx}} \qquad (2.20)$$

It follows that:

$$\frac{c_1}{c_x} = \frac{l_1 K_{b1}}{l_x K_{bx}} \qquad (2.21)$$

The number of receptors that have bound the first or second ligand will be:

$$\frac{c_1}{C} = \frac{l_1 K_{b1}}{1 + l_1 K_{b1} + l_x K_{bx}} \tag{2.22}$$

$$\frac{c_x}{C} = \frac{l_x K_{bx}}{1 + l_1 K_{b1} + l_x K_{bx}} \tag{2.23}$$

If the concentration of one of the ligands is set constant and the other varies, then we will obtain the following graph of the dependence of the concentration of either ligand that is bound by the receptor (c_1 or c_x) on total target ligand concentration l_x at constant concentration of the testing ligand. This is the target ligand binding isotherm (Fig. 2.4).

This graph shows that when the labeled competitor binds stronger than the target to the receptor, there is a range of sensor insensitivity at low target concentrations because the target cannot replace the competitor from the complex. The substitution occurs at higher concentrations and if the competitor changes the parameters of its fluorescence in a binding-release process, this allows determination of the concentration of the target. When the target concentration becomes too high, all the receptor sites become occupied with the target and all of the competitor is released, so the sensor becomes insensitive again. Note that as in the case of sensors based on a direct response to target binding, the range of target concentrations that can be detected is within the same two orders of magnitude.

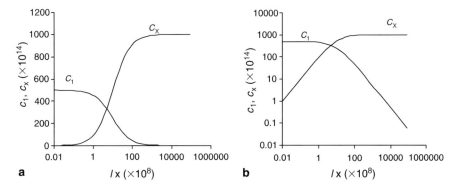

Fig. 2.4 Concentrations of the bound competitor and target ligands (c_1 and c_x) as a function of total target ligand concentration l_x at constant competitor concentration l_1 in semi-logarithmic (a) and double-logarithmic (b) coordinates. The graphs were composed for the following parameters: $N = 6.0 \times 10^9$, $K_{b1} = 1.0 \times 10^8$ M^{-1}, $K_{bx} = 1.0 \times 10^7$ M^{-1}, $l_1 = $ const $= 1.0 \times 10^{-8}$ M and l_x changing from 1.0×10^{-10} to 8.4×10^{-4} M

2.3.4   Interactions in a Small Volume

We now consider the case when the volume V, in which the detection is made, is so small that on binding the ligands to receptors their concentration in this volume essentially decreases (S.N. Bobrovnik, to be published). As we will see below, the present tendency in sensor design is the dramatic decrease of the testing volume (Chapter 9) and this case may represent a real situation in a number of applications. Sensing inside the living cells and using the sensing devices reaching picoliter (10^{-12} l) detection volumes (Chapter 11), may present such cases.

As in the case discussed above, we consider the system composed of N receptors immobilized on planar support, whereas the tested target remains in solution. In the absence of ligands, all receptors are unoccupied and their efficient concentration is N/V, where V is the sample volume. Consider the case of equimolar (1:1) reversible binding, so that at equilibrium n receptors are occupied; they bind n ligand molecules. Then the number of free receptors will be $N - n$ and their apparent concentration in solution will be $(N - n)/V$. An apparent concentration of ligand-receptor complex will be n/V. The free ligand concentration, being initially l/V (l is total amount of ligand molecules), changes on binding to the receptor and becomes $l - n/V$.

The mass action law can be applied for this case in the following form:

$$K_b = \frac{\dfrac{n}{V}}{\left(\dfrac{N-n}{V}\right)\left(l-\dfrac{n}{V}\right)} = \frac{nV}{(N-n)(lV-n)} \tag{2.24}$$

Equation (2.24) can be rewritten as an expression for f as a function of l ($f = n/N$):

$$f = \frac{NK_b + lVK_b + V - \sqrt{\left(NK_b + lVK_b + V\right)^2 - 4NlVK_b^{\,2}}}{2K_bN} \tag{2.25}$$

Equation (2.25) represents the *ligand binding isotherm*. Its graph is a sigmoid dependence similar to that in Fig. 2.1. It may be interesting to analyze how the isotherm of ligand binding will change with the decrease of detection volume. The results are presented in Fig. 2.5.

We observe that with the decrease of solution volume the ligand binding isotherm transforms significantly. It becomes narrower and shifts to higher concentrations. The origin of this effect is in ligand redistribution between its free and bound forms. When the sample volume becomes smaller, a higher ligand concentration is needed to occupy the same amount of binding sites as in a large volume. The miniaturization of sensor technologies requires accounting for these effects.

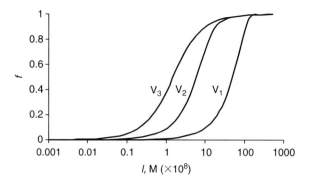

Fig. 2.5   The influence of solution volume on ligand binding isotherm (the $f(l)$ function). The graphs were composed for the following parameters: $N = 1 \times 10^{-10}$ M, $K_b = 1 \times 10^8$ M^{-1} and the solution volumes $V_1 = 0.1$ ml, $V_2 = 1$ ml and $V_3 = 10$ ml

2.4   Kinetics of Target Binding

The sensing commonly starts from the ligand and receptor physically separated in space and in order to interact they first have to approach each other. It can be the motion of one partner, ligand, if the receptor is immobilized on the surface of a spotted array or of both partners if the receptor molecules are free to diffuse in an analyzed solution. In sensor technologies, the knowledge of ligand-receptor association-dissociation kinetics is important in several aspects. For correct applications of conditions of equilibrium binding we always need to know when this equilibrium is established and this determines the necessary incubation time. If the binding is irreversible, we need to wait until all the ligand is bound. Some sensing technologies allow target determination in a kinetic regime.

In kinetic analysis of sensor performance the problem is *to find the amount of the ligand bound to the receptor as a function of time t* after an initial application of a sample containing the ligand to the sensor system. The rates of ligand binding reactions are described by differential equations, which in the simplest case (1:1 stoichiometry) can be written in the form:

$$dB(t) / dt = k_1[N - B(t)]c_L(t) - k_2 B(t), \tag{2.26}$$

where $B(t)$ is the number of bound ligand molecules and N is the total number of receptor molecules. Here $c_L(t)$ is the local concentration of the ligand solution at the reaction surface. It is the function of $B(t)$ and this relation is the subject of modeling (Klenin et al. 2005). In general, the reaction rate as a function of time can be obtained numerically as a solution of a non-linear integral equation and its analytical solutions are available only for some special cases. Various sensor geometries of practical interest can be modeled but the practically important solution is experimental. The output signal should reach a steady-state value as a function of time.

Depending on the receptor-target pair and on the immobilization of receptors (their presence in solutions, attachment to nanoparticles or immobilization on flat or porous surface) the rate-limiting step in the formation of their complexes can be different. The target-receptor mutual diffusion imposes an upper limit on the rates of all these reactions. Meanwhile, some intermolecular interactions can be coupled with conformational isomerizations in ligand or target molecules and their rates can be slower than the diffusion. This can be a *multistep process* of search in a conformational space.

In the case of one of the interacting components (usually the receptor) being immobilized on the flat surface, the kinetics of ligand binding will become slower not only because only one component has the freedom to diffuse, but also because the kinetics of ligand binding becomes time-dependent. This fact can be explained by the following. At time zero, $t = 0$, all receptor binding sites are empty and every interaction with them may lead to binding. At $t > 0$, some of the sites become occupied and the search for a smaller number of unoccupied sites slows down the process. The ligand migration along the receptor sites is usually much slower than free diffusion in a solution because the desorption-sorption process at each step occurs with overcoming higher energy barriers (Sadana and Madugula 1993).

The observed strong tendency towards miniaturization of sensor systems raises new challenges. Due to the small binding area and high target concentration that are usually applied in a small volume, the kinetics of target binding depends on this concentration. For micro flow channel devices the transport equations have to be solved and diffusion processes simulated to guarantee perfect mixing.

The importance of *reaching the conditions of thermodynamic equilibrium* in target binding has been discussed by many authors. They attribute the loss of reproducibility and many systematic errors in assays to the fact that these conditions were not satisfied. Both calculation and experiment show that for sensors immobilized on the surface of planar arrays, the target binding occurs two to three orders of magnitude slower than on binding in solutions. Meanwhile, both immunosensors and DNA hybridization arrays are made by immobilization on solid surfaces, which increases the time of reaching the equilibrium to many hours and even many days (Carletti et al. 2006). A smaller decrease in the rate of binding can be achieved if the sensors are immobilized on microspheres (Sekar et al. 2005). Both these array formats will be discussed in this book.

The characteristic rate of binding reaction depends on the rates of two processes: (a) transport of the analyte from the bulk compartment to the reaction area (reaction compartment) and (b) the subsequent binding process. In immunosensors, the mass transport plays the major role. In contrast, in DNA arrays the process of recognition of correct sequence, in which a huge amount of target-receptor sequences hybridize simultaneously in limiting the kinetics of binding, plays the major role.

Different strategies for decreasing the assay time can be applied by knowing the origin of the rate-limiting step: (a) decreasing the volume of the tested system, (b) using inert polymers or nanoparticles to provide a 'molecular crowding' effect, (c) the application of microwaves for local heating. Still, the issue of binding kinetics remains very actual. With the knowledge of the origin of such retardation, the binding kinetics can be optimized with the acceleration of target binding by several orders of magnitude (Kusnezow et al. 2003). This is especially important for the microarray analysis of complex biological samples.

2.5 Formats for Fluorescence Detection

Fluorescent sensing similarly to other sensing technologies is the method that is based on the measurement of the *amount of bound target* and, based on these data, the data on the *amount of total target* in the tested system is obtained. It is important that fluorescence response operates with parameters that show the difference between ligand-free and ligand-bound forms of receptor, and the presented above relations based on mass action law can be simply re-written with introduction of these parameters.

The detailed description of different parameters that are used in fluorescence detection technologies and the analysis of their various applications will be presented in the next chapter. For the present discussion, we only indicate that they are derived from several types of measurements. Usually at fixed wavelengths, the intensity (in one response channel), intensity ratio (in two channels), anisotropy (or polarization) and the lifetime are recorded. The cases in which a fluorescence parameter provides the full-scale response changing from zero to a very high value are rare. In all practically important cases, we deal with the overlap of at least two signals generated by fluorescence reporters indicating the receptor sites with free and bound ligands. The simplest intensity sensing provides a *highly linear response*. The applications of anisotropy and lifetime sensing are different, they result in interesting *non-linear effects*.

2.5.1 Linear Format

In fluorescence sensing the dependence $f(l)$ has to be transformed into the dependence of the *fluorescence response function* on ligand concentration, which can be used as a calibration curve for a sensor device. Let the fluorescence signal F be proportional to the number of bound ligand molecules (which is the same as the number of occupied receptors). Its value will be $F = n\phi$, where ϕ – is the proportionality factor. At saturation, when $N = n$, we have $F = F_{max}$. Since $f = n/N = F/F_{max}$ and the function $f(l)$ can be determined by calibration, the values of l in the tested system can be obtained from the *linear fluorescence response* to ligand binding by the sensor.

In the case where the background fluorescence and/or the fluorescence of the ligand-free sensor with the intensity F_{min} contributes to a fluorescence signal, this parameter has to be accounted for in the determination of the ligand concentration, l. The transformed Eq. (2.5) can be presented as:

$$l = K_d \left(\frac{F - F_{min}}{F_{max} - F} \right) \qquad (2.27)$$

Usually, in intensity sensing the linear format is observed over many orders of magnitude. There may be the cases, however, when the fluorescence reporters are connected by an electronic conjugation to produce an enhanced collective effect (See Chapter 6). In these cases, the response function may become complicated.

A popular method in intensity sensing that allows compensating for the variation of sensor concentration is the introduction of a reference channel with a ligand-independent intensity F_{ref}. Dividing the numerator and denominator of Eq. (2.26) by F_{ref}. we obtain:

$$l = K_d \left(\frac{R - R_{min}}{R_{max} - R} \right) \tag{2.28}$$

In contrast to Eq. (2.27), Eq. (2.28) contains only the intensity ratios that are independent of sensor concentration. $R = F/F_{ref}$, $R_{min} = F_{min}/F_{re}$ and $R_{max} = F_{max}/F_{ref}$. Such measurements are called *ratiometric*. The most convenient in fluorescence detection is the application of the ratios of fluorescence intensities at two excitation or two emission wavelengths.

In sensor technologies there are the cases when the two sensor forms, ligand-free and ligand-bound, are highly emissive and they differ in positions of excitation or emission spectra. Thus, the increase of intensity at one wavelength is coupled with its decrease at the other wavelength. The spectra of these forms are usually overlapped and in these cases accounting for such variation can be avoided if we select the wavelength for measuring the response to binding, F, at the wavelength of maximal variation of intensity and the reference intensity, F_{ref}, at the crossing point of two spectra (*isoemissive point*), where this intensity does not change. We can then easily use Eq. (2.26).

It can be more practical to use a broader scale of intensity ratio variations by measuring the intensities at two points of its maximal change, usually at the band maxima corresponding to ligand-free and ligand-bound forms of the sensor. We can again consider the intensity at one wavelength as a 'sensor signal' and at the other wavelength as the 'reference' but we have to introduce the factor that accounts for intensity redistribution between two bands. This is the ratio of intensities of free and bound forms, F_F/F_B, at a 'reference' wavelength λ_2:

$$l = K_d \left(\frac{R - R_{min}}{R_{max} - R} \right) \left(\frac{F_F(\lambda_2)}{F_B(\lambda_2)} \right) \tag{2.29}$$

In the case of multivalent receptors these equations become more complicated and the reader may find them in the literature (Yang et al. 2003).

If the sensing is based on the detection of changes in excitation spectra, then the F_F/F_B ratio has to be substituted by the ratio of 'brightnesses' at the reference wavelength ($\varepsilon_F\Phi_F/\varepsilon_B\Phi_B$), which is the ratio of molar absorbances, ε, multiplied by correspondent quantum yields, Φ (Lakowicz 2007).

2.5.2   *Intensity-Weighted Format*

In Chapter 3 we will discuss the methods that are frequently applied in sensing and for which the fluorescence response function in a general case is not linear. They are the methods of *anisotropy* and *lifetime* sensing. When measured in a homogeneous population of dyes, these parameters are concentration-independent. The two states of the sensor, ligand-bound and ligand-free, may differ significantly in these parameters and, based on this difference, the target concentration has to be determined. In the applications of these methods an essential non-linearity in response function appears. This is because each state (bound or unbound) displays its own intensity and anisotropy decays. Their fractional contributions depend on the relative intensities of their correspondent forms and the additivity law is valid only for the intensities. Therefore the parameters derived in anisotropy and lifetime sensing appear to be weighted by fractional intensities.

Fluorescence anisotropy is extremely sensitive to those intermolecular interactions in which a small rotating unit becomes a rigidly coupled associate of a much larger size. The bound and free ligands are detected because they possess different values of anisotropy r, r_f of free and r_b of ligand-bound receptor. The measured anisotropy is obtained as being formed by the contributions of these two forms weighted by fluorescence intensities of these forms:

$$r = F_f r_f + F_b r_b \tag{2.30}$$

This means that if the intensity of one of the forms is zero, such an anisotropy sensor is useless since it will show anisotropy of only one of the forms. Meanwhile, if the test conditions are selected so that fluorescence intensity is not changed on ligand binding, $F_f = F_b$, then the fraction of the bound ligand can be estimated in a simple way:

$$f = \frac{r - r_f}{r_b - r_f} \tag{2.31}$$

The account of fractional intensity factor $R = F_b/F_f$ (the ratio of intensities of bound and free forms) leads to a more complicated function:

$$f = \frac{r - r_f}{\left(r - r_f\right) + R(r_b - r)} \tag{2.32}$$

In different polarization assays the measured steady-state anisotropy is usually a weighted average of the low r values of free ligand and the large anisotropy of high molecular volume ligand–receptor complexes and the fluorescence intensities as the weighting factors that can vary in both directions. The extent of non-linearity that can appear is illustrated in Fig. 2.6, in which the intensity changes on ligand binding are linear but a strong non-linearity is observed in anisotropy.

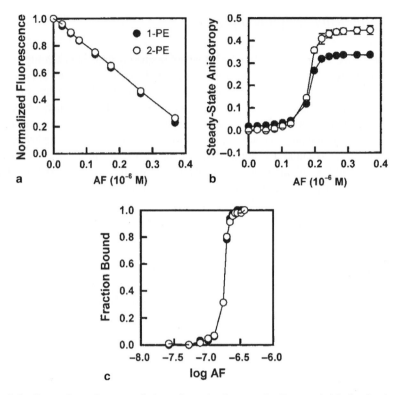

Fig. 2.6 Comparison of response in intensity and anisotropy for fluorescein binding by the anti-fluorescein antibody (AF). (a) Normalized steady-state fluorescence from fluorescein (1.0 μm) as a function of added AF, showing the linear effect of quenching. (b) Steady-state fluorescence anisotropy as a function of added AF. Because both intensity and anisotropy change, the effect becomes non-linear. (c) The fraction of AF-bound fluorescein as a function of total AF concentration plotted on a logarithmic scale. The data points represent the results from one- (•) and two-photon (O) excitation. Reproduced with permission from (Baker et al. 2000)

A similar situation is observed in lifetime sensing, which will be discussed in Section 3.3. Fluorescence decays as a function of time (in an ideal case, exponentially) and this decay can be described by initial amplitude α and lifetime τ_F for each of the two free (with index F) and bound (with index B) forms. If both of these forms are present in the emission, we observe the result of the additive contributions of two decays:

$$F(t) = \alpha_F \exp(-t / \tau_F^{\,F}) + \alpha_B \exp(-t / \tau_F^{\,B}) \tag{2.33}$$

The pre-exponential factors α_F and α_B are related to concentrations of these forms. If for one of the forms $\alpha = 0$, there will be no sensing, since irrespective of target concentration we will observe the response only the form, which is emissive. Meanwhile, if $\alpha_F = \alpha_B$, the sensor response will be determined by the ratio of $\tau_F^{\,F}$ and $\tau_F^{\,B}$ values. In a general case, the ratio of concentrations of free and occupied receptors will be determined not only by α_F and α_B values and correspondent lifetimes τ_F^F and τ_F^F. They have to be weighted by correspondent brightnesses, which in the simplest description are the products of molar absorbances ε_F or ε_B and quantum yields Φ_F or Φ_B (Lakowicz 1999):

$$\frac{\alpha_B}{\alpha_F} = \frac{\varepsilon_B}{\varepsilon_F} \frac{\Phi_B \tau_F^{\,F}}{\Phi_F \tau_F^{\,B}} \frac{[LR]}{[L]} \tag{2.34}$$

It follows from Eq. (2.34) that the ratio of pre-exponential factors (α_B/α_F) represents the ratio of the concentrations of ligand bound and free forms ([LR]/[L]) only if there is no change in absorbance at the excitation wavelength ($\varepsilon_B = \varepsilon_F$) or of the product of quantum yield and lifetime.

Thus, the anisotropy and fluorescence decay functions change in a complex way as a function of target concentration Both ligand-free and ligand-bound forms of the sensor should provide measurable but different contributions to reporter signal. The same condition should be observed if the reporter signal is collected from dissociating competitor. Therefore in any of the forms fluorescence intensity should never be zero. Species that fluoresce more intensely contribute disproportionally more strongly to the measured parameters. Simultaneous measurements of intensities allows accounting for this effect. It is also essential to note that neither the choice of reference signal in intensity-ratiometric measurements nor the choice of reference levels in anisotropy or lifetime sensing changes the range of target concentrations to which the sensor can respond. In the state of equilibrium it is determined mainly by the target-receptor affinity.

Sensing and Thinking 2: How to Provide the Optimal Quantitative Measure of Target Binding

The first lesson that should be obtained from the considerations presented above is that the range of potential target concentrations is extremely large and covers in molar units 19–20 orders of magnitude. The second lesson is that no single sensor can cover such a vast range in target detection and no single strategy can even be used for this. There may be a possibility to fit the range of sensor sensitivity by the dilution or concentration of the sample but this cannot be done in many practical cases. In all these cases strict limitations exist. They are determined by the target-receptor affinity.

Only in the case of *very strong binding* can the binding be considered as *irreversible*. For the determination of such targets the concentration of receptors should be higher than the upper range of target concentrations in a, usually extracted from the tested system, small sample volume. Then, an irreversible binding will provide the measure of target concentration. Irreversibility means that the sensor is unable to measure the target continuously. It cannot be re-used or can be reused only after regeneration. Reversible binding in these cases is also possible but at very high dilutions, namely, at concentrations approaching the K_d values.

If the *affinity is lower*, then the major operation principle becomes the *reversible binding*. Thermodynamic analysis and the application of the mass action law allows obtaining the optimal sensor parameters, such as affinity and concentration (or density) of receptors, for the determination of the target in the desired concentration range. This range is very narrow and covers only two orders of magnitude, roughly from 0.1 K_d to 10 K_d. It can be extended by the application of an array of sensors with receptors possessing different K_d values. The gradual K_d increments should cover the whole range of potential target concentrations. In order to achieve the broadest dynamic range of measurement, the receptor concentration should be lower than the expected target concentrations.

This is the first reason to use an array of sensors with different receptors instead of a single type of receptor. The other is the *cross-reactivity* that is commonly present in real situations and that cannot easily be eliminated by better sensor design. This is the case, for instance, when the non-specific ligand, when present in the concentrations much higher than the specific target, produces the same effect of binding. Designing a variety of receptors that bind to analytes with varying affinities allows providing the 'pattern recognition' (Sandanaraj et al. 2007). The third reason for the application of the series of receptors derives from the fact that by detecting one compound one is never able to characterize a system of even medium complexity. The challenges that are put forward are the whole genome and whole proteome, requiring the analysis of many thousands of different targets.

Questions and Problems

1. Respond quickly, how many zeptomoles are in 100 pmol? How many attomoles are in 10 nmol? How much should you dilute 50 μM of concentration to obtain the concentration of 10 aM? 100 zM? What is the dimension of a cube filled with 1 fl of water? What volumes will you get by separating 5 pl into 100 equal portions?
2. Exercise in transferring the number of molecules (denoted as N or n) into moles. These values should be divided by solution volume, V and then divided by Avogadro's number 6×10^{23}. For example, if the sensor is composed of 6×10^{10} receptors and it is immersed into a volume of 1 ml, then the number of moles of receptors will be $N = 6 \times 10^{10}/10^{-3} \times 6 \times 10^{23} = 1 \times 10^{-10}$.
3. Let the sensor be based on receptors with $K_d = 100$ μM. Can we measure at equilibrium the target concentrations of 1 μM? 100 μM? 100 nM?
4. The system contains analyte displaying a specific and relatively strong ($K_d = 10$ nM) binding. Its concentration varies in the range of 1–100 nM. A nonspecific ligand is also present in this system in unspecified concentrations ranging between 1 and 100 μM. Can the analyte concentration be accurately determined if for this ligand K_d is 5 μM? 100 nM?
5. What is the dynamic range of a pH sensor with a single titrating group? How can we increase this range by incorporating into the sensor groups with variable titration properties? To provide a smooth scale of pH sensitivity in the pH range 2–10, how many of these groups are needed?
6. Why is weak binding commonly associated with low selectivity? Provide an explanation based on thermodynamic considerations. Can there be a strong but low-selective binding?

7. Ca^{2+} ions are present in resting cells on the level of 100 nM and Mg^{2+} ions on the level of 1 mM. Imagine that your sensor is based on a receptor that does not exhibit selectivity between these ions but you can vary K_d in broad ranges. Can you measure Mg^{2+} in the presence of Ca^{2+} or Ca^{2+} in the presence of Mg^{2+}? If yes, select the sensors with optimal K_d and estimate the selectivity of such measurements.

8. Analyze the practical example of calculating the occupancies of the sensor binding sites by two ligands in the case of their simultaneous presence in the tested system. Assume that for two ligands dissolved in a defined volume V we know the concentrations and the binding constants. Let $l_1 = 2 \times 10^{-8}$ M, $K_{b1} = 1 \times 10^8$ M^{-1}, $l_x = 5 \times 10^{-8}$ M, $K_{bx} = 3 \times 10^8$ M^{-1}, the number of sensor binding sites 6×10^{12} and the volume, V, is 1×10^{-3} l (i.e., 1 ml). The efficient concentration of the ligand binding sites is then $6 \times 10^{12}/1 \times 10^{-3} \times 6 \times 10^{23} = 1 \times 10^{-8}$ M. From Eqs. (2.14) and (2.15) we find the number of those sites that bind the first ligand: $n_1 = 4 \times 10^{-9} \times 10^{-3} \times 6 \times 10^{23}$ $= 24 \times 10^{11}$. The number of receptors that bind the second ligand will be $n_{x1} = 3.49$ $\times 10^{-9} \times 10^{-3} \times 6 \times 10^{23} = 20.94 \times 10^{11}$.

9. Consider the previous case in which we change the concentration of the first ligand (increase it to $l_2 = 3 \times 10^{-8}$ M), whereas the concentration of the second ligand is left unchanged. Then the number of receptors that have bound the first ligand will be n_2 $= 5.08 \times 10^{-9} \times 10^{-3} \times 6 \times 10^{23} = 30.54 \times 10^{11}$ and the number of receptors that bind the second ligand will be $n_{x2} = 2.88 \times 10^{-9} \times 10^{-3} \times 6 \times 10^{23} = 17.28 \times 10^{11}$. The known values of n_{x1} and n_{x2} can be substituted together with the known values $l_1 = 2 \times 10^{-8}$ M, $K_{b1} = 1 \times 10^8$ M^{-1}, $l_2 = 3 \times 10^{-8}$ M into the system of equations (2.16). Their numerical solution yields: $l_x = 4.9997 \times 10^{-8}$ M, $k_{bx} = 3.00016 \times 10^8$ M^{-1}. Thus, we have found the values l_x and K_{bx}, which differ very little from those values that were taken for the calculation of the number of ligand-receptor complexes, i.e., $l_x = 5 \times 10^{-8}$ M, $K_x = 3 \times 10^8$ M^{-1}.

10. What is faster, the binding to a receptor in solution or the binding to the same receptor immobilized on a surface? By how much? Explain the difference in kinetic mechanisms.

11. Why do the kinetics of target binding to a receptor immobilized on a solid surface become dependent on its concentration?

12. In what sensor technologies and for what reason is the response 'weighted'? How do weighting parameters appear?

References

Anonymous (1980) Guidelines for data acquisition and data quality evaluation in environmental chemistry. Analytical Chemistry 52:2242–2249

Baker GA, Pandey S, Bright FV (2000) Extending the reach of immunoassays to optically dense specimens by using two-photon excited fluorescence polarization. Analytical Chemistry 72:5748–5752

Bobrovnik S (2005a) New capabilities in determining the binding parameters for ligand-receptor interaction. Journal of Biochemical and Biophysical Methods 65:30–44

Bobrovnik SA (2005b) Determining the parameters for receptor-ligand interaction by serial dilution method for the case when the ligand and receptor are in a pre-existing mixture. Ukrainskii Biokhimichnii Zhurnal 77:138–145

Bobrovnik SA (2007) The influence of rigid or flexible linkage between two ligands on the effective affinity and avidity for reversible interactions with bivalent receptors. Journal of Molecular Recognition 20:253–262

Carletti E, Guerra E, Alberti S (2006) The forgotten variables of DNA array hybridization. Trends in Biotechnology 24:443–448

Demidov VV, Frank-Kamenetskii MD (2004) Two sides of the coin: affinity and specificity of nucleic acid interactions. Trends in Biochemical Sciences 29:62–71

Eaton BE, Gold L, Zichi DA (1995) Let's get specific: the relationship between specificity and affinity. Chemistry & Biology 2:633–638

Gaugain B, Barbet J, Capelle N, Roques BP, Le Pecq JB (1978) DNA Bifunctional intercalators. 2. Fluorescence properties and DNA binding interaction of an ethidium homodimer and an acridine ethidium heterodimer. Biochemistry 17:5078–5088

Hong S, Leroueil PR, Majoros IJ, Orr BG, Baker JR Jr., Banaszak Holl MM (2007) The binding avidity of a nanoparticle-based multivalent targeted drug delivery platform. Chemistry & Biology 14:107–115

Kellner R, Mermet J-M, Otto M, Valcarcei M, Widmer HM (2004) Analytical chemistry. Wiley-VCH, New York

Kitov PI, Bundle DR (2003) On the nature of the multivalency effect: a thermodynamic model. Journal of American Chemical Society 125:16271–16284

Klenin KV, Kusnezow W, Langowski J (2005) Kinetics of protein binding in solid-phase immunoassays: Theory. Journal of Chemical Physics 122: 214715

Klotz IM (1983) Ligand-receptor interactions - what we can and cannot learn from binding measurements. Trends in Pharmacological Sciences 4:253–255

Klotz IM, Hunston DL (1971) Properties of graphical representations of multiple classes of binding sites. Biochemistry 10:3065–3069

Kusnezow W, Jacob A, Walijew A, Diehl F, Hoheisel JD (2003) Antibody microarrays: An evaluation of production parameters. Proteomics 3:254–264

Lakowicz JR (1999) Principles of fluorescence spectroscopy. Kluwer, New York

Lakowicz JR (2007) Principles of fluorescence spectroscopy. Springer, New York

Li CY, Zhang XB, Han ZX, Akermark B, Sun L, Shen GL, Yu RQ (2006) A wide pH range optical sensing system based on a sol-gel encapsulated amino-functionalized corrole. Analyst 131:388–393

Marvin JS, Corcoran EE, Hattangadi NA, Zhang JV, Gere SA, Hellinga HW (1997) The rational design of allosteric interactions in a monomeric protein and its applications to the construction of biosensors. Proceedings of the National Academy of Sciences of the United States of America 94:4366–4371

Munson PJ (1983) A computerized analysis of ligand binding data. Methods of Enzymology 92:543–576

Pivovarenko VG, Klueva AV, Doroshenko AO, Demchenko AP (2000) Bands separation in fluorescence spectra of ketocyanine dyes: evidence for their complex formation with monohydric alcohols. Chemical Physics Letters 325:389–398

Sadana A, Madugula A (1993) Binding kinetics of antigen by immobilized antibody or of antibody by immobilized antigen: influence of lateral interactions and variable rate coefficients. Biotechnology Progress 9:259–266

Sanchez SA, Gratton E (2005) Lipid-protein interactions revealed by two-photon microscopy and fluorescence correlation spectroscopy. Accounts of Chemical Research 38:469–477

Sandanaraj BS, Demont R, Thayumanavan S (2007) Generating patterns for sensing using a single receptor scaffold. Journal of the American Chemical Society 129:3506–3507

Sekar MM, Bloch W, St John PM (2005) Comparative study of sequence-dependent hybridization kinetics in solution and on microspheres. Nucleic Acids Research 33:366–375

Tetin SY, Hazlett TL (2000) Optical spectroscopy in studies of antibody-hapten interactions. Methods 20:341–361

Yang RH, Li KA, Wang KM, Zhao FL, Li N, Liu F (2003) Porphyrin assembly on beta-cyclodextrin for selective sensing and detection of a zinc ion based on the dual emission fluorescence ratio. Analytical Chemistry 75:612–621

Chapter 3
Fluorescence Detection Techniques

Fluorescence is the phenomenon of the *emission* of a light quanta by a molecule or material (*fluorophore*) after initial electronic *excitation* in a light-absorption process. After excitation, a molecule resides for some time in the so-called *excited state* and its fluorescence emission can be observed usually with a lower energy (longer wavelength) than the excitation. The time range of fluorescence emission (*fluorescence lifetime*) depends on both the fluorophore and its interactions with the local environment. Thus, for organic dyes it is in the picosecond (ps) to nanosecond (ns) time range, typically 10^{-8}–10^{-11} s. Fluorescence is a part of a more general phenomenon, *luminescence*. The latter includes the emission of species excited in the course of chemical reactions (*chemiluminescence*), biochemical reactions (*bioluminescence*) or upon oxidation/reduction at an electrode (*electrochemiluminescence*). Important for sensing is also emission with a long lifetime from triplet state (*phosphorescence*). The duration of these types of luminescence can be much longer than the fluorescence. For semiconductor nanocrystals it can be tens of nanoseconds; for organometallic compounds and lanthanide complexes – hundreds of nanoseconds, up to milliseconds (ms).

Several parameters of fluorescence emission can be recorded and all of them can be used in sensing (Fig. 3.1). Fluorescence intensity F can be measured at the given wavelengths of excitation and emission (usually, band maxima). Its dependence on emission wavelength, $F(\lambda_{em})$ gives the fluorescence *emission spectrum*. If this intensity is measured over the excitation wavelength, one can obtain the fluorescence *excitation spectrum* $F(\lambda_{ex})$. *Emission anisotropy*, r (or the similar parameter, polarization, P) is a function of the fluorescence intensities obtained at two different polarizations, vertical and horizontal. Finally, emission can be characterized by the *fluorescence lifetime* τ_F, fluorescence-detected excited-state lifetime what is often called. All of these parameters can be determined as a function of excitation and emission wavelengths. They can be used for reporting on sensor-target interactions and a variety of possibilities exist for their employment in sensor constructs.

In this chapter we will discuss different possibilities of using these parameters of fluorescence emission and corresponding detection methods for the design of operational fluorescence sensors.

A.P. Demchenko, *Introduction to Fluorescence Sensing*,
© Springer Science + Business Media B.V. 2009

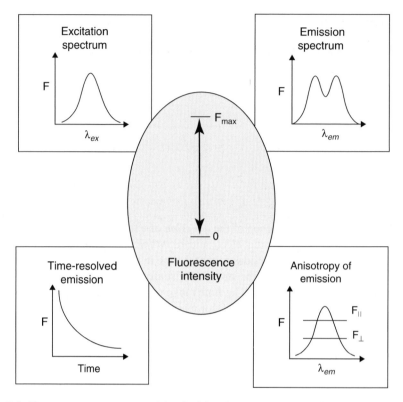

Fig. 3.1 Fluorescence parameters used for obtaining the sensor response. The simplest parameter is the fluorescence intensity F measured in the steady-state spectrum at particular excitation (λ_{ex}) and emission (λ_{em}) wavelengths. F is measured in relative units and is sensitive to all quenching effects. Dependence $F(\lambda_{ex})$ gives the excitation and $F(\lambda_{em})$ the emission spectrum. Positions of their maxima may be sensitive to intermolecular interactions of reporting dyes. The rate of emission decay gives the reverse function – the fluorescence lifetime. Recording fluorescence emission at two perpendicular polarizations gives emission anisotropy

3.1   Intensity-Based Sensing

The change from light to dark is easily observed and recorded. This change can be efficiently reproduced on a molecular level with a recording of the response from fluorescent dyes that switch from an emissive to a dark non-emissive state. Thus, everybody knows fluorescein for its bright fluorescence emission. The fluorescence of this dye can be almost completely quenched on its binding to the anti-fluorescein antibody. This provides the reader with an estimate of the dynamic range of intensity sensing of intermolecular interactions: from extremely bright to almost totally dark. If fluorescein is attached through a flexible linkage to another dye, tetramethylrhodamine, the opposite effect on the binding to anti-fluorescein antibody can be shown.

Two dyes make a non-fluorescent stacking dimer and binding to an antibody results in disrupting this dimer with dramatic enhancement of fluorescence by a rhodamine dye (Kapanidis and Weiss 2002). Thus, for various sensing applications, modulation of fluorescence intensity can be observed in both directions, the *quenching* and the *enhancement* and in the broadest possible ranges.

3.1.1 Peculiarities of Fluorescence Intensity Measurements

The proper selection of excitation and emission wavelengths is important for intensity sensing. Commonly, the fluorescence spectrum is shifted with respect to the excitation spectrum to longer wavelengths and this shift is called the *Stokes shift* (Fig. 3.2). The highest fluorescence intensity will be observed if the excitation and emission are provided at the wavelengths of the correspondent band maxima. The complicating factor is the *light scattering* that occurs at the excitation wavelength, the spectral profile of which corresponds to the profile of incident light. The light-scattering complications are especially strong if the studied sample contains

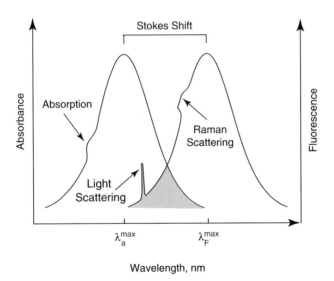

Fig. 3.2 Typical absorption and fluorescence emission spectra. The fluorescence spectrum is located at longer wavelengths with respect to the absorption spectrum. The area between the absorption and emission spectrum is shadowed. The distance between their band maxima is called the Stokes shift. It is usually expressed as a difference in energy on a wavenumber scale v, v (cm^{-1}) = $10^7/\lambda(nm)$. The fluorescence excitation spectrum usually matches the absorption spectrum and the excitation wavelength is usually selected close to this band maximum. The major light-scattering band is observed at the wavelength of excitation. The Raman scattering band is shifted by the value of the energy of the vibrations, it is seen only if the fluorescence intensity is low

macromolecules or particles, the size of which is larger than the wavelength of light. Therefore, excitation and emission wavelengths should always be separated and dyes with a large Stokes shift are preferred. The *Raman scattering* bands present in any solvent are shifted with respect to the main scattering band and overlap the fluorescence spectrum. This may produce complications in fluorescence measurements if the fluorescence intensity is very low. In a practical sense, the excitation and emission wavelengths should be selected to optimally match the wavelengths of the radiation of the light source (if it is a laser) and of the optimal sensitivity of the detector.

The measurement of fluorescence intensity at a single wavelength (usually at the emission band maximum) is the simplest and still very sensitive method to obtain information from fluorescence reporters. In many cases it is relatively easy to provide the coupling of this enhancement/quenching response with a sensing event. Intensity change is the reflection of change of the fundamental parameter of emission, the quantum yield. The *fluorescence quantum yield*, Φ, is the ratio of the number of quanta emitted as fluorescence to the number of all absorbed quanta. Fluorescence intensity at any wavelength $F(\lambda)$ is proportional to the fluorescence quantum yield Φ. The latter is determined by the ratio of radiative (k_r) and the sum of all non-radiative (k_{nr}) rate constants of an excited-state decay:

$$\Phi = \frac{k_r}{k_r + \sum k_{nr}} = \frac{1}{1 + \sum k_{nr}/k_r} \tag{3.1}$$

In highly fluorescent dyes, the emission determined by k_r should occur much faster than all the 'dark' processes occurring with the rates Σk_{nr}. Many dyes that can be used for providing the response to target binding can change their fluorescence emission on the variation of their weak noncovalent interaction with the local environment and this change is mediated by non-radiative rate constants k_{nr}. A change in fluorescence intensity from very high values (that correspond to quantum yield $\Phi \sim 100\%$) to zero or almost zero values between the target-free and the target-bound forms can be used in the background of the operation of *intensity-based molecular sensors* exploring the quenching effects.

3.1.2 How to Make Use of Quenching Effects

In intensity sensing, one of the sensor forms (target-bound or unbound) should possess the highest intensity and the other form should become 'dark' due to substantial quenching of the fluorescence of the reporter dye. Switching between these emissive and dark states of the dye can be realized in different ways (Demchenko 2005b):

1. *The quenching can be intrinsic to the reporter molecule.* This means that the reporter dye molecule itself can switch between 'light' and 'dark' states. This is possible but the dye molecule should be properly constructed. For instance, if the

intrinsic mechanism of quenching is the *intramolecular electron transfer* (IET), then the dye should contain electron-donor and electron-acceptor fragments (de Silva et al. 2001). There are many possibilities to make such constructions (see Section 6.1) and they are extensively explored, especially in ion sensing.

2. *The quenching requires intermolecular interactions.* Fluorescent dyes in their excited states may be electron-rich or electron-poor compared to the target molecules, so on their complexation, an electron-transfer quenching can be observed. An example is the detection of dopamine based on 2,7-diazapyrenium dications. In this system quenching is observed on complexation of a target electron-rich dioxyarene fragment with an electron-deficient excited dye molecule (Cejas and Raymo 2005). The same type of quenching by nucleic acid bases can be applied for their detection (Torimura et al. 2001). The quenching can be observed on dye interaction with heavy and transition metal ions.

Fluorescence quenching can be connected with *conformational flexibility*, which may exist in some dye molecules. Should the rigidity in the environment of these dyes increase, a dramatic enhancement of fluorescence can be achieved (McFarland and Finney 2001). Classical examples of this type of behavior are triarylmethane dyes, such as Crystal Violet and Malachite Green. They are nonfluorescent in liquid media but become strongly fluorescent on absorption to rigid surfaces and on binding to proteins.

3. *The quenching involves the participation of solvent.* The formation-disruption of hydrogen bonds with solvent molecules and different solvent-dependent changes of dye geometry can be observed in many organic dyes. Dramatic quenching in water (and to a lesser extent in some alcohols) may occur due to the formation by water molecules of traps for solvated electrons. In addition, the solvent can influence on dye energetics, particularly on the inversion of n (non-fluorescent) and π (fluorescent) energy levels. According to the *Kasha rule*, fluorescence occurs from the lowest state in energy. If the non-emissive state becomes the lowest, fluorescence is quenched. These and other factors can also strongly modulate fluorescence intensity (de Silva et al. 1997). Practically, for sensing applications, this means that exposing or screening the fluorescence reporter from the solvent may influence dramatically the emission intensity and can provide the necessary response.

From this short discussion, one can derive the existence of a variety of possibilities for exploring fluorescence quenching effects. Some of them may operate together. The researcher has a lot of choice for constructing a sensor with a response based on the principle of intensity sensing.

3.1.3 Quenching: Static and Dynamic

Fluorescence quenching differs not only by its intrinsic mechanism but also by the way in which it is applied (Lakowicz 2007). The quencher can form a long-lasting contact with the reporter dye, so that its emission is abolished. The emission of the same closely located dyes that do not interact with the quencher may not change. Such quenching is called *static*. It decreases fluorescence intensity but not the lifetime. This is because the dyes interacting with the quencher do not emit light and those which do not interact emit normally.

The other limiting case is the *dynamic quenching*. Here the effect of quenching competes with the emission in time and is determined by the diffusion of a quencher in the medium and its collisions with the excited dye. In this case, the relative change of intensity, F_0/F, is strictly proportional to the correspondent change of the fluorescence lifetime, τ_0/τ, where F_0 and τ_0 correspond to conditions without quencher.

Fluorescence lifetime is the average time spent in the excited state as detected by fluorescence emission. It is the reverse function of the rate of the excited-state depopulation:

$$\tau_0 = 1 / (k_r + \Sigma k_{nr}) \tag{3.2}$$

The *Stern-Volmer relationship* establishes the correlation of these changes with the quencher concentration [Q] by introducing the rate constant proportional to this concentration:

$$F_0 / F = \tau_0 / \tau = 1 + k_q \tau_0 [Q] = K_{SV} [Q] \tag{3.3}$$

Here k_q is the rate constant of quenching, which in the case of 100% efficient quenching is equal to the quencher diffusion constant (being typically in the range of 10^9–10^{10} $M^{-1}s^{-1}$). $K_{SV} = k_q \tau_0$ is called the Stern-Volmer quenching constant. This can be determined from the slope of the graph displaying dependence of intensity or lifetime on [Q], expressed by Eq. (3.3).

If the lifetime does not change in proportion to the change of intensity or if the quenching effect in intensity exceeds the diffusional limit, this indicates the contribution of *static quenching*. In the case of all the quenching being static, a dye bound to a quencher appears immediately (much faster than the decay) in a dark nonemissive state. The unbound fluorophores exhibit their natural lifetimes and the slope of F_0/F vs [Q] yields a binding constant.

The static and dynamic quenching can also be distinguished by the temperature effects. If the quenching is static, it decreases with the increase of temperature because of the disruption of molecular complexes that produce quenching. On the contrary, the dynamic quenching increases because of an increase of the quencher diffusion rate.

Both static and dynamic quenching can be used for providing the fluorescence response. In sensing technology, the target or the competitor can be used as a dynamic quencher and its concentration can be determined from Eq. (3.3). This condition could be better applied to detecting small molecules, the diffusion of which is fast. Meanwhile, more efficient in many cases could be the employment of such sensor constructs that use static quenching produced by intramolecular or intermolecular complexation.

3.1.4 Non-linearity Effects

When fluorescence intensity is measured at low dye concentrations and low intensities of incident light and dye distribution in the studied system do not depend upon its concentration, this intensity should be strictly proportional to dye concentration.

Non-linearities can appear due to several factors. They are:

(a) Depending on *sample volume and geometry*. In the case of high *incident light absorption* in the system, the light passing through the sample becomes weaker and excites fluorescence at a lower level (inner filter effect). In the cases of high fluorophore concentration, the fluorescent light emitted from illuminated volume can be *re-absorbed* in a dark volume instead of passing to a detection device. The involvement of these factors can be reduced by reducing the optical path or sample volume.

(b) Depending on the *properties of the dye*. Some dyes form *non-fluorescent* dimers and aggregates and this formation is concentration-dependent. In addition, at short distances between the dye molecules, excitation can migrate between the dyes so that the dye absorbing light can transfer its energy to neighboring dyes and these non-fluorescent species serve as the light traps. This explains why one cannot load many dye molecules (fluorescein or rhodamine derivatives) on a single protein molecule (e.g., antibody), such complexes become non-fluorescent. The mechanisms of these processes will be explained in Section 6.2. Energy migration to non-fluorescent associates can be reduced by using the dyes with large Stokes shifts.

However, there is a useful application of this phenomenon. *Self-quenching* effects may bring useful information in the cases when observed phenomenon involves a strong change in the local dye concentration. For instance, the dyes trapped at high concentrations within phospholipids vesicles (liposomes) can be non-fluorescent and when vesicle integrity is disrupted, the dye comes out from the vesicle and being diluted, dramatically increases its fluorescence (Domecq et al. 2001). Hence, the control for vesicle integrity.

(c) Occurring at *high excitation light intensities*. They are explained by the fact that in this case more molecules appear in the excited state and the ground state becomes depopulated. Since fluorescence intensity is proportional to ground-state concentration, the so-called *light quenching* effect appears (Lakowicz 2007).

All these effects do not appear at common excitation and emission conditions and commonly used dye concentrations but they have to be accounted for if the researcher is working outside this range.

3.1.5 Internal Calibration in Intensity Sensing

Calibration in fluorescence sensing means the operation, as a result of which at every sensing element (molecule, nanoparticle, etc.) or at every site of the image, the fluorescence signal becomes independent of any other factor except the concentration of the bound target. The problem of calibration in intensity sensing is very important because commonly the fluorescence intensity is measured in relative units that have no absolute meaning if not compared with some standard measurement. When the fluorescence spectrum is recorded, the intensity at one wavelength is compared with intensities at other wavelengths; therefore the spectrum has an informative value. However, one cannot compare the fluorescence intensities of two samples measured today and 1 year later because of the difference in experimental conditions. One cannot compare the fluorescence intensities at two neighboring spots in the sensor array in which the amount of sensor molecules is different. A similar problem exists in cellular imaging: the image formed of light intensities depends not only on the target distribution but also on the distribution of the sensors themselves.

A fluorescence signal *is obtained in relative units* (the quanta counts or the analog signal obtained on their averaging by the detection system). In this case the fluorescence response depends not only upon the sensor – target binding but also upon such technical parameters as the intensity of the incident light, geometry of the instrument, detector sensitivity, monochromator slit widths, etc. It is sensitive to every fluctuation in the light source intensity or in the sensitivity of the detector. Thus, the recorded changes of intensity always vary from instrument to instrument and the proper reference, even for compensating these instrumental effects, is difficult to apply.

In the common spectrofluorimeter, the measured signal can be compared with the signal coming from the reference sample. This reference could be the sensor molecules in a given concentration calibrated with the addition of given concentrations of the target. By variation of the target concentration, a calibration function can be obtained. It is then used for determining the target concentration from the read-out of the fluorescence response from the sample. This approach is hard to realize on a microscopic level or in a simultaneous analysis of different sensor-target compositions. We need to analyze why this difficulty appears with the aim of finding proper solutions.

(a) It is difficult to achieve a *strictly defined concentration* (or density) of sensor molecules in an analyzed volume. It is also common that both the sensing properties and the reporting abilities degrade in time, due either to inactivation or to photobleaching.

(b) It is difficult to provide a *separate reference* to every sensor unit in case there are arrays of them and each of them is sensing a different target.

(c) Even if this is done, it is hard to *equalize the concentrations* of sensor molecules in a sensor and in a reference element or to provide a correction for their difference.

Thus, if our aim is obtaining quantitative information about the concentrations of many different targets simultaneously, we have to develop a special methodology. We have to have the requirement that each element of the array should become *self-calibrating*. This means that its response should contain not only the signal reporting on the target binding but also a different signal (or signals) that could allow accounting for or compensation of all the effects that influence the fluorescence parameter(s) besides the target binding.

In the simplest case of reversible binding with stoichiometry 1:1, the target analyte concentration [A] can be obtained from the measured fluorescence intensity F as:

$$[A] = K_d \left(\frac{F - F_{min}}{F_{max} - F} \right) \quad (3.4)$$

Here, F_{min} is the fluorescence intensity without binding and F_{max} is the intensity if the sensor molecules are totally occupied. K_d is the dissociation constant (introduced in Section 2.2). The differences in intensities in the numerator and denominator allows compensation for the background signal and the obtained ratio can be calibrated in the target concentration. But since F, F_{min} and F_{max} are expressed in relative units, they have to be determined in the same test and under exactly the same experimental conditions. This is difficult and often not possible.

The most efficient and commonly used method is the introduction of the *reference reporter*. This additional dye can be introduced into a sensor molecule (or into a support layer, the same nanoparticle, etc.) so that it can be excited together with the reporter dye and emit light at a different wavelength but it should not change its signal on target binding. This should provide an additional *independent channel of information* that could serve as a reference (Fig. 3.3). If the reference is properly selected, then one can observe two peaks in fluorescence – one from a reporter with a maximum at λ_1 and the other from the reference with a maximum at λ_2. Their intensity ratio can be calibrated in the concentration of the bound target. Thus, if we divide both the numerator and the denominator of Eq. (3.4) by $F_{ref}(\lambda_2)$, the intensity of the reference measured in the same conditions, we can obtain a target concentration from the following equation, which contains only the intensity ratios $R = F(\lambda_1)/F_{ref}(\lambda_2)$, $R_{min} = F_{min}(\lambda_1)/F_{ref}(\lambda_2)$ and $R_{max} = F_{max}(\lambda_1)/F_{ref}(\lambda_2)$:

$$[A] = K_d \left(\frac{R - R_{min}}{R_{max} - R} \right) \quad (3.5)$$

This approach to quantitative target assay has found many applications. As an example, it was used in the construction of double-labeled molecular biosensors based on glutamine and glucose binding proteins (Ge et al. 2004; Tolosa et al. 2003). In other research, for the design of a ratiometric Cl⁻ sensor, a molecular hybrid was synthesized that contained two covalently linked dyes (Jayaraman et al. 1999). They can be excited simultaneously but exhibit separate emission spectra. One of them is strongly quenched by Cl⁻ ions, while the other does not interact with Cl⁻ and serves as a reference for the sensor concentration. The sensor designs with the introduction of additional dyes as the references are also used in cation and pH sensing (Koronszi et al. 1998).

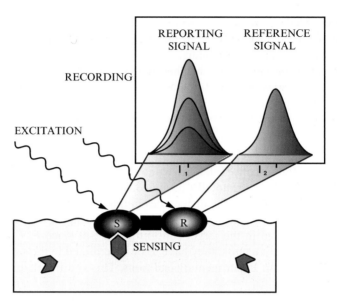

Fig. 3.3 The scheme illustrating obtaining self-calibrated signals with the aid of reference dyes. The two dyes, sensing (S) and referencing (R), are excited and detected simultaneously. The sensing dye provides the variation of intensity on the target binding but not the reference dye

The intensity-responding probes are very popular in intracellular measurements of Ca^{++} ions (Minta et al. 1989) and to obtain a more accurate representation of ion concentrations, the simultaneous injection into the cell of Ca^{++}-responsive and Ca^{++}-insensitive dyes is used (Oheim et al. 1998). Of course, this is not the best solution, since the intracellular distribution of the two dyes can be different. A better possibility is to use the nanoparticles that incorporate two dyes, one responsive and the other insensitive to target binding. This approach has been suggested for the measurement of pH, oxygen and ion concentrations in different test media and also within living cells.

The *separate detection of the two signals*, one from the reporter dye and the other from the reference, can be provided based not only on the difference of their fluorescence band positions but also on the difference in anisotropy (Guo et al. 1998) or lifetime (Guo et al. 1998; Liebsch et al. 2001). The measured anisotropy or lifetime is a sum of intensity-weighted anisotropies or lifetimes of contributing species (see Sections 3.2 and 3.3), thus if in the presence of the reference dye the intensity of the reporting dye changes, this change can be recorded as an anisotropy or lifetime change. This type of calibration can be used even if the reporter and the reference dyes possess strongly overlapping fluorescence spectra. Intensity calibration in the lifetime domain has an advantage in studies in highly light-scattering media.

Summarizing, we outline what is achieved and what is not with the introduction of a reference dye. The two dyes, responsive and non-responsive to target binding, can be excited and their fluorescence emission detected simultaneously, which compensates for the instability of instrumental factors. In principle, the results should be reproducible on instruments with a different optical arrangement, light source intensity, slit widths, etc. In addition, if the two dyes are distributed very similarly in the illuminated volume, the two-band ratiometric signal can be calibrated in the target concentration. This calibration, in some range of target concentrations, will be insensitive to the concentration of sensor (and reporter dye) molecules.

It should be remembered however, that the sensor molecules can exhibit degradation and, additionally, the dye molecules can photobleach as a function of time. These effects may provide time-dependent and target-independent changes of the measured intensity ratios. In addition, because the sensitivity to quenching (by temperature, ions, etc.) is different for reporter and reference dyes and they emit independently, every effect of fluorescence quenching unrelated to target binding will interfere with the measured result. This can make the sensor non-reproducible in terms of obtaining precise quantitative data, even in serial measurements with the same instrument. In these cases, other detection methods, which will be reviewed below, are preferable.

3.1.6 Intensity Response as a Choice for Fluorescence Sensing

As we observed above, in intensity sensing there are many good possibilities to choose the dye and its response mechanism, satisfying the requirements for a strong change of fluorescence intensity. After choosing the proper dye and optimizing its variation of fluorescence intensities, the researcher can concentrate on other parameters that are important for the operation of the sensor. There are convenient for observation excitation and emission spectra, high molar absorbance at the wavelength of excitation, high quantum yield of one of the forms (bound or unbound) of the sensor and its strong change on the binding (see Section 4.1). It may be important to choose a dye with a strong Stokes shift, since this will allow not only reducing the light-scattering effects but also, using broad-band filters for detection, which will increase the recorded intensity and thus the sensitivity of detection. It is important also to choose a dye with a high photostability, which will not photobleach during the measurement and will provide the necessary reproducibility in serial repetitive measurements.

Concluding, we also note that the weak point of the intensity sensing technique is the necessity to resolve the *problem of calibrating* the sensor element. This problem exists in all sensor applications; be it macromolecules in solutions, nanoparticles, waveguide tips or multi-sensor arrays. The intensity depends upon the concentration of sensor molecules, excitation wavelength, optical density at this wavelength and, of course, on the emission wavelength, which still further complicates the calibration. It is difficult to obtain a linear response in fluorescence intensity operating with fiber-optical waveguides. But what is really difficult or even impossible is to calibrate or compensate the effects of light-absorption and light-scattering. This can be a serious problem, especially if the studied medium is heterogeneous, such as the colloid suspension, the solution of high-molecular-weight polymer or the suspension of living cells. If the target molecule is of large size and if it can aggregate during the measurement, this may create irresolvable problems.

If the sensor molecules are incorporated into living cells, then an endogenous *autofluorescence* becomes also an essential problem (Billinton and Knight 2001). In addition, even if the target binding is completely reversible and both emission intensity levels without target and with the target saturation can easily be determined, there is always a problem of time-dependent degradation of the sensor affinity. Finally, any time-dependent processes occurring with the fluorophore, especially photobleaching, will make the sensor non-reproducible even in serial measurements with the same instrument and when all the above-listed conditions are kept constant or properly corrected. These difficulties justify the strong efforts of the researchers to develop fluorescence dyes and sensing methods that allow excluding or compensating these factors.

Discussing the disadvantages of sensing based on fluorescence intensity, we must not forget its *important merits*. One such merit is the simplicity in instrumentation, in the measurement and in adjustability of this instrumentation for use in integrated sensing devices and micro-arrays. Another merit is the highest possible absolute sensitivity, since in other methods the application of higher resolution in spectra or resolution in time and anisotropy results in loss of absolute intensity. If the interfering factors, such as background fluorescence and light scattering, are not very important, the integral intensity over the whole emission band can be collected. Thus the sensors based on the registration of ΔF changes (sometimes called *intensiometric* sensors) can be recommended for use in qualitative or semi-quantitative detection technologies, where high sensitivity is very desirable but high precision of quantitative measurement is of secondary importance.

3.2   Anisotropy-Based Sensing and Polarization Assays

Linearly polarized light is light that propagates with its electric vector keeping a unique direction in space. Such light waves can excite the dyes only if they are in definite orientation. If these dyes are immobile on a time scale of fluorescence decay, they also emit polarized light. However, due to rotational diffusion, they may randomly change their orientation, so that being initially anisotropic, with time their orientation becomes isotropic. This leads to *depolarization* of emitted light. If the dye represents a fluorescent reporter and in the course of a sensing event the size of the rotating unit increases or local viscosity increases, then the rotation will slow down and there will be an increase of *fluorescence polarization* and *anisotropy*. The time window for this rotation is also important. It is given by the fluorescence lifetime τ_F. When τ_F increases, our sensor will then have more time to rotate and anisotropy (polarization) will decrease. Thus, the *rotation rate* and the *fluorescence emission rate* determine the response in anisotropy (Fig. 3.4).

The recording of anisotropy requires measurement of the *ratio of fluorescence intensities obtained at two different polarizations* of the emission light, *vertical* and *horizontal* (Guo et al. 1998; Jameson and Croney 2003). Primarily fluorescence polarization measurements were used to study the rotational mobility of labeled macromolecules, the presence of their flexible fragments, association-dissociation of protein subunits and of intermolecular interactions influencing this mobility. Later on, different *polarization assays* were developed based on the same principle: the slowing down of the rotational mobility of the sensor molecule. Their development led to *anisotropy-based sensor arrays*, in which small rotating sensor units are immobilized on support and target binding is detected as the decrease of the rate of its rotation.

3.2.1   Background of the Method

The measurement of *steady-state anisotropy r* is simple and needs two polarizers, one in excitation and the other in emission beams. When the sample is excited by vertically polarized light (indexed as $_V$) and the intensity of the emission is measured at vertical (F_{VV}) and horizontal (F_{VH}) polarizations, one can obtain r from the following relation:

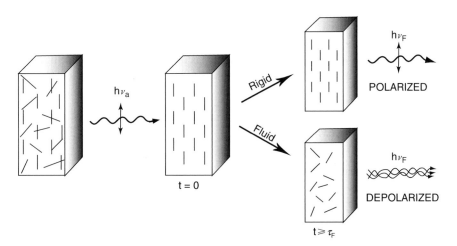

Fig. 3.4 The connection between light depolarization and rotational mobility. Excitation by polarized light selects from the dye molecules in different orientations those oriented in a particular direction (usually, vertical). In rigid environments these oriented molecules emit polarized light. If they are able to rotate during the lifetime of the emission, their emission becomes depolarized

$$r = \frac{F_{VV} - G \times F_{VH}}{F_{VV} + G \times 2F_{VH}} = \frac{1 - G \times (F_{VH} / F_{VV})}{1 + G \times 2(F_{VH} / F_{VV})}, \qquad (3.6)$$

where G is an instrumental factor. *Polarization P* is the parameter, which is also frequently used for characterizing polarized emission. Though it is less convenient, it was historically the first parameter introduced and provides essentially the same information. Its relation to r is simple:

$$r = 2P / (3 - P) \qquad (3.7)$$

Equation (3.6) shows that r in fact is a *ratiometric parameter* that does not depend upon the absolute intensity of emission light because the variations of intensity proportionally influence the F_{VV} and F_{VH} values. Therefore the anisotropy, similarly to the lifetime (Section 3.3), is weakly sensitive to the variation of reporter dye concentration or to its change in the course of experiment (e.g., to photobleaching). Thus, we may consider that anisotropy and polarization of emission that are derived from the intensities collected at the same wavelength at vertical (F_{VV}) and horizontal (F_{VH}) polarizations are the *intrinsic properties* of the dye molecule in a particular environment.

In the simplest case, when both the rotation and the fluorescence decay can be represented by single-exponential functions, the range of variation of anisotropy (r) is determined by the variation of the ratio of the fluorescence lifetime (τ_F) and *rotational correlation time* (φ) describing the dye rotation:

$$r = \frac{r_0}{1 + \tau_F / \varphi} \qquad (3.8)$$

where r_0 is the *fundamental anisotropy*. It can be defined as the limiting anisotropy obtained in the absence of rotational motion. It is an intrinsic parameter of a fluorophore that is determined by the angle between its absorption and emission dipoles.

3.2.2 Practical Considerations

The dynamic range of anisotropy sensing is determined by the difference of this parameter observed for a free sensor, which is the rapidly rotating unit and the sensor-target complex that exhibits a strongly decreased rate of rotation (Fig. 3.5). The reporter dye should be rigidly attached to the sensor for reflecting this change of mobility.

The polarization experiment allows obtaining *two-channel information* on a sensing event from a single reporter dye. The range of variation of r values in this event could extend from 0 to r_0 (which is commonly around 0.3–0.4) and should not depend upon the instrumental factors or the dye concentration.

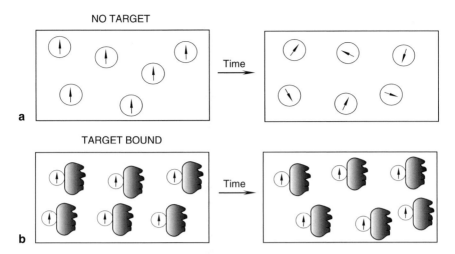

Fig. 3.5 Illustration of the basic principle of fluorescence anisotropy sensing (polarization assays). Dye molecules with their absorption transition vectors (arrows) aligned parallel to the electric vector of linearly polarized light (along the vertical page axis) are selectively excited. Due to rapid rotational diffusion, the distribution in their orientations becomes randomized prior to emission, resulting in low fluorescence anisotropy. The binding of a large, slowly rotating target molecule or particle results in highly polarized fluorescence. The difference between the two values of anisotropy provides a direct readout of the target binding

As follows from Eq. (3.8) and the above discussion, the variation of anisotropy can be observed in two cases, on the variation of the *rotational mobility* of the fluorophore (the change of φ) and on the variation of the *emission lifetime* τ_F. At given τ_F the rate of molecular motion determines the change of r, so that in the limit of slow molecular motion ($\varphi >> \tau_F$) r approaches r_0 and in the limit of fast molecular motion ($\varphi << \tau_F$) r is close to 0. This determines the dynamic range of the assay.

Since the change of anisotropy on target binding can be achieved by changing two parameters, φ and τ_F, it is essential that their ratio should exhibit the most significant difference between target-bound and target-free states of the sensor. It is possible and it is highly preferable to realize the case when the change of φ and τ_F occurs in opposite directions. Meanwhile, it is not uncommon that the target binding, resulting in increasing φ due to the increase of the size of rotating unit, also increases τ_F due to dye immobilization or screening from the quenching effect of water. In order to provide a high sensitivity for the assay, these situations need to be avoided.

Anisotropy sensing is well compatible with the methodology of *direct reagent-independent sensing* (Section 1.6). The attachment of the dye at the sensor-target contact areas in this case is not required, which increases the possibilities in the application of this method. In competitor displacement assays (Section 1.3) the binding of a small competitor molecule is easily detected as the increase of anisotropy. However, the advantages of this method are often difficult to realize in heterogeneous assays, since for the sensor attached to the surface the rotational mobility can already be lost and the anisotropy on binding the target may not exhibit an additional increase. Therefore, in such cases the fluorophore has to be attached to a flexible linker to maintain its own degree of rotational freedom relative to the sensor molecule, so that the loss of its mobility can report about binding.

Another problem is the resolution of anisotropy sensing in the case of *high molecular weight targets*. In order to achieve a good resolution in anisotropy, the rotational correlation time of the labeled target should correspond to the fluorescence lifetime. Using the well-known Stokes-Einstein relation for isotropic Brownian rotational diffusion, one can estimate the rotational correlation time of the solute molecule or particle (assuming its spherical shape) on the basis of molecular mass (*M*):

$$\varphi = \frac{\eta V}{kT} = \frac{\eta M}{RT}(v+h) \qquad (3.9)$$

Here k is the Boltzmann constant, T is the absolute temperature in Kelvin, η is the viscosity of the solution, V is the molecular volume, R is the ideal gas constant, v is the partial specific volume in cm^3/g and h is the hydration (typically for proteins, 0.2 g of H$_2$O/g). Assuming the typical v value about 0.73, temperature 20°C and the viscosity of water, one can use a simplified version of Eq. (3.9) for a protein molecule approximated as unhydrated sphere:

$$\varphi(\text{ns}) = 3.05 \times 10^{-4} \times M \qquad (3.10)$$

It follows that to achieve the best sensing signal in assay for a protein of molecular mass 40,000 Daltons the fluorophore lifetimes should be approximately 12 ns with no account of hydration and 19 ns on the assumption that h = 0.4. Asymmetry of molecule or particle will further increase this value.

From Eq. (3.9), using Eq. (3.8) one can evaluate the fluorescence lifetime that will be optimal to detect the variation of anisotropy in the necessary range. The calculations based on Eq. (3.9) correlate reasonably well with experimental data. Figure 3.6 presents the graphs built on this correlation (Guo et al. 1998).

From these estimates it can immediately be seen that common organic dyes with fluorescence lifetimes of the order of 1 ns are suitable for detection in the anisotropy response of only small and rapidly rotating sensors (or competitors) and do not properly fit the size of even small proteins. To label larger molecules, such as antibodies, luminophores with longer lifetimes are needed. They should satisfy the best sensing conditions, so that $\varphi < \tau_F$ before the binding and $\varphi > \tau_F$ after the binding of target.

3.2.3 Applications

Like other methods of fluorescence sensing, anisotropy sensing is based on the existence of two states of the sensor, so that switching between them depends on the

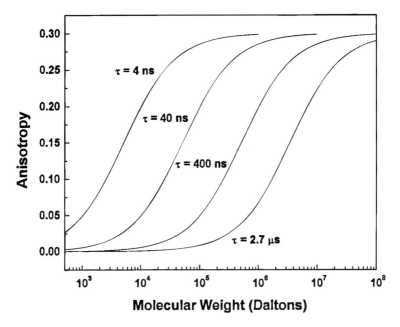

Fig. 3.6 Molecular-weight-dependent anisotropy for a protein-bound luminophore with luminescence lifetimes of 4, 40, 400 and 2,700 ns. The curves are based on Eqs. 3.8 and 3.9 assuming an aqueous solution at 20°C with a viscosity of 1 cP and (v + h) = 1.9 (Guo et al. 1998)

concentration of the bound target. In these states, the dye reporters exhibit different possibilities to emit light and to rotate. The major application of this technique in fluorescence sensing is to provide a response to target binding by the change of the *rotational mobility* of the fluorescent reporter. If the free fluorescent sensor molecule is relatively small, it rotates rapidly and displays a low value for anisotropy. On target binding the size of this rotating unit increases, producing a sharp rise in anisotropy. Due to these features, anisotropy sensing can be applied both in competitor displacement and direct sensing formats (see Sections 1.3 and 1.5).

The most extended application is found in sensing techniques based on antigen-antibody recognition. The so-called "*polarization immunoassays*" are based on measurements of steady-state anisotropy in both heterogeneous and homogeneous assay formats. In a *homogeneous* fluorescence polarization immunoassay, for the detection of antigens or antibodies (Nielsen et al. 2000), either the antigen or the antibody can be determined in solution with the labeled fluorescent partner. On formation of their complex, the steady-state anisotropy increases due to an increase of the size of the rotating unit. The fluorescence polarization assay can also be used to determine the antigen-antibody affinity (Smallshaw et al. 1998).

Anisotropy measurements are also used in *competitive immunoassays* that have become common not only in research but also in routine clinical analysis. Primarily, the complex of the sensor antibody with the target analog labeled with fluorescent dye is formed. Consequently, the added target produces a change of fluorescence anisotropy by displacing the competitor from the binding site. Since the free competitor is a smaller rotating unit than its complex with the sensor, a decrease in anisotropy is detected. This procedure can be performed in a solution as a *homogeneous immunoassay* as it does not require any separation steps. Similarly to other competitive assays, in solutions it cannot be extended to measure several targets simultaneously.

An example of the successful application of the anisotropy detection technique in ion sensing can be the detection of picomolar to nanomolar concentrations of Zn^{2+} by application of aminobenzoxadiazole dye covalently attached to genetically modified metalloprotein carbonic anhydrase (Thompson et al. 1998). The results demonstrate that the free transition metal ions can be determined at trace levels in aqueous solution using inexpensive instruments.

Anisotropy sensing is an efficient method for detecting the binding of fluorescent ligands (and displacement) to *imprinted polymers* (Hunt and Ansell 2006). Here, in comparison to intensity sensing, no separation of bound and unbound ligand is needed.

In *drug discovery*, the fluorescence polarization (anisotropy) technology became a routine technique because of its convenience and low price (Owicki 2000). Originally, the polarization assays were developed for the single-cuvette analytical instruments but the technology was rapidly converted to high-throughput screening assays when commercial plate readers with equivalent sensitivity became available (Burke et al. 2003). These assays include different targets, such as kinases, phosphatases, proteases, G-protein-coupled receptors and nuclear receptors (Burke et al. 2003; Sportsman et al. 2003).

There is no limit for the generalization of the anisotropy sensing technique to the large-format proteomics arrays, using probably the most advanced capture molecules, nucleic acid aptamers (see Section 5.5). Aptamers can be immobilized at one point by a chain terminal on solid support and can rotate around this point of attachment with the observed decrease of anisotropy (McCauley et al. 2003). The target binding decreases the rate of its rotation.

Based on this principle, a biosensor in a chip configuration that is capable of quantifying several proteins with relevance to cancer was demonstrated (McCauley et al. 2003). A rapid, homogeneous aptamer-based bioanalysis was recently reported for the sensitive detection of immunoglobulin E (IgE) in the low-nanomolar range with a high specificity (Gokulrangan et al. 2005).

Summarizing, we outline three possibilities for using the fluorescence response in anisotropy (polarization) in sensing:

(a) When anisotropy increases with the increase of the molecular mass of the rotating unit
(b) When anisotropy increases due to the increase of local viscosity producing higher friction on the rotating unit
(c) When anisotropy increases due to a fluorescence lifetime decrease being coupled to any effect of dynamic quenching

All these effects are interrelated and observed within a rather narrow range of related parameters. Nevertheless, the possibilities for the application of this technique are very broad.

3.2.4   Comparisons with Other Methods of Fluorescence Detection

Anisotropy sensing allows a direct response that is independent of sensor concentration and allows both direct and competitor substitution sensing. In the background of this independence is the independence of measured anisotropy (polarization) on absolute fluorescence intensity. In this respect we have to stress the following.

(a) Two information channels in this case are derived from a single dye molecule exhibiting a single electronic transition. Thus, the two parameters, (F_{VV} and F_{VH}) are strongly coupled.
(b) If additional perturbations appear, such as variations of lifetime or the appearance of additional degrees of rotational freedom, F_{VV} and F_{VH} values change in a different manner, producing a change of anisotropy that overlaps the response obtained on target binding. In particular, any factor that causes the decrease of fluorescence lifetime (produces quenching) increases the anisotropy.

The *dynamic range* of the fluorescence response is determined by the "window" in anisotropy values between the readings obtained with the free sensor and the sensor with the bound target (Jameson and Croney 2003); the possibility to increase this range is offered by long-lifetime luminophors (see Section 5.2).

The scaling of *bound target concentration* is performed between these two limiting (low and high) values and the position of the whole scale depends on the fluorescence lifetime τ_F. In the case of the presence of fluorescence quenchers, the whole scale will be shifted, producing a systematic error. Due to this fact, the self-calibration that can be achieved in anisotropy sensing may not always be sufficient.

It should be emphasized that the weak point of this detection technique is its great *sensitivity to light-scattering effects*, especially to those that may result in a non-specific aggregation of the sample during the sensing procedure. This occurs because the scattered light is always 100% polarized and its contribution cannot be removed in steady-state measurements if there is a spectral overlap between scattered and fluorescent light. To avoid the light-scattering artifacts, dyes with a large Stokes shifts should be preferably used. This allows increasing the gap between excitation (and also light-scattering) and emission wavelengths.

3.3   Lifetime-Based Fluorescence Response

Fluorescence is the process developing over time after initial electronic excitation. It decays as a function of time in the subnanosecond-nanosecond time range. It is hard to imagine such short periods of time; they are outside our everyday experience. But the life of molecules in these periods is very busy. Molecules move, rotate, collide and participate in different reactions. Fluorescence emission helps us to reveal the mechanisms of all these events. The duration of fluorescence also depends on them. If we wish to make a sensor, our target should also participate in these processes.

3.3.1 Physical Background

Usually, molecules emit light independently, so their decay function is essentially concentration-independent. It is an intrinsic property of the dye, of its interactions, dynamics and participation in reactions. In an ideal case of identical independent fluorescence emitters, the fluorescence decay is a *single exponential function of time t*:

$$F(t) = F_0 \exp(-t/\tau_F) \tag{3.11}$$

Here, F_0 is the initial intensity at time $t = 0$. The time constant τ_F is called the *excited-state (fluorescence) lifetime*, it is the reverse function of the *fluorescence decay rate*, k_F. In many real cases, the decay is not exponential but it is often possible to characterize it by averaged lifetime $< \tau_F >$. The necessary condition in lifetime sensing is the possibility of detecting the strong difference of $< \tau_F >$ between analyte-bound and analyte-free forms of the sensor.

Lifetime sensing is based on two different principles:

(a) *Modulation of lifetime by a dynamic quencher.* Molecular oxygen is a strong quencher of long-lifetime emission operating by diffusional mechanism. The decrease of τ_F occurs gradually with oxygen concentration. This change obeys the Stern-Volmer relationship (Eq. 3.3), which allows determination of the oxygen concentration. High mobility combined with potent quenching ability are unique features of oxygen molecules.

(b) Shifting of the sensor to a *discrete form with a different lifetime*. The presence of the distinct difference in lifetime between the two forms of the sensor, free (A) and bound (B) is necessary for the application of this type of lifetime sensing. Belonging to the same dye, these two forms can be excited at the same wavelength. When excited, they emit light independently and the observed nonexponential decay has to be decomposed into two different individual decays with lifetimes τ_F^A and τ_F^B:

$$F(t) = \alpha_A \exp(-t/\tau_F^A) + \alpha_B \exp(-t/\tau_F^B) \tag{3.12}$$

In relation to molecular sensing, the interest to lifetime detection techniques is due to the fact that fluorescence decay is characteristic of the fluorophore and of its environment and *does not depend upon fluorophore concentration*. Moreover, the results are not sensitive to the optical parameters of the instrument, so that the attenuation of the signal in the optical path does not distort it. The light scattering also produces much fewer problems, since the scattered light decays on a very fast time scale and does not interfere with fluorescence decay observed at longer times. Most often, the decays do not depend on emission wavelength (with the exception of excited-state reactions), which allows one to collect them from the whole spectra and achieve relatively high sensitivity.

3.3.2 Technique

Fluorescence decay functions can be obtained with the aid of special instrumentation. It allows observing the decay directly using the single photon counting or stroboscopic techniques or by observation of the demodulation and phase shift of harmonically modulated excitation light. These methods are well described in the literature (Lakowicz 2007; Valeur 2002). In brief, in the *phase-modulation* technique the excitation light is modulated with a high frequency (100–200 MHz) and the detector records the shift of the phase of the modulated signal (the phase angle) and the decrease in its modulation depth in fluorescence compared to that in excitation light. The slower the fluorescence decay (the longer lifetime), the larger the phase shift and the smaller the modulation. In *single-photon counting* the excitation occurs by a light source producing short pulses and single quanta are randomly selected from every pulse to form the experimental emission decay function (Fig. 3.7). The emission decay function can be obtained by mathematical transformation of the primary data obtained by phase-modulation. Thus, the methods are equivalent although their technical realization is different.

Lifetime measurement requires a more complicated and expensive technique than that required for steady-state measurements, so the prospects for this method depend upon the development of simple and cheap ways for its realization. The most expensive part of this instrumentation is the light source and the application of cheap pulsed semiconductor lasers can make this method very attractive. Until recently, these lasers were available only in the "red" spectral region, where a very limited range of dyes was available in sensor applications. With the introduction of inexpensive "blue"-emitting diode lasers, we expect a new increase of interest to lifetime sensing. The other extension of this methodology is the development of assays that involve the long-lifetime luminescence of rare earth ions or explore the reactions with their participation (Maliwal et al. 2001).

3.3.3 Time-Resolved Anisotropy

Time-resolved detection can be combined with the measurements of anisotropy to give *time-resolved anisotropy r(t)*. Technically, this can be achieved by incorporating the polarizer and analyzer into a time-resolved fluorimeter. The time resolution

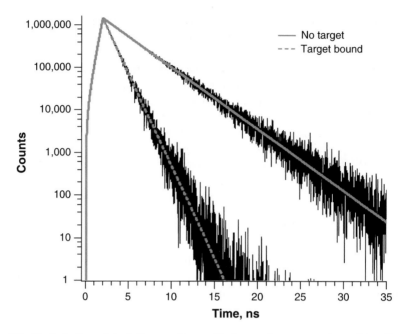

Fig. 3.7 Typical effect of the influence of target binding on fluorescence decay. Short fluorescence decay (small τ_f) changes to long decay (larger τ_f) on target binding (it can also be the opposite)

allows the elimination of all the light-scattering effects, to which the steady-state anisotropy is sensitive. If the two forms of a sensor, A and B, differ by anisotropy decay, then the observed decay will be composed by contributions provided by these individual forms:

$$r(t) = r_0 [F_A \exp(-t / \varphi_A) + F_B \exp(-t / \varphi_B)]. \tag{3.13}$$

Similarly to steady-state anisotropy, $r(t)$ is an 'intensity-weighted' parameter in sensing (see Section 2.5). Such systems can yield unusual anisotropy decays that show decline to a minimum at short times and increase at long times (Lakowicz 2007).

3.3.4   Applications

Examples of the application of lifetime sensing were presented for popular fluorescence probes, used for measuring cation concentration (such as Calcium Green) and pH (such as SNAFL-2 (Szmacinski and Lakowicz 1994). Recently, one of the pH-sensitive dyes was incorporated into polymeric hydrogel and it was shown that lifetime sensing provides reliable results in light-scattering media (Kuwana and Sevick-Muraca 2002).

The other important application is sensing the compounds that produce the decrease of τ_F in molecular collisions. An example is chloride sensing in living cells. The mechanism of the response of many cellular chloride sensors is based on collisional quenching by a Cl^- ion that decreases both the intensity and the lifetime. Using a SPQ (6-methoxy-N-[3-sulfopropyl]quinolinium) chloride-sensitive probe, a comparative study of intensity-sensing and lifetime-sensing was performed and the strong advantages of lifetime sensing were demonstrated (Szmacinski and Lakowicz 1994).

3.3.5 Extension to Reporter Response Based on Phosphorescence

Phosphorescence exhibits much longer emission lifetimes than fluorescence. It is commonly observed at cryogenic temperatures but in special conditions (incorporation of dye into a rigid matrix and removal of mobile oxygen) it can also be observed at room temperature with lifetimes covering a broad range of 10^{-6}–10^2 s. The rigidity of the dye environment can be provided by sol-gel and polymeric materials. The long duration of the phosphorescence emission is due to the fact that it is a spin-forbidden transition from the triplet excited state to the ground state. Phosphorescence spectra are usually insensitive to environment perturbations whereas the lifetimes may change by orders of magnitude. This makes phosphorescence very attractive for development sensor techniques based on decay-time measurements (Sanchez-Barragan et al. 2006).

Other attractive features of phosphorescence include the *simplicity of time-resolved detection* techniques compared to fluorescence, easy discrimination of the light-scattering background and contaminating fluorescence due to stronger Stokes shift and the longer lifetime of phosphorescence emission. The latter advantage makes phosphorescence sensors attractive for intracellular studies. The quantum yield of phosphorescence is usually low but can be increased substantially by incorporation into the matrix surrounding phosphorescent dye with heavy atoms, such as iodine. Optical sensors with phosphorescence response are expected to undergo rapid development with the development of new dyes and nano-composites hosting them.

Present applications of phosphorescence lifetime sensors are developed mostly for the sensing of oxygen. Oxygen is a potent collisional quencher of phosphorescence and this effect is very strong due to the long duration of this emission in comparison to the oxygen diffusion rate. The current applications include sensing pesticides, antibiotics and CO_2 (Sanchez-Barragan et al. 2006), see Section 8.3.

3.3.6   Comparison with Other Fluorescence Detection Methods

Before attempting to apply lifetime sensing to detecting a particular target, one has to understand clearly that cases where one of the forms of reporter fluorophore (with a bound or unbound target) is quenched statically (becomes non-fluorescent) and that are very good for intensity measurement cannot be analyzed with the lifetime sensing technique. This is due to the fact that the static quenching changes the number but not the lifetimes of emitting species. If this happens, the target binding will remain unnoticed. In contrast to static quenching, collisional quenchers produce a dramatic change in the lifetime. These features explain why the best results in the application of lifetime sensing were obtained for determining analytes that are strong collisional quenchers of fluorescence (oxygen, chloride, sulfur dioxide and NO). Meantime, the effects of dynamic molecular collisions with the target are different from that of target binding: in these cases one can observe a gradual change of lifetime as a function of the concentration of the quencher instead of intensity redistribution between two lifetimes belonging to forms with a bound and unbound analyte.

These facts indicate the necessity of synthesizing the "lifetime-sensing" dyes, for which the response in $< \tau_F >$ to environment changes could be significant. Very beneficial for lifetime sensing could be the cases when the sensor in unbound form emits light with a short but resolved $< \tau_F >$ and on analyte binding the lifetime increases substantially. This condition can be realized with careful selection of the dye and the site of its binding (Szmacinski and Lakowicz 1994).

Thus, lifetime sensing can be recommended as a more advanced substitute for intensity sensing in cases where the local sensor concentration varies and cannot be measured or controlled, e.g., in microscopy. It also has advantages over intensity and anisotropy measurements in all cases where the stray and scattered light is an important problem (Guo et al. 1998; Kuwana and Sevick-Muraca 2002). It is reasonable to use this method in the cases where there are no substantial spectral changes on analyte binding and the wavelength-ratiometric methods cannot be applied.

3.4   Excimer and Exciplex Formation

When a molecule absorbs light, its properties change dramatically. It may participate in reactions that are not observable in the ground state. Particularly, it can make a complex with ground-state molecules like itself. These excited dimeric complexes are called the *excimers*. The excimer emission spectrum is very different from that of monomers; it is usually broad, shifted to longer wavelengths and does not contain vibrational structure. Particularly, this is the case of the *excimer of pyrene*.

The excited-state molecule can also make complexes with molecules different in structure. Usually, these unexcited partners are electron-donor molecules like amines. These complexes are called *exciplexes* (the term excimer comes from excited dimer and the exciplex from excited complex). The formation of excimers and exciplexes is reversible: after emission they break apart, so that they may form again on excitation. There are many possibilities to use these complex formations in fluorescence sensing. We just need to make a sensor in which the free and target-bound forms of the sensor differ in the ability of the reporter dye to form excimers and exciplexes. The fluorescence spectra will then report on the sensing event.

The choice of dyes forming excimers with attractive spectroscopic properties is not large. Pyrene derivatives are chosen by many researchers due to the very distinct spectra of the monomers and excimers. The structured band of monomer is observed at about 400 nm, whereas that of the excimer is broad, structureless and long-wavelength shifted; it is located at 485 nm. Both forms possess the same excitation spectrum in the near-UV, which is not convenient for cellular studies. However, long lifetimes (~300 ns for monomer and ~40 ns for excimer) allow easy discrimination of background emissions (that do not extend longer than 3–5 ns) in a time-resolved experiment.

The formation of excimers and exciplexes requires close location and proper orientation between the partners (for excimer the formation of a cofacial sandwich between two heterocycles rich in π-electrons is usually needed). Since these interactions appear in the excited state and do not exist in the ground state, the complexes cannot be distinguished in excitation spectra and both forms can be excited at the same wavelengths. If the excimer is not formed, we observe an emission of the monomer in the fluorescence spectra and upon its formation, a characteristic emission of the excimer appears.

3.4.1 Application in Sensing Technologies

In chemical sensing a number of molecular constructs have been suggested that employ excimer formation. The idea was to change on target binding the *ground-state configuration* of sensor molecules in such a way that on excitation they can form excimers but only in the presence of a target. Thus, to obtain a sensor for sodium ions, a pair of dioxyanthracene molecules was included into a polyether ring. On their binding, the configuration of molecule changes and the excimer emission begins to be observed (Collado et al. 2002). Similar ideas were realized in the construction of sensors for other cations (Yuasa et al. 2004).

An efficient ratiometric fluorescent sensor for silver ions based on a pyrene-functionalized heterocyclic receptor was suggested. It forms an intramolecular sandwich complex via a silver ion-induced self-assembly that results in a dramatic increase in fluorescence intensity of the excimer and a dramatic decrease of monomer fluorescence (Fig. 3.8). The intensity ratio of the excimer and monomer emissions (at 462 and 378 nm) is an ideal measure of Ag^+ ion concentrations (Yang et al. 2003).

The performance of excimer sensors has been improved on the incorporation of excimer-forming reporter dyes into a cyclodextrin cavity (Yamauchi et al. 1999). Excimer-forming pyrene-conjugated oligonucleotides can be used for the detection of DNA and RNA sequences (Mahara et al. 2002). If incorporated into thermosensitive polymer, they can serve as molecular thermometers (Chandrasekharan and Kelly 2001).

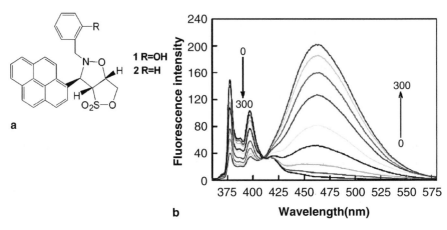

b   **Wavelength(nm)**

Fig. 3.8 The structure (**a**) and fluorescence emission spectra (**b**) of pyrene derivative (**1**) forming excimers in the presence of silver ions. Addition of these ions in concentrations of 0, 4.0, 10, 15, 20, 40, 75, 150 and 300 μm results in the decrease of intensity of the normal emission with typical bands at 378 and 397 nm (excitation 344 nm) and the appearance of a red-shifted structureless maximum centered around 462 nm, typical of a pyrene excimer (Reproduced with permission from Yang et al. 2003)

The change of excimer fluorescence on hybridization was used for the quantification of DNA in solution (Kostenko et al. 2001). Oligonucleotide conjugates bearing two pyrene residues at the 5′-phosphate of oligonucleotide exhibited excimer fluorescence intensity that is highly sensitive to duplex formation: the binding of the bis-pyrenylated oligonucleotides to their DNA and RNA targets leads to a ten-fold increase of fluorescence. This depends linearly on the concentration of the target DNA and permits the quantification of DNA in solution.

A prototype of a double-labeled protein sensor for lipids has been described (Sahoo et al. 2000). An attachment of pyrene groups to two engineered Cys residues resulted in the observation of strong excimer fluorescence. The excimer band disappeared upon lipid binding. Formation of excimers is used in the studies of protein aggregation in pathology (Thirunavukkuarasu et al. 2008).

Aptamers (see Section 5.5) that use two pyrene substituents can be efficiently used for detecting specific proteins in complex biological fluids (Yang et al. 2005).

3.4.2 Comparison with Other Fluorescence Reporter Techniques

There are special requirements for the application of excimer and exciplex sensing techniques. First, double labeling is commonly needed (with the exception of when the target itself can form exciplex with the dye incorporated into the sensor). Second, the researcher is limited in the selection of reporter dyes. Usually, pyrene derivatives are used for this because of the unique property of this fluorophore to form stable excimers with fluorescence spectra and lifetimes that are very different from that of monomers. Other dyes are commonly not applicable or less convenient for this application. Finally, a conformational change is needed in the sensor molecule to couple or separately remove the dye monomers on target binding.

If these requirements are properly addressed, the user can benefit from the advantages of this technique. With the observations at two emission wavelengths one can achieve *two-channel sensor operation* that will be independent of sensor concentration and will allow self-calibration of the fluorescence response. The recorded signal will be insensitive to instrumental factors and to the sensor (and reporter) concentration. Meantime, we have to keep in mind that monomers and excimers (or exciplexes) are independent emitters and that the non-specific influence of quenchers may be different for the two forms.

3.5   Förster Resonance Energy Transfer (FRET)

Two or more dye molecules with similar excited-state energies can exchange their energies due to a dipole-dipole resonance interaction between them. One molecule, the *donor*, can absorb light and the other, the *acceptor*, can emit it (Clegg 1996; Selvin 2000). When this communication between excited and unexcited molecules is strong, we observe that on excitation of the donor its emission is quenched and, instead, the emission of the acceptor is increased. When this coupling is absent, only the emission of the donor is observed. This approach is frequently used in sensing. Meantime, it usually needs labeling with two dyes serving as donor and acceptor. Only in rare, lucky cases can the intrinsic fluorescent group of sensor or target molecules be used as one of the partners in FRET sensing.

The acronym FRET is frequently presented as 'Fluorescence Resonance Energy Transfer'. This is not correct, since this general mechanism of energy transfer in the excited state does not necessarily involve fluorescence. It can be observed regardless the mechanism of excitation and emission (phosphorescence, bioluminescence and chemiluminescence). Therefore, according to recommendations approved by IUPAC (Braslavsky 2007), we will use the term '*Förster Resonance Energy Transfer*' in recognition of the scientist who first suggested this mechanism.

3.5.1   Physical Background of the Method

The FRET effect depends strongly upon the *distance between the donor and acceptor*. If the distance is short and one of the partners (donor) is excited, it can transfer the energy of electronic excitation to the other partner (acceptor), resulting in the latter emitting fluorescence. FRET has to be distinguished from the effect of the re-absorption of the light quanta emitted by one dye by the other dye molecule: re-absorption depends on the geometry of the system but FRET does not. Even at a relatively long distance, the dyes may interact through space by dipole-dipole resonance interactions and the resonance provided by this dipole coupling allows the energy to migrate from donor to acceptor without emission.

FRET can take place if the emission spectrum of the donor overlaps with the absorption spectrum of the acceptor (Fig. 3.9) and they are located at separation distances within or less than 1–10 nm from each other. The *efficiency of energy transfer E* can be defined as the number of quanta transferred from the donor to the acceptor divided by all the quanta absorbed by the donor. According to this definition, $E = 1 - F_{DA}/F_D$, where F_{DA} and F_D are the donor intensities in the presence and absence of the acceptor. Both have to be normalized to the same donor concentration. If time-resolved measurements are used, then knowledge of donor concentration is not required and $E = 1 - <\tau_{DA}>/<\tau_D>$, where $<\tau_{DA}>$ and $<\tau_D>$ are the average lifetimes in the presence and absence of the acceptor (Wu and Brand 1994).

The energy transfer efficiency exhibits a very steep dependence on the distance separating two fluorophores, R:

$$E = R_0^{\,6} / (R_0^{\,6} + R^6) \tag{3.14}$$

Equation (3.14) determines this steep distance function of observed intensity ratios of donor and acceptor emissions. Here R_0 is the parameter that corresponds to a distance with a 50% transfer efficiency (the Förster radius), expressed in Ångstroms. It is characteristic for a given donor-acceptor pair in a particular medium. It depends strongly on the quantum yield of the donor in the absence of acceptor Φ^D and on the overlap of the fluorescence spectrum of the donor with the absorption spectrum of the acceptor (Fig. 3.9), expressed as normalized spectral overlap integral J:

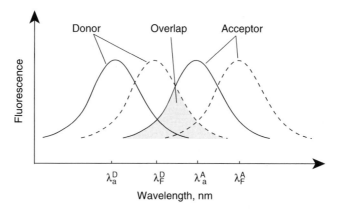

Fig. 3.9 The absorption (or excitation) and fluorescence emission spectra of two fluorophores exhibiting Förster resonance energy transfer (FRET). The light-absorbing and emitting at shorter wavelengths fluorophore (donor) can transfer its excitation energy to another fluorophore (acceptor) absorbing and emitting at longer wavelengths. For an efficient transfer, the absorption spectrum of the acceptor should overlap the emission spectrum of the donor (shaded area). At close donor-acceptor distance, the excitation of the donor results in emission of the acceptor and if the distance is large, the donor itself will emit. This produces distance-dependent switching between two emission bands

$$R_0{}^6 = 9.78 \cdot 10^3 (\kappa^2 n^{-4} \Phi^D J) \tag{3.15}$$

Here n is the refractive index of the medium between donor and acceptor. The spectral overlap integral J is a function of the fluorescence intensity of the donor, F_D and molar absorbance of the acceptor, ε_A, as a function of wavelength, normalized against the total donor emission:

$$J = \int F_D(\lambda)\varepsilon_A(\lambda)\lambda^4 d\lambda \Big/ \int F_D(\lambda)d\lambda \tag{3.16}$$

In Eq. (3.16) the integration should be made over the whole spectrum. κ^2 is the orientation factor that depends on the relative orientation of the donor and acceptor dipoles; it can assume values from 0 to 4. The lowest value will be if they are oriented perpendicularly and the maximum value corresponds to coaxial aligned dipoles. For random orientations of donor and acceptor, $\kappa^2 = 2/3$ if they move rapidly on a fluorescence time scale and $\kappa^2 = 0.475$ if they are rigidly fixed. Thus, transfer efficiency depends on the relative orientation of donor and acceptor dyes. If the lifetime is long and rotations of segments containing reporter dyes are fast, then the approximation $\kappa^2 = 2/3$ is justified.

In the case of a rigidly fixed donor and acceptor the situation is less definite and the estimation of donor-acceptor distances may lead to great errors. In designing the FRET-based sensors there are cases when even at short distances FRET does not occur. Such a case was described on the interaction of a conjugated polymer (donor) with the double-helical DNA containing dye ethidium bromide as an acceptor. The rigid structure of this complex and the orthogonal orientation of transition dipole moments prevented the detection of DNA hybridization by FRET (Xu et al. 2005b). However in a practical sense, variations of both the distance and orientation may substantially increase the dynamic range of the fluorescence response.

The overlap integral J is the parameter that often determines the choice of donor and acceptor for particular needs. ESIPT can occur if the donor and acceptor are the same molecules (*homo-transfer*). Such a transfer is typical for dyes in concentrated solutions. If it is required, the Stokes shift should be minimal and if not required – maximal. In sensor technologies, if we need to avoid homo-transfer, we have to keep the sensor molecules spatially separated. Proper manipulation of this phenomenon may allow achieving an amplification effect (Section 6.2) in case this transfer is combined with *hetero-transfer* (when donor and acceptor are different dyes).

FRET based on hetero-transfer is the most frequently used and a number of donor-acceptor pairs are selected for this (see Section 4.1). By a proper choice of this pair one can optimize the excitation and emission wavelengths and also the distance dependence of the effect (see Fig. 3.10). One can observe that these distances are comparable with the dimensions of many biological macromolecules and their complexes (Clegg 1996; Selvin 2000).

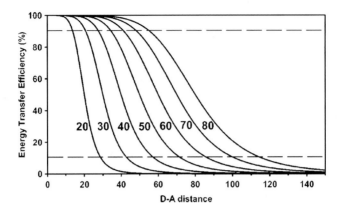

Fig. 3.10 The dependence of the dynamic range of Förster resonance energy transfer (FRET) on the donor-acceptor (D-A) distance for different R_0 values given in Ångstroms. The dotted lines delineate the regime of maximum sensitivity of response to distance variation for each D-A pair with different R_0, which is commonly between 0.7 and 1.5 R_0 (Reproduced with permission from Kapanidis and Weiss 2002)

The effect of FRET is not limited to organic dyes. It can be observed between a large-sized noble metal and semiconductor (Quantum Dot) nanoparticles or fluorescent proteins (Chapter 4). Moreover, it occurs if an acceptor is a plain conducting surface. In these cases FRET can extend to longer distances and display a $1/R^4$ function, thus extending the possibilities of its observation beyond the traditional distance limitations. These issues will be discussed at length in Section 6.2.

3.5.2   FRET Modulated by Light

In some sensing technologies and especially in cellular imaging, it can be important to compare two signals or images, with and without FRET, with the same composition and configuration in the system. In these cases, one can play with the light-absorption properties of the FRET acceptor. It can be *photobleached*, so that its absorption spectrum disappears. This leads to the disappearance of FRET.

Another very elegant approach was suggested recently (Giordano et al. 2002) and received the name "*photocromic FRET*". As FRET acceptors, so-called *photochromic compounds* can be used having the ability to undergo a reversible transformation between two different structural forms in response to illumination at appropriate wavelengths. These forms may have different absorption (and in some cases, fluorescence) spectra. Thus, they offer the possibility of reversible switching of the FRET

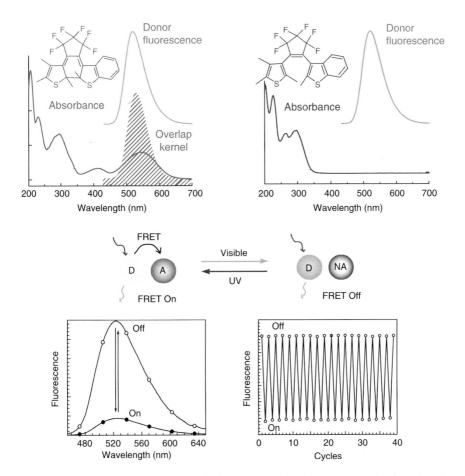

Fig. 3.11 Photochromic FRET. The chemical structures depicted correspond to the photochromic dithienylethene in the colored closed form (left) and colorless open form (right). The absorption spectrum of the latter overlaps well with the emission spectrum of the donor; the kernel of the overlap integral (striped) corresponds to the lucifer yellow dye as the donor. Ultraviolet light induces the photochromic transition to the closed form (On) and visible (green) light reverses the process to the open form (Off) (Reprinted with permission from Jares-Erijman and Jovin 2003)

effect between "on" and "off" states without any chemical intervention, just by light (Fig. 3.11).

The best compounds for serving this purpose are probably spiropyrans. Their molecules exist in closed spiro forms absorbing at wavelengths shorter than 400 nm. They undergo a light-driven molecular rearrangement to an open merocyanine form with absorbance at 500–700 nm (Bahr et al. 2001). A family of these photo-switchable acceptors has been synthesized.

Sensing technology benefits from these possibilities. The photo-switching quenching phenomenon provides the light-activated control of the excited states and, consequently, controls any subsequent energy or electron-transfer processes.

In imaging the living cells, the switching on and off of FRET allows one to obtain the necessary controls on the distribution of donor molecules and their fluorescence parameters (Jares-Erijman and Jovin 2003). Emission of not only the dye molecules but also of the fluorescent nanoparticles (Quantum Dots) can be modulated by the photo-switching phenomenon (Medintz et al. 2004).

3.5.3 Applications of FRET Technology

The attractive feature in the application of FRET in sensing technologies is the possibility of *ratiometric* spectral observation of the sensing event. However, for the detection of a two-wavelength response (one band in emission belonging to the FRET donor, the other – to the acceptor) one needs to satisfy very specific requirements, which are the double-labeling by different dye molecules and a substantial change in the wavelength-ratiometric signal during the sensing event. If the acceptor is non-fluorescent, one would only record the donor quenching with all the disadvantages arising from single-wavelength intensity recording.

The most straightforward application of FRET in sensing can be realized in conditions where the distance or orientation between the donor and the acceptor changes in the sensing event. This can be achieved in two different cases.

1. Both donor and acceptor fluorophores belong to the sensor and a conformational change in the sensor on analyte binding causes the change in their relative distance.
2. One fluorophore label belongs to the sensor and the other – to the analyte or its analog (in displacement assay) and upon their binding, the two fluorophores come into close proximity (Fig. 3.12).

Thus, should the analyte absorb light, it can be used as a FRET acceptor and the problem is reduced to selecting a proper donor that can be used as a reporting element of the sensor. Of course, the labeled analyte analog can be used in a displacement assay.

Since detectable switching between donor and acceptor can be achieved at separation distances up to 5–10 nm, the distance changes within this limit are explored in sensing. The most typical examples of such applications are:

(a) Quantitative determinations of *enzyme activities* with double-labeled substrates. This application is straightforward. If the covalent bond belonging to a link between two dyes is cleaved, the emission of the acceptor observed before the cleavage disappears and the emission of the donor only is observed. This allows wavelength-ratiometric measurements of enzyme kinetics of proteases (Gershkovich and Kholodovych 1996), phosphatases (Takakusa et al. 2003) and nucleases (Ghosh et al. 1994). Meantime, since the fluorogenic molecules are modified (destroyed) during the assay, this procedure cannot be formally considered as "sensing" but as a traditional biochemical analysis.

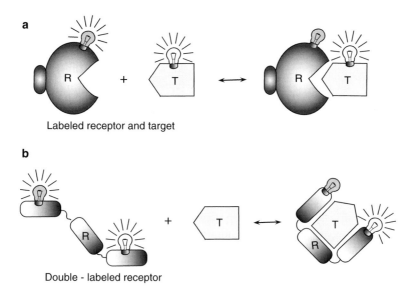

Fig. 3.12 Two frequently used possibilities for the application of FRET. (**a**) Both sensor and analyte are labeled, with different dyes. In a sensing event, FRET appears as a result of their approach within a critical distance. (**b**) The analyte is not labeled but the sensor is labeled with different dyes at two different sites. In the analyte-free sensor conformation, the distance between two dyes is longer than critical. On analyte binding, the two dyes appear in close proximity due to conformational change. The opposite is also possible – the conformational change on analyte binding increases the distance between the dyes outside the critical value for FRET

(b) *Competitor displacement assays* (Xu et al. 2005a). FRET with either a fluorescent or quenched acceptor is frequently used in displacement assays. Thus, in one of the proposed sensors for maltose based on the *E. coli* maltose binding protein (Medintz et al. 2003), a complex between the protein labeled with Quantum Dot (as a donor) and a competitor, β-cyclodextrin labeled with another dye serving as the acceptor, is formed. The substitution of labeled β-cyclodextrin with an unlabeled target (maltose) results in the disappearance of FRET. Other examples refer to essays for enzyme inhibitors in which Trp residues of enzymes are FRET donors and the dye-labeled inhibitor analogs as competitors that allow detection of the binding of different inhibitors. This approach was applied for acidic protease (Epps et al. 1989) and metalloproteinase stromelysin (Epps et al. 1999).

(c) *'Remote sensing'* based on the change of the mutual distance between two labels on a sensor molecule. FRET can be applied to a sensor protein composed of a single polypeptide chain exhibiting conformational change as well as to protein composed of two or several subunits, if the sensing event involves changing the distance between them. This strategy was used for the detection of cAMP in smooth muscle cells by labeling the catalytic and regulatory subunits of cAMP-dependent protein kinase with two different dyes (Adams et al. 1991).

Peptides that undergo a conformational change upon ligand binding have been transformed into peptidic FRET sensors. For example, Zn^{2+}-binding domains are characteristic for a number of proteins that participate in protein – nucleic acid recognition. Their structural motives, the so-called zinc fingers, are disordered in the absence of zinc and attain a regular three-dimensional structure upon coordination of a zinc atom to Cys and His residues. Thus the ion-dependent ordering can be reproduced on relatively short peptides. Labeling by two dyes allowed constructing a fully synthetic fluorescent zinc sensor (Godwin and Berg 1996). The binding of Zn^{2+} produces the conformation ordering in this flexible molecule, so that its distant parts can be brought into close proximity and FRET can be induced between two bound dyes located distantly in the sequence. These signal transduction mechanisms will be discussed at length in Section 6.3.

(d) *FRET imaging* inside the living cells. This approach offers the important advantage of being independent of fluorescence reporter concentration. A number of constructs based on pairs of visible fluorescent proteins were suggested (Tsien 1998). These results will be discussed in more detail in Chapter 11.

Double labeling is often a serious complication. However, this approach expands the possibilities to construct those sensors for which single-dye molecular reporters are not applicable.

3.5.4 FRET to Non-fluorescent Acceptor

The FRET donor should necessarily be emissive, since the transfer occurs during the lifetime of the excited state. In contrast, the FRET acceptor can be non-emissive and can provide only the quenching effect on the donor emission. FRET to emissive acceptor can allow observing both donor and acceptor bands in a two-band ratiometric emission. If the acceptor is only the quencher, this very attractive feature is lost. This situation has a different benefit, however. It is normally very hard to select the wavelength at which only the donor can be excited without noticeable excitation of the acceptor. If the acceptor is non-emissive, its direct excitation will not influence the fluorescence spectrum. This is probably the reason why in the first versions of the 'molecular beacon' specific DNA binding assay, FRET quenching was used (Marras et al. 2002; Tyagi and Kramer 1996; Tyagi et al. 2000).

There is one more possibility to realize FRET quenching. If the sensing event produces the shift of the absorption band of the acceptor, this will also influence the FRET efficiency and result in the variation of the fluorescence intensity of the FRET donor (Takakusa et al. 2003). Thus, many non-fluorescent pH indicators display pH-dependent absorption spectra in the visible with their different positions depending on ionization state. Such pH-dependent variations of absorbance can be transformed into a fluorescence signal by selecting as a donor the dye with a fluorescence spectrum overlapping the absorption spectrum of one of the forms of such an acceptor with no or a much weaker overlap of the other form.

3.5.5 *Comparison with Other Detection Methods*

FRET is increasingly popular in the studies of recognition between macromolecules and in the construction of biosensors based on this recognition. It operates with two reporter dyes (FRET donor and acceptor), which allows introducing the geometry parameter into fluorescence sensing and observing the sensing effects based on geometrical changes extending to a relatively long distance, up to 5–8 nm. The efficiency of FRET depends strongly on the distance and typically decreases to the sixth power of the distance between the two dyes. If in the sensing event the distance between the donor and acceptor changes, then the change of the relative contribution of emission intensity produced by the donor and acceptor to fluorescence spectrum can be observed. The two correspondent bands are usually well separated on the wavelength scale and the switching between them in the course of the sensing event can be registered as the ratio of their intensities.

For a better understanding of the applicability of the FRET technique to fluorescence sensing, one important remark should be made. While the FRET reaction allows the fluorophores to "talk" at sufficient distances and respond to variations of these distances, their responses to other factors influencing the fluorescence band intensities are to a significant extent independent. For instance, the quenching of the fluorescence of the donor also quenches the acceptor emission but the quenching of the acceptor emission does not influence the emission of the donor. Hence, when intermolecular interactions of fluorophores change on analyte binding, then one may expect a strong variation of fluorescence response unrelated to donor-acceptor distances and this possibility has to be considered and is in fact used in sensor design.

Based on the response in FRET, different sensor constructs can be designed. Based on this principle, a single-stranded dual fluorescently labeled DNA molecule that adopts a stem-loop conformation in its nonhybridized state with the close proximity of a FRET donor and acceptor dyes can be constructed. It exhibits a conformational change on hybridization with an increase of dye-to-dye distance and the appearance of a new band belonging to the acceptor is recorded (Ueberfeld and Walt 2004). It can be the re-arrangement of subunits in the sensor protein (Adams et al. 1991) and also a conformational change within the subunits. Since the distance dependence of FRET is very steep, efficient switching between two emissions can be achieved.

3.6 Wavelength-Shift Sensing

Most of the information that the human brain obtains from the surrounding world is acquired by visual perception. Colored vision and characterization of visual objects in real color is an important contribution to this process. Therefore, it is not surprising that in spectroscopy and in its combination with microscopy, attempts to obtain information in real color in the visible range of the spectrum are in the minds of many researchers. This situation was not changed with the application of electronic photometric devices, since the spectral shift information is often more easily obtained and more unambiguously interpreted than the information obtained in the studies of anisotropy and lifetimes. Therefore, the fluorescent dyes that change the color of emission are in great demand. In many respects the application of these dyes may substitute sensing technologies based on excimer formation and FRET, since single labeling is always better than the labeling with two dyes.

Usually, the dyes that exhibit the changes in emission spectra are called *wavelength-ratiometric* because they allow a convenient quantitative characterization of the changes in emission spectra by an evaluation of the intensity ratio at two selected wavelengths. These are usually the points on the wavelength scale at which the most significant changes of fluorescence intensity are observed. According to the effects that are produced in the original spectra, the ratiometric probes can be classified into two groups. One group combines the *wavelength-shifting* probes, in which the ratiometric effect is produced by the spectral shifts of a single band displayed in the emission spectrum. Their properties and applications are overviewed in this section. The other group of ratiometric dyes, the *two-band ratiometric*, allows achieving the interplay of intensities of two emission bands. Such reporters will be discussed in the next section.

3.6.1   The Physical Background Behind the Wavelength Shifts

The background of wavelength shifts in excitation or emission bands is the change of energy of correspondent electronic transitions that can be influenced by intermolecular interactions. A number of textbooks (Demchenko 1986; Lakowicz 2007; Suppan and Ghoneim 1997; Valeur 2002) provide a detailed analysis of these mechanisms and here we provide only their simplified description.

It has to be recollected that any molecules that are able to absorb light possess *discrete energy levels* in the *ground state* (the electronic state in which it resides at equilibrium, without excitation) and in the *excited state* (the electronic state in which it appears on the absorption of light quanta). A molecule can absorb and emit the light quanta only of the energy that corresponds to the *energy gap* between these states. This gap is inversely proportional to the wavelength of the band maximum. In spectroscopy, the *wavenumber scale* ν (expressed in cm^{-1} units) is commonly used as an energy scale, so that ν $(cm^{-1}) = 10^7/\lambda$ (nm). The spectrum recorded on a wavelength scale is in fact a reverse function of absorbed or emitted energy.

The energies of both ground and excited states are influenced by intermolecular interactions: the stronger the interactions, the lower the level on an energy scale. This is true for the energies of both ground and excited states. Since not only the energy but often the spatial distribution of electrons changes dramatically on excitation, these states interact differently with surrounding molecules. If the interaction with the surroundings is stronger in the ground state, then, when this interaction increases, the difference in energy between the states increases. The molecule then absorbs and emits light of higher energies and the spectrum becomes shifted to higher energies, i.e., to shorter wavelengths. On the other hand, if the interaction energy is stronger in the excited state, then on increase of this interaction the spectra shift to longer wavelengths. For an illustration, one may refer to the classical Jablonski diagram which, for the purposes of our discussion, was modified and simplified (without showing vibrational levels, high-order electronic states and intramolecular relaxations). It is presented in Fig. 3.13.

In highly fluorescent organic dyes the lowest ground-state and excited-state energy levels are formed by π-*electrons*. Usually, the π-electronic system becomes more *polarized* in the excited state and, by becoming a stronger dipole, it interacts more strongly with the molecular environment. As a result, the energy gap between the ground and excited states becomes smaller and the spectra shift to longer wavelengths. There are possibilities to increase this effect by introducing chemical substitutions into the π-electronic system. If we attach an electron-donor group (such as dialkylamino-) to one side of the molecule and an electron acceptor (such as carbonyl) to its opposite side, the charge separation in the excited state becomes greater. The molecule will become a strong electric dipole, its interactions with surrounding dipoles will become stronger and the spectra will exhibit stronger shifts. Such structures are often called '*push-pull structures*'. The structures of the so-called '*polarity-sensitive*' dyes follow this principle. The *higher the polarity* of the medium, the stronger the shift of fluorescence spectra to *longer wavelengths*.

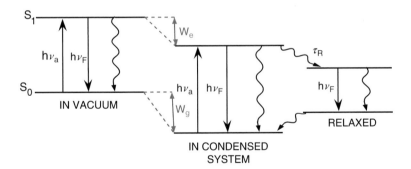

Fig. 3.13 A simplified Jablonski diagram of ground S_0 and excited S_1 energy levels and transitions between them. The vertical upward arrow shows the excitation and the downward arrows show emissive (straight) and non-emissive transitions to the ground state. In condensed media, the energies of ground and excited states are decreased due to electronic interactions with the environment by solvation energies W_g and W_e correspondingly. Their values are commonly not equal, and this determines the direction of spectral shift. In addition, in polar media there occurs an establishment of equilibrium in interactions of dye and surrounding dipoles (dipolar relaxation). As a result, the energy gap between S_0 and S_1 states decreases and the spectra shift to longer wavelengths

3.6.2   The Measurements of Wavelength Shifts in Excitation and Emission

The shift of the whole fluorescence band results in an increase of intensity at one of its edges and a decrease at the other. As the information channels for collecting the fluorescence intensities one can use *two selected wavelengths* at the two edges, in which the effect is the strongest. Those are the wavelengths of the maximal slope of the spectrum (the wavelengths of the maximum and minimum of its first derivative). Taking the intensity ratio at these points, one may then benefit from two-channel ratiometric measurements. Meantime, it should be kept in mind that the intensities at the edges of the spectra are about one half lower than at the band maxima and the presence of impurities in the sample and light-scattering effects may influence the short and long-wavelength wings differently.

Wavelength shifts in excitation spectra are not frequently used in sensing. They are measured mostly for electrochromic dyes (e.g., aminostyryl derivatives), which are strongly sensitive to a variation of local electric fields (Section 4.1). Based on it, the binding of functional dyes to cell membranes and their phospholipids analogs (liposomes) allows the detection of variations in membrane potential (Gross et al. 1994). The collective effects of charges that can be generated in phospholipid bilayers may be used in the development of nano-scale and whole-cell biosensors.

Wavelength-ratiometric measurements of spectral shifts in emission are used more frequently. Some dyes exhibit very significant (up to 100 nm and more) shifts in wavelength positions of their fluorescence spectra in response to a change in their interaction with the molecular environment (Valeur 2002). These shifts are usually due to a combination of effects of *polarity* with the effects of intermolecular *hydrogen bonding* (Vazquez et al. 2005). Thus, if the dye is incorporated into the sensor molecule in such a way that its interaction with the environment changes as a result of a conformational change in the sensor or due to direct contact with the target, the event of sensing can be detected by the spectral shift. This principle is used in the design of both chemical sensors and biosensors.

It could be ideal to observe the environment-dependent spectral shifts without change or with an insignificant change of integral intensity, since this will provide a high signal-to-noise ratio and a broad range of variation of intensity ratios. In reality, this condition is difficult to realize because the environment-dependent quenching is the common property of these dyes. The increase of their dipole moment on excitation produces an increase in interactions in the excited state with the environment and results in the shift of the spectra to the red (Demchenko 1991). In the background of these shifts is the reaction of dielectric relaxations of the environment dipoles. They are fast in liquid environments and result in rapid acquisition of equilibrium. In practice, this relaxation is associated with the activation of a number of quenching factors (Demchenko 2005b). This is why the shift to longer wavelengths is often associated with decrease in intensity (Fig. 3.14). This makes the ratiometric measurements based on wavelength shifts imprecise and subject to systematic errors.

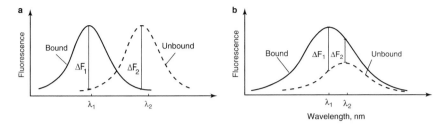

Fig. 3.14 Sensing with the wavelength-shifting fluorophores by recording the intensity ratios at two wavelengths, λ_1 and λ_2. (**a**) The ideal case, when the analyte-free and analyte-bound form (located at shorter wavelengths) of the sensor exhibit similar fluorescence intensities. (**b**) Often observed reality when in unbound form, in addition to shift, the spectrum is quenched and broadened

3.6.3 Wavelength-Ratiometric Measurements

In principle, some of the important problems observed with the intensity measurements can be avoided if the sensing response is provided by the observation of a substantial shift of the fluorescence band on the wavelength scale. This could allow instead the measurement of one parameter, $F(\lambda)$, obtaining variations of *two intensity values at two different wavelengths*, λ_1 and λ_2 and taking the *ratio of their intensities*, $F(\lambda_1)/F(\lambda_2)$. In the dynamic range of sensing this ratio will depend on the concentration of the bound analyte but will not depend on the concentration of the sensor molecules. Photobleaching of the fluorophore that decreases the concentration of the responsive sensors will then also be without influence on the result (if, of course, the photodegradation products are non-fluorescent).

If the spectrum shifts (for instance, to the red), then the intensity at the short-wavelength slope decreases and increases at the long-wavelength slope (see Fig. 3.14) and this can allow convenient measurement of intensity ratios at two fixed wavelengths. The choice of λ_1 and λ_2 in the case of spectral shifts can be made in two different ways. One is to choose λ_1 at the point of maximal increase of intensity and λ_2 at the point of maximal decrease. Then the dynamic range of change of $F(\lambda_1)/F(\lambda_2)$ will be the largest and this change can be calibrated as a function of the concentration of the bound analyte.

The other possibility is to choose λ_1 at the point of maximal change of intensity and λ_2 at the point, in which the two spectra cross and the intensity does not change at all (*isoemissive point*). Then λ_1 can be chosen either at the blue (short-wavelength) or red (long-wavelength) slope, apart from λ_2 and in the region of the maximal change of spectra. Since commonly the fluorescence spectra are not symmetrical but exhibit a shallower slope at longer wavelengths, then the changes at the blue side are usually stronger and the dynamic range of response broader. However, in some cases, if the Stokes shift is small, the readings at shorter wavelengths can be subjected to a stronger light scattering perturbation.

Formally, there is no advantage of wavelength-ratiometric sensing over intensity sensing with a molecular reference. In both cases the ratio of two intensities at two wavelengths, λ_1 and λ_2, is obtained. The only difference is that instead of being a permanent value, the intensity at the second (reference) channel (at λ_2) increases or decreases in a converse manner to its change in the first channel, at λ_1. If for recording the reference signal we select the wavelength at which all the calibrated spectra cross one another (isoemissive point), then we can determine the analyte concentration using Eq. (3.5). In a more general case, when λ_2 is a different wavelength (e.g., it is the maximum of the second band), then we have to include the factor that accounts for this intensity redistribution, which is the ratio of the intensities of the free and bound forms at λ_2:

$$[A] = K_d \left(\frac{R - R_{min}}{R_{max} - R} \right) \left(\frac{F_F(\lambda_2)}{F_B(\lambda_2)} \right) \tag{3.17}$$

3.6.4   Application in Sensing

In a number of studies, the fluorescence shifts were detected in response to the change in intermolecular interactions and their ratiometric response were used in molecular sensing. Thus, a strong red shift of the fluorescence spectra of Ca^{2+}-binding proteins (parvalbumin, troponin C) labeled with acrylodan was demonstrated on interaction with calcium ions (Prendergast et al. 1983). This shift was accompanied by a significant decrease of intensity. It was suggested that a conformational change occurs as a result of the ion binding and the covalently bound dye located in a low-polar environment becomes exposed to the protein surface. This idea was further developed for Ca^{2+} sensing by labeling single-Cys mutants of troponin C (Putkey et al. 1997). It was also observed that an ICT dye attached to a zinc finger responds to zinc binding (Walkup and Imperiali 1996), since the binding induces the formation of structured peptide conformation with the screening of a fluorescence reporter from the solvent.

Some progress was also achieved in *fatty acid* and *saccharide* sensing. A significant spectral shift (by 35 nm to the red) was observed for acrylodan selectively attached to a Lys residue of a fatty acid binding protein upon binding fatty acids (Richieri et al. 1992). This occurs due to a displacement of the dye from the hydrophobic fatty acid binding pocket. The blue shifts were observed on binding the ligand to the maltose binding protein (Gilardi et al. 1994). A wavelength-shift sensor for saccharides was made based on two donor-acceptor substituted diphenylpolyenes, in which the dimethylamino group served as an electron donor and the boronic acid group functioned both as electron acceptor and as the sensor for saccharides (Di Cesare and Lakowicz 2001). These authors achieved a blue shift and an increase of intensity on analyte binding and recorded a wavelength-ratiometric response. They observed the decrease in fluorescence lifetime on sugar binding and claimed that the lifetime sensing may be a more convenient observation in this case.

The prototype of the *direct molecular immunosensor* has been described (Bright et al. 1990). For its construction, the complex of human serum albumin (HSA) with an anti-HSA antibody was formed and subjected to labeling by a fluorescent dansyl derivative. This allowed labeling the antibody outside the antigen-binding site at unspecified locations of amino groups. After removal of HSA and immobilization on solid support, the antibody became reactive to HSA binding by the blue shift of dansyl emission and by an increase of its intensity.

The outlined cases above are rather uncommon, in which the environment-sensitive dyes respond to analyte binding by a fluorescence shift that is strong enough for ratiometric measurement. There are many examples where these shifts were undetected and the measurements were reduced to single-wavelength intensity measurements. This problem is clearly seen in the work of de Lorimier et al. (de Lorimier et al. 2002), in which a very detailed study of 11 bacterial periplasmic binding proteins modified with 8 environmentally sensitive fluorophores was conducted. The common spectroscopic effects on binding the analyte ligands were only the changes in fluorescence intensities, whereas the spectral shifts were either insignificant or even totally absent. Only in two cases, of Glu/Asp and glucose binding proteins with the bound acrylodan dye, were the shifts (13 and 21 nm correspondingly) sufficient for ratiometric measurements. This insensitivity can only be understood if no significant change of polarity of the fluorophore environment occurs on analyte binding and also if the fluorophores remain highly hydrated and retain hydrogen bonds with water molecules.

3.6.5 Comparison with Other Fluorescence Reporting Methods

Ideally, wavelength-shift detection is optimal for realizing the concept of *direct sensing*. It would be ideal to obtain a response to intermolecular interactions by spectral shifts only, without the change or with insignificant change of band intensity. In reality, this condition is difficult to realize because the same interactions may result in quenching. Whereas in intensity detection the strong change in intensity on analyte binding is a necessary requirement for sensor operation, in wavelength-shift sensing it becomes a complicating factor.

It is difficult to predict these quenching effects. They are the decrease of energy separation between the locally excited and intramolecular charge transfer (ICT) states up to the level inversion, quenching by electron transfer to solvent traps, quenching via newly formed intermolecular hydrogen bonds and increased thermal quenching due to a smaller gap between ground and excited-state energies. Intramolecular flexibility in fluorophores may contribute strongly to the quenching effect. These quenching effects follow quite different regularities with respect to the spectral shifts (the two effects are often not correlated).

On the increase of polarity with environment-sensitive dyes together with the long-wavelength shift, one can commonly observe an increase in the width of the spectrum. As a result, the combination of three factors, the shift, the change of intensity and the broadening, results in a shifted component that will be strongly quenched, so that in reality one will observe the spectrum depicted in Fig. 3.13b.

It will exhibit only a small shift but a strong decrease in intensity. In these cases, the application of common environment-sensitive dyes is not justified and the fluorophores specially designed for intensity measurements are preferable.

3.7 Two-Band Wavelength-Ratiometric Sensing with a Single Dye

The previous section should give the reader a full impression of the advantages of the operation of fluorescence reporters based on the change in the fluorescence spectrum in an intramolecular process and on the difficulties in achieving this. Therefore, researchers try to develop fluorescence reporters in which the sensitivity to the change of intermolecular interactions could be greatly increased. Ideally, it would be to possess one type of dye molecules (this will eliminate double-labeling), which serve as reporters so perfectly, that in the sensing event switch completely between the two ground-state or excited-state forms. These forms should give separate bands in excitation or emission spectra to produce a dramatic, easily recordable change of color.

From basic photophysics we may derive that in order to obtain the switching *in excitation spectra* we need to operate with two or more ground-state forms of the reporter dyes and to couple the sensing event with the transitions between them. To obtain two switchable bands *in fluorescence spectra*, one ground-state form is enough (and preferable) but there should be an excited-state reaction generating new species. This reaction should proceed between discrete energy states and both the reactant and the product of this reaction should emit fluorescence with the shifts in energy.

3.7.1 Generation of a Two-Band Ratiometric Response by Ground-State Isoforms

The switching of a fluorescence signal can be performed between two *ground-state forms* that may be isomers, different charge-transfer forms or forms differing in ionization of a particular group. If the two such forms are fluorescent, they should exhibit different excitation spectra. Though wavelength-ratiometric recording in excitation is less convenient than in fluorescence and needs two light sources or re-tuning of a single source between two wavelengths, there are many applications of such approaches in sensing and many suggestions for their technical realization.

The Ca^{2+}-chelating dyes Fura 2 and Indo-1 (Grynkiewicz et al. 1985) serve as good examples of the influence of analyte binding on the ground-state *intramolecular charge transfer* (ICT) behavior of stilbene-like dyes. The ion binding results in the redistribution of absorbed quanta between long and short-wavelength excitation bands. The appearance on Ca^{2+} binding of a new short-wavelength band at 405 nm (490 nm for a free dye) is the result of an IST reaction that occurs on the interaction of a chelated Ca^{2+} ion with the electron-donor nitrogen atom.

In principle, if the excitation spectra are so different, fluorescence spectra should be different too. But only in the case of Indo-1 the two-band response is also observed in fluorescence. The story is that in the excited states of these dyes, due to the redistribution of electronic density, there appears a positive charge close to the bound ion. This results in ion ejection and only the ion-free form remains present in emission. In contrast to other molecular Ca^{2+} sensors, in Indo-1 the electrostatic repulsion in the excited state between the nitrogen atom and the bound ion is not sufficient for its ejection from the binding site and this interaction generates a new band in emission. Indo-1 is excited in the near-UV (about 345 nm) and all the attempts to synthesize the Ca^{2+} probe with similar properties but with the excitation and emission spectra shifted to the visible range were unsuccessful. It was only reported that the positioning of the ion-binding group at the electron-acceptor site of coumarin dye reversed the effect (Bourson et al. 1993).

Notably, the application of the principle of reporter operation based on the interplay of ground-state isoforms is presently limited to the indication of pH changes or sensing the ions. Probably, only the ions can provide such a strong perturbation of the ground state that could allow switching between the two forms; this effect is hard to achieve on recognition of neutral molecules. There are many reports on the construction of chemical sensors for other, besides calcium, ions based on the ICT mechanism (de Silva et al. 1997). A new seminaphtho-fluorescein platform suggested for ion sensing (Chang et al. 2004) uses the switching between two ground-state tautomers of the dye that produces different excitation and emission spectra under the influence of the bound ion. In all these cases, switching between two ground-state isoforms of the same dye may be considered as switching between the distinct species that possess their characteristic excitation spectra.

3.7.2   Excited-State Reactions Generating a Two-Band Response in Emission

When the effects of two-band switching occur in emission spectra, this offers additional advantages over excitation spectra. In this case, for target detection one does not have to change or re-tune the excitation light source and ratiometric fluorescence detection can be accomplished using the simplest bandpath filters. This advantage can be realized with the use of dyes possessing strong electron-donor and electron acceptor substituents and exhibiting localized *intramolecular electronic charge transfer* (ICT) states.

The excited-state ICT is well described in the literature (Suppan and Ghoneim 1997) and will be discussed in Section 6.1. It is a basically fast and reversible transition between two excited-state forms, as illustrated in Fig. 3.15. A response to the sensing event of the dyes exhibiting these reactions is the modulation of the transition from initially excited A^* form to the product B^* form. Usually, in ICT the initially low-polar A^* state is excited and the reaction results in the generation of a highly polar and strongly interacting with the environment charge-transfer state. The shift in excited-state equilibrium to the level of lower energy guides the spectroscopic effect: the stronger the interaction with the environment of one of the excited-state forms, the stronger the correspondent band in the fluorescence emission.

The ICT reaction can often generate the *two-band emission* in aromatic dyes possessing a strong *electron-donor* and *electron acceptor* substituents (Rurack 2001; Yoshihara et al. 2003) and the switching between them can be in the background of efficiently operating sensors (Collado et al. 2002; Malval et al. 2003). However, the significant and often unpredictable quenching effects that are often observed for a separated-charge form complicate the development of this approach.

The possibilities of functional substitutions in some of these dyes are also limited, since they may dramatically alter the fluorescence properties. Thus in bianthryl (the dimer of anthracene) the two bands belonging to initially excited and ICT emissions are well separated and have a different shape, with a sufficiently high quantum yield. Variations of their relative intensities suggest bianthryl to be a good reporter of the polarity and mobility of its environment (Demchenko and Sytnik 1991a, b), which

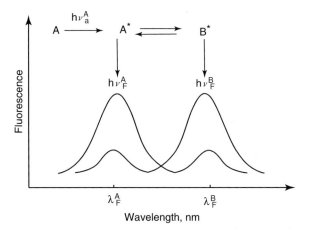

Fig. 3.15 Illustration of the principle of two-band fluorescence ratiometric sensing. Fluorescent dye A on absorption of light quanta hv_{abs}^a emits light quanta hv_f^a from the excited state A^* and also exhibits transition to another excited state B^* that emits quanta of energy hv_f^b at a different wavelength, usually with a strong red shift. The effect of sensing switches the population of excited-state species between A^* and B^*, which produces the redistribution of emission intensity between A^* and B^* bands. The sensing event triggers such redistribution

might be used in sensing. However, this dye can not be functionalized, since any chemical substitution changing the symmetry between two anthracene rings results in the disappearance of the ability to switch between the two emission bands.

Recently, as much as 11 different possibilities to generate two-band fluorescence emission were counted (Inoue et al. 2002) but none of them seems to have good prospects in sensor applications. The reason for this is the often occurring coupling of a basic ICT reaction with other reactions, such as isomerizations, with the loss of properties that could be attractive for sensing.

3.7.3 Excited-State Intramolecular Proton Transfer (ESIPT)

In aromatic molecules that contain *hydrogen bond donor* and *acceptor* groups located at a close distance, protons can migrate from one group to the other giving rise to a *tautomer* form that emits fluorescence with a substantial shift of spectrum to longer wavelengths. For sensing applications, the molecules exhibiting ESIPT are especially attractive because the switch between the emission of the initially excited normal (N^*) form and of the tautomer (T^*) form can be modulated by the target. Meantime, without understanding the mechanisms of ESIPT reactions, it is hard to provide this switch. Basically, these reactions are very fast (subpicosecond) and the needed simultaneous presence of N^* and T^* bands in emission could be due to two factors:

(a) *Kinetic*. The equilibrium in the ESIPT reaction is shifted towards the T^* form. Meantime, the N^* and T^* forms are separated by an energy barrier that makes the reaction slow and comparable in rate with the emission. The barrier appears due to the presence of conformers in configurations unfavorable for the transfer. In protic environments, the intermolecular hydrogen bond partners that compete with H-bonds on the ESIPT pathway can be involved. This slows down the ESIPT reaction and the emission from the N^* form occurs in the course of its transition to the T^* form.

(b) *Thermodynamic*. In this case the ESIPT reaction can remain very fast but the reverse reaction on a time scale faster than the emission allows establishing the equilibrium between the N^* and T^* forms, so that two emissions of comparable intensities can be observed, in line with the scheme of Fig. 3.15.

Remarkably, in the latter case the *fluorescence decay* functions along the spectra are composed of two components: long decays that are identical for the two forms and fast decays that change their sign on transition from the A^* to B^* band and reflect the ESIPT reaction (Shynkar et al. 2003). The identity of long lifetimes shows that the two forms are in equilibrium. This also means that any fluorescence quenching produces an identical decrease of lifetime and intensity of both forms, which allows the ratiometric response to be unchanged.

In an attempt to find the optimal system for realizing the principle of two-band ratiometric sensing, the present author and his colleagues have suggested to explore the properties of fluorophores exhibiting an ESIPT reaction (Demchenko 2006; Demchenko et al. 2003). To our knowledge, the 3-hydroxychromone derivatives (3HCs) are the only class of organic dyes exhibiting ESIPT (shown in Fig. 3.16), in which the observation of two separate bands in emission can be achieved without the involvement of two or more ground-state or excited-state conformers or due to the slow (on the scale of fluorescence lifetimes) rate of this reaction. In 3HCs, the proton-donor 3-hydroxyl and proton-acceptor 4-carbonyl are attached to a rigid skeleton and coupled by a hydrogen bond (for more detailed information on their structures see Section 4.1). The ESIPT reaction is much faster than the emission and the observation of two emission bands belonging to an initially excited normal (N^*) form and the reaction product the tautomer (T^*) form is due to a rapidly established dynamic equilibrium between these forms (Shynkar et al. 2003).

An increase of the dipole moment of the N^* form by proper substitution in the chromone ring allows one to make not only the position of the N^* band but also the ratio of band intensities (I_N^*/I_T^*) highly sensitive to interaction with the environment (Klymchenko et al. 2001a, b, 2003b). More detailed information about these dyes and their applications is presented in Section 4.1 and other sections of this book. The essential aspect is that all the perturbations of spectra that are commonly observed as the band shifts are transformed into dramatic variations of band intensities, thus manifesting a strong amplification effect.

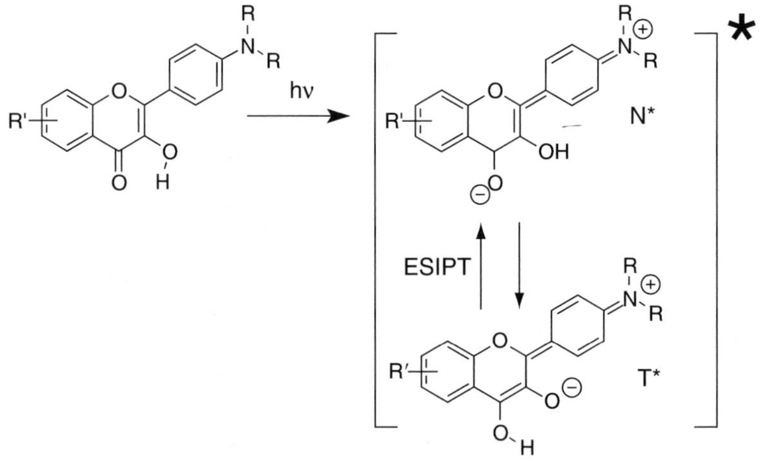

Fig. 3.16 The structure of a typical 3HC dye and its excited-state transformations between N^* and T^* forms (Klymchenko et al. 2003a)

3.7.4 Prospects for Two-Band Ratiometric Recording

The two-band wavelength-ratiometric recording is the most convenient way of fluorescence reporting. Using only the single type of labeling allows obtaining an internally self-calibrated response signal. The importance of such calibration for microarray technology and other applications has recently been indicated (Demchenko 2005a). The arguments were presented that this approach has strong advantages over other methods that allow internal calibration – intensity sensing with the reference, anisotropy, lifetime sensing and FRET (Table 3.1).

Based on this analysis, the conclusion can be made that only in lifetime sensing does the single-channel response allow obtaining an *internally calibrated signal*. In anisotropy sensing the two (vertical and horizontal) polarizations provide the necessary *two channels* and in FRET to fluorescent acceptor and in wavelength-shifting response these two channels are selected as *intensities at two wavelengths*. In two-band ratiometric sensing, the ratio of the intensities at two wavelengths is also obtained but because the signal comes from a single type of the dye (in contrast to FRET) and the forms emitting at two wavelengths have the same lifetimes (in contrast to wavelength-shifting), the internally calibrated signal is resistant to any uncontrolled quenching effect; for 3HC dyes the fact that the two spectroscopic forms are quenched proportionally with the retention of *the same intensity ratio* under the influence of collisional quenchers or the temperature has been confirmed experimentally (Oncul and Demchenko 2006; Tomin et al. 2007).

These facts provide a strong concern for the prospects of the application of the new molecular sensor technologies based on the switching in light intensities between the two bands. This new technique exhibits extreme simplicity of record-

Table 3.1 Relevance of different fluorescence detection methods to the application in direct sensing arrays

Method of detection	Number of read-out channels	Insensitivity to instrumental factors	Insensitivity to sensor concentration[a]	Insensitivity to uncontrolled quenching
Intensity	1	No	No	No
FRET	1	No	No	No
	2	Yes	Yes	No
Lifetime	1	Yes	Yes	No
Wavelength shifting	2	Yes	Yes	No
Anisotropy	2	(Yes)	Yes	No
Two-band ratiometry[b]	2	Yes	Yes	Yes

[a]In the responsive range of analyte concentrations. [b]In the case of fast reversible excited-state reaction.

ing the intensity at two fluorescence band maxima combined with the possibility of obtaining strongly enhanced and internally calibrated ratiometric responses. It features a broad-range applicability on both molecular, nanoscale and whole cell levels.

Sensing and Thinking 3: The Choice of Fluorescence Detection Technique and Optimization of Response

The choice of the mechanism of fluorescence response, responsive dye and the method of observation is always a compromise in two directions. One is the compromise in choosing between the brightest dyes that are optimal for fluorescence intensity sensing and 'responsive' dyes that are needed for more complicated methods of observation, based on the change in lifetime, formation of excited-state complexes, band wavelength shifts or generation of two fluorescence bands.

The other dilemma is between the simplicity and information content of the response, which also depends on the selected detection technique. The choice in this respect is rather limited but still exists. In addition to intensity sensing, one may choose between anisotropy and lifetime measurements and use spectroscopic detection based on spectral shifts or the interplay of two bands used in FRET, excimer sensing or sensing by generating the excited-state reaction.

Intensity sensing is the simplest technique but it features many disadvantages, mainly due to the absence of absolute scale and the difficulties in applying the internal calibration. The internally calibrated signal can be provided in anisotropy and lifetime sensing. In principle, this can be done in a two-wavelength ratiometric recording with the application of two dyes (excimers and FRET) or a single dye exhibiting a ground-state or excited-state reaction. Every one of these techniques needs proper selection of the reporter dyes. Many requests therefore should be addressed to synthetic chemists and photochemists.

Questions and Problems

1. Why is the wavenumber scale more correct for the presentation and analysis of spectra than the wavelength scale? Convert into wavenumbers 365, 436 and 630 nm. Convert into wavelengths 14,820, 19,600 and 24,300 cm^{-1}. Convert into wavenumbers the wavelength shift by 10 nm of the fluorescence band at 400 nm and at 800 nm. Convert into wavelengths the wavenumber shift by 200 cm^{-1} at 12,500 cm^{-1} and at 24,200 cm^{-1}. Estimate in cm^{-1} the bandwidths of 50 nm for the bands located in the blue (450 nm) and red (650 nm) ranges of the spectra.

2. Let the reporter dye be attached to the surface of the sensor molecule and appear on the sensor-target interaction at their contact interface. Count all possible effects that can be used to provide its quenching/unquenching response.

3. Explain how to introduce the reference signal into an intensity sensor to make its internal calibration possible. What output readings should then be obtained and, based on them, how does one calculate the target concentration?

4. What are the possibilities to distinguish static and dynamic quenching? Explain the physical processes behind them.

5. How can you distinguish between two interpretations of the quenching effect observed in intensity on exposing your sensor to the test sample? (A) the target binds to the sensor producing the quenching or (B) there is no target in the sample but some substance producing dynamic quenching by collisions with the sensor. How could the experiments in the steady state, using sample dilution or time-resolved studies, resolve this issue?

6. In anisotropy sensing should the free sensor or its recognition segment necessarily rotate? What rotation rate is acceptable (in comparison with the rate of fluorescence decay)? What other possibilities, except providing the sensor, target or competitor rotation can be used in anisotropy sensing?

7. What is the difference between polarization and anisotropy? Calculate the polarization values for anisotropy $r_0 = 0$; 0.1; 0.2; 0.3; 0.4.

8. On the target binding let a total quenching of fluorescence be observed so that the intensity goes down from a high-level value to zero. What lifetime changes will be observed? Will the lifetime sensing in this case be efficient?

9. Is a geometry change in a sensor molecule necessary to provide the response in excimer formation? Is this change always needed for a response in FRET?

10. Let the distance between the donor and acceptor on interaction with the target change from 5 to 10 Å and R_0 is 50 Å. Will this sensor be efficient?

11. Explain the relative advantages and disadvantages in the application of FRET to fluorescent or to non-fluorescent acceptors.

12. Let all the conditions for optimal FRET observation be satisfied but the donor is quenched on interaction with the target. Will this system work as efficiently as a FRET sensor?

13. Let the donor and acceptor in FRET be identical dye molecules. Can FRET be efficient between them? If so, on what conditions? Propose the construction of a sensor based on this principle.

14. Explain why and in what conditions the sensor response, based on the principle of FRET between a labeled sensor and a labeled competitor, will be more advantageous than intensity sensing with the labeling of the competitor only.

15. What is the dynamic range of the wavelength-ratiometric response, ideally and in reality? Make the estimates by applying reasonable values of spectral shifts.

16. Suggest the sensor design based on a combination of the wavelength-ratiometric response and FRET.

17. Why is internal calibration needed? Compare the solutions suggested for the internal calibration of the fluorescence response in the detection techniques discussed in this chapter.

References

Adams SR, Harootunian AT, Buechler YJ, Taylor SS, Tsien RY (1991) Fluorescence ratio imaging of cyclic AMP in single cells. Nature 349:694–697

Altschuh D, Oncul S, Demchenko AP (2006) Fluorescence sensing of intermolecular interactions and development of direct molecular biosensors. Journal of Molecular Recognition 19:459–477

Bahr JL, Kodis G, de la Garza L, Lin S, Moore AL, Moore TA, Gust D (2001) Photoswitched singlet energy transfer in a porphyrin-spiropyran dyad. Journal of the American Chemical Society 123:7124–7133

Billinton N, Knight AW (2001) Seeing the wood through the trees: a review of techniques for distinguishing green fluorescent protein from endogenous autofluorescence. Analytical Biochemistry 291:175–197

Bourson J, Pouget J, Valeur B (1993) Ion-responsive fluorescent compounds. 4. Effect of cation binding on the photophysical properties of a coumarin linked to monoaza- and diaza-crown esters. Journal of Physical Chemistry 97:4552–4557

Braslavsky SE (2007) Glossary of terms used in photochemistry 3(rd) edition (IUPAC Recommendations 2006). Pure and Applied Chemistry 79:293–465

Bright FV, Betts TA, Litwiler KS (1990) Regenerable fiber-optic-based immunosensor. Analytical Chemistry 62:1065–1069

Burke M, O'Sullivan PJ, Soini AE, Berney H, Papkovsky DB (2003) Evaluation of the phosphorescent palladium(II)-coproporphyrin labels in separation-free hybridization assays. Analytical Biochemistry 320:273–280

Cejas MA, Raymo FM (2005) Fluorescent diazapyrenium films and their response to dopamine. Langmuir 21:5795–5802

Chandrasekharan N, Kelly LA (2001) A dual fluorescence temperature sensor based on perylene/exciplex interconversion. Journal of the American Chemical Society 123:9898–9899

Chang CJ, Javorski J, Nolan EM, Shaeng M, Lippard SJ (2004) A tautomeric zinc sensor for ratiometric fluorescence imaging: application to nitric oxide-release of intracellular zinc. Proceedings of the National Academy of Sciences of the United States of America 101:1129–1134

Clegg RM (1996) Fluorescence resonance energy transfer. In: Wang XF, Herman B (eds) Fluorescence imaging spectroscopy and microscopy. Wiley, New York, pp. 179–252

Collado D, Perez-Inestrosa E, Suau R, Desvergne JP, Bouas-Laurent H (2002) Bis(isoquinoline N-oxide) pincers as a new type of metal cation dual channel fluorosensor. Organic Letters 4:855–858

Demchenko AP (1986) Ultraviolet spectroscopy of proteins. Springer, Berlin/Heidelberg/New York

Demchenko AP (1991) Fluorescence and dynamics in proteins. In: Lakowicz JR (ed) Topics in Fluorescence Spectroscopy. Plenum Press, New York, pp. 61–111

Demchenko AP (2005a) The future of fluorescence sensor arrays. Trends in Biotechnology 23:456–460

Demchenko AP (2005b) Optimization of fluorescence response in the design of molecular biosensors. Analytical Biochemistry 343:1–22

Demchenko AP (2006) Visualization and sensing of intermolecular interactions with two-color fluorescent probes. FEBS Letters 580:2951–2957

Demchenko AP, Sytnik AI (1991a) Site-selectivity in excited-state reactions in solutions. Journal of Physical Chemistry 95:10518–10524

Demchenko AP, Sytnik AI (1991b) Solvent reorganizational red-edge effect in intramolecular electron transfer. Proceedings of the National Academy of Sciences of the United States of America 88:9311–9314

Demchenko AP, Klymchenko AS, Pivovarenko VG, Ercelen S, Duportail G, Mely Y (2003) Multiparametric color-changing fluorescence probes. Journal of Fluorescence 13:291–295

de Lorimier RM, Smith JJ, Dwyer MA, Looger LL, Sali KM, Paavola CD, Rizk SS, Sadigov S, Conrad DW, Loew L, Hellinga HW (2002) Construction of a fluorescent biosensor family. Protein Science 11:2655–2675

de Silva AP, Gunaratne HQN, Gunnaugsson T, Huxley AJM, McRoy CP, Rademacher JT, Rice TE (1997) Signaling recognition events with fluorescent sensors and switches. Chemical Reviews 97:1515–1566

Di Cesare N, Lakowicz JR (2001) Wavelength-ratiometric probes for saccharides based on donor–acceptor diphenylpolyenes. Journal of Photochemistry and Photobiology A, Chemistry 143:39–47

Domecq A, Disalvo EA, Bernik DL, Florenzano F, Politi MJ (2001) A stability test of liposome preparations using steady-state fluorescent measurements. Drug Delivery 8:155–160

Epps DE, Schostarez H, Argoudelis CV, Poorman R, Hinzmann J, Sawyer TK, Mandel F (1989) An experimental method for the determination of enzyme-competitive inhibitor dissociation constants from displacement curves: application to human renin using fluorescence energy transfer to a synthetic dansylated inhibitor peptide. Analytical Biochemistry 181:172–181

Epps DE, Mitchell MA, Petzold GL, VanDrie JH, Poorman RA (1999) A fluorescence resonance energy transfer method for measuring the binding of inhibitors to stromelysin. Analytical Biochemistry 275:141–147

Gershkovich AA, Kholodovych VV (1996) Fluorogenic substrates for proteases based on intramolecular fluorescence energy transfer (IFETS). Journal of Biochemical and Biophysical Methods 33:135–162

Ghosh SS, Eis PS, Blumeyer K, Fearon K, Millar DP (1994) Real time kinetics of restriction endonuclease cleavage monitored by fluorescence resonance energy transfer. Nucleic Acids Research 22:3155–3159

Gilardi G, Zhou LQ, Hibbert L, Cass AE (1994) Engineering the maltose binding protein for reagentless fluorescence sensing. Analytical Chemistry 66:3840–3847

Giordano L, Jovin TM, Irie M, Jares-Erijman EA (2002) Diheteroarylethenes as thermally stable photoswitchable acceptors in photochromic fluorescence resonance energy transfer (pcFRET). Journal of the American Chemical Society 124:7481–7489

Godwin HA, Berg JM (1996) A fluorescent zinc probe based on metal-induced peptide folding. Journal of the American Chemical Society 118:6514–6515

Gokulrangan G, Unruh JR, Holub DF, Ingram B, Johnson CK, Wilson GS (2005) DNA aptamer-based bioanalysis of IgE by fluorescence anisotropy. Analytical Chemistry 77:1963–1970

Gross E, Bedlack RS, Loew LM (1994) Dual-wavelength ratiometric fluorescence measurement of the membrane dipole potential. Biophysical Journal 67:208–216

Grynkiewicz G, Poenie M, Tsien RY (1985) A new generation of Ca2 + indicators with greatly improved fluorescence properties. Journal of Biological Chemistry 260:3440–3450

Guo XQ, Castellano FN, Li L, Lakowicz JR (1998) Use of a long lifetime Re(I) complex in fluorescence polarization immunoassays of high-molecular weight analytes. Analytical Chemistry 70:632–637

Hunt CE, Ansell RJ (2006) Use of fluorescence shift and fluorescence anisotropy to evaluate the re-binding of template to (S)-propranolol imprinted polymers. Analyst 131:678–683

Inoue Y, Jiang P, Tsukada E, Wada T, Shimizu H, Tai A, Ishikawa M (2002) Unique dual fluorescence of sterically congested hexaalkyl benzenehexacarboxylates: mechanism and application to viscosity probing. Journal of the American Chemical Society 124:6942–6949

Jameson DM, Croney JC (2003) Fluorescence polarization: past, present and future. Combinatorial Chemistry & High Throughput Screening 6:167–173

Jares-Erijman EA, Jovin TM (2003) FRET imaging. Nature Biotechnology 21:1387–1395

Jayaraman S, Biwersi J, Verkman AS (1999) Synthesis and characterization of dual-wavelength Cl-sensitive fluorescent indicators for ratio imaging. American Journal of Physiology 276: C747–757

Kapanidis AN, Weiss S (2002) Fluorescent probes and bioconjugation chemistries for single-molecule fluorescence analysis of biomolecules. Journal of Chemical Physics 117:10953–10964

Klymchenko AS, Ozturk T, Pivovarenko VG, Demchenko AP (2001a) A 3-hydroxychromone with dramatically improved fluorescence properties. Tetrahedron Letters 42:7967–7970

Klymchenko AS, Ozturk T, Pivovarenko VG, Demchenko AP (2001b) Synthesis and spectroscopic properties of benzo- and naphthofuryl-3-hydroxychromones. Canadian Journal of Chemistry-Revue Canadienne De Chimie 79:358–363

Klymchenko AS, Duportail G, Mely Y, Demchenko AP (2003a) Ultrasensitive two-color fluorescence probes for dipole potential in phospholipid membranes. Proceedings of the National Academy of Sciences of the United States of America 100:11219–11224

Klymchenko AS, Pivovarenko VG, Ozturk T, Demchenko AP (2003b) Modulation of the solvent-dependent dual emission in 3-hydroxychromones by substituents. New Journal of Chemistry 27:1336–1343

Koronszi I, Reichert J, Heinzmann G, Ache HJ (1998) Development of submicron optochemical potassium sensor with enhanced stability due to internal reference. Sensors and Actuators B 51:188–195

Kostenko E, Dobrikov M, Komarova N, Pyshniy D, Vlassov V, Zenkova M (2001) 5'-bis-pyrenylated oligonucleotides display enhanced excimer fluorescence upon hybridization with DNA and RNA. Nucleosides Nucleotides Nucleic Acids 20:1859–1870

Kuwana E, Sevick-Muraca EM (2002) Fluorescence lifetime spectroscopy in multiply scattering media with dyes exhibiting multiexponential decay kinetics. Biophysical Journal 83:1165–1176

Lakowicz JR (2007) Principles of fluorescence spectroscopy. Springer, New York

Liebsch G, Klimant I, Krause C, Wolfbeis OS (2001) Fluorescent imaging of pH with optical sensors using time domain dual lifetime referencing. Analytical Chemistry 73:4354–4363

Mahara A, Iwase R, Sakamoto T, Yamana K, Yamaoka T, Mirakami A (2002) Bispyrene-conjugated 2''-O-methyloligonucleotide as a highly specific RNA-recognition probe. Angewandte Chemie-International Edition 41:3648–3650

Maliwal BP, Gryczynski Z, Lakowicz JR (2001) Long-wavelength long-lifetime luminophores. Analytical Chemistry 73:4277–4285

Malval J-P, Lapouyade R, Leger JM, Jany C (2003) Tripodal ligand incorporating dual fluorescent ionophore: a coordinative control of photochemical electron transfer. Photochemical and Photobiological Sciences 2:259–266

Marras SA, Kramer FR, Tyagi S (2002) Efficiencies of fluorescence resonance energy transfer and contact-mediated quenching in oligonucleotide probes. Nucleic Acids Research 30:e122

McCauley TG, Hamaguchi N, Stanton M (2003) Aptamer-based biosensor arrays for detection and quantification of biological macromolecules. Analytical Biochemistry 319:244–250

McFarland SA, Finney NS (2001) Fluorescent chemosensors based on conformational restriction of a biaryl fluorophore. Journal of the American Chemical Society 123:1260–1261

Medintz IL, Goldman ER, Lassman ME, Mauro JM (2003) A fluorescence resonance energy transfer sensor based on maltose binding protein. Bioconjugate Chemistry 14:909–918

Medintz IL, Trammell SA, Mattoussi H, Mauro JM (2004) Reversible modulation of quantum dot photoluminescence using a protein-bound photochromic fluorescence resonance energy transfer acceptor. Journal of the American Chemical Society 126:30–31

Minta A, Kao JP, Tsien RY (1989) Fluorescent indicators for cytosolic calcium based on rhodamine and fluorescein chromophores. Journal of Biological Chemistry 264:8171–8178

Nielsen K, Lin M, Gall D, Jolley M (2000) Fluorescence polarization immunoassay: detection of antibody to Brucella abortus. Methods 22:71–76

Oheim M, Naraghi M, Muller TH, Neher E (1998) Two dye two wavelength excitation calcium imaging: results from bovine adrenal chromaffin cells. Cell Calcium 24:71–84

Oncul S, Demchenko AP (2006) The effects of thermal quenching on the excited-state intramolecular proton transfer reaction in 3-hydroxyflavones. Spectrochimica Acta A. Molecular and Biomolecular Spectroscopy 65:179–183

Owicki JC (2000) Fluorescence polarization and anisotropy in high throughput screening: perspectives and primer. Journal of Biomolecular Screening 5:297–306

Prendergast FG, Meyer M, Carlson GL, Iida S, Potter JD (1983) Synthesis, spectral properties, and use of 6-acryloyl-2-dimethylaminonaphthalene (Acrylodan). A thiol-selective, polarity-sensitive fluorescent probe. Journal of Biological Chemistry 258:7541–7544

Putkey JA, Liu W, Lin X, Ahmed S, Zhang M, Potter JD, Kerrick WG (1997) Fluorescent probes attached to Cys 35 or Cys 84 in cardiac troponin C are differentially sensitive to Ca(2+)-dependent events in vitro and in situ. Biochemistry 36:970–978

Raymond FR, Ho HA, Peytavi R, Bissonnette L, Boissinot M, Picard FJ, Leclerc M, Bergeron MG (2005) Detection of target DNA using fluorescent cationic polymer and peptide nucleic acid probes on solid support. BMC Biotechnology 5:10

Richieri GV, Ogata RT, Kleinfeld AM (1992) A fluorescently labeled intestinal fatty acid binding protein. Interactions with fatty acids and its use in monitoring free fatty acids. Journal of Biological Chemistry 267:23495–23501

Rurack K (2001) Flipping the light switch 'on' - the design of sensor molecules that show cation-induced fluorescence enhancementwith heavy and transition metal ions. Spectrochimica Acta Part A 57:2161–2195

Sahoo D, Narayanaswami V, Kay CM, Ryan RO (2000) Pyrene excimer fluorescence: a spatially sensitive probe to monitor lipid-induced helical rearrangement of apolipophorin III. Biochemistry 39:6594–6601

Sanchez-Barragan I, Costa-Fernandez JM, Valledor M, Campo JC, Sanz-Medel A (2006) Room-temperature phosphorescence (RTP) for optical sensing. Trac-Trends in Analytical Chemistry 25:958–967

Selvin PR (2000) The renaissance of fluorescence resonance energy transfer. Nature Structural Biology 7:730–734

Shynkar V, Mely Y, Duportail G, Piemont E, Klymchenko AS, Demchenko AP (2003) Picosecond time-resolved fluorescence studies are consistent with reversible excited-state intramolecular proton transfer in 4′-dialkylamino-3-hydroxyflavones. Journal of Physical Chemistry A. 109: 9522–9529

Shynkar VV, Klymchenko AS, Piemont E, Demchenko AP, Mely Y (2004) Dynamics of intermolecular hydrogen bonds in the excited states of 4 '-dialkylamino-3-hydroxyflavones. On the pathway to an ideal fluorescent hydrogen bonding sensor. Journal of Physical Chemistry A 108:8151–8159

Shynkar VV, Klymchenko AS, Duportail G, Demchenko AP, Mely Y (2005) Two-color fluorescent probes for imaging the dipole potential of cell plasma membranes. Biochimica Et Biophysica Acta 1712:128–136

Smallshaw JE, Brokx S, Lee JS, Waygood EB (1998) Determination of the binding constants for three HPr-specific monoclonal antibodies and their Fab fragments. Journal of Molecular Biology 280:765–774

Sportsman JR, Daijo J, Gaudet EA (2003) Fluorescence polarization assays in signal transduction discovery. Combinatorial Chemistry & High Throughput Screening 6:195–200

Suppan P, Ghoneim N (1997) Solvatochromism. Royal Society of Chemistry, Cambridge, UK

Szmacinski H, Lakowicz JR (1994) Lifetime-based sensing. In: Lakowicz JR (ed) Topics in fluorescence spectroscopy. Plenum Press, New York, pp. 295–334

Takakusa H, Kikuchi K, Urano Y, Kojima H, Nagano T (2003) A novel design method of ratiometric fluorescent probes based on fluorescence resonance energy transfer switching by spectral overlap integral. Chemistry 9:1479–1485

Thirunavukkuarasu S, Jares-Erijman EA, Jovin TM (2008) Multiparametric fluorescence detection of early stages in the amyloid protein aggregation of pyrene-labeled alpha-synuclein. Journal of Molecular Biology 378:1064–1073

Thompson RB, Maliwal BP, Feliccia VL, Fierke CA, McCall K (1998) Determination of picomolar concentrations of metal ions using fluorescence anisotropy: biosensing with a "reagentless"
enzyme transducer. Analytical Chemistry 70:4717–4723

Tolosa L, Ge X, Rao G (2003) Reagentless optical sensing of glutamine using a dual-emitting
glutamine-binding protein. Analytical Biochemistry 314:199–205

Tomin VI, Oncul S, Smolarczyk G, Demchenko AP (2007) Dynamic quenching as a simple test
for the mechanism of excited-state reaction. Chemical Physics 342:126–134

Torimura M, Kurata S, Yamada K, Yokomaku T, Kamagata Y, Kanagawa T, Kurane R (2001)
Fluorescence-quenching phenomenon by photoinduced electron transfer between a fluorescent
dye and a nucleotide base. Analytical Sci 17:155–160

Tsien RY (1998) The green fluorescent protein. Annual Review of Biochemistry 67:509–544

Tyagi S, Kramer FR (1996) Molecular beacons: probes that fluoresce upon hybridization. Nature
Biotechnology 14:303–308

Tyagi S, Marras SA, Kramer FR (2000) Wavelength-shifting molecular beacons. Nature
Biotechnology 18:1191–1196

Ueberfeld J, Walt DR (2004) Reversible ratiometric probe for quantitative DNA measurements.
Analytical Chemistry 76:947–952

Valeur B (2002) Molecular fluorescence. Wiley-VCH, Weinheim

Vazquez ME, Blanco JB, Imperiali B (2005) Photophysics and biological applications of the
environment-sensitive fluorophore 6-N,N-Dimethylamino-2,3-naphthalimide. Journal of the
American Chemical Society 127:1300–1306

Walkup GK, Imperiali B (1996) Design and Evaluation of a Peptidyl Fluorescent Chemosensor
for Divalent Zinc. Journal of the American Chemical Society 118:3053–3054

Wu PG, Brand L (1994) Resonance Energy-Transfer - Methods and Applications. Analytical
Biochemistry 218:1–13

Xu H, Wu HP, Huang F, Song SP, Li WX, Cao Y, Fan CH (2005a) Magnetically assisted DNA
assays: high selectivity using conjugated polymers for amplified fluorescent transduction.
Nucleic Acids Research 33:e83

Xu QH, Wang S, Korystov D, Mikhailovsky A, Bazan GC, Moses D, Heeger AJ (2005b) The fluorescence resonance energy transfer (FRET) gate: a time-resolved study. Proceedings of the
National Academy of Sciences of the United States of America 102:530–535

Yamauchi A, Hayashita T, Nishizawa S, Watanabe M, Teramae N (1999) Benzo-15-crown-5
fluoroionophore/cyclodextrin complex with remarkably high potassium ion sensitivity in
water. Journal of the American Chemical Society 121:2319–2320

Yang RH, Chan WH, Lee AWM, Xia PF, Zhang HK, Li KA (2003) A ratiometric fluorescent sensor for Ag-1 with high selectivity and sensitivity. Journal of the American Chemical Society
125:2884–2885

Yang CJ, Jockusch S, Vicens M, Turro NJ, Tan W (2005) Light-switching excimer probes for
rapid protein monitoring in complex biological fluids. Proceedings of the National Academy
of Sciences of the United States of America 102:17278–17283

Yuasa H, Miyagawa N, Izumi T, Nakatani M, Izumi M, Hashimoto H (2004) Hinge sugar as a
movable component of an excimer fluorescence sensor. Organic Letters 6:1489–1492

Chapter 4
Design and Properties of Fluorescence Reporters

A variety of fluorescent and luminescent materials in the form of molecules, their complexes and nanoparticles, are available for implementation as the response units into sensing technologies. In this chapter we will concentrate on their properties and evaluate their applicability for particular tasks. In this respect, organic fluorescent dyes remain of primary importance. Meantime, we observe increasing competition in research and applications from the side of luminescent metal ion chelating complexes, from fluorescent polymer molecules and especially, from different kind of nanoparticles.

We observe that with the incorporation into polymeric and silica nanoparticles the dyes attain new, important properties, including a dramatically increased brightness. Semiconductor nanocrystals (Quantum Dots) are unique in their fluorescence properties and possibilities in applications as sensing reporters. Noble metal particles of a very small size are not only fluorescence quenchers; they are light-absorbing and fluorescent themselves. Conjugated polymers are formed by coupling into a polymer chain of light-absorbing and fluorescent monomeric units. Finally, the fluorescent proteins that can be synthesized inside living cells are capable of spontaneous formation of fluorescent moiety inside these protein structures. They produce a revolutionary advancement in cell imaging and show good prospects in sensing and reporting on cellular processes.

4.1 Organic Dyes

Organic dyes are the most commonly used molecules as reporters in fluorescence sensing. Their advantages are not only easy availability and low price but also, most importantly, their versatility. The number of fluorescent synthetic organic products is so great that a researcher can easily select the proper dye corresponding to a particular need in terms of spectroscopic properties and chemical reactivity.

In the past, the field of organic dye synthesis was strongly stimulated by the needs of color photography and dye laser technologies. These needs were reduced with the development of electronic digital photography and solid-state laser techniques, so that the implementation into molecular probe and sensor technologies became the major field of the application of new organic dyes. The amount of newly synthesized dyes grows exponentially and it is not possible to describe in a small section of this book even the most important of them. Therefore, we will concentrate on their general properties that are essential for sensing applications and describe several of their classes that are frequently used. Many original publications, reviews and web-based resources will be of help to the reader. Regarding information on particular dyes, one can refer to the web sites of distributors of fluorescent dyes. The recently established public database (http://www.fluorophores. org) is also useful in choosing dyes with the desired properties.

4.1.1 General Properties of Fluorescence Reporter Dyes

Organic fluorescent dyes offer an immense variation of chemical, photochemical and spectroscopic properties. Some of them are listed in Table 4.1 to help the reader with a brief navigation of these properties and compare them with those of other luminescent emitters discussed in this chapter.

In view of the tremendous diversity of structures, chemical reactivities and spectroscopic properties, one needs certain practical, useful criteria for dye selection that could be formulated in a most general way. The parameters characterizing the dyes and important for sensor applications are suggested below.

Table 4.1 The properties of organic dyes used in fluorescence sensing technologies

Parameter	Property
Size	0.5–2 nm
Absorption spectra	Variable and often narrow, asymmetric and sometimes with vibronic structure, sensitive to intermolecular interactions
Molar absorbance, ε	Variable, usually in the range 10^3–10^5 M^{-1} cm^{-1}
Emission spectra	Broad (~40–70 nm half-width) and asymmetric, position may exhibit strong sensitivity to intermolecular interactions
Stokes shifts	Variable, commonly ~10–100 nm
Excited-state lifetime, τ_F	Short, ~0.5–5 ns (with few exceptions)
Fluorescence quantum yield, Φ	Variable, strongly dependent upon the medium
Two-photonic cross-section, α	Strongly variable and uncorrelated with ε
Fundamental emission anisotropy, r^0	Variable, up to 0.3–0.4
Photostability	Variable to poor
Chemical resistance	Variable and often low
Availability of chemical modification/functionalization	Variable to high
Cell toxicity	Variable, often low. High phototoxicity may be observed in some cases

1. High *molar absorbance*, ε. This is the parameter describing the ability of the molecule to absorb light at a particular wavelength. The absorbance E measured at any wavelength on a spectrophotometer is proportional to the concentration c (in mol/l) and path length in a sample l (in cm). In this relation, ε plays the role of a coefficient of proportionality, so that $E = \varepsilon c l$.

The value of ε^{max} at the maximum of the absorption band characterizes the light-absorbing power of the dye. Together with the quantum yield, Φ, this parameter is the major contributor to the dye '*brightness*', which determines the absolute sensitivity of fluorescence detection (Wetzl et al. 2003). Usually, for organic dyes, ε^{max} varies from 3 ÷ 5 × 10³ up to 2 ÷ 3 × 10⁵ l mol⁻¹ cm⁻¹. It should be as high as possible. Using dyes with lower ε^{max} values is not reasonable and much higher values are hardly achievable.

2. High *quantum yield*, Φ. The quantum yield of any process is the ratio of the number of quanta participating in this process to the total number of absorbed quanta. For fluorescence, it is the ratio of the number of light quanta emitted as fluorescence to the total number of quanta that were absorbed. In addition to fluorescence, the dyes can participate in different nonemissive reactions. Thus, Φ is used as a measure of efficiency of fluorescence compared to other processes occurring during the excited-state lifetime, such as static and collisional quenching, transition to the triplet state or photochemical transformations.

Primarily, Φ depends on the dye structure but it can be modulated dramatically by the dye's environment. The Φ values may vary from very low numbers (in relative units) to almost 100%. The sensing technique, in which the change of fluorescence intensity is recorded as an effect of quenching (Section 3.1), is based on variations of the quantum yield.

3. Optimal *excitation wavelength*. In order to achieve the highest brightness, one has to excite fluorescence at wavelengths close to the absorption band maximum. Since the dye brightness depends on the product of the molar absorbance $\varepsilon(\lambda_{ex})$ (at the applied excitation wavelength) and the fluorescence quantum yield Φ, it is important that the excitation wavelength, λ_{ex}, should be chosen optimally with this account.

Other factors are important for selecting the dyes with an excitation band at a particular wavelength, such as the availability of light sources, necessity to avoid autofluorescence in living cell studies, etc. To avoid cells' autofluorescence, excitation at wavelengths longer than 400–460 nm is required. For diagnostic sensors implemented through the human skin, one has to account that the skin has a very low transparency throughout the whole visible range, due to absorption by porphyrins and skin pigments. Such absorption is dramatically reduced in the near-IR, until the vibrational overtones of water appear, so the range 800–1,000 nm is optimal for this application.

4. Optimal emission wavelength. The range of operational wavelengths should be selected taking into account the wavelength dependence of the sensitivity of the detector. If the detector is a human eye, then the emission should be in the visible range of 400–750 nm. Many spectrofluorimeters available on the market are equipped with photodetectors with the cutting edge range of 700 nm, which is

not sufficient for many dyes. However, the special "red-sensitive" photomultipliers can be supplied by special request. Different red-infrared detectors using semiconductor elements have become available in recent times.

5. Large *Stokes shift* – the shift between absorption and emission band maxima on the wavenumber scale. This property is desirable but not necessary in many applications. It should be noted that the most popular as fluorescent tags and labels, fluorescein and rhodamine derivatives possess a rather small Stokes shift. Light scattering, which enables excitation and emission aside from the band maxima, can be a difficult problem with such dyes. Meantime, a stronger separation between absorption and emission bands allows reducing the light-scattering effects and provides possibilities for a more efficient collection of emitted light, using broad-band filters or larger monochromator slits.

Larger Stokes shift allows reducing homo-FRET (excitation energy transfer between the same dye molecules), thus allowing a high-density attachment of the dyes to macromolecules and their incorporation into nanocomposite structures (see Section 4.3). In some sensing technologies, the dyes possessing strong Stokes shifts behave within these high-density units as independent emitters with a suppressed concentration-dependent depolarization and quenching.

6. Optimal *fluorescence lifetime*, τ_F. This parameter depends on the dye and cannot be longer than the radiative lifetime, which is the fundamental property of a particular dye. The lifetime also depends on the immediate environment of the dye, so depending on the application, it can be selected long or short by choosing the dye and its environment.

A long lifetime is easy to detect and analyze. It is usually associated with a high quantum yield (the absence or small involvement of quenching effects). However, long lifetimes (longer than 10^{-7} s, such as that of pyrene) are quenched by oxygen diffusing in the medium, so it is sometimes difficult to control this parameter. In anisotropy sensing (polarization assays), to provide the strongest effect the lifetime should fit the time scale of the molecular rotations. Its long values are needed to detect rotations of large molecular units.

Meantime, for some applications, a short τ_F is preferable. This is so in flow cytometry, in which a small number of dyes are under intensive excitation and the response may be limited by the duration of the excitation-emission cycle. It must also be noted that due to a fundamental limitation, a high ε^{max} cannot be achieved with a long τ_F.

7. High *photostability*. Each fluorescent dye can be degraded (photo-bleached) after some number of excitation-emission cycles (Eggeling et al. 1998). Photobleaching is a consequence of a higher dye chemical reactivity in the excited states. Degradation occurs due to some photochemical reaction that often involves molecular oxygen and is coupled with the production of singlet oxygen. Fluorescein molecules can survive between 10^4 and 10^5 excitation-emission cycles before decomposing and commonly, the dyes are distinguished as being more photostable or less photostable than fluorescein.

In sensing technologies that do not require exposure to intensive light, photo-stability is not a big problem. It may not be important in spectroscopic studies, in which the narrow excitation slits and the fast recording of spectra are used. In flow cytometry and microplate reading, the photo-bleaching is generally not a problem because the sample remains in the laser beam for a very short period of time. Otherwise, oxygen can be removed from the system, which requires an oxygen-free atmosphere or consumption of oxygen in any coupled reaction. Such a possibility may not exist in the fluorescence microscopy of living cells and this often limits obtaining cellular images to a short time scale. Photostability is needed in a single-molecule detection, where the number of quanta detected from a single molecule has to be maximized to increase the statistical accuracy of the detection based on the intensity or lifetime.

8. Optimal *solubility – penetration – reactivity* in the used system. Each dye has particular physical and chemical properties (polarity, charge distribution, reactivity to form covalent bonds, ability to participate in hydrogen bonding and other noncovalent interactions). These properties determine the dye distribution in a heterogeneous system (e.g., on liquid-solid interfaces, in cell cytoplasms or biomembranes) and determines the extent of the labeling of particular macromolecules or organelles. In addition, the dye has to be chemically stable and its fluorescence should be insensitive or weakly sensitive to factors like pH or temperature, if this sensitivity is not required.

The reason for describing these criteria at such a length is due to the fact that the ideal dyes simply do not exist and the researcher has to compromise some properties to make other properties ideal for a particular application. Therefore, the first question to respond to is concerning a **particular role that the dye has to play in the sensing process**. It could be the labeling of a sensor, of a competitor or of a sample pool containing the target (the dye indicates its location or attachment to particular structure). These are the simplest applications to which many dyes are well suited. If the reporting function is needed (the parameters of dye emission should be changed on interaction with the target), then additional criteria should be introduced.

4.1.2 Dyes for Labeling

There are several classes of dyes that conform to the dye selection criteria in the best possible way. If a 'responsive' function is not needed, the optimal dyes for applications such as labels and tags are the modified *fluorescein* and *rhodamine* derivatives and also *cyanine* dyes. Synthetic chemistry suggests many improvements in dye properties. The 'classical' dyes together with their most efficient improvements are discussed below.

Figure 4.1 presents the structure of fluorescein and the spectra of rhodamine 123, which is typical in this respect.

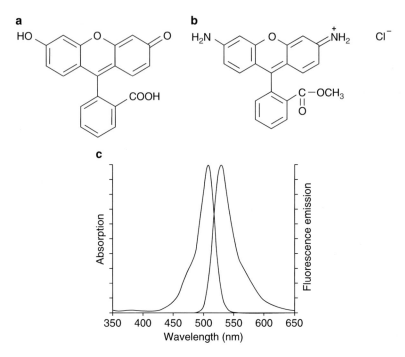

Fig. 4.1 The structure of fluorescein (a) and rhodamine 123 (b) and the normalized absorption and fluorescence spectra of rhodamine 123 in methanol (c) (Haugland 2005)

Fluoresceins. The structure of parent fluorescein dye is presented in Fig. 4.1. These dyes are widely used for labeling in the form of amine-reactive and SH-reactive derivatives (see Fig. 7.6). They possess a relatively high absorbance, a high fluorescence quantum yield (even in water) and a high aqueous solubility. Additionally, due to the fact that their excitation band maximum closely matches the 488 nm spectral line of the most popular argon-ion laser, these dyes are very well accepted in confocal laser-scanning microscopy and flow cytometry applications. Meantime, fluorescein dyes possess many essential drawbacks. They possess a low photostability and also pH-sensitivity of the fluorescence spectra with a significant reduction of intensity below pH 7. They are subjective to strong concentration-dependent quenching, which does not allow one to obtain a high density of labels. Oregon Green dyes are fluorinated analogs of fluorescein, which, in the conditions of a high degree of labeling, are quenched to a much lesser extent. They are more photostable and are pH-insensitive in the physiological pH range (Haugland 1996).

Rhodamines (see Fig. 4.1). Just like fluoresceins, rhodamine dyes are considered a popular choice of dyes for labeling. They absorb and emit light at a longer wavelength and may be used in techniques employing them as acceptors for FRET from fluorescein dyes. Their excitation band maxima are located at about 520 nm, which is close to the 514 nm spectral line of the argon-ion laser, making them both popular and important in studies involving confocal microscopy and flow cytometry.

Selected as an example, rhodamine 123 is a cell-permeant, cationic, green-fluorescent dye with a good solubility in water. When used in biological labeling it readily penetrates into the cell and is sequestered by active mitochondria without cytotoxic effects. Because of its positive charge its binding to mitochondria depends on the membrane potential of this organelle. Because of a small Stokes shift (Fig. 4.1), this dye is self-quenched when it accumulates in mitochondria in high concentrations and thus provides a 'slow' response to its membrane potential (Plasek and Sigler 1996).

Alexa dyes. These organic dye molecules are obtained by sulfonating aminocoumarin, rhodamine, or carbocyanine dyes (Haugland 1996). The introduction of negatively charged sulfonic acid groups makes these molecules more water-soluble and decreases their inherent tendency to form aggregates. This also allows performing bioconjugation in the absence of organic solvents. The excitation and emission wavelength ranges of Alexa dyes cover the entire spectrum from ultraviolet to red with the retention of such properties of parent compounds as high molar absorbance. They are less subjective to self-quenching and their photostability is even higher (Panchuk-Voloshina et al. 1999).

BODIPY dyes (Fig. 4.2). The variations of their synthetic pathways and structures have been recently described (Loudet and Burgess 2007). In contrast to Alexa dyes, BODIPY molecules are uncharged. These dyes, emitting in a broad wavelength range (500–700 nm), can be selected. They are insensitive to polarity and solvent pH (Haugland 1996, 2005). They possess a high molar absorbance, $\varepsilon \approx 80,000\,\mathrm{cm^{-1}\,M^{-1}}$ and a quantum yield Φ approaching 100% (even in water!). Their

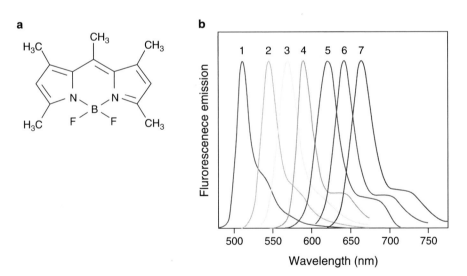

Fig. 4.2 The BODIPY (4,4-difluoro-4-bora-3a,4adiaza-s-indacene) dyes (Haugland 1996). (a) The structure of parent fluorophore. (b) Normalized fluorescence emission spectra of (1) BODIPY FL, (2) BODIPY R6G, (3) BODIPY TMR, (4) BODIPY 581/591, (5) BODIPY TR, (6) BODIPY 630/650 and (7) BODIPY 650/665 dyes in methanol

emission wavelength maxima ranges from 510 to 675 nm. The application of these dyes is extremely versatile. They are used to generate fluorescent conjugates of proteins, nucleotides, oligonucleotides and dextrans, as well as to prepare fluorescent enzyme substrates, fatty acids, phospholipids, lipopolysaccharides, receptor ligands and polystyrene microspheres. Due to their relatively long lifetime (4 ns and longer) they are particularly useful for fluorescence polarization-based assays.

Acrydine and ethidium dyes. These cationic dyes are frequently used for staining double-helical DNA. These dyes bind with a stoichiometry of one dye per 4–5 base pairs of DNA and the binding mode is an intercalation between the nucleic acid bases with little or no sequence preference. *Intercalation* is the noncovalent inserting of molecules between the base pairs of the DNA duplex, in which the dye molecule is held stacked perpendicular to the helix axis. *Ethidium bromide* (Fig. 4.3) and its analogs exhibit approximately a 30-fold enhancement of emission intensity on their binding. Screening of fluorophore moiety from the quenching effect of water is in the origin of such an enhancement.

Cyanine dyes. These dyes belong to a large class of organic compounds that is in extensive use in photography and dye laser technology. A great interest in them, from the side of biological labeling and sensing techniques, is due to their high molar absorbance, relatively high photostability, good water solubility and usually

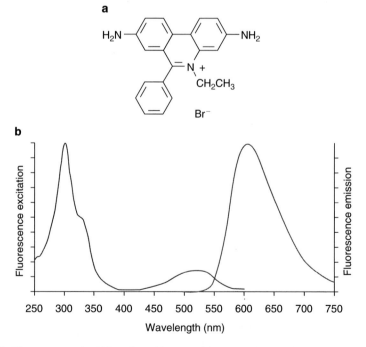

Fig. 4.3 The structure of ethidium bromide (a) and its absorption and fluorescence spectra on binding to double-stranded DNA (b)

high quantum yield (Mujumdar et al. 1993). Their common disadvantage is a small Stokes shift, short lifetime (commonly less than 1 ns) and the unavoidable presence of a positive charge. Their absorption and fluorescence spectra cover a broad wavelength range extending to red and near-IR.

These molecules are composed of a polymethine chain with two heterocyclic units at its terminals (Fig. 4.4). The length of this chain determines the spectral range of absorption and emission: with its increase by one vinylene unit (CH = CH) the spectra shifts to longer wavelengths by about 100 nm (Ishchenko et al. 1992). The shortest monomethine dyes, which include the asymmetric cyanines thiazole

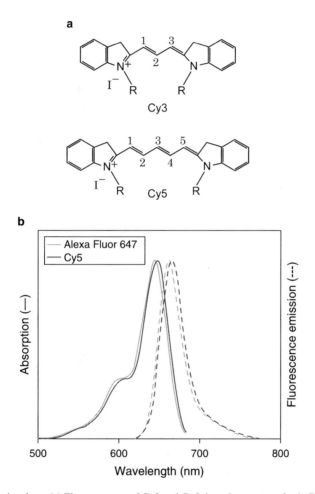

Fig. 4.4 Cyanine dyes. (a) The structures of Cy3 and Cy5 dyes that are popular in DNA labeling. R are the sites of substitutions that are used in covalent labeling. Usually, they are not identical. (b) The comparison of the absorption and fluorescence spectra of Cy5 with that of Alexa Fluor 647, the product of Invitrogen. The spectra have been normalized to the same intensity for comparison purposes (Haugland 2005)

orange (TO), oxazole yellow (YO), or dimers of both TO and YO (TOTO, YOYO) are used for staining nucleic acids; they do not fluoresce in solution but possess intense fluorescence when bound to them.

Polymethine cyanines include the frequently used dyes Cy3 and Cy5 (see Fig. 4.4) that possess an increased chain length and are known for binding nucleic acids with much greater affinity. Cy3 can be maximally excited near 550 nm and emits an orange light, whereas Cy5 has a narrow excitation peak at 650 nm and produces an intense signal in the far-red region of the spectrum. Their remarkable brightness is due to a high molar absorbance that can exceed 10^5 M^{-1} cm^{-1}. The extraordinary strong increase in fluorescence intensity upon binding to nucleic acids is a result of rigid fixation of their trans-conformation. In some of them, such an increase of quantum yield can be 100-fold, from less than 1% to nearly 100%. Polyfluorination allows reducing their aggregation in an aqueous media and enhancing their fluorescence quantum yield and photostability (Renikuntla et al. 2004). They are frequently used in DNA and RNA microarray hybridization techniques.

There are two important ways in which cyanine dyes can reversibly bind to double-strand DNA: by *intercalation* or by binding in the minor groove. As a rule, monomethine dyes act as intercalators, whereas the increase in the polyme-thine chain length allows their binding to the minor groove. As a rule, intercalat-ing monomethine dyes do not exhibit specificity to DNA sequences. The longer, polymethine dyes can serve as minor groove binders and display sequence specificity.

At groove binding, the cyanine dyes can interact between themselves forming aggregates. There can be face-to-face dimers called H-aggregates that possess a new absorption band shifted to shorter wavelengths with respect to the monomer band. For some cyanine dyes an end-to-end aggregation in the minor groove of the DNA is observed, which results in the appearance of a new very narrow absorption band shifted to longer wavelengths (Fig. 4.5). These structures are called *J-aggregates.*

These aggregates are useful for the DNA staining in living cells using two-pho-ton excitation (Guralchuk et al. 2007). Though the major application of J-aggregate forming dyes is in detection of nucleic acids, they can be used in other applications, such as in measuring membrane potentials in mitochondria (Salvioli et al. 1997). In both cases, electrostatic effects modulate the formation of aggregates.

The dramatic change of optical properties on the J-aggregation is due to a very high polarizability of their Π-electronic system along the polymethine chain. It is a rare case when such a high polarizability is observed already in the ground state, so that it gives rise to extraordinary strong attraction forces between these dye molecules. In J-aggregates the appearance of narrow absorption bands that are red-shifted as compared to the monomer transition is due to the presence of *excitons* (delocalized excitations) producing the so-called 'exciton splitting' (Davydov 1971). So, the dramatic transformation of spectra on aggregation forms the basis for the application of J-aggregates in sensing technologies (Losytskyy et al. 2002).

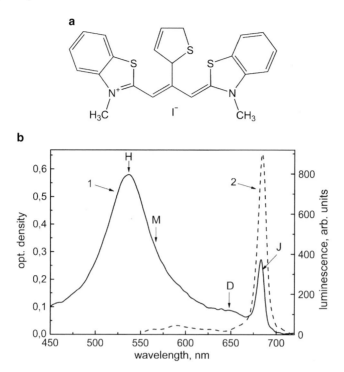

Fig. 4.5 Cyanine dye forming J-aggregates on interaction with DNA. (a) The structure of cyanine dye L-21. (b) Absorption (1) and fluorescence (2) spectra in the molecular form and in the form of J-aggregates when it binds to double-helical DNA (Reproduced with permission from Guralchuk et al. 2007)

4.1.3   The Dyes Providing Fluorescence Response

Here we concentrate on the dye properties that are required for performing a fluorescence response in sensing systems. They should optimally correspond to the detection techniques described in Chapter 3 and allow achieving the broadest dynamic range and highest sensitivity of detection. Only in sensing based on anisotropy (polarization) can the role of the dye reporter be passive to respond to rotations of large molecules or their fragments. In detection based on FRET, the response can be coupled to the variation of the distance between two dyes. In all other cases, the dyes have to respond to the formation/breaking of specific or nonspecific noncovalent interactions with their neighboring molecules or groups of atoms.

The special requirements for such responsive dyes are the following:

1. *Strong variation* in one or several fluorescence parameters described in Chapter 3 that can be quantified. The response can be provided by the formation of specific complexes of the dyes (e.g., with the quenchers), by formation of their dimers (e.g., excimers) or aggregates. Alternatively, they can participate in interactions

that can be described by such variables as polarity, viscosity or hydrogen-bonding ability.
2. *Strong discrimination* in response to the measurable parameter against the response to interfering parameters (e.g., polarity probes should not be highly sensitive to H-bonding and the reverse). Alternatively, the possibility of the multiparametric analysis of the complex response may be provided (Klymchenko and Demchenko 2003).
3. *Insensitivity of response* to different interfering factors, such as concentrations of co-solvents or ions, temperature, pH, etc. if these factors are not the targets of the assay.

The generation of the reporting signal from responding dyes involve different signal transduction mechanisms that will be described in Chapter 6 and the dye properties should be optimally selected for realizing these mechanisms.

4.1.4 The Environment-Sensitive (Solvatochromic) Dyes

Many organic dyes respond to weak changes of intermolecular interactions by the shifts of their fluorescence spectra. This extremely valuable property is used in the design of probes and sensors (see Section 3.6). Below we overview the representatives of several classes of dyes, the response of which is based on this principle.

The polarity-responsive properties are already present in the amino acid *tryptophan*, which is a natural constituent of many proteins (Demchenko 1986). But small values of these effects together with an inconvenient UV excitation do not allow its extensive use in fluorescence sensing. In contrast, there are fluorophores that provide record values in solvent-dependent spectral shifts in the visible range. Like tryptophan and more than tryptophan they have to change their dipole moment on excitation that changes their interaction with surrounding molecules and results in the shifts of spectra. This concept was introduced by Gregorio Weber and the first dyes that found many applications, mostly in the studies of structure, dynamics and interactions in proteins and biomembranes, were the naphthalene sulfonate derivatives such as ANS and TNS (Fig. 4.6). This principle was realized in the design of dyes of the latter generation represented in Fig. 4.6 by Nile Red and Prodan.

Prodan is one of the most solvent-sensitive among the commonly used probes operating on the principle of the solvent-dependent band shift. Its absorption spectrum is observed at 350–370 nm and the fluorescence spectrum ranges from 416 nm for toluene to 505 nm in methanol (Fig. 4.7).

Recently, a long-wavelength emitting an analog of Prodan, Anthradan, was synthesized. In its structure the naphthalene group was replaced by anthracene, which led to the shifts of both the excitation and emission spectra by about 100 nm with a retention of high sensitivity to polarity and hydrogen bond formation in the protic environment (Lu et al. 2006). A number of dye derivatives used for the covalent

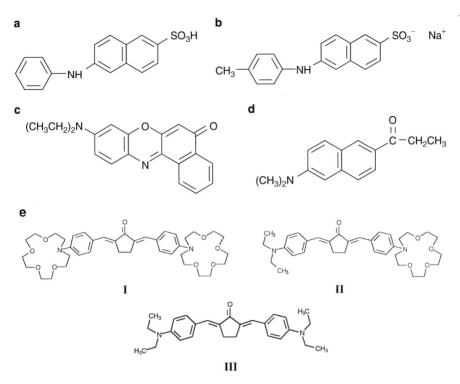

Fig. 4.6 The structures of typical environment-responsive dyes. (a) 2,6-ANS (6-(phenylamino)-2-Naphthalenesulfonic acid). (b) 2,6-TNS (2-(p-toluidinyl)naphthalene-6- sulfonic acid). (c) Nile Red. (d) Prodan (6-propionyl-2-dimethylaminonaphthalene). (e) Ketocyanine dye 2,5-bis[4-N,N-dimethylaminobenzylidene]cyclopentanone-1 and its crown-substituted derivatives

labeling of proteins (dansyl, acrylodan, NBD, IAEDANS, etc. (Haugland 1996)) exhibit strong polarity-dependent shifts. They were developed for the labeling of proteins. Nowadays they are used on an extended scale, for covalent labeling to amino or sulfhydryl groups of any molecular sensors (see Section 7.3).

Meantime, the progress in the development of environment-sensitive dyes exhibiting the spectral shifts is slow. Most of them demonstrate often undesirable fluorescence quenching in some environments. The dipole moment in the excited state can be dramatically increased by substituents that leads to an intramolecular charge transfer (ICT) state and this state is often most significantly quenched. Thus, for one of the rod-shaped ICT dyes called '*fluoroprobe*' the fluorescence maximum shifts from 407 nm in non-polar hexane to 697 nm in polar acetonitrile (Hermant et al. 1990) and its maleimide derivative has been synthesized for labeling amines, thiols and other reactive groups in natural and synthetic macromolecules (Goes et al. 1988). The extreme solvent sensitivity of this molecule is connected with disadvantages, such as the dramatic fluorescence quenching in all protic media.

The other highly solvent-sensitive type of fluorescent dyes are *ketocyanines* (Kessler and Wolfbeis 1991), the terminal groups of which are strong electron donors

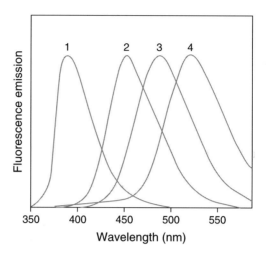

Fig. 4.7 Solvent effects on the fluorescence of Prodan, the popular wavelength-shifting fluorescent dye (Haugland 2005). The recorded and peak-normalized fluorescence spectra were recorded in cyclohexane (1), dimethylformamide (2), ethanol (3) and water (4)

in ICT and the central carbonyl – an electron acceptor, which can also serve the proton acceptor in intermolecular hydrogen bonding with an additional solvent effect (Pivovarenko et al. 2000). Without any additional sensor groups the whole ketocyanine molecule can serve as a polarity sensor. The solvent-dependent change of fluorescence color covers almost the whole visible spectrum – from blue to red. Below (see Fig. 4.8) we will discuss their properties in more detail, in relation to hydrogen bond sensing.

Ketocyanine dyes are effective tools for the studies of the molecular structure of solvent mixtures and allow observing and qualifying the effects that are usually hidden in the spectra of other fluorescence probes. The terminal groups can be functionalized. For instance, with the addition of azacrown groups they become ion chelators (Doroshenko et al. 2001) and biomembrane probes (Doroshenko et al. 2002).

4.1.5 Hydrogen Bond Responsive Dyes

The generation of the fluorescence reporting signal based on the modulation of intermolecular *hydrogen bonding* could be a very important possibility for fluorescence reporting. In Section 3.1 we indicated that the formation/disruption of these bonds is an important mechanism of fluorescence quenching. Enhancement of fluorescence in protic environments was also reported (Uchiyama et al. 2006). Here we will concentrate on the ability of some dyes to report on the formation of their intermolecular hydrogen bonding by strong spectral shifts. Surprisingly, those are the same dyes that were known for decades as 'polarity sensors'. The electron acceptor carbonyl groups present in their structures are also the proton acceptors in

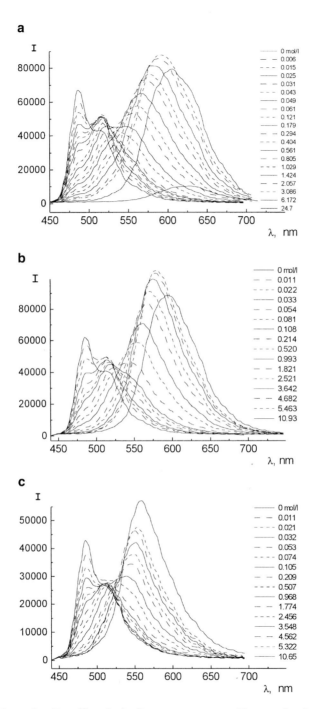

Fig. 4.8 Hydrogen bonding effects in the fluorescence spectra of ketocyanine dye. The spectra are obtained in toluene (the narrow structured spectrum at shorter wavelengths) on the addition of various concentrations of methanol (a) or butanol-1 (b) and 2-methylpropanol (c). Panel on right indicates molar concentrations of alcohols. Extreme curves refer to pure solvents. (Reproduced with permission from Pivovarenko et al. 2000)

hydrogen bonds with external proton donors, such as hydroxyls. Many highly polar solvents, such as alcohols and water, are able to form hydrogen bonds with the dyes being the proton donors. The dyes containing the H-bond proton acceptor groups, such as the carbonyl group of Prodan (Fig. 4.6), exhibit additional spectral shifts due to this bonding in the same direction as the increase of polarity, to longer wavelengths (Fig. 4.7). In Prodan almost a half of this shift is due to hydrogen bonding (Balter et al. 1988). Water molecules are among the strong proton donors, so the sensing effect can be coupled with the change in dye hydration.

Hydrogen bonding can provide a substantial influence on the energies of ground and excited states and therefore produce shifts of the absorption and emission spectra of organic dyes. Such shifts can be accompanied by the change of the shape of the fluorescence spectra, for instance, by the loss of the fine vibrational structure in the spectra of dyes that have it. *Ketocyanine* dyes can be viewed as very efficient hydrogen bond sensors, in which these changes are clearly observed. As can be seen in Fig. 4.8, an addition of millimolar amounts of alcohol to aprotic solvent produces dramatic changes in these spectra (Pivovarenko et al. 2000). In their origin is the formation of 1:1 (with K_b = 30–50 M^{-1}) and 1:2 (with K_b = 0.3–3 M^{-1}) complexes of solvent hydroxyls with central carbonyl of the dye.

4.1.6 Electric Field Sensitive (Electrochromic) Dyes

The request for sensing the *electric fields* at the distances of molecular dimensions is great. The electric field sensitive dyes can respond to electrostatic potential at interfaces, biomembrane potential, surface potential of protein molecules or nanoparticles, etc. They can be efficiently used in sensing technologies, in which the sensing effect is coupled with relocation with a nearby located charge. Their response is based on *electrochromism* (also known as the *Stark effect*), which is the phenomenon of shifts of electronic (absorption and fluorescence) spectra under the influence of electric fields. The interaction with the field changes the energies of the ground and excited states thus inducing the spectral shifts (Bublitz and Boxer 1997).

The *electrochromic dye* senses the integrated electric field at the site of its location whenever this field is applied externally in a macroscopic device or internally, on a molecular level, by a nearby charge. This allows some averaging and integration of produced field effects. The 'mesoscopic' approach, considering the fluorophore as a point dipole, electric field as a vector $\vec{F}$ that averages all the fields influencing the fluorophore and its surrounding as the medium with effective dielectric constant ε_{ef}, can be used for the description of electrochromism (Stark effect) in the simplest dipole approximation. The direction and magnitude of the shift, Δv_{obs}, is proportional to the electric field vector $\vec{F}$ and the change of the dipole moment associated with the spectroscopic transition $\Delta\vec{\mu}$:

$$h\Delta v_{obs} = -\left(1/\varepsilon_{ef}\right)\cdot\left|\Delta\vec{\mu}\right|\cdot\left|\vec{F}\right|\cos(\theta) \qquad (4.1)$$

where θ is the angle between the $\Delta\vec{\mu}$ and $\vec{F}$ vectors. It follows that in order to show maximal sensitivity to electrostatic potential, the probe dye should exhibit a substantial change of its dipole moment $\vec{\mu}$ on electronic excitation, which implies a substantial redistribution of the electronic charge density. Furthermore, it should be located in a low-polar environment (low ε_{ef}) and oriented parallel (cos $\theta = 1$) or anti-parallel (cos $\theta = -1$) to the electric field.

Styryl dyes and particularly 4-dialkylaminostyrylpyridinium derivatives with electron-donor and electron-acceptor substituents at the opposite ends of their rod-shaped aromatic conjugated moieties are among the best known electrochromic dyes (Gross et al. 1994). They exhibit the following transformation to the excited state (Fig. 4.9):

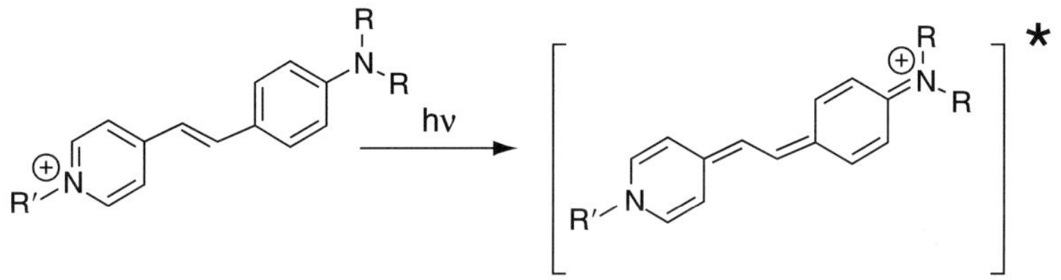

Fig. 4.9 Translocation of charge in the excited state of 4-dialkylaminostyrylpyridinium dye di-ANEPPS

The change of the dipole moment in such dyes on electronic excitation is dramatic. This change can be modulated in broad ranges by the electric field. Therefore these dyes exhibit rather significant spectral shifts that can be modulated by nearby charges. These shifts are best observed in the excitation spectra, where the intensities at the slopes of these spectra can be recorded in a ratiometric manner. The shifts in the fluorescence spectra are much smaller due to the involvement of dielectric relaxations during the fluorescence lifetime. Strong electrostatic effects can be observed in this way in biomembranes and their phospholipid analogs. In biomembranes the electrostatic fields produced by charged groups, adsorbed ions and oriented dipoles of lipids are functionally important. They can be modulated by different factors with a clear observation of the response of the incorporated dyes. Therefore, it can be expected that electrochromic dyes will find an application in sensing devices that involve charged interfaces or sensing surfaces of nanoparticles.

It was demonstrated recently that the sensing response from styryl dyes can be obtained based on the phenomenon of *second harmonic generation* (Millard et al. 2004). Observation of this nonlinear effect needs intensive optical excitation and the presence in the medium of highly organized system of polarizable nonsymmetric molecules. Styryl dyes conform to these requirements.

Thus, three major effects, *polarity*, *electric fields* and *hydrogen bonding* can induce strong shifts of fluorescence spectra. They can be used for probing the properties of different media (liquids, solids, their interfaces), see Section 10.2. Here we stress that if the sensing event is coupled with the change of reporting dye molecular environment, this event can be recorded with these dyes. The dye is needed to be incorporated into a sensor molecule in such a way that its interactions with the environment change as a result of conformational change in the sensor or due to direct contact with the target. This event can be detected as the *spectral shift*. This principle is used both in chemical sensor (Alonso et al. 2002; Moschou et al. 2006) and in biosensor (de Lorimier et al. 2002) design.

4.1.7 Supersensitive Multicolor Ratiometric Dyes

From the overview presented above, we derive that organic fluorescent dyes are unique in the ability of reporting the changes of all major types of weak noncovalent intermolecular interactions. Meantime, for many applications these effects in spectra are not strong enough and since all of them result in spectral shifts, it is often hard to identify their origin. In recent years, the efforts of the present author and his collaborators were concentrated on finding the ways for providing a dramatic amplification of the reporter signal. We started exploring the idea of coupling the modulation by these interactions of excited-state energy with a reaction producing a more dramatic spectral change.

This idea is visualized schematically in Color Plate 1. The balance in any reaction in the state of equilibrium between concentrations of reactant and product depends on the difference in free energies of correspondent states. If some, even weak, intermolecular interactions of reactant or product molecules change this energy balance, the correspondent balance in relative concentrations will also change. The stronger interacting state (possessing decreased energy) will become more populated, just in accordance with the Boltzmann law. Imagine then, that the reactant and the product are the excited-state species. The reactant state is populated upon absorption of a light quantum and the equilibrium with product is established on a time scale that is short compared to the fluorescence lifetime. If the two states emit fluorescence, the intensities of these emissions depend on the relative populations of correspondent states. Thus, we achieve a strong *amplification effect in response* – a small change in energy dramatically changes the relative intensities of the two emissions.

This simple idea requires practical realization. The requirements for this are very stringent. First, the equilibrium between the two states should be dynamic. If they interconvert rapidly, their lifetimes will be averaged and the only easy possibility to measure their relative populations is to observe relative intensities of the two fluorescence bands representing these species. Therefore, the two forms have to be highly emissive and the correspondent fluorescence bands well separated, as shown in Color Plate 2. However, this is not enough. The two species should exhibit different sensitivity to different types of intermolecular interactions in order to efficiently change the energy balance.

The designed organic dyes belonging to the family of 3-hydroxychromones (3HCs) satisfy these requirements (Demchenko 2006; Demchenko et al. 2003; Klymchenko and Demchenko 2003). Due to rigidity of their skeleton, 3HCs (and their derivatives 3-hydroxyflavones, 3HFs) are the dyes in which the observation of two separate bands in emission is not due to the presence of two or more ground-states or excited-state structural isoforms, which produce the two-band spectra in some other dyes. The second (long-wavelength) band is generated due to the *excited-state intramolecular proton transfer* (ESIPT) reaction (see Section 6.1) along the intramolecular hydrogen bond. The structural modifications of the parent fluorophore allow the variation of the spectroscopic properties of these dyes in very broad ranges.

In Fig. 4.10 the dyes of a two series, based on 3-hydroxyflavone (**1, 3m–3–3j**) and 2-benzofuranyl-chromones (**2, 4m–4–4j**) are shown. In these series of 3HC derivatives such an increase occurs in the direction **3m→3→3j** or **4m→4→4j**. By proper substitutions, the gradual increase of the excited-state dipolar moment is achieved, shifting the range of wavelength-ratiometric sensitivity to lower polarities.

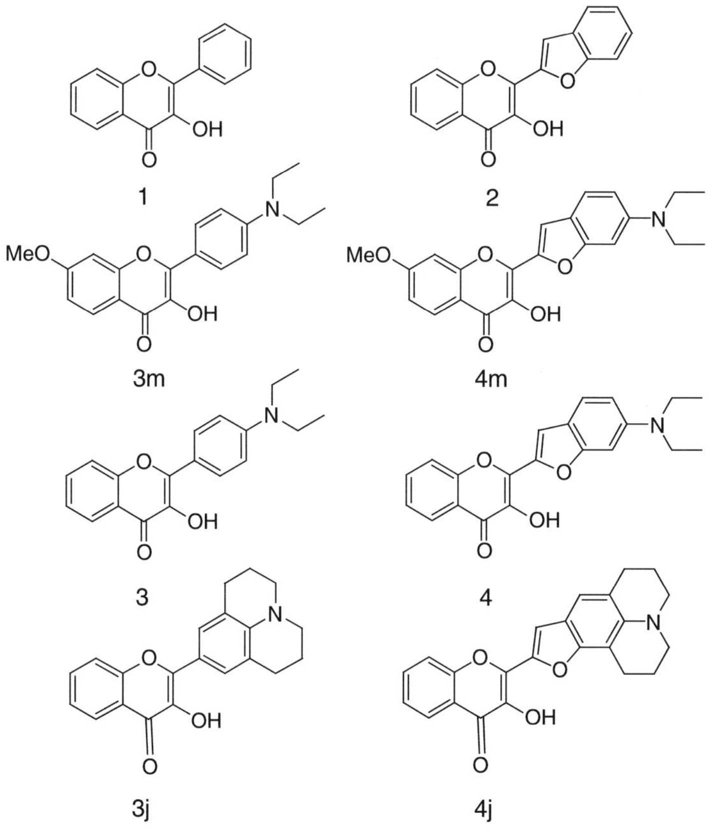

Fig. 4.10 Two series of 3-hydroxychromone dyes exhibiting a dramatic variation of wavelength-ratiometric response to the polarity of the molecular environment (Klymchenko et al. 2003b; Yesylevskyy et al. 2005)

The choice of dyes for obtaining an optimal response in particular environments can be complemented by the dyes of other series, in which furanyl (Klymchenko and Demchenko 2004), naphthofuranyl (Klymchenko et al. 2001b) or thiophenyl (M'Baye et al. 2007) groups are attached in position 2 of the chromone heterocycle. Moreover, the compounds, in which oxygen in position 1 is substituted by a group containing nitrogen (3-hydroxy quinolones) were synthesized. They possess an unique sensitivity to the variation of intermolecular hydrogen bonding (Yushchenko et al. 2006).

The basic mechanism of excited-state transformations of 4′-diethylamino-3-hydroxyflavone (dye **3** in Fig. 4.10) is presented in Fig. 4.11. For these dyes the ESIPT reaction is *much faster than the emission* and the observation of the two emission bands belonging to the initially excited normal (N^*) form and the reaction product tautomer (T^*) form is due to a rapidly established dynamic equilibrium between these forms (Shynkar et al. 2003). This is in line with the general scheme presented in Color Plate 1.

Thus, an increase of the dipole moment of the N^* form by proper substitution in the chromone ring (Klymchenko et al. 2001a; Yesylevskyy et al. 2005) allows to not only make the position of the N^* band but, especially, the *ratio of band intensities* (I_N^*/I_T^*) highly sensitive to the interaction with the environment (Fig. 4.12). These

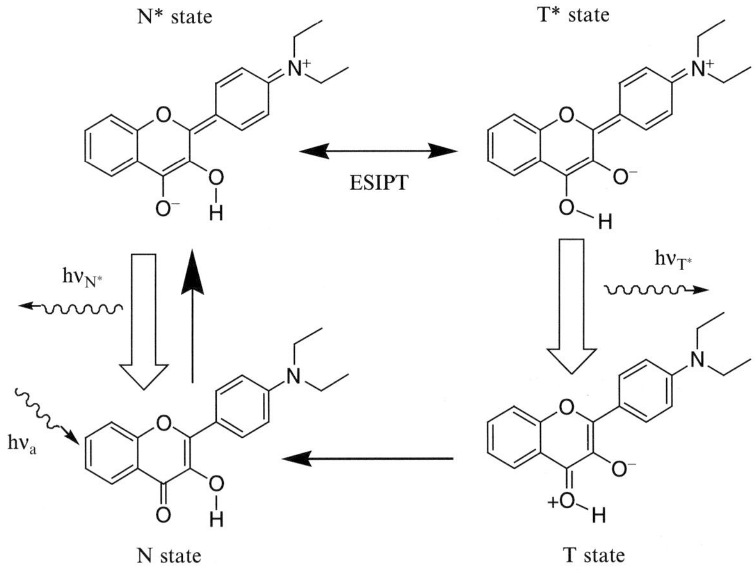

Fig. 4.11 ESIPT reaction in 3-hydroxychromone derivative 4′-dialkylamino-3-hydroxyflavone. The high excited-state dipole moment of the N^* form is provided by the dialkylamino group (electron donor) and carbonyl group (electron acceptor). In the tautomer T^* form, due to proton transfer, the charge distribution is more symmetric and the hydrogen bonding with external proton donors is different. Therefore the sensitivity of the two N^* and T^* forms to all types of intermolecular interactions is different, producing different variations in positions and intensities of correspondent fluorescence bands

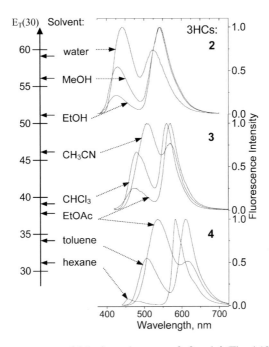

Fig. 4.12 Fluorescence spectra of 3-hydroxychromones **2**, **3** and **4** (Fig. 4.10) showing that the two-band ratiometric response of these dyes is adapted to different ranges of solvent polarities – polar, medium-polar and low-polar, respectively. This fact is illustrated by the measurements of spectra in solvents representing different polarities in accordance to polarity scale $E_T(30)$ presented as an ordinate to the left (Reproduced with permission from Klymchenko et al. 2003b)

dyes possess an absorption spectra at 400–460 nm with $\varepsilon = 30,000 \div 40,000$ and Φ up to 70%. Separation between N^* and T^* bands in emission depends upon the environment. Usually it is significant and in some cases exceeds 100 nm. Notably, due to the rapid establishment of ESIPT equilibrium, the factors that induce the quenching of one of the forms do not change the two-band ratiometric response. Photostability of these dyes is lower than that of fluorescein but quite sufficient for common spectroscopic measurements. For these dyes the switching between the two different colors of emission, blue-green and orange-red is commonly observed, in accordance with the artwork of Color Plate 2.

Subsequent studies have shown one more remarkable property of 3HC dyes. By observing the spectroscopic changes in protic environments we started to distinguish and analyze an additional H-N* band. Being characteristic for these environments only, it was attributed to intermolecular *hydrogen-bonded forms* that involve a 4-carbonyl group (Shynkar et al. 2004). A new important step was the discovery of the strong *electrochromism* of these dyes (Klymchenko and Demchenko 2002a), which opened a new dimension in their study and applications. Thus, it was shown that the positions of the two emission bands and relative variations of their intensities are highly sensitive to properties of their microenvironment such as polarity, hydrogen bonding and local electric fields. In particular:

- In *polarity sensing* they exhibit an ultrasensitive two-band ratiometric response to the polarity of the environment, up to complete switching between N^* to T^* bands in emission in a narrow range of variation of this parameter. This range can be adjusted by chemical substitutions (Klymchenko et al. 2003b) and shifted from very low (Ercelen et al. 2002) to very high (Klymchenko et al. 2004b) polarities. The origin of this effect, in line with the mechanism described in Color Plate 1, is a much stronger stabilization of the N^* state in the polar media, as a result of which this state becomes increasingly populated at equilibrium. The wavelength-ratiometric response is observed of such dramatic magnitude that only the use of a number of dyes can cover the whole polarity range. The results on a more detailed characterization of solvent polarity with the use of one of these dyes can be found in Section 10.2.
- In the *sensing of the hydrogen bonding* potential of the environment (which in biomolecular structures can also be a measure of hydration), a strong response to the presence of proton donor groups is typical for these dyes. In contrast to many other carbonyl-containing fluorescent dyes, in 3HCs no formation or disruption of H-bonds occurs in the excited state, so the probe can sense the true concentration of the H-bond partners existing in the ground state (Shynkar et al. 2004). Sensitivity to hydrogen bonding can be eliminated by proper chemical substitution that provides a sterical screening of this carbonyl group (Klymchenko et al. 2002b).
- In *electric field sensing* a very strong two-band ratiometric response to the magnitude and direction of the local electric field on a molecular scale was demonstrated (Klymchenko and Demchenko 2002) (the so-called electrochromic modulation of the ESIPT reaction) and this effect was applied for the determination of the electrostatic potential in biomembranes (Klymchenko et al. 2003a). The origin of this response is an electrochromic (due to internal Stark effect) influence on the relative energies of the N^* and T^* states.

In a general sense, the sensitivity to many different types of intermolecular interactions is not an attractive feature in sensing. But in the case of 3HCs there is a possibility for deriving in the spectral response of a single fluorophore a number of spectroscopic parameters (the wavelengths of the absorption and of two emission band maxima together with intensity ratios). Due to this feature we were able to develop an approach for distinguishing the effects of *polarity*, *electronic polarizability* and *hydrogen bonding* (Klymchenko and Demchenko 2003). The effects of the *electric field* can be evaluated by an analysis of the shifts in the excitation spectra and by comparing the data obtained for 3HF probes located in different orientations to this field (Klymchenko et al. 2003a).

Thus, the molecules of the 3HC family can respond to several, the most important, types of intermolecular interactions. The common response to these interactions, the *spectral shift*, can be amplified – being *transformed into the ratio of the two band intensities*. By chemical modifications, the 3HC dyes can be transformed into a variety of functional units. They were suggested for the probing of biomembrane structures by locating the dyes at different orientations and depths

(Klymchenko et al. 2002). More sophisticated functional dyes sense the variations of the dipole potential in membranes of living cells (Shynkar et al. 2005) and are able to detect the early steps of apoptosis (Shynkar et al. 2007). Derivatives of these dyes for covalent labeling of amino and SH-groups were also obtained (Klymchenko et al. 2004), which can be used in various types of protein and peptide biosensors. Their azacrown ether derivatives exhibit dramatic spectral responses to the binding of bivalent cations (Roshal et al. 1998, 1999). Therefore these reporters can provide a link between chemical sensing and biosensing. Some of their applications will be highlighted in different sections of this book.

Why an apparent increase in a number of bands in fluorescence spectra has led to a so remarkable result? That is, the ability of obtaining from a single type of dye a record number of valuable parameters characterizing its environment and recording them with record sensitivity. Most important in this success is the appearance of *additional channels* for acquiring and the transduction of information from the dye. Generated in an excited-state reaction, these additional channels are almost independent. One can find an analogy with the transition from black-and-white (single-channel) to color (three-channel) television. The color television rapidly occupied the whole world. Is the same future expected for multicolor dyes?

4.1.8   The Optimal FRET Pairs

Förster resonance energy transfer (FRET) is the excited-state reaction that requires the presence of two dyes. One (the donor) is initially excited and it then transfers its excitation energy to the other (acceptor) that may either emit or not emit light (Section 3.5). FRET to a nonfluorescent acceptor has the advantage of the absence in necessity to account for the direct excitation of the acceptor. If the acceptor is highly fluorescent, its direct excitation may contribute to the observed emission but the two wavelengths at the emission band maxima of the donor and of the acceptor can be used for ratiometric detection. In both cases, the best FRET donors are the dyes with high molar absorbances and quantum yields.

The needed spectroscopic matching between donor and acceptor consists in an overlap of the emission spectrum of the donor and absorption spectrum of acceptor expressed as the overlap integral J (Eq. 3.14). The data on Förster distances R_0 for typical FRET pairs are presented in Table 4.2. These data are solvent-dependent and should be used only for rough estimates.

4.1.9   Phosphorescent Dyes and the Dyes with Delayed Fluorescence

Phosphorescence is the emission from a triplet state. It usually exhibits strong Stokes shifts, long emission lifetimes (up to milliseconds and seconds) and strong

Table 4.2 Typical FRET pairs and Förster distances (compilation of literature data)

Donor	λ_a(nm)	Acceptor	λ_a(nm)	R_0 (nm)
Fluorescein	494	Tetramethylrhodamine	555	5.5
Dansyl	340	FITC	494	3.3–4.1
Pyrene	339–345	Coumarin	400	3.9
IAEDANS	336	Fluorescein	494	4.6
IAEDANS	336	IANBD	465	2.7–5.1
Carboxyfluorescein	492	Texas Red	595	5.1
Fluorescein	494	Fluorescein	494	4.4*
BODIPY FL	505	BODIPY FL	505	5.7*
AEDANS	336	Dabcyl	453	3.3**

Notes: *Homotransfer; **Transfer to nonfluorescent acceptor.

temperature-dependent quenching. Therefore in sensing techniques only phosphorescent dyes are used that display strong room-temperature phosphorescence. Using phosphorescent dyes as the sensor reporters has important advantages, since a strong Stokes shift allows avoiding the background emission and light-scattering effects, whereas the long lifetimes allow the additional possibility of removing fluorescent background by using the time-discrimination or lifetime-based techniques.

Among the dyes exhibiting room-temperature phosphorescence are eosin and erythrosine derivatives. A much larger number of dyes can become phosphorescent if the dye is protected from quenching by a rigid environment, if the heavy atoms are used to increase the population of the fluorescent state by the so-called '*Kasha heavy atom effect*' and by removing oxygen from the medium.

In view of these requirements, the best way of the application of phosphorescent dyes is their incorporation into porous sol-gel materials or into inorganic or organic polymers and the optimal detection technique is the emission lifetime (Sanchez-Barragan et al. 2006). The latter can exhibit variation by many orders of magnitude.

Since phosphorescent dyes are, as a rule, also fluorescent, the *ratio of fluorescence to phosphorescence emissions* can be used as an important reporter parameter in sensing. Commonly, the two emissions can be easily separated based on spectral positions and in time domain: fluorescence is a 'short-lived' and phosphorescence is a 'long-lived' emission. This difference can be used in oxygen sensing in solutions, since oxygen is a collisional quencher of long-lived emission (Section 10.3).

Phosphorescent complexes can also be obtained by incorporating the *noble metal ions* platinum, palladium and others into heterocyclic organic structures such as porphyrins. Their properties and applications will be discussed in the next section.

Delayed fluorescence is the emission with spectral parameters as that of fluorescence and duration as that of phosphorescence. It can be obtained by intra- and intermolecular triplet-triplet annihilation and, in rare cases, by thermal activation. Its application in sensing and imaging technology is still limited (Benniston et al. 2007).

4.1.10   Combinatorial Discovery and Improvement of Fluorescent Dyes

A strong demand for new fluorescent dyes for different applications meets with the problem of difficulty in predicting their properties, even in simple cases. From the data analyzed in this section we derive that for every particular application the dye properties have to be optimized in many directions and neither empirical knowledge nor the aid of quantum mechanics calculations may be a good help for that. There are three central obstacles to the discovery of new dyes for probing and sensing based on their rational design (Finney 2006):

(a) Complexity of underlying photophysical phenomena. The excited-state processes are different from those occurring in the ground states and not so strongly related to the chemically derived structures. They involve a multiple relaxation phenomena that are hard to predict. Because of that, fluorescence properties are hardly predictable.
(b) The still common one-molecule-at-a-time dye synthesis together with its purification and characterization has low productivity.
(c) For the evaluation of dye properties, a labor-intensive characterization needs to be used that involves spectroscopic characterization and comparison with the known standards.

The *combinatorial approaches* may offer solutions to overcome these difficulties. Optimization can be achieved by exploring the structural space around the known core fluorophore, keeping the line along the needed properties. This approach simplifies the synthetic challenge and offers many related compounds for comparative analysis. Then the products can be evaluated in the simplest way, disregarding most of the quantitative aspects. This can be done with a multi-sample plate reader or by a simple evaluation of the fluorophore function. Any interesting hits can be then characterized in more detail. A number of solid-phase and solution-phase synthetic procedures are now well developed and their application offers good results. Though the discovery of completely new fluorophores with this approach is not addressed, it allows optimizing many properties together. In addition to spectroscopy, there can be the chemical reactivity or affinity to particular structures. The library design is critical in this endeavor.

4.1.11   Prospects

It is expected that organic dyes will maintain playing the major role in the design of the response units of future chemical sensors and biosensors. They offer tremendous possibilities of choice in application to sensor technologies due to their huge number, diversity of their spectroscopic properties, possibilities of variation of these properties by chemical modifications, facile methods of their synthesis and easy spectroscopic control of their properties. They can respond to the changes of all possible fluorescence parameters (positions of spectra, intensities, anisotropies and lifetimes) and allow realizing for different basic photophysical mechanisms involving the transfers of electronic charge, proton and excited-state energy.

Computer design together with the application of rapidly progressing methods of quantum chemistry calculations will aid to the improvement of the required properties of the dyes. It is also very attractive that the straightforward and facile synthetic routes may lead to direct chemical sensors by coupling the dyes with binders for ions and other small molecules. Their labeling of proteins and nucleic acids and also of polymers and nanoparticles becomes an easy routine procedure.

The versatility and practically unlimited possibilities for chemical modifications of organic dyes allow the researcher to make the proper choice for any particular application. One has to recognize however that on a general scale, a scientifically motivated design strategy of fluorescent dyes is still lacking and their *selection is largely empirical.* Quantum chemical calculations can predict (and with difficulty) only the excited-state energies, the charge distributions and dipole moments but not the quantum yields or lifetimes. For optimizing them, an empirical rule suggests to choose the most planar molecules without rotating segments and to select among most efficient analogs. Therefore the most efficient of presently used approaches is the selection of an optimal dye within large libraries of synthetic products.

4.2   Luminescent Metal Complexes

In sensing technologies there is a strong need to overcome the limitations in lifetime values and extend to the time ranges of microseconds and milliseconds. Longer life-times should allow observing molecular motions of large molecules and particles and using simpler instrumentation for lifetime analysis. They offer a simple way to eliminate the light-scattering effects and short-living background fluorescence, thus increasing significantly the sensitivity and precision of the analysis. Two types of metal complexes can achieve the long-lifetime emission – the coordination complexes of lanthanide ions and of noble metal ions of the platinum group.

The *lanthanides* usually exist as trivalent cations. Only four of them can emit light of detectable intensity in the visible region: terbium (Tb^{+3}), europium (Eu^{+3}), samarium(Sm^{+3}) and dysprosium (Dy^{+3}) and only the first two exhibit a relatively high emission intensity. This emission comes from formally forbidden transitions of *f* electrons, which being shielded from external perturbations by the filled 5s and 5p orbits do not interact strongly with the environment. Therefore, the emission of these ions is almost insensitive to intermolecular interactions and because of that, there is no observed quenching on collisions with molecular oxygen.

This property is in strong contrast with the long-living emission of noble metal complexes (ruthenium, rhenium osmium and palladium). When excited, these complexes convert to the triplet state and emit phosphorescence. This explains the fact that they are more subjective to different quenching effects and the collisions in this state with oxygen molecules provide a strong quenching.

4.2.1 Structure and Spectroscopy of Complexes of Lanthanide Ions

Strictly speaking, lanthanide emission is not a fluorescence (it does not result from singlet-singlet transitions) and therefore a broader term '*luminescence*' has to be applied in this case. They exhibit different types of electronic transition, the most important of which are forbidden (by quantum mechanical rules) and therefore faint intra-configuration *f-f* transitions. These transitions are easily recognizable by a series of sharp peaks in the emission spectra and the major problem with the direct application of these ions is in their very low molar absorbance ($\varepsilon \sim 1\,M^{-1}\,cm^{-1}$). Therefore, for overcoming this drawback the lanthanide ion was incorporated into the chelating complex formed by the light-absorbing organic heterocycles (Fig. 4.13).

The operation of such complexes is similar to the natural *systems of photosynthesis*. The high efficiency of light absorption in these systems is provided by the so-called antenna pigments, which transfer the excitation energy to a much smaller number of reaction centers. Here also, aromatic heterocycles surrounding the ion chelating groups, called sensitizers or antennas, are excited primarily. Then they transfer the excitation energy to lanthanide, which becomes the source of a much more efficient emission than on direct excitation. This allows increasing their molar absorbance to the values above $10,000\,M^{-1}\,cm^{-1}$ (Table 4.3), which is sufficient for many applications in fluorescence sensing.

Such complexation allows solving several other problems that arise in constructing the sensors. Chelator groups stabilize the lanthanide ion and shield it from undesired interactions. They can be the sites of conjugation with other molecules, which is needed for constructing the sensors.

Synthetic strategies for the incorporation of lanthanide ions into molecular complexes are well developed. They use pre-organized macrocycles (such as cryptands and crowns), pre-disposed macrocycles (such as substituted cyclens) and prodands.

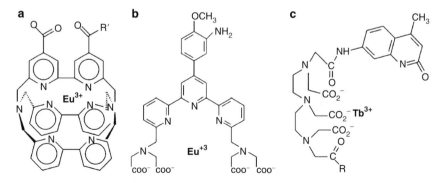

Fig. 4.13 The typical structures of chelating complexes with europium and terbium ions (Wang et al. 2006a). For cryptates (left) and terpyridine (center), the chelate and antenna are logically the same entity. The complex may be composed of a chelating group and an antenna (right)

Table 4.3 The properties of lanthanide ion chelating complexes

Parameter	Property
Size	~1–2 nm
Absorption spectra	The broadband in the range 337–420 nm, depending on the ligand
Molar absorbance, ε	Very small (~1 M^{-1} cm^{-1}) upon direct excitation. Highly increased up to ~10,000 M^{-1} cm^{-1} in chelating complex
Emission spectra	Series of strongly separated, narrow (~10 nm half-width) bands
Tunability of emission spectra	Possible only in a very limited range (several nanometers)
Stokes shifts	Large, 100–200 nm
Excited-state lifetime, τ_F	Very long, up to ~1 ms, normally insensitive to oxygen quenching; sensitivity may appear due to quenching of ligand emission
Photoluminescence quantum yield, Φ	Variable, from ~0.05 to ~0.4; dramatic quenching by coordinated water
Fundamental anisotropy, r^0	Zero
Photostability	Relatively high but not ideal
Chemical stability	Often low, due to low binding energy in chelating complex. Can be increased by selecting proper ligands
Availability of chemical modification/ functionalization	High, depending upon the ligand
Cell toxicity	Definitive data are absent

Different structures formed by the self-assembly process from complex heterocycles such as substituted benzimidazole pyridines (Bunzli 2006) can be also efficiently used. Modified calixarenes (see Section 5.1) were also suggested as ligands (Oueslati et al. 2006). This allows realizing many possibilities for modification, functionalization and incorporation into the desired system. In addition, all these complexes should be able to play their major role: to minimize the vibration-induced quenching and protect the ion from interaction with the solvent.

Consequently, lanthanide chelates exhibit a *broad excitation spectra* owing to the organic ligands and a *sharply spiked emission spectra* resulting from the lanthanide ions.

Indirect excitation of lanthanides via an antenna effect proceeds in three steps (Bunzli and Piguet 2005). First, light is absorbed by aromatic ligands surrounding the ion, leading to a singlet excited state, then a conversion occurs to its triplet (phosphorescent) state, from which the energy is transferred to an emitting lanthanide ion (Fig. 4.14). The singlet excited state of the ligand may also transfer its energy but due to the short lifetime of this state, this process is not efficient.

The *long lifetimes* (τ) of lanthanide-chelator complexes that may extend from ~100 ns to milliseconds are their primarily attraction points that offer many advantages

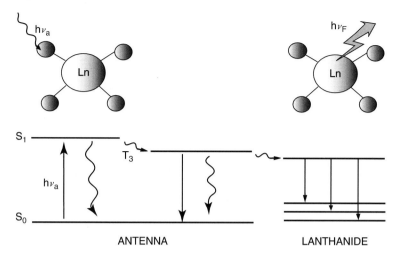

ANTENNA   LANTHANIDE

Fig. 4.14 Schematic representations of excited-state processes in lanthanide complexes. The light quanta are absorbed by aromatic ligands (antenna) and after conversion from a singlet (S_1) to a triplet (T_3) state the energy is transferred to a lanthanide ion, which emits light

(Aoki et al. 2005), including the ability to eliminate background fluorescence and light-scattering by time-resolved detection (or even by technically simple time-gated detection) by collecting the emission after the microsecond time delay. If both lanthanide complexes with τ values in the µs-ms time range and conventional fluorescent dyes with $\tau_F \sim 1$–5 ns are present in the emission, the lanthanide emission can be easily distinguished.

The *luminescence spectra* of lanthanide ions are very characteristic (Pandya et al. 2006). They are composed of several discrete *narrow-shaped bands* (less than 10 nm in width) that exhibit very strong Stokes shifts. Thus, for terbium chelates with excitation at about 340 nm a series of emission bands is observed, starting from about 500 nm and extending to longer wavelengths giving a predominantly green light emission (Fig. 4.15).

Regarding europium ions, *multiple peaks of their emission* are located in the yellow-orange spectral region (Fig. 4.16). Thus, the strong spectral separation between excitation and emission bands is the second, beside the long lifetime, attractive property of lanthanide ion complexes.

The quantum yields of lanthanide probes can be as high as 0.1–0.4 but often are much lower, ~0.05. A severe quenching by water can be observed (Bunzli and Piguet 2005). In contrast to organic dyes (in which the quenching usually occurs due to an electron transfer to traps formed in bulk water) the quenching here is due to coordinating water molecules. The latter interacting with an inner coordinating sphere dissipates energy via O–H vibrations. Thus, a special effort has been made for achieving the high emissive properties of these complexes in water (Charbonniere et al. 2007) by constructing a saturated co-ordination sphere. Finally, this allowed realizing different simple and sensitive bioassays (Yuan and Wang 2006).

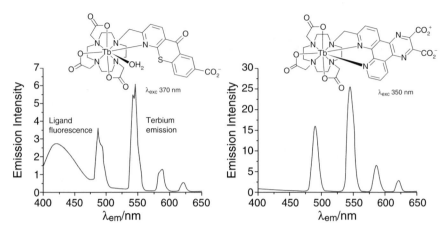

Fig. 4.15 The emission spectra for selected terbium complexes (298 K, 0.1 M NaCl), showing residual ligand fluorescence from the aza-thiaxanthone chromophore (left) (Pandya et al. 2006)

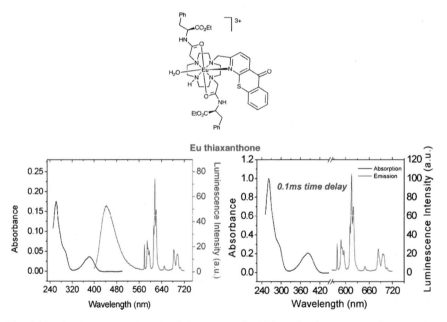

Fig. 4.16 The absorption and total emission spectra (λ_{ex} 384 nm) for the aqua–europium complex, highlighting the short-lived ligand-based fluorescence and the long-lived europium luminescence, the latter being observed uniquely when a 100 μs time delay is used (right) (Reprinted with permission from Pandya et al. 2006)

4.2.2 Lanthanide Chelates as Labels and Reference Emitters

Lanthanide chelates are very efficient labels. This is not only because of their intrinsic properties, including a high brightness and many possibilities of their covalent and non-covalent attachments but also because they do not exhibit concentration quenching and therefore can be used for loading macromolecules and nanoparticles in very high concentrations (Kokko et al. 2007).

Lanthanide complexes are ideal as reference dyes in fluorescence sensors based on two-wavelength ratiometric recording (see Section 3.1). Their broad excitation spectra allow exciting them by the same wavelength as the reporting dyes and providing a stable signal that is well separated from that of the reporter. The response of this reference should be independent of target binding and of any external perturbation and easily distinguishable from the reporter signal by emission wavelength or lifetime. The low dependence of the emission spectra on the temperature and the presence of quenchers make lanthanide chelates very attractive for this purpose.

4.2.3 Dissociation-Enhanced Lanthanide Fluoroimmunoassay (DELFIA)

The formation-dissociation of lanthanide complexes result in a dramatic (by orders of magnitude) change in their luminescence emission intensity and lifetime by the on-off switching of the antenna effect, so the idea of the application of this property in sensing is straightforward. Moreover, the chelating complexes can be both strong and weak (easily dissociating) and this fact suggests very elegant ideas in applications, particularly in immunoassays. In DELFIA and related technologies (Hemmila and Laitala 2005) the non-luminescent lanthanide Eu^{3+} and Tb^{3+} chelates are used as labels for antibodies. The complexes are formed in such a way that metal ions are loosely bound within them and can dissociate into the medium. After performing an immunoassay, in which the antibodies are bound by the antigens, the lanthanide ions dissociate from their ligands. During the next step, they are transferred to a development solution, where they form new strongly emitting complexes with the high-affinity chelators that are present in this solution. These complexes can be easily detected and quantified.

This technique and its versions are very popular both in research and in clinical analysis because of high sensitivity and reproducibility. The recent comparative studies of several sensing technologies for detecting cAMP have shown that DELFIA has the highest sensitivity (Gabriel et al. 2003). We cannot forget, meantime, that this and related techniques are the techniques based on the concept of a heterogeneous assay that needs separation and reagent addition steps. Simplification of these techniques by transforming them to homogeneous format is a great challenge.

4.2.4 Switchable Lanthanide Chelates

For producing direct sensing, one can use many possibilities to incorporate recognizing and transducing units (Fig. 4.17). Because the primary light absorption sites in lanthanide complexes are the chelating aromatic groups (sensitizers) and the emission properties are determined by lanthanide, there are basically two application

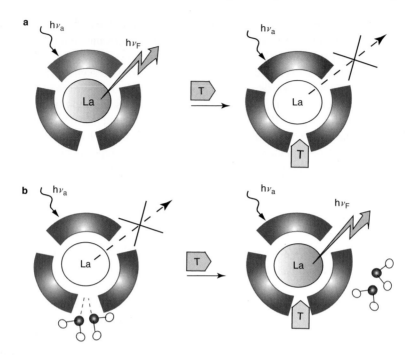

Fig. 4.17 Illustrations of the mechanisms in the background of the modulation of the lanthanide emission on a reversible binding of the target (T): (a) Influence on the excited-state transitions of the antenna via PET mechanism. (b) Direct influence on the lanthanide ion by screening the access of the lanthanide ion to water molecules

sites for inducing the changes in emission intensity and lifetime that could be used to produce a sensing response (Pandya et al. 2006):

(a) *Perturbations of sensitizers.* The reporting signal can be generated by the influence on two (singlet and triplet) excited states of sensitizer groups. The photoinduced electron transfer (PET) can be used as a well-known mechanism of excited-state quenching, in which an electron is transferred from an electron donor moiety to an acceptor moiety. With lanthanide complexes in PET the electron donor group can be covalently attached to a chelating heterocycle, so that coupling-decoupling of this quencher can provide the response in intensity of emission (Terai et al. 2006).

(b) *Perturbation of the lanthanide ion.* It can be provided via coupling/decoupling the sensor operation with its access to water molecules that are extremely potent quenchers (Bunzli and Piguet 2005). When these quenching water molecules are displaced, for instance, by reversible binding of anion, a high emission intensity is restored with a correspondent increase of lifetime. A better protection from the interaction with water can be achieved by incorporation of the whole complex to detergent micelles and to different nanostructures and this offers new possibilities to impose the sensor response.

Lanthanide complexes with ionizable groups incorporated into the antenna may respond to the change of pH not only by variation of emission intensity but also by the redistribution of intensity between emission bands. This opens the way for the construction of probes that are valuable for luminescent ratiometric pH imaging studies (Woods and Sherry 2003).

4.2.5 Transition Metal Complexes that Exhibit Phosphorescence

Long-living emission can be observed from a different type of metal-ligand complexes, formed by ruthenium (Ru), osmium (Os) and rhenium (Re) ions. Their well-described and most frequently used examples are the complexes of Ru^{2+} with ligands containing pyridine groups, such as tris-(2,2′-bipyridine) chelate, $Ru(bpy)_3^{2+}$ (Balzani et al. 2000). The structure of one of such complexes and its absorption and emission spectra are presented in Fig. 4.18.

Despite the visual similarity in the formation of ion-heterocyclic chelates, such an emission is quite different from that of the lanthanide complexes in its origin. It is the *phosphorescence* in the *metal-ligand charge transfer* (MLCT) complexes. Therefore, their spectroscopic properties differ and resemble the absorption and emission spectra of typical organic dyes. Absorption spectra are broad. For Ru^{2+} complexes they are located around 470 nm with the half-width ~100 nm. Similarly, broad emission spectra are observed at 610–670 nm, demonstrating a large Stokes shift.

The molar absorbances of these complexes are at the level of ~10,000–30,000 M^{-1} cm^{-1}, which is sufficient for different applications in sensing. Since phosphorescent emission occurs from triplet state, its duration is long and can be observed on a scale of several microseconds. This emission is highly sensitive to temperature and concentrations of oxygen in the medium.

The long lifetime of the ruthenium chelate complexes allows using them as *reference emitters* in two-channel wavelength-ratiometric sensors based on the principle of intensity sensing with the reference (Section 3.1). This was done with double labeling in glucose sensor (Ge et al. 2004) and in glutamine sensor (Lakowicz et al. 1999) based on bacterial periplasmic glucose and glutamine binding proteins correspondingly. In these sensors the environment-sensitive dye acrylodan was the responsive dye reacting to the target binding by quenching its emission. Ruthenium bis-(2,2′-bipyridyl)-1,10-phenanthroline-9-isothiocyanate served as a nonresponsive long-lived reference. This allowed recording the ratio of the emission intensities of organic dye Acrylodan and ruthenium (at 515 and 610 nm) as a reporting signal, which allows increasing substantially the accuracy of the sensor operation.

The other important application of metal chelate complexes is in fluorescence to luminescence *anisotropy sensors* and *polarization immunoassays*. These complexes exhibit large fundamental anisotropy values, allowing the examination of microsecond rotational dynamics. The problem here (as discussed in Section 3.2) is to develop luminescence probes with lifetimes that are comparable to the rotational correlation times of the antibody, the antigen and the bioconjugates that they

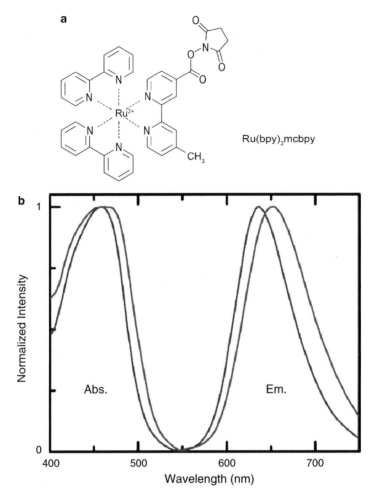

Fig. 4.18 An example of a chelating complex incorporating ruthenium ions and containing a reactive group for covalent binding. (a) The structure of Ru bis(2,2′-bipyridine)-4′-methyl-4-carboxybipyridine N-succinimidyl ester, (Ru(bpy)$_2$mcbpy). (b) Normalized absorption and emission spectra of (Ru(bpy)$_2$dcbpy) in aqueous buffer, pH 7.4 and when conjugated to IgG (slightly shifted to the red) (Reproduced with permission from Piszczek 2006)

form. Ruthenium or rhodium chelate complexes with lifetimes 1,000-fold longer than of the common dyes allow achieving this goal.

4.2.6   Metal-Chelating Porphyrins

Among organic heterocyclic compounds that can incorporate the transition metal ions with the formation of phosphorescent complexes are the porphyrins (see Fig. 5.6). They possess a tetrapyrrolic structure and may contain different side substituents. These compounds widely occur in nature forming various heme derivatives and chlorophyls. Many of these structures can be produced now by chemical synthesis. Interesting for fluorescence sensor applications are *metalloporphyrin* complexes of platinum (Pt) and palladium (Pd). They display a narrow and intensive absorption band at 360–400 nm with molar absorbance $1–5 \times 10^5$ M^{-1} cm^{-1} (Soret band) and also the absorbance in the 500–550 nm region (Q-bands). The red (600–750 nm) phosphorescence emission exhibits lifetimes in the range from 0.01 to 1 ms (Papkovsky and Ponomarev 2001). These features make these reporters useful as probes for a number of analytical applications, particularly those employing time-resolved fluorescent detection (Papkovsky and O'Riordan 2005).

A number of analytical systems have been developed that are based on the use of phosphorescent porphyrin probes (Papkovsky et al. 2000). Among them are the fiber-optic phosphorescence lifetime-based oxygen sensors constructed on the basis of hydrophobic platinum-porphyrins and also biosensors for metabolites (glucose; lactate) based on the detection of oxygen consumption by glucose oxydase. Water-soluble platinum and palladium-porphyrins can also be used as labels for ultrasensitive time-resolved phosphorescence immunoassays.

The techniques of labeling proteins and nucleic acids with porphyrins as the probes have been developed (O'Riordan et al. 2001). They can be used for the labeling of oligonucleotides that are applied in hybridization assays (Burke et al. 2003). Upon hybridization with complementary sequences bearing common organic dyes, the metal chelate probes display a strong quenching due to close proximity effects, which allows establishing separation-free hybridization assays. An attractive field of the application of phosphorescent metalloporphyrins is the stains and labels in time-resolved luminescence microscopy (Soini et al. 2002).

4.2.7 Prospects

The most important trends in the application of lanthanide chelating complexes and of charge-transfer complexes of transition metals are related by their strong Stokes shifts and long luminescence lifetimes. Both these factors strongly facilitate the recording of the output signal and, especially, its discrimination from background emission. More efforts are needed for their implementation in direct sensing technologies and the first steps have already been made.

Their applications in two-wavelength ratiometric sensors in which the ion chelates serve as references or FRET donors are attractive. Moreover, due to their long lifetimes these emitters can respond to rather slow motions of molecules and nanoparticles, which can be the mechanism of the report on the binding of targets of relatively large size. In sensor operation, different quenching effects can be used as the mechanism of on/off switching.

The applications of metal chelates as FRET donors are the most promising. Being coupled to molecules or nanoparticles with a bright but short-living fluorescence they can provide to these acceptors the property of long-lifetime emission (this issue will be discussed in Section 6.2). The time-resolved detection in this case allows achieving important benefits: not only the light-scattering and background emission are suppressed but also the short-living emission of the directly excited acceptor does not contribute to the detected signal.

4.3 Dye-Doped Nanoparticles and Dendrimers

In conventional fluorescence sensor techniques, single dye molecules are used as reporters. The application of supramolecular structures containing multiple dyes can increase the output signal and thus achieve lower limits of detection. A convenient way to create these structures is to produce nanoparticles of inorganic and organic polymer origin that are 'soaked' with dyes.

There may be different additional benefits for such an incorporation. One, of course, is the isolation of the dyes from molecular contact with the test medium and thus making them insensitive to the changes of composition in this medium. Important in this respect is the possibility to avoid degradation of dyes in aggressive media and in living cells. The dyes in solid environments usually exhibit a much higher fluorescence intensity, quantum yield and lifetime than in liquid media. This is due to a stronger restriction imposed by the solid matrix on the rotation of dye segments, intermolecular collisions and dielectric relaxations – all the factors that quench the fluorescence.

Due to a high number (hundreds to thousands) of dye molecules incorporated into one unit, the particles feature a much higher brightness than individual small molecules. The larger surface of nanoparticles may allow an attachment instead of one, of several macromolecular recognition units. The attachment to their surface of various functional groups (such as avidin, biotin, antigenic peptides, ligand-binding proteins, etc.) allows their facile bioconjugation, immobilization and incorporation into various assays.

The dyes can be located inside the nanoparticles but they can also be attached to their surfaces by physical adsorption and by a covalent attachment to surface-exposed groups. The porous nanoparticles are the most attractive in this respect, since they exhibit the highest surface-to-volume ratio and the cavities that are suitable for efficient dye incorporation. This provides the possibility for a high density of dye molecules in the particle that may allow achieving a dramatically increased brightness, by two to four orders of magnitude. All this contributes to the new rapidly growing field of nanosensors (Riu et al. 2006).

4.3.1 The Dye Concentration and Confinement Effects

The most general motivation for scientists to develop and use organic and inorganic (silica) nanoparticles doped with dyes is the increasing of the brightness by assembling a large number of dye molecules in a small unit. Meantime, being located at short distances, the dye molecules can loose the properties of independent emitters. Being excited, they start to interact in accordance with the FRET mechanism (Section 3.5). In many sensor applications, referred to throughout this book, FRET is realized between two different dyes – donor and acceptor. This case can be called *hetero-FRET*. But there can be FRET between the same dye molecules, *homo-FRET*. Their mechanism is the same; it requires an overlap of the emission spectrum of the donor with the absorption spectrum of the acceptor. So the same dye molecules can be both donors and acceptors and, if the Stokes shift is small, they can exchange energies even being at large distances, up to 5–7 nm (Runnels and Scarlata 1995). The FRET effects can be observed already in the case of high-density labeling of proteins or nucleic acids or when the dyes are confined in concentrated solutions within phospholipid vesicles (liposomes). In dye-doped nanoparticles these effects can be especially important and therefore they are discussed here.

The primary and most general influence of homo-FRET is on *emission anisotropy* (Jameson and Croney 2003). If one dye molecule absorbs light and some other being in a different orientation emits, the emission in the case of efficient homo-FRET becomes totally depolarized. High polarization can be restored only on excitation at the red edge of the absorption band but this case is not very interesting for sensor applications. Thus, polarization measurements in the case of homo-FRET are not efficient.

In the case of polarity-sensitive dyes located in rigid environments, homo-FRET may result in *shifts of spectra* to longer wavelengths (Demchenko 1986). If these dyes appear in different locations, for instance, in the core and on the surface of a particle, the emission will occur from those dyes, the absorption and emission spectra of which are shifted to the red. This is because of interplay of FRET overlap integrals: the dye with long-wavelength-shifted absorption and emission spectra is less efficient as the donor and more efficient as the acceptor.

Meantime, the most important process that can be observed at high concentrations of dye molecules is the *concentration-dependent quenching*, or 'self-quenching'. It is known that when the protein or nucleic acid molecule is modified by attaching the increasing number of fluorescent dyes, then very often its fluorescence instead of linear increase passes through the maximum and falls down. For some dyes, the self-quenching is a limiting factor in the design of devices with a high sensor and fluorophore density. This happens due to FRET to non-fluorescent traps.

Such traps can be formed by *non-fluorescent associates* of dye molecules, usually the dimers (Fig. 4.19). Association to dimers often results in static fluorescence quenching, commonly by the PET mechanism (Johansson and Cook 2003). Additional effects of quenching can be due to the formation of exciplexes (Bhattacharyya and Chowdhury 1993). These non-fluorescent species absorb light from other dyes by the homo-FRET mechanism but do not emit it (serving as light traps), which results in quenching in the whole system. Involvement of these effects depends strongly on the properties of the dye. Fluoresceins and rhodamines exhibit high homo-FRET efficiency and also self-quenching, pyrene and perylene derivatives – high homo-FRET but little self-quenching and luminescent metal complexes may not exhibit homo-FRET at all because of their very strong Stokes shifts.

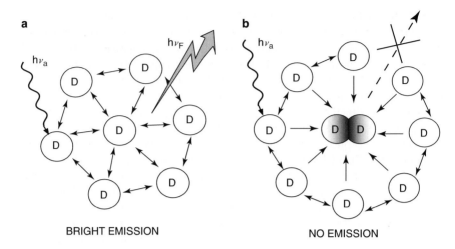

Fig. 4.19 Illustration of homo-FRET and self-quenching in dimers. (a) The dyes (D) exchange their excitation energy and are brightly emitting. (b) The presence of a non-fluorescent dye dimer (DD) acting as the trap for excitation energy results in quenching of all the dyes located at a close distance to it

The location of a number of dye molecules exhibiting both homo-FRET and hetero-FRET allows realizing many interesting possibilities that will be discussed in detail in Section 6.2. One of them is to shift gradually the emission spectrum by a proper choice of acceptor dye without altering the light absorption properties. This may be useful for multiplex assays (single excitation and several emissions for several sensors). Moreover, impregnation of nanoparticles with a number of different dyes serving as both donors and acceptors will make the particles efficient wavelength converters with emissions over the whole visible range extend to near-infrared. The other possibility is to increase the brightness of the acceptor by transferring the excited-state energy from the donors. In this case, the donors may 'light-up' the acceptor, increasing in this way the sensitivity of its response.

4.3.2 Nanoparticles Made of Organic Polymer

The techniques for preparing polymer nanoparticles are well developed. They can be obtained from a variety of synthetic polymers: polystyrene, polymethyl methacrylate, polyacrilic acid, polyvinyl chloride, etc. Usually this is done by mini-emulsion polymerization (Landfester 2006). *Mini-emulsions* are the specially formulated heterophase systems, in which the stable nanodroplets of one phase are dispersed in a second, continuous phase. Each of the nanodroplets can be considered as a nanoscopic individual batch reactor. Because of this confinement, a whole variety of reactions and processes resulting in both organic and inorganic nanoparticles can be performed and the size of the polymerized particles controlled. Surfactants are used for the stabilization of this nano-scale heterophase and an ultrasound treatment helps for dispersing these particles. With this technique, different kinds of polymerization reactions, such as radical, anionic and enzymatic polymerization, as well as polyaddition and polycondensation, can be carried out. This permits the formulation of a variety of polymers, copolymers, or hybrid particles that have not been previously synthesized in other heterophase processes. Miniemulsion technology is highly suited for the encapsulation of various organic and inorganic, solid or liquid materials, including fluorescent dyes, into the particles.

Polystyrene latexes that form stable dispersions in water and other polar media are known and have been used in research for a long time. With the incorporation of different dyes, they can be made highly fluorescent. One of the synthetic schemes involves the formation of a reaction mixture composed of styrene, cross-linker divinyl benzene and a hydrophilic co-monomer amino ethyl methacrylate hydrochloride. This allows the synthesis of 20–50 nm in size narrowly dispersed fluorescent cross-linked nanoparticles incorporating pyrene. Characterized by a steady-state fluorescence spectra, these nanoparticles show high luminescent intensity and no detected pyrene desorption. The surface of the nanoparticles can be modified by amino and amidino functional groups introduced by the co-monomer and the initiator 2, 2′-azobis (2-amidinopropane) dihydrochloride. Fluorescent dye N-(2,6-diisopropylphenyl)perylene-3,4-dicarboximide was incorporated into the co-polymer nanoparticles formulated from styrene and acrylic acid or styrene and aminoethyl methacrylate hydrochloride (Holzapfel et al. 2005). The resulting latexes were stable and showed a monodisperse size distribution within the range 100–175 nm. These functionalized fluorescent particles are commonly used as markers for living cells and the cell uptake was visualized using fluorescence microscopy.

A simple technique of *doping the nanoparticles* after their synthesis can also be applied. It involves swelling the particles in organic solvent containing the dye and then drying. Being transferred to water, these particles will retain hydrophobic dyes but a restriction exists for their use in low-polar solvents. Preparation of polymer nanoparticles that form stable dispersions in nonpolar media can also be made. Several synthetic routes for that have been described in the literature, particularly, for the incorporation of pyrene (Tauer and Ahmad 2003).

Many possibilities exist for the functionalization of the particle surface by the attachment of peptides, oligonucleotides, biotin and other molecules serving as tags or as recognition units. This allows making polymer nanoparticles highly soluble, fluorescent and reporting.

Special attention in recent literature has been paid to the production of particles that contain *luminescent metal chelates*, such as Eu(III) beta-diketonates complexes. As we observed in Section 4.2, such complexes have a long decay time, a large Stokes shift and very narrow emission bands in comparison with organic fluorescent dyes. Upon incorporation into the polymer matrix these properties are preserved (Huhtinen et al. 2005).

The template-directed polymerization that is often called 'molecular imprinting' (Section 4.7) is a very promising approach for the preparation of selective and high-affinity binding sites for different targets. Thus, the polymer nanoparticle can serve not only as a support and reporting element but also as a recognition functionality.

4.3.3 Silica-Based Nanoparticles

The labeled silica nanoparticles are attractive alternatives to organic polymer particles in serving the same purpose: confining, concentrating and making functional the dye molecules. Their advantage is a higher stability in organic solvents and at variations of pH. Compared to natural and some synthetic polymers, the silica particles are more hydrophilic and stable in biological milieu; they are not subjected to a microbial attack. Their separation and fractionation is easy by sedimentation due to the high density of silica ($1.96 \, g/cm^3$). Their surface modification and bioconjugation are also not very difficult: the reactive end groups can be added during the synthesis. The decoration of their surface with amino groups that is the most important for conjugation is described in detail (Liu et al. 2007). The surface of silica particles can be modified in different ways including the attachment of enzymes (Qhobosheane et al. 2001).

The *sol-gel matrices* are excellent supports for the immobilization of dyes and metal-chelating complexes due to their inherent mechanical and chemical stability, optical transparency and ease of formation. Fluorescent nanoparticles composed of dye-doped silica synthesized by reverse micro-emulsion allow achieving a high density of dye molecules. The particles exhibit a higher fluorescence quantum yield and a stronger resistance against photobleaching. The dye properties may improve substantially on their immobilization. The higher polarity than that of synthetic polymers and the higher stability against the change of environment conditions together with the presence of a negative charge allows efficient retention in silica matrix of fluorescent dyes, especially of those bearing a positive charge.

The developed technologies allow selecting a preferential location of reporter dyes – in the rigid core, at the porous periphery or at the particle surface (Ow et al. 2005). These possibilities can be realized not only with organic dye molecules but also with luminescent metal complexes (Lian et al. 2004; Qhobosheane et al. 2001). Sol-gel matrices are typically used to provide a microporous support matrix, in which the analyte-sensitive fluorophores are entrapped and into which smaller analyte species may diffuse and interact. The versatility of the process of producing these particles facilitates tailoring their physicochemical properties and optimizing sensor performance.

Due to the large number of dye units housed in the relatively small volume of a single nanoparticle, the dye-doped silica nanoparticles present considerable advantages with respect to the same dyes used in molecular form. In the same excitation conditions one can obtain a fluorescence signal, which is in some cases 10^4 to 10^5 times stronger (Wang et al. 2006b).

Being highly luminescent and extremely photostable, the dye-doped silica nanoparticles have been developed for ultrasensitive bioanalysis and diagnosis (Yao et al. 2006).

4.3.4   Dendrimers

Dendrimers are unique molecules of high complexity and order with a precise and gradable range of dimensions within the range of several nanometers. They are highly branched spherically shaped polymeric structures that emanate from a central core (Grayson and Frechet 2001). A typical dendrimer contains a core monomer, from which many branches stem in all directions producing a spherical shape (Fig. 4.20). It is hard to believe that a rather simple synthetic chemistry can provide the materials of such regularity and beauty.

Each branch can be further expanded by sequential branching with the addition of other layers of monomers, forming in this way the higher order dendrimer '*generations*' (Gorman and Smith 2001). Since these molecules are produced in an iterative reaction sequence, they *lack polydispersity*, so that each molecule has the same number of atoms and the same shape. Sometimes, because of the different polarity of the inner volume and surface, these structures are called 'single-molecular micelles'. The typical and most frequently used dendrimer is water-soluble poly(amidoamine), PAMAM, exposing amino groups on its surface (Table 4.4).

Dendrimers are currently attracting the interest of many scientists because of their unusual chemical and physical properties and the wide range of potential applications. Many possibilities for such applications exist in sensing technologies. Because of containing selected chemical units in predetermined sites, the dendrimers

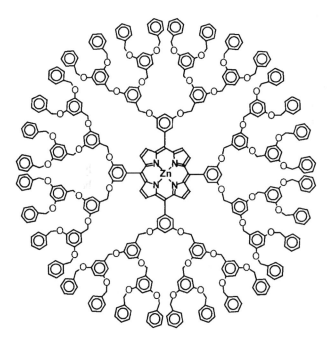

Fig. 4.20 The structure of a typical dendrimer containing porphyrin core and benzene rings forming the surface (Grayson and Frechet 2001)

Table 4.4 Physical properties of PAAMAM dendrimers (Scott et al. 2005)

Generation	Number of surface groups	Number of tertiary amines	Molecular mass (Da)	Diameter (nm)
0	4	2	517	1.5
2	16	14	3,256	2.9
4	64	62	14,215	4.5
6	256	254	58,048	6.7
8	1,024	1,022	233,383	9.7
10	4,096	4,094	934,720	13.5

are able of *encapsulating in their internal dynamic cavities* the ions or neutral molecules, including fluorescent dyes (Balzani et al. 2000; Shcharbin et al. 2007) and luminescent metal ions (Balzani et al. 2000). In addition, their surface is exposed for the covalent attachment of fluorescent dyes (Oesterling and Mullen 2007) that can make chelating complexes within the dendrimer interior.

Luminescent moiety can also serve as a dendrimer core, around which the branches can grow. Dendrimers formed on the basis of [Ru(bpy)(3)](2+) as a core, exhibit the characteristic [Ru(bpy)(3)](2+)-type luminescence that can be protected from external quenchers by the dendrimer branches and sensitized by attached dyes on the periphery of a dendrimer (Balzani et al. 2000).

The dendrimers started to find an application similar to that of organic polymers and silica. Researchers use their ability to exhibit a bright emission, which can be modulated by the joint incorporation of multiple dyes into the structure. The advantage in this case is the well-defined shape and chemical organization of these structures that allows shielding the dye molecules from their direct contacts with each other, avoiding their quenching. Meantime, dendrimeric structures are rather flexible and this allows penetration from the outside into their interior of small quencher molecules. In addition, a broad range of functionalization is possible for the introduction of target-recognizing groups into these structures.

There is also a new type of intrinsically fluorescent dendrimers, which is even more attractive. The dendrimers can be formed of *aromatic groups with a high level of their conjugation* (Balzani et al. 2003). These dendrimers resemble in many properties the conjugated polymers (Section 4.6) with the advantage of their structures being well organized and symmetric. Using modern techniques it is possible to design and synthesize the dendrimers containing a variety of light absorbing and emitting groups. Even more, the effect of light harvesting (Section 6.2) can be easily obtained with these structures. This makes them prospective transducers and reporters in sensor technologies.

4.3.5 Applications of Dye-Doped Nanoparticles in Sensing

Organic polymer and silica-based nanoparticles as well as dendrimers, with the incorporation of different dyes have found a broad range of applications in sensor technologies.

The *brightness and resistance to photobleaching* were the primary criteria for the application of organic polymer nanostructures. Thus, they were suggested for application as fluorescent tags for DNA and in the cell imaging – based on the trapping of chelated europium ions in polystyrene nanoparticles. They are used as labels in time-resolved bioassays for prostate-specific antigens (Harma et al. 2007). In DNA detection (Huhtinen et al. 2004) it was easy to achieve the detection limits on the level of ~1 pg/ml for proteins and 6.1×10^4 copies of DNA.

The application of polymer nanoparticles in ion sensing can be illustrated by the reported fluorescent sensor for the detection of Cu^{2+} ions at the nanomolar level in water (Meallet-Renault et al. 2004). In this application, the sensor was constructed by associating a BODIPY fluorophore and a copper chelator (cyclam) in ultrafine polymer nanoparticles and the response was based on the ability of the copper ion to quench the fluorescence emission.

Regarding fluorescent dye-doped silica nanoparticles, they are also actively used in different sensing technologies, particularly in *fluorescence immunoassays*. They offer an extremely high sensitivity, especially when they are impregnated with luminescent metal chelating complexes. This sensitivity is explored for cell quantitation based on antigen-antibody recognition (Wang et al. 2006b). When the luminescent silica nanoparticles were used for leukemia cell recognition, the antibody was first immobilized onto the particle through silica chemistry and the cell binding was detected by an optical microscopy imaging (Qhobosheane et al. 2001). Using these antibody-coated nanoparticles, the leukemia cells were identified easily, clearly and with a high efficiency.

The larger-size fluorescent *polymeric microspheres* are actively used in analytical assays using flow and image cytometry (Szollosi et al. 1998). In the case of the detection of fluorescent labeled molecules, they can be used both as support and recognition units with the application of homogeneous assay methodology. Single-nucleotide polymorphism (SNP) genotyping directly from human genomic DNA samples is another area of their efficient applications (Rao et al. 2003).

The dendrimers containing a high amount of fluorescent label also find application in DNA hybridization assays, improving or replacing the organic dye-based assays (Caminade et al. 2006). An immunoassay based on the composed antibody-conjugated PAMAM-dendrimer-gold Quantum Dot complex was also suggested (Triulzi et al. 2006). Due to the possibility of an easy and efficient modification of their surface, different groups recognizing small molecules and ions can be attached to this surface. Due to their membrane permeation ability (Khandare et al. 2005), PAMAM dendrimers, especially those of higher generation, have good prospects for *in vivo* biological applications.

Finally, it has to be stressed that due to the close proximity of dye molecules incorporated into these particles and extensive homo-FRET between them, the fluorescence response to the sensing event becomes collective. This is clearly seen, for instance, when the particles incorporate a dye responsive to pH. All the dyes together respond to a change of pH producing a strongly amplified effect (Kim et al. 2006).

4.3.6 Summary and Prospects

Commonly, fluorescence reporters are the molecules or their ensembles that use only one or several dye molecules for providing the response. In this section, we demonstrated the new and very attractive possibilities that appear from the incorporation of organic dyes and luminescent metal chelating complexes into and onto nanoparticles of mineral or organic origin. We could observe that, in addition to the improvement of chemical stability and photostability due to full or partial protection of dye moieties from the solvent, these composite particles attain new properties. They exhibit an extremely bright emission allowing 100–10,000-fold increase over common organic dye molecules. Their large surface areas allow multi-point chemical modifications. The chemistry of dye incorporation is determined by the host particle composition and can be the same for different dyes. In addition, the dyes in these nanostructures can locate in such distances that the FRET between them becomes efficient but self-quenching is avoided.

One can operate with these possibilities in many ways. Homo-FRET can make the particles respond cooperatively to quenching of a single dye. If this effect is not desired, it can be suppressed by choosing the dyes with a strong Stokes shift. The FRET hetero-transfer may allow increasing the brightness of the acceptor dye, to modulate the absorption and emission properties in broad ranges and to achieve the "antenna" effect. These features dramatically increase the sensitivity in analytical applications and this fact has to stimulate the development of new more efficient assays. Eventual use of these assays will depend not only on optical properties of these reporters but also on the ease with which they can be synthesized and manipulated.

The large and bulky size of polymer particles (10–100 nm) may limit some of the applications. In all cases when this factor is not crucial, being the brightest emitters, they show the brightest prospects. Such prospect is expected in cellular studies and in whole-body imaging, where loading the particles with two-photonic dyes will allow achieving extreme brightness and resolution (Bertazza et al. 2006). The protection of cells from the toxic effects of these dyes and protection of dyes from chemical and photochemical decomposition will appear as additional benefits.

4.4 Semiconductor Quantum Dots and Other Nanocrystals

Quantum Dots (QDs) are the luminescent nanoparticles with unique optical and spectroscopic properties including the ability of efficient absorption and emission of light in the visible range of spectra. They are emerging as a new class of fluorescent reporters with properties and applications that are not available with traditional organic dyes or nanocomposite structures doped with them (Costa-Fernandez et al. 2006; Grecco et al. 2004; Sapsford et al. 2006). Their novel properties have opened new possibilities in many areas including ultrasensitive chemical analysis and cellular imaging.

4.4.1 The Properties of Quantum Dots

Quantum Dots are the *semiconductor nanocrystals* in which, due to their small size, the motion of elementary electric charges (electrons) are confined in all three spatial directions. In large-scale semiconductors, the electrons are not free to move. This freedom appears on the absorption of light quanta and this may generate an electric current. When semiconductor particles become small, specific quantum effects change their properties. In absorbing and emitting light, they start to behave like atoms.

So these particles absorb light and emit fluorescence without the involvement of $\pi \rightarrow \pi^*$ electronic transitions, which is a completely different way from that of traditional dyes. These special spectroscopic properties appear due to an effect called '*quantum confinement*' that emerges when the particle size is smaller than the so-called Bohr exciton radius. In these conditions, the semiconductor valence and conduction bands break into quantized energy levels. The radiative recombination of an exciton occurs during a relatively long lifetime (>10 ns) and leads to the emission of a photon in approximately the same time range as the fluorescence of organic dyes. For displaying these new properties the particles should contain a small finite number of atoms that on excitation could form a small number (of the order of 1–100) of *conduction-band electrons* that become delocalized but remain confined. Their charge is compensated by positively charged *valence-band holes* forming together the *excitons* (the bound pairs of conduction band electrons and valence band holes). Like atoms and small molecules, these pairs generate discrete quantized energy levels, the transitions between which lead to very characteristic absorption and emission spectra.

Because these transitions are not coupled with atomic vibrations, the emission spectra of the Quantum Dots are narrow and symmetric. The separation between quantized ground-state and excited-state energy levels and therefore the positions of absorption and emission bands are directly related to the size of the QDs. The smaller they are, the fewer energy levels can be occupied by electrons and the greater is the distance between individual levels. Therefore, the decrease in their size results in blue shifts of their emission. These particles can be as small as 2–10 nm (corresponding to 10 to 50 atoms) in diameter and a total of 100 to 100,000 atoms within the QD volume. This is roughly the size of a typical globular protein.

Many QDs are the composites made of a narrow dispersed highly crystalline CdSe core overcoated with a shell of a few atomic layers of a material with a larger band gap, such as ZnS and CdS, on top of the nanocrystal core. Thus, the typical so-called core-shell structures are formed (Fig. 4.21). Such constructions are beneficial because the surface defects in the crystal structure act as temporary "traps" for the electron or hole, preventing their radiative recombination and thus reducing the quantum yield. The shell also protects surface atoms from oxidation and other chemical reactions. With the proper shell design, it becomes possible to obtain photoluminescence quantum yields close to 90% and to increase photostability by several orders of magnitude relative to conventional dyes.

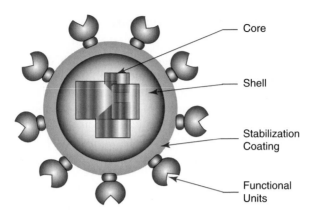

Fig. 4.21 Schematic of the overall structure of a Quantum Dot nanocrystal conjugate. The layers represent the distinct structural elements and are drawn roughly to scale

Table 4.5 The properties of Quantum Dots used in sensing technologies

Property	Quantum dots
Size	~2–10 nm
Absorption spectra	Extremely broad, extending from the first long-wavelength band continuously to the UV
Molar (particle) absorbance, ε	Very high, up to 10^7 M^{-1} cm^{-1}
Emission spectra	Narrow (~25–30 nm half-width) and symmetric, without red-side tailing
Tunability of emission spectra	Size-dependent, available in a broad range from UV to IR
Stokes shifts	Small, ~15 nm from absorption onset
Excited-state lifetime, τ_F	Usually longer than that of organic dyes, ~10–20 ns
Photoluminescence quantum yield, Φ	Variable, strongly depends on surface coating, can be ~90%
Photostability	Excellent
Chemical resistance	Excellent
Availability of chemical modification/function-alization	Rather poor
Cell toxicity	Variable, insufficiently studied
Two-photonic cross-section, α	High

The properties of semiconductor Quantum Dots are summarized in Table 4.5. The comparison with similar data, presented for organic dyes in Table 4.1, shows that the QD features are really unique. Their *absorption spectra* are extremely broad extending from a long-wavelength band far into the UV region with a gradual increase of ε, whereas their fluorescence spectra are amazingly narrow. This allows exciting fluorescence at any shorter wavelengths that are far aside from the emission band and thus avoiding light-scattering problems. Attractive is also the possibility to use for excitation the low-cost UV lamp with a broad-band filter or a cheap 405 nm diode laser. With a single light source, the QDs emitting at different wavelengths throughout the whole visible range of the spectrum can be excited. This allows multicolor detection of several targets.

The *emission spectrum* can be engineered by controlling the QD geometrical size, shape and strength of the confinement potential. The larger the particle, the redder (the more towards the long-wavelength end of the spectrum) the fluorescence. The smaller the Quantum Dot, the bluer (the more towards the short-wavelength end) it is (Color Plate 3). Depending upon the size, the core-shell CdSe-ZnS particles can change emission wavelength from 480 nm (2 nm in diameter) to 660 nm (with diameter 8 nm). Correspondingly, for the redder emitting CdTe-CdSe QDs the variation of emission wavelengths is from 650 nm (4 nm diameter) to 850 nm (diameter 8 nm) (Medintz et al. 2005).

Since the positions of fluorescence spectra can be modulated in broad ranges by the composition and size, these particles offer unique possibilities in sensing, mainly as labels and tags. They may allow the development of multi-color multi-analyte assays with a simple single-wavelength excitation based on QDs with essentially the same composition. A number of "multiplexed" assays based on this principle were suggested for DNA and RNA hybridization (Liang et al. 2005; Shepard 2006).

Quantum Dots exhibit *excellent brightness* due to high molar absorbance (reaching the values of several millions) and high quantum yields, typically of 20–70%, depending on surface coating. Such brightness cannot be easily achieved with organic dyes even with their high-density coupling with synthetic polymers and proteins or their sorption on nanoparticles. This is because of the concentration-quenching effects observed in common dyes and discussed above; in QDs these effects are absent. Excellent chemical stability and photostability are also their characteristic features.

Regarding fluorescence at two-photon excitation, QDs are dramatically superior to organic dyes (Raymo and Yildiz 2007), which makes them very prospective for cellular imaging or even imaging of the whole tissues using near-infrared excitation. Meantime, their cyto-toxicity is not studied in sufficient detail (Shiohara et al. 2004), it will definitely depend upon the properties of covering shells.

4.4.2 Stabilization and Functionalization of Quantum Dots

Naked QDs with their hydrophobic surfaces are insoluble in water and in other highly polar solvents. Their surface is not amendable for easy conjugation with functional groups needed for molecular sensing. Therefore, many efforts are being made for modifying this surface for making it friendlier to synthetic chemistry (Dubois et al. 2007).

There are several methods currently applied for QDs stabilization and functionalization (Sapsford et al. 2006):

(a) The use of surface *covalent modification chemistry* to generate $-NH_2$, $-SH$ and $-COOH$ groups on the particle surface, which could be further modified to provide attachment of target recognition molecules. This method is the most commonly used for the attachment of different recognition units, such as DNA or antibodies.

(b) The use of *coordination chemistry* that may allow, with the exposed ZnS shell, the interaction of the Zn atom with polyhistidines and of sulphur atoms with thiols. For this purpose, the peptides containing His and Cys residues can be used. They can be further modified by attaching the molecules with a specific function.

(c) Finally, the *electrostatic interactions* with oppositely charged polymer molecules can form the surface layer around the particle. Alternatively, the amphiphilic polymers can be used, which will screen the hydrophobic QD surface and expose polar groups to the solvent. A simple method for the preparation of CdSe-ZnS quantum dots that are highly fluorescent and stable in an aqueous solution was reported using calix[4]arene carboxylic acids as the surface coating agents (Jin et al. 2006).

The application of designed peptides for coating the QDs can combine all these approaches since it allows realizing all types of interactions with the core particles plus using exposed reactive groups for additional functionalization (Iyer et al. 2006). Peptides have the advantage of being easily customized and can yield all necessary functions: (a) protect the core/shell structure and maintain the original QD photophysics, (b) solubilize QDs, (c) provide a biological interface and (d) allow the incorporation of multiple functions (Sapsford et al. 2006). A number of self-assembled QD-peptide conjugates were suggested for selective intracellular delivery (Delehanty et al. 2006).

The formation of the lipid monolayer as a capping functionality can also be made for intracellular imaging and intracellular detection of nucleic acids (Dubertret et al. 2002). Thus, the aim to get the QD particles coated in such a way that would allow the application of common organic chemistry to make these particles functional is achievable and can be reached in different ways.

A relatively large size of Quantum Dots is commonly not a problem in the application of different sensing technologies. Each QD particle may carry several recognition units represented by DNA or proteins (Medintz et al. 2004). They can be integrated into microfluidic cells and other integrated devices (Sapsford et al. 2004).

4.4.3 Applications of Quantum Dots in Sensing

During the first steps of their application, the QD conjugates have started to be used as simple replacements for conventional dye conjugates when their unique performance characteristics allow achieving better results. Their high brightness helps lowering the limit of detection. The large surface areas allow simultaneous conjugation of many biomolecules to a single QD particle. Advantages conferred by this approach are obvious. One may achieve increased affinity for targets due to multipoint binding, the potential for obtaining cooperative binding and the possibility of simultaneous recognition of several targets.

Such a traditional trend can be seen in technologies that involve detection by antibodies. The streptavidin-coated QDs can be used by their incorporation into a generic three-layer sandwich approach using (a) an antibody against a specific target, (b) a biotinylated secondary antibody and (c) a streptavidin-coated QD labeling this secondary antibody (Goldman et al. 2006). The same approach can be used with ligand-binding proteins, etc. This type of application is quite similar to that of 'nonresponsive' dyes. The fluorescent particles play a passive role, in which only their presence in the system is detected. This cannot satisfy the researchers that try to find possibilities to make QDs directly sensitive to intermolecular interactions.

The problem here is an insignificant knowledge of QD properties. Very little is known on the variation of the different parameters of their emission in response to different kind of intermolecular interactions in view of various possibilities of forming their surface shells. Presently the properties of QDs are not studied in such detail as that of organic dyes and no focused attempts have been made to date to increase this response by optimizing the composition of core and coating layers. Meantime, some fragmental research in this direction shows promise. Thus, it was reported (Algar and Krull 2007) that the spectroscopic properties of CdSe/ZnS core-shell QDs change dramatically in a pH-dependent manner when linear thio-alkyl acids of variable chain lengths are used as capping ligands. The variation of luminescence of thioalkyl acid capped QDs is a complex function of the dielectric constant and electrostatic or hole-acceptor interactions with ionized ligands. These effects have to be further explored.

Sensing the target binding and the response in the change of the QD emission intensity can be connected by the photoinduced electron transfer (PET) mechanism (See Section 6.1), as depicted in Fig. 4.22. The sensor for fatty acids was suggested as a proof-of-principle for this mechanism (Aryal and Benson 2006). The electron donor is a ruthenium complex and the electron transfer occurs through the fatty acid binding protein serving as a recognition unit, so the binding results in quenching.

Attachment to CdSe-ZnS core-shell QDs of the strongly electron-donor indolyl group resulted in dramatic quenching of their fluorescence (Raymo and Tomasulo 2006). Via some conformational change this effect can be coupled with the target binding by the proper recognition unit. The QD reporting function has been suggested as one of the promising possibilities for inducing in this way (Yiidiz et al. 2006). It is based on the electrostatic adsorption of cationic quenchers on the surface of anionic quantum dots. The adsorbed quenchers suppress efficiently the emission of QDs due to the presence of a photoinduced electron transfer. Based on this mechanism a reporting signal could be generated indicating receptor-target interactions. The QD emission is restored in the presence of target receptors able to bind the quenchers and prevent the electron transfer.

Wavelength shifts of the emission spectra of Quantum Dots have been observed by several authors but have still not been studied in detail. Particularly, they were

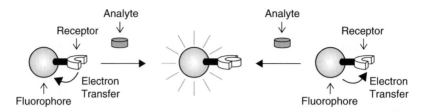

Fig. 4.22   Fluorescence sensors utilizing Quantum Dots by employing mechanisms of excited-state electron or energy transfers. The association of the receptor with a complementary analyte can suppress either an electron or an energy transfer process and switch on the fluorescence of the adjacent fluorophore (Reprinted with permission from Raymo and Yildiz 2007)

detected for CdSe/ZnS QD conjugates with antibodies and aptamers upon the binding of bacteria (Dwarakanath et al. 2004). Strong blue shifts were reported together with the enhancement of fluorescence intensity.

Quantum Dots have started to be actively used in sensing technologies based on *Förster resonance energy transfer* to or from an external agent that provides or mediates the sensor response acting as a FRET partner adsorbed or covalently attached to a semiconductor particle. Signaling may be obtained by directly detecting the luminescence from a QD and/or the conjugate (Somers et al. 2007). The approach based on the QD donor and dye acceptor has found many applications: in DNA and RNA hybridization assays, in immunoassays, in the construction of biosensors based on ligand-binding proteins and enzymes. Usually QDs are conjugated with sensor molecules and organic dyes and are used either for target labeling or for the labeling of a target analog in a competitor displacement assay.

The donor-acceptor pairs for FRET can be easily selected from QDs of different sizes and from the series of dyes possessing gradual variations of their fluorescence spectra, such as Alexa Fluor dyes. The choice between QDs emitting at 565, 605 and 655 nm as the energy donors and dyes with absorbance maxima at 594, 633, 647 and 680 nm as the energy acceptors provides a lot of possibilities for the selection of optimal donor-acceptor pairs (Nikiforov and Beechem 2006). This allows composing the systems for different homogeneous assays based on the formation of complexes avidin-biotin and antigen-antibody. Competitive binding assays can also be developed in such systems.

Quantum Dots also have the prospective for applications in immunoassays. As examples, the cancer cell marker protein her2 (Huang and Choi 2007), diphtheria toxin and tetanus toxin have been detected successfully using QD-labeled antibodies (Hoshino et al. 2005). An example of an advanced QD-based immunosensor could be the technique for detecting explosive 2,4,6-trinitrotoluene (TNT) in aqueous media (Goldman et al. 2005). The hybrid sensor consists of anti-TNT specific antibody fragments attached to a hydrophilic QD via a metal-affinity coordination. A dye-labeled TNT analogue pre-bound at the antibody binding site quenches the QD photoluminescence via proximity-induced FRET. Analysis of the data collected at the increasing of the dye-labeled analogue to QD ratios provided an insight into understanding how the antibody fragments self-assemble on the QD. An addition of soluble TNT displaces the dye-labeled analogue, eliminating FRET and resulting in a concentration-dependent recovery of QD photoluminescence.

The peculiarities of Quantum Dots as the components of hybrid composite structures that dramatically improve the properties of fluorescence reporters are discussed in Section 6.2. Due to their very high brightness but very broad excitation band their most efficient application is as FRET donors with much lesser prospects as acceptors.

4.4.4   Nanobeads with Quantum Dot Cores

For many applications the capping of QDs with small molecules such as mercaptoacetic acid is not efficient, since they can be easily degraded by hydrolysis or oxidation of the capping ligand. Therefore, the coating of silica on ZnS-CdSe QDs is a straightforward solution among many of them that were suggested (Gerion et al. 2001). Coating QDs with a silica layer can provide improved stability without affecting their optical properties. In addition to capping the single QDs with a silica monolayer, multiple QDs can also be encapsulated into silica nanobeads (Gao and Nie 2003).

The range of applications implying the potential of such nanobeads to incorporate QDs of different size and, therefore, different fluorescence spectrum, is tremendous. Each particle can incorporate several different QDs that may compose a very specific 'color barcode'. Such coding is needed in microsphere-based array technology (Section 9.4). The DNA hybridization studies demonstrate that the coding and target signals can be simultaneously read at the single-bead level, which allows applying this technique in high throughput screening and in medical diagnostics.

The encapsulation of Quantum Dots into monodisperse polystyrene latex nanobeads has also been suggested (Han et al. 2001). The swelling method was used for tagging 1.2 μm beads with different combinations of QDs of various colors to create QD barcodes. It was demonstrated that the use of six colors and ten intensity levels can theoretically encode up to one million types of biomolecules.

4.4.5   Porous Silicon and Silicon Nanoparticles

Silicon, which is the basic material of microelectronics, upon photoexcitation does not demonstrate any emission. But microcrystalline silicon that is often called porous silicon has a bright luminescence. Such a difference is attributed to the nanostructured composition of porous silicon. It possesses tiny pores that range from less than 2 nm to micrometer dimensions. Easy fabrication allows obtaining this material with a high surface to volume ratio (as much as $500 \, m^2 \, cm^{-3}$). Films made of porous silicon have found their own application in sensor technologies (Mizsei 2007; Zhang et al. 2007). Here we will concentrate on the properties of silicon nanocrystals, which can be presented as a new version of Quantum Dots. Their inertness and bio-compatibility together with high brightness has recently attracted the attention of researchers.

To date, there are only a few studies on these prospective materials (Veinot 2006). Physicists try to understand why photoluminescence cannot occur in infinite crystal. This is because Si, in contrast to CdS or CdSe, is an indirect bandgap semiconductor in which the lowest point of the conduction band and the highest point of the valence band occur at different wavevectors in reciprocal space. Hence, the optical transition is dipole-forbidden. Luminescence properties appear when the size of the structure forming the elements (particle size or the size of walls of porous silicon) is reduced to nanoscale dimensions, below the Bohr exciton radius (~5 nm). Then the quantum confinement effects emerge that reduce the conduction band to a discrete energy level and make electronic transition 'weakly' allowed (Buriak 2002).

Like other semiconductor QDs, the size-calibrated particles exhibit an emission spectra spanning the visible range but these spectra are much broader and exhibit stronger dependence on modifications of their surface. The small (1.2–2 nm) particles, commonly used in experiment, exhibit blue emission at 440–460 nm. Water-soluble derivatives of Si nanoparticles were obtained (Veinot 2006), they were used in cellular imaging.

4.4.6   Other Fluorescent Nanocrystal Structures

Recent studies have discovered many alternatives to the classical semiconductor Quantum Dots with regard to shape and chemical composition. Nanostructured materials can adopt different shapes, such as nanotubes, nanorods, nanobelts and nanowires (Huang and Choi 2007). Such variation of shapes opens a new dimension in variations of sensor properties. It was found that QDs of an elongated shape, "Quantum Rods" with diameters ranging from 2 to 10 nm and with lengths ranging from 5 to 100 nm demonstrate superior properties to that of QDs, in particular, a higher brightness (Fu et al. 2007). Quantum Rods emit linearly polarized fluorescence light and this property can be potentially used in anisotropy (polarization) assays. Whereas QDs can allow improving the sensitivity of biological detection and imaging by at least 10- to 100-fold, Quantum Rods offer even superior possibilities for that. The labeled with Quantum Rods antibodies can specifically recognize the cancer cell markers on the cellular surface.

It is interesting that the particles that exhibit quantum confinement effects and a bright emission can be obtained from pure carbon. They possess spectral features and properties comparable to surface-oxidized silicon nanocrystals and obtained the name 'Carbon Dots' (Sun et al. 2006). The materials are chemically inert but can be surface-functionalized easily with polymers containing carboxylic groups for specific or nonspecific binding with nucleic acids and proteins. Their emission is observed at wavelengths around 500 nm (when excited at 400 nm) with a quantum yield 4–10%. Its mechanism is probably the radiative recombination of excitons.

The other new discoveries in the fluorescent nanoworld involve 'nanodiamonds' (Chao et al. 2007). They are nanosized diamond powders containing atomically dispersed nitrogen atoms. The nitrogen defect centers absorb light strongly at ~560 nm and emit fluorescence efficiently at ~700 nm with the lifetime 11.6 ns and a quantum yield approaching 100%. These particles do not photobleach. Some possibilities, yet unexplored in detail, exist for changing the color of their emission.

4.4.7   Prospects

Quantum Dots are emerging as a new class of fluorescent labels with improved brightness, resistance against photobleaching and size-dependent fluorescence emission in different colors. In comparison with the dye-doped nanoparticles they are more homogeneous in size and in different properties and exhibit a much narrower emission spectra. This feature, together with their availability in different sizes, determines their prospect in multiplex assays. The new generations of QDs have a far-reaching potential for the study of intracellular processes at the single-particle level and with two-photon excitation. They offer many applications in high-resolution cellular imaging, long-term *in vivo* observation of cell trafficking, tumor targeting and diagnostics.

Compared to many years of the application of fluorescent organic dyes, the employment of Quantum Dots into sensing technologies started only recently and by trying the products that were already in use by physicists. Then the new materials better satisfying the demands appeared and very rapid progress is observed. Electron and energy transfer processes can be designed to switch the QD luminescence in response to molecular recognition events. On the basis of these operating principles, the presence of target analytes can be transduced into detectable luminescence signals.

Luminescent QD-based chemosensors are starting to be developed to detect small molecules, monitor DNA hybridization, assess protein–ligand complementarities, test enzymatic activity and probe pH distributions (Raymo and Yildiz 2007). Fundamental research on these nanoparticles continues and it is highly probable that in the near future the material nanotechnology will offer to the sensor community the light-emitting reporters with advanced properties that are very difficult to predict.

4.5   Noble Metal Nanoparticles and Molecular Clusters

Gold and *silver nanoparticles* are used with increasing activity in different sensor technologies mainly because they exhibit a strong size-dependent ability to absorb light in the visible range of spectrum and because they are extremely potent fluorescence quenchers. This ability can be used in combination with strong emitters such as dyes and fluorescent nanoparticles. Their features as fluorescence emitters are also observed. Quite recently, their ability to enhance the fluorescence of other emitters has attracted the attention of researchers. Similar properties were also described for *copper* but the latter are used to a much lesser extent. The features of noble metal materials exhibit a new dramatic change when the particle size goes down to several tens of atoms. Such metal clusters attain molecule-like properties with a strong luminescence.

4.5.1   Light Absorption and Emission by Noble Metal Nanoparticles

The light absorption by these particles is due to their *confined plasmon resonances*. For monodisperse particles of a relatively small size (~10 nm) the plasmon absorption spectra are the narrow bands located at ~410 nm for silver and at ~540 nm for gold. They are resistant to photobleaching and their absorbances are relatively large (~10^5 cm^{-1} M^{-1}).

In contrast to semiconductor Quantum Dots, the light absorption and emission of noble metal nanoparticles are related to the motion of *free electrons* (Eustis and El-Sayed 2006). The electrons can easily travel through the material and their free path in gold and silver is about 50 nm. Therefore, in particles smaller than this, no scattering is possible from the bulk and all interactions are with the surface. When the wavelength of light is much longer than the particle size, the standing resonance conditions are established and the light in resonance with the surface plasmon oscillation causes oscillation of free electrons in the metal (Fig. 4.23). This phenomenon is known as surface plasmon resonance.

The observed color originates from the strong light absorption by the metal nanoparticle when the frequency of the electromagnetic field becomes resonant with the coherent motion of the electrons (the *surface plasmon absorption*). The resonance conditions and thus the particle color can be modulated by the size and shape of the particle and also, most importantly, by interactions at the surface. It also depends on the material. Noble metals such as copper, silver and gold have a strong visible-light

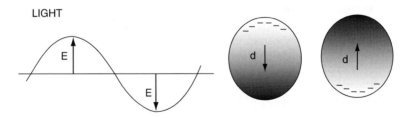

LIGHT

Fig. 4.23 Light-induced oscillations in noble metal nanoparticles giving rise to surface plasmon absorption. Electrons oscillate with the same phase as the electric field (E) oscillation produced by an incident light wave. A much heavier ionic core does not follow these oscillations, which creates an oscillating net charge difference on the particle surface. In the simplest case, this process can be approximated by dipoles oscillating with the frequency of light

plasmon resonance, whereas most of the other transition metals show only a broad and poorly resolved absorption band in the ultraviolet region.

The properties of gold, silver and copper nanocrystals of different shapes have been studied in detail (Burda et al. 2005). With particle elongation as well as with the increase of their size the spectra shift to longer wavelengths.

4.5.2 Preparation and Stabilization

Many different techniques have been developed to generate the metal nanoparticles. The two general strategies can be called 'top down' and 'bottom up' (Eustis and El-Sayed 2006). *Top down* are the methods that use for the generation of nanostructures the removal of particles from bulk material, whereas the *bottom up* methods use the reduction of ions in solutions and their assembling to generate nanostructures. For the production of small size particles the bottom up methods are the most popular. They do not need complicated instrumentation and the major problem is the necessity to stop the particle growth at a particular point to produce monodisperse particles of desired size.

Different methods use different reduction mechanisms: thermal, chemical, electrochemical, photochemical, etc. An agent should be employed to stop the growth of particles on a nanoscale level. Capping materials, such as polymers and detergents are used for that. The nanoparticles can be made in hydrogels and small metal clusters in dendrimers.

4.5.3 Gold and Silver Nanoparticles as Fluorescence Quenchers

A strong electromagnetic field generated at the surface of metal nanoparticles influences the spectroscopic properties of different dyes if they are located in close proximity. When the distance of their location is shorter than ~5 nm, then their strong interaction with the metal surface results in fluorescence quenching. This is due to photoinduced electron transfer, i.e., donation of an excited electron to the contacting surface. At longer distances the mechanism of quenching could be FRET, which requires the overlap of the dye emission spectrum with the absorption spectrum of the particle. The distance dependence of quenching was studied by many authors and a more detailed description of these mechanisms of quenching can be found in Section 6.1.

The quenching of dye fluorescence by gold or silver nanoparticles is used in a number of sensor technologies, for instance, in a single-mismatch detection in DNA with the application of gold-quenched fluorescent oligonucleotides (Dubertret et al. 2001). Here a hybrid material composed of a single-stranded DNA (ssDNA) molecule, a 1.4 nm diameter gold nanoparticle and a dye that is highly quenched by the nanoparticle through a distance-dependent process. The fluorescence of this hybrid increases by several thousand times as it binds to a complementary ssDNA. This composite molecule can be considered as a special type of molecular beacon with a sensitivity enhanced up to 100-fold. In competitive hybridization assays, the ability to detect a single mismatch is eightfold greater with this probe than with other molecular beacons.

Metal nanoparticles are also efficient in quenching the emission of Quantum Dots. As an example, the detection of avidin can be provided by a competitor assay, in which streptavidin-functionalized QDs are used. They emit intensive fluorescence. Avidin-functionalized gold nanoparticles quench fluorescence by interacting with them. Avidin in solution competes with streptavidin-labeled QDs for the biotin-gold particles and removes the quenching (Oh et al. 2005).

When the distance between the excited dye and the metal particle is more than 5 nm, the electron transfer quenching becomes a low probability but the electric field is still strong enough to influence the dye emission. In these conditions, at a distance of 10–20 nm, a strong fluorescence enhancement can be observed. We will discuss the origin of this phenomenon in Section 8.6.

4.5.4 Nanoparticles and Molecular Clusters as Emitters

Whereas the quenching abilities of noble metal nanoparticles are well-known and extensively explored, their intrinsic fluorescence is still waiting for such a level of exploration. The intrinsic fluorescence of metal nanoparticles is commonly considered to be weak but some authors have reported about their rather strong emission. In particular, nanoparticles with a shape-dependent absorption spectra and highly intense fluorescence were observed in silver (Maali et al. 2003).

Recently, the luminescent silver nanoparticles stabilized by a two-armed polymer with a crown ether core were described (Gao et al. 2004). The emission of these particles was blue-shifted and the intensity was significantly increased due to the conjugation between the crown ether embedded in the polymer and the particles.

4.5.5 Metal Nanoclusters

The metal particles with their size reduced to several or several tens of atoms attain new properties. They are already too small to have the continuous density of states necessary to support a plasmon characteristic of larger free-electron metal nanoparticles, so they start to behave like molecules. They possess *discrete electronic states* and electronic transitions between these states. The characteristic length scale here is the Fermi wavelength of an electron, which is the de Broglie wavelength of an electron at the Fermi energy. It is ~0.5 nm for gold and silver. The electronic transitions allow observing strong visible fluorescence. In view of their very small size, comparable with that of organic dyes and of their biocompatibility, these nanoclusters represent an attractive alternative to other types of nanoparticles in labeling and, especially, in imaging.

The absorption and fluorescence spectra of small gold nano-clusters are structure-less with a moderate Stokes shift. They cover the whole visible range of spectrum extending from near-UV to near-IR. Much like semiconductor Quantum Dots these clusters have the size-tunable emission bands that shift to longer wavelengths when increasing the number of atoms in the cluster (Zheng et al. 2007). Thus, for gold clusters the band maxima are at 385 nm for Au_5, 510 nm for Au_{13}, 760 nm for Au_{23} and 866 nm for Au_{31}. Their quantum yield ranges between 10% and 70% and the lifetime between 3 and 7 ns. Their photostability is high.

Gold and silver clusters can be incorporated into PAMAM dendrimers (Zheng et al. 2007), although the methods of their functionalization by incorporating into different media need further development. Recently, the labeling by gold nanoclus-ters incorporated into PAMAM dendrimers was suggested for immunoassays aimed at detecting antibodies (Triulzi et al. 2006).

The water-soluble, near-IR-emitting silver nanoclusters were encapsulated into DNA molecules (Fig. 4.24). They were shown to exhibit an extremely bright and photostable emission on the single-molecule and bulk levels. It was concluded that these species represent significant improvements over existing dyes and that these very small species may be alternatives to much larger and strongly intermittent semiconductor Quantum Dots (Vosch et al. 2007).

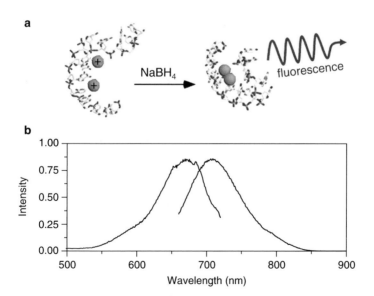

Fig. 4.24 Near-IR-emitting Ag nanoclusters. (a) Schematic of the IR-emitting DNA C12–Ag_n formation. After complexation of the C12 DNA with silver cations, the mixture is reduced with $NaBH_4$ and the near-IR-emitting Ag nanocluster is studied. (b) Normalized excitation and emission spectra of the studied species (Reproduced with permission from Vosch et al. 2007)

4.5.6 Prospects

Colloid gold particles have already been used for a long time as the stains for electron microscopy. With the decrease of their size below ~50 nm, the surface plasmon oscillations activate oscillations of free electrons in a confined space with the generation of optical absorption and emission bands. Gold and silver nanoparticles that demonstrate these properties have shown themselves as very potent fluorescence quenchers. In combination with fluorescent dyes and other types of nanoparticles this allows designing different sensor constructs with an efficient response in fluorescence intensity. In addition, the optical properties of these particles are sensitive to the association between them and modifications of their surface, which has also found application in sensing. With the further decrease of the particle size to tens of atoms they attain molecule-like properties. Such nanoclusters possess a bright emission, which, depending on the number of atoms, possess absorption and emission spectra variable over the whole visible range. Having the solution for the problems of their stabilization and functionalization they will have a strong potential for application in sensing and, especially, in imaging technologies.

4.6   Fluorescent Conjugated Polymers

Conjugated polymers are the organic synthetic polymers possessing collective effects of electronic excitation and emission (McQuade et al. 2000). They are polyunsaturated compounds with alternating single and double bonds (or aromatic units) along the polymer chain (Hoeben et al. 2005), see Fig. 4.25. The key parameter that determines the optical properties of conjugated polymers is their polarizable π-electronic system extended along the conjugated backbone. Being electron-rich and with a high level of π-electronic conjugation between each repeat unit, such a polymeric chain is able to absorb and emit visible light.

The interaction between electronic orbitals is responsible for the polymer semiconductor properties resulting in the formation of a valence band and a

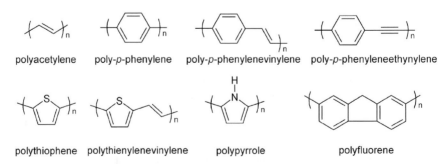

Fig. 4.25 Chemical structures of several most frequently used conjugated polymers

conduction band. In the ground state, the valence band is filled with electrons. The conduction band is devoid of electrons and it becomes populated in the excited state. It is important that these properties are collective, which results in collective optical and conducting effects. Due to their high molar absorbance ($\varepsilon \sim 10^6$ M^{-1} cm^{-1}) these polymers are very good light harvesters and strong fluorescence emitters (Achyuthan et al. 2005).

4.6.1 Structure and Spectroscopic Properties

Many conjugated polymers are fluorescent. Their fluorescence efficiency depends on delocalization and polarization of the electronic structure. The π-electrons of each monomer interact strongly with that of neighboring monomers, up to their significant delocalization over the polymer chain. This brings these polymers the property of electronic conductance and even allows considering them as 'molecular wires'. Upon excitation, there appear conduction-band electrons forming together with the positively charged nuclear sites, the collective excitations called *excitons*. Excitons can migrate along the polymer chain and, in densely packed aggregates, between the chains. The mechanism of this migration involves both the through-space dipolar coupling and the strong mixing of electronic states.

Fluorescence quenching is essentially the termination of this migration (Fig. 4.26). This produces a collective effect involving not only the site of the quencher location but the whole polymer or its significant part. The situation, in which a large number of otherwise brightly fluorescent monomeric units are simultaneously quenched by one quencher molecule is often referred to as '*superquenching*' (Thomas et al. 2007).

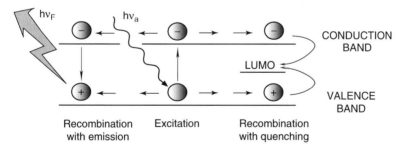

Fig. 4.26 Diagram illustrating the mechanisms of electronic excitation, exciton transport along a conjugated polymer chain, fluorescence emission and quenching by electron-hole recombination. The excitation promotes the electron to a high-energy conduction band with the generation of a positively charged 'hole' in the ground state. An electron-hole pair generated on excitation forms the exciton that can migrate along the polymer chain with an emissive transfer to a ground state. An external quencher can easily interfere into this process and induce the non-emissive recombination if its LUMO (lowest unoccupied molecular orbital) band located lower in energy than the excited state of a polymer

This effect is interesting. With the increase of the chain length the fluorescence intensity becomes higher (and therefore the effect of any external quencher stronger) until a certain limit is reached that is determined by exciton mobility and its lifetime. With a further increase of the chain length, the intensity does not grow and the quenching effect, if it exists, decreases. This means that the superquenching effect stops at some particle size of the polymer chain. Electron acceptors are commonly very strong fluorescence quenchers in these systems. Therefore, the presence of only one such group interacting with one element of the polymer chain may quench the fluorescence of the whole chain (or of its significant part if it is too long), as depicted in Fig. 4.27.

If a thin film is made of such a polymer, one may observe a dark spot on the film, due to the possibility of exciton migration both within and between the chains, which manifests the superquenching. In comparison with corresponding 'molecular excited states' that could be observed with uncoupled monomers, a greater than million-fold amplification of the sensitivity to fluorescence quenching can be achieved. Such a reaction is reversible if the polymer and the quencher are free to dissociate. 'Superenhancement' is also possible if the initially present quencher is removed in some event such as competitor displacement.

The emission of conjugated polymers depends on chain conformation, which makes them very different from other brightly luminescent nanostructures discussed above. The changes in chain conformation from extended to bent or tilted may change both the absorption and fluorescence spectra. Planar conformation extends the chain conjugation and favors the high fluorescence quantum yield providing a

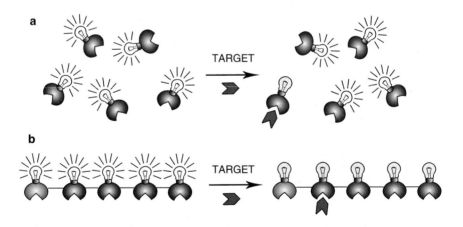

Fig. 4.27 A cartoon explaining the mechanism of 'superquenching'. (a) Molecular sensors involving monomeric dyes. Target binding producing the quenching induces a response in a single monomeric unit. Other units remain emissive and the effect is small. (b) The same sensors are linked by a conjugate polymer chain. The binding of a single target molecule to a single recognition site on a chain switches off completely the fluorescence of the whole chain

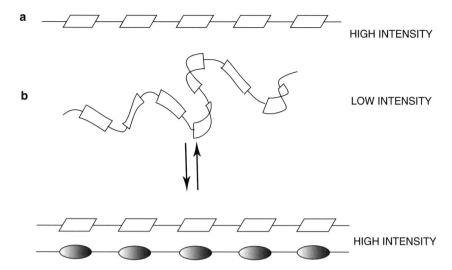

Fig. 4.28 Scheme illustrating the influence of the conformation of a conjugated polymer on its fluorescence. (a) Planar conformation. The extent of conjugation is high. Fluorescence intensity is high, so if the quenching exists it extends to a significant length. The spectra may be shifted to longer wavelengths. (b) Non-planar conformation. The extent of conjugation is low and the emission can be to a significant extent localized. The intensity is expected to be low and the quenching effect localized. If a planar conformation is re-established in a sensing event (e.g., by binding to DNA), then the high intensity can be restored

broad dynamic range for realizing strong superquenching (Fig. 4.28). This property has found application in homogeneous DNA assays in which extended conformation is achieved for a cationic conjugated polymer on its electrostatic binding with DNA. More on these assays can be found in Sections 6.3 and 10.4.

The change in chain conformation may produce also the shifts in emission wavelength. This is because the planar conformation favoring extended conjugation length allows achieving the strongest Stokes shift. The monomers in this conformation are more efficient excitation energy acceptors than the others and therefore they are the best emitters. Thus, when the energy is transferred to these sites, the spectrum shifts to longer wavelengths. When the planarity is destroyed as a result of conformation change, the spectra shift to shorter wavelengths.

4.6.2   Possibilities for Fluorescence Reporting in Sensor Design

There are many possibilities to use conjugated polymers in sensing. The first possibility is to combine the sensing effect with superquenching. The frequently used term 'superenhancement' is in fact 'super-dequenching', since the polymer does not allow any additional enhancement over its natural emission. The strong increase in emission intensity is achieved simply by the removal of the quencher and allowing displaying the natural brightness. The quencher could be the target itself, the competitor or a designed group in the sensor unit.

The other possibility that can be realized with the conjugated polymers is to modulate the extent of conjugation. In this case, the quencher can be present permanently. If the conjugation is high, its effect will extend over a large distance. Then the conformational change in the polymer chain may reduce the conjugation so that the quenching effect will be reduced with a resulting fluorescence enhancement. Moreover, with the change in conjugation the color of the fluorescence can change and this effect can be recorded in the form of a ratiometric response.

In all these cases, the greatly enhanced sensitivity of the polymer over the monomeric compound is due to facile energy migration along the polymer backbone and this can be modulated by target binding. In contrast to a sensor with a monomeric fluorescence reporter, in which every site must be occupied for complete quenching, the polymer only needs to have a small fraction of target binding sites occupied to produce complete quenching (check Fig. 4.27).

Such unique properties made conjugated polymers very popular among researchers that develop novel sensing techniques. A simple chemical structure of many of them together with many possibilities to modify it by attaching functional groups adds to this popularity. Among the frequently used polymers are polyacetylene, polythiophene, polypyrolle, polyaniline, polyparaphenylene vinylene (PPV) and others. All the core elements of these polymer chains are hydrophobic (see Fig. 4.25). The attachment of side groups allows modulating the polymer charge and hydrophobicity in broad ranges. Biosensing in water needs a good polymer solubility, which is provided by the introduction of charged groups. Such modification makes them fluorescent polyelectrolites. There is a whole range of possibilities for introducing into their structures different functional elements. For instance, the introduction of crown ester groups makes them potent ion sensors (Section 5.1) and of cationic groups – the DNA sensors (Section 10.4).

An example of the exploration of the superquenching effect can be the conjugated polymer-based probes for proteases (Wosnick et al. 2005b). A set of carboxylate-functionalized poly(phenylene ethynylene) chains was synthesized, in which the carboxylic acid groups are separated from the polymer backbone by oligo(ethylene glycol) spacer units, which made the polymer water soluble. The carboxylate groups are appended to each repeating unit allowing covalent binding of peptides that are protease substrates. The attachment of the strong fluorescence quencher dinitrophenyl group at the peptide terminal quenches the polymer emission. The presence of a protease enzyme is detected by splitting the peptide leading to disconnecting the quencher and the subsequent fluorescence enhancement.

The sensitivity of the fluorescence response to the conformation of the conjugated polymer has started to be actively used in sensing techniques involving nucleic acids. If DNA or RNA is a target or when the recognition element changes its conformation, this can be detected by electrostatic binding of a polycationic conjugated polymer. In Section 6.3 this mechanism will be analyzed in more detail.

4.6.3   Nanocomposites Based on Conjugated Polymers

Conjugated polymers being active participants in different supramolecular constructs allow realizing different signal transduction and reporting mechanisms. They endow these constructs the ability to exhibit collective effect in PET and FRET quenching. A designed combination of this polymer with another partner serving as a quencher or FRET acceptor opens many interesting possibilities in sensing. One of them is a combination of the conjugated polymer with organic dye. The dye can be either directly attached to the polymer chain or co-incorporated into the nanostructure with an efficient collective effect in FRET. The latter is very efficient when it occurs from an excited polyanionic conjugated polymer to a cationic cyanine dye (Tan et al. 2004).

Interesting is the possibility of combining different conjugated polymers in one unit. Their chemistry is well developed and involves the formation of not only the polymer films or latexes but also of well organized nanoparticles – dendritic conjugated polymers (Zhang and Feng 2007). The encapsulation of several conjugated polymers into one unit leads to new spectroscopic properties. A very small proportion of polymers that may serve as the FRET acceptors allows observing only their emission with the excitation of the major component excited as the donor (Wu et al. 2006). These experiments demonstrated that the fluorescence brightness of the blended polymer nanoparticles can be much higher than that of inorganic quantum dots and dye-loaded silica particles of similar dimensions. Different nano-scale support media such as silica can be used for the assembly of polymer-coated particles (Wosnick et al. 2005a), in which different routes of the reporter signal transduction can be activated.

An exceptional ability of gold nanoparticles to quench the cationic conjugated polymer was demonstrated (Fan et al. 2003). During fabrication, these particles can be stabilized with citrate, so they can contain the negative charge. In the bound form they produce the quenching effect in such a way that one gold nanoparticle can 'superquench' several polymer chains.

A promising field of the application of conjugated polymers is the sensing technology based on thin films, especially in gas sensing (Section 9.1) and the development of artificial 'noses' and 'tongs' (Section 12.5). The proper polymer design allows avoiding the quenching in aggregates. For instance, poly(phenylene ethynylene)s with specially designed substitutions are highly emissive in aggregated forms due to the formation of intermolecular excimers (Kim et al. 2005).

A series of non-aggregating carboxylate-functionalized poly(phenylene ethynylene)s have been synthesized for immobilization via electrostatic adsorption onto Eu^{3+}-polystyrene microspheres (Liao and Swager 2007). This system is shown to constitute a ratiometric system that measures fluorescence quenching with high fidelity.

4.6.4   Prospects

Conjugated polymers are very interesting and prospective tools for designing fluorescence sensors. A key advantage of sensors utilizing these polymers over the devices using the elements composed of monomeric dyes is their potential to exhibit collective properties. These properties are sensitive to minor perturbations produced by occupation by the target of a very small number of binding sites. Their backbone itself can emit light and demonstrate high sensitivity to different perturbations. Due to the conjugation of many fluorophores linked in the chain, they provide the effect of superquenching. Its simplest application can be the introduction of an external quencher that allows sensing the binding of a target by coupling or decoupling this dye with the polymer.

More complicated mechanisms of response involve the conformational change in a polymer producing a strong effect on its fluorescence. Due to these unique properties, the conjugated polymers demonstrate a very promising tendency of development. Their implementation as reporters is expected to be very efficient, especially in the form of thin films and in sensing of vapors. Their decoration with polar and charged groups increases dramatically the range of their applications, which is manifested by their successful application in DNA sensing.

4.7   Visible Fluorescent Proteins

Fluorescence emission from common natural proteins can be observed only in the near-UV due to the presence in their structure of amino acid residues of Tyr and Trp (Demchenko 1986). Coenzyme groups of NAD and FAD, being excited at 335–360 nm, can emit fluorescence at longer wavelengths but these compounds are abundant and do not allow identification or imaging of particular structures in the living cell. Therefore, the discovery of proteins that can be synthesized and folded inside the cell with the spontaneous formation of a fluorescent site produced revolutionary change in the techniques of cell imaging and intracellular sensing. These proteins are intrinsically fluorescent requiring no exogenously added proteins, substrates or co-factors. The green fluorescent protein (GFP) was the first in these series. Its introduction into research stimulated finding its new genetic variants and the improvement of their properties.

4.7.1   Green Fluorescent Protein (GFP) and Its Colored Variants

The green fluorescent protein (GFP), a naturally fluorescent protein derived from the jellyfish *Aequorea Victoria*, was the first in these series. This protein can be synthesized in any cell as a genetically engineered product. The mechanisms of protein fluorophore formation and engineering has been established (Miyawaki et al. 2003). Upon the folding, the fluorescent group is formed spontaneously by the cyclization and oxidation of the protein's own sequence Ser–Tyr–Gly located at positions 65–67 in the primary structure (Fig. 4.29). Due to this important feature, no additional genes or any reactants are needed to be introduced inside the cell.

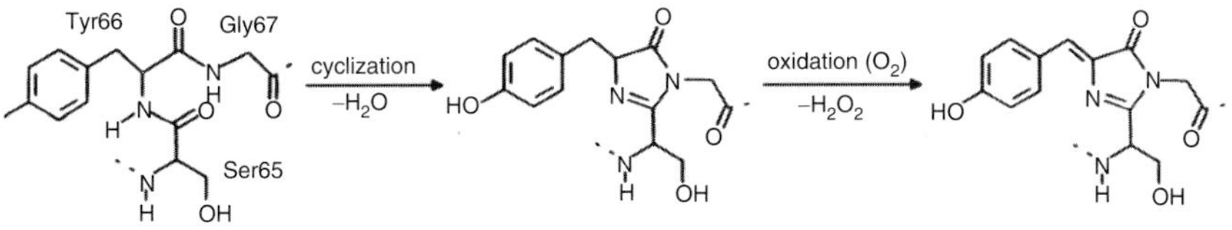

Fig. 4.29 The mechanism of intramolecular biosynthesis of GFP fluorophore. The first step is a nucleophilic attack of the amino group of Gly-67 onto the carbonyl group of Ser-65. Subsequent elimination of water results in the formation of an imidazolidinone ring. In a second step the C_α–C_β bond of Tyr-66 is oxidized to give a large delocalized π-electronic system p-hydroxyben-zylideneimidazolinone

Moreover, the GFP gene can be fused with the genes of other proteins. This allows the location of different proteins inside the cell and visualize the structures formed by them.

It is important that in GFP the fluorescent group is located deeply inside the protein molecule and has no contact with the solvent. It is located in the central helix and is surrounded by 11 β strands forming the so-called β-barrel (Color Plate 4).

Many GFP variants have been obtained by now. They differ in protein and fluorescent group structure and allow the selection of their light-absorption and fluorescence properties (Tsien 1998). Now they are available in different colors and include blue (BFP), cyan (CFP) and yellow (YFP) fluorescent proteins (Shaner et al. 2004). In recent developments, fluorescent GFP-like proteins were obtained from different sources (*Anthozoa*, *Hydrozoa* and *Copepoda* species), demonstrating the broad evolutionary and spectral diversity of this protein family.

Naturally these proteins are tetrameric and their monomeric versions have been engineered for applications that involve fusion with other proteins. An important step towards their improvement was obtaining the species that become fluorescent or change the colors of emission upon illumination at specific wavelengths (Miyawaki et al. 2005). Further molecular characterization of the structure and maturation of these proteins is in progress, aimed at providing information for the rational design of variants with desired fluorescence properties. It was found that the fluorescence quantum yield Φ values for these proteins are generally relatively good but depend on the final structure of the fluorescent group and the mutations around it. Usually Φ ranges from 0.17 for BFP to 0.79 for the wild-type of GFP but it is sensitive to pH, temperature, oxygen concentrations and other environment conditions (Giepmans et al. 2006).

Very promising in this respect is the red fluorescent protein from *Anthrozoa* corals (DsRed), mainly due to its strongly red-shifted fluorescence spectrum. With the molar absorbance at $57,000\,M^{-1}\,cm^{-1}$ and the quantum yield at 79% it competes with GFP and shows superior properties in terms of photostability and independence of spectra on pH in the range 5.0–12.0 (Shrestha and Deo 2006). Excited in the range 558–578 nm, DsRed and its mutants display fluorescence with the band maxima ranging between 583 and 605 nm. With these properties, it becomes valuable as a fluorescent tag in various cell biology applications.

4.7.2 Labeling and Sensing Applications of Fluorescent Proteins

Due to the ability to be co-synthesized as fusion proteins with many proteins that are naturally synthesized and functioning inside the living cells, GFP and its mutant variants serve as unique tools in cell imaging. Presently they are the only fully genetically encoded fluorescent probes available to the researcher. The benefits of using such markers and sensors are obvious because they obviate the issues of intracellular delivering and targeting. Mutagenic studies have given rise to diversified and optimized variants of fluorescent proteins, which have never been encountered in nature (Chudakov et al. 2005).

GFPs are used as tools in numerous applications. They include biological labeling to track and quantify individual or multiple protein species, probing to monitor protein–protein interactions and sensing to describe intracellular events and detecting different targets. However, their relatively large size (~26–28 kDa) may create problems by interference with the expression and folding of fused protein partners and affecting interactions with other molecules. The GFP mutants and circularly permuted GFP variants have been used to develop fluorescent probes that sense physiological signals such as membrane potential and concentrations of free calcium (Miyawaki et al. 2005).

4.7.3 Other Fluorescent Proteins

In addition to GFP and its relatives, there are other proteins that contain strongly fluorescent pigments. They can be isolated from species containing photosynthetic systems. These are algae *phycobiliproteins* whose function is to absorb energy of light and transfer it to chlorophyll. They possess an absorption maxima at 545–650 nm with the molar absorbance as large as $2.4 \times 10^6 \, M^{-1} \, cm^{-1}$. Their Stokes shift is not very large, 10–25 nm but the quantum yield may reach 85%. Such a high quantum yield is achieved because the protein matrix provides not only the location sites for a number of fluorescent pigments but also their separation allows them not to come into contact and not to produce self-quenching. Due to their high brightness and relatively long-wavelength emission, these proteins have found important application primarily in flow cytometry, where they can be combined with different dyes for the multicolor labeling of cells. Meantime, at present it is not possible to synthesize them in fully assembled fluorescent forms inside the living cells as is done with GFP-type proteins. This limits their application to *in vitro* testing.

Due to their high brightness and long-wavelength emission, phycobiliproteins and, in particular, *B-phycoerythrin*, are used in FRET-based immunoassays, in which they make complexes with antibodies (Petrou et al. 2002). They are also useful as acceptors in sensing technologies based on FRET. Their major disadvantage in these applications is their large molecular size (1–1.5 times the molecular mass of IgG), which results in their slow diffusion leading to a long incubation time. Presently, in this application the nanoparticles are expected to be more prospective, regarding availability, stability and price.

4.7.4 Finding Simple Analogs of Fluorescent Proteins

The discovery and very successful applications of fluorescent proteins has stimulated the search of naturally formed fluorophores of peptide origin. Fluorescent proteins have shown that spontaneous reactions between amino acid residues in a peptide chain with fluorophore formation are possible. This justifies the search for simpler peptide sequences that can give rise to self-assembling and self-catalizing reactions leading to the formation of fluorescent species. The focused search within a constructed ~500,000-member solid-phase combinatorial library of Trp-containing Ala-flanked peptide hexamers ended with successful findings (Juskowiak et al. 2004). It was demonstrated that several non-fluorescent peptides selected from the library after UV irradiation and oxidation in the air attain intensive visible fluorescence.

The other trend in this line is mimicking the environment of GFP fluorophore analogs in order to suppress the free rotation around an aryl-alkene bond that is restricted in the β-barrel protein environment. Some of the compounds synthesized along this line showed relatively bright fluorescence (Wu and Burgess 2008).

This research has strong potential to continue.

4.7.5 Prospects

An exciting new development in fluorescent imaging and sensing has appeared with the introduction of naturally fluorescent proteins. The interest towards these proteins is great, especially to those that can be fused with other proteins on a genetic level producing hybrids that can be synthesized inside the living cells. These hybrids can be unique intracellular markers and tags. Their applications as the sensors for small compounds (such as sugars and ions) have also been demonstrated successfully. Meantime, in this respect their application will probably not go far beyond intracellular research. The single cells with incorporated smart fluorescence proteins will become ideal whole-cell biosensors. Size reduction and also increasing the stability and brightness of their emission together with the development of sensor properties are the primary goals.

Sensing and Thinking 4: Which Reporter to Choose for Particular Needs?

The design and optimization of fluorescence reporter units remains a key determinant for the future success of sensing technologies. Presently, organic dyes dominate as fluorescence reporters. Their advantage is their small size and extreme diversity of spectroscopic properties. Many synthetic procedures exist for modulating these properties and for the incorporation of these dyes into different sensor constructs. Many dyes possess an extremely valuable attribute to respond by the changes in parameters of their emission to the changes in intermolecular interactions in their environment and to participate directly in molecular sensing events. However, with each step of technology development, more flexibility is being required from these dyes together with increased brightness and photostability and often the traditional dyes are simply unable to meet the necessary standards.

In the cases when the small size of the reporter unit is not important and a direct response to a target binding event is not requested or may be provided by the design of complex hybrid structures, the fluorescent nanoparticles have a great advantage. They can be much brighter and less susceptible to chemical damage and photodamage. These properties are achieved by different mechanisms. There could be a high local concentration of dyes incorporated into silica or polymeric particles together with immobilization and protection from external quenching. There can also be plasmon effects in the case of metal nanoparticles and an exciton confinement effect in the case of semiconductor Quantum Dots. In the latter cases a remarkable size-dependent variation of wavelengths of their emission can be achieved. This variation is even broader with silver and gold nanoclusters.

The weak point of all fluorescent nanoparticles is a difficulty in providing a strong response of their fluorescence to external stimuli that could be explored in sensing. We observe how these properties can be attained with proper modifications of these particles and, especially, with the design of hybrids between them and organic dyes. The number of combinations here is enormous.

Fluorescent proteins are the unique tools for cell imaging and sensing inside the living cells. Despite tremendous success in imaging, their application in sensing is rather modest and limited to several examples. The problem here is similar to that of nanoparticles. Very limited possibilities were explored to make these proteins responsive, so a strong effort is needed in this direction.

Thus, in addition to traditionally used organic dyes, new materials have been developed and successfully applied in sensing technologies. In some aspects they are already superior to the dyes and can be endowed with new functionalities, for instance, they can be made magnetic. The success in the use of these materials will depend strongly not only on their fluorescence reporter properties but also on the ease with which they can be synthesized and manipulated.

Questions and Problems

1. List the most important properties of fluorescence reporter dyes. In what directions should they be optimized for application in sensor technologies described in Chapter 3?
2. The dye brightness and quantum yield are connected with measured fluorescence intensity via an unknown factor of proportionality. What are the possibilities to evaluate them?

3. Working with very small amounts of dye, it is frequently needed to determine its concentration from absorbance and evaluate absorbance from concentration. Estimate the dye absorbance in concentrations 10^{-5} and 10^{-6} M. Estimate the dye concentrations for the measured absorbances 0.5 and 0.05. Assume that the dye molar absorbance is 40,000 M^{-1} cm^{-1} and the measurements are made in a cell with a 1 cm path length.

4. What is photobleaching? On what physical and chemical factors does it depend and what are the possibilities to reduce it? Can it find a useful application?

5. Let the reporter dye be incorporated onto the surface of a sensor molecule. On sensor-target interaction let it be located at their contact interface. Specify all possible effects that could be used to provide its quenching/enhancement response.

6. Explain the mechanisms by which the fluorescence spectra may shift on the wavelength scale. Suggest several examples of the application of this effect in sensing.

7. What are the basic requirements for a dye to be sensitive to polarity? To hydrogen bonding? To electric fields?

8. Explain the idea of signal enhancement by transforming the spectral shift into an intensity ratio of two fluorescent bands.

9. In view that the optical cross-section (that determines molar absorbance) cannot be larger than the fluorophore geometrical cross-section and that the quantum yield cannot be higher than 100%, evaluate the possibilities for the improvement of presently used organic dyes.

10. Why do lanthanide ions need to be incorporated into coordination complexes? Suggest different structures of coordination shells.

11. Suggest the sensor design in which a conjugate between a lanthanide complex and a dye-impregnated polymeric particle are used. How will this sensor work?

12. Explain the correlation between the size of Quantum Dots and the positions of their optical spectra.

13. Explain how to make the Quantum Dot responsive as the detection unit in direct sensing (avoiding sandwich assays).

14. Is it easy to achieve FRET between Quantum Dots of different sizes? What are the major problems? How are these problems resolved if, instead of Quantum Dots, other types of nanoparticles are used as FRET partners.

15. What is common and what is different between the superquenching effect in a conjugated polymer and the concentration quenching of organic dyes?

16. Explain how to protect gold and silver nanoparticles from oxidation and, simultaneously, make them amendable to functional modifications.

17. Is it easy to achieve FRET between two GFP-like fluorescent proteins of different colors? What is the difference from the Quantum Dot case?

18. Assume you have a responsive element (dye, nanoparticle) and an irresponsive element (producing a constant level of emission) and you wish to make a responsive FRET sensor out of them. To which positions (donor, acceptor) will you place the responsive and irresponsive elements?

References

Achyuthan KE, Bergstedt TS, Chen L, Jones RM, Kumaraswamy S, Kushon SA, Ley KD, Lu L, McBranch D, Mukundan H, Rininsland F, Shi X, Xia W, Whitten DG (2005) Fluorescence superquenching of conjugated polyelectrolytes: applications for biosensing and drug discovery. Journal of Materials Chemistry 15:2648–2656

Algar WR, Krull UJ (2007) Luminescence and stability of aqueous thioalkyl acid capped CdSe/ZnS quantum dots correlated to ligand ionization. A European Journal of Chemical Physics and Physical Chemistry 8:561–568

Alonso M-T, Brunet E, Juanes O, Rodríguez-Ubis J-C (2002) Synthesis and photochemical properties of new coumarin-derived ionophores and their alkaline-earth and lantanide complexes. Journal of Photochemistry and Photobiology A-Chemistry 147:113–125

Aoki S, Zulkefeli M, Shiro M, Kohsako M, Takeda K, Kimura E (2005) A luminescence sensor of inositol 1,4,5-triphosphate and its model compound by ruthenium-templated assembly of a bis(Zn2+-cyclen) complex having a 2,2′-bipyridyl linker (cyclen = 1,4,7,10-tetraazacyclododecane). Journal of the American Chemical Society 127:9129–9139

Aryal BP, Benson DE (2006) Electron donor solvent effects provide biosensing with quantum dots. Journal of the American Chemical Society 128:15986–15987

Balter A, Nowak W, Pawelkiewicz W, Kowalczyk A (1988) Some remarks on the interpretation of the spectral properties of Prodan. Chemical Physics Letters 143:565–570

Balzani V, Ceroni P, Gestermann S, Kauffmann C, Gorka M, Vogtle F (2000) Dendrimers as fluorescent sensors with signal amplification. Chemical Communications:853–854

Balzani V, Credi A, Venturi M (2003) Molecular logic circuits. A European Journal of Chemical Physics and Physical Chemistry 4:49–59

Benniston AC, Harriman A, Llarena I, Sams CA (2007) Intramolecular delayed fluorescence as a tool for imaging science: synthesis and photophysical properties of a first-generation emitter. Chemistry of Materials 19:1931–1938

Bertazza L, Celotti L, Fabbrini G, Loi MA, Maggini M, Mancin F, Marcuz S, Menna E, Muccini M, Tonellato U (2006) Cell penetrating silica nanoparticles doped with two-photon absorbing fluorophores. Tetrahedron 62:10434–10440

Bhattacharyya K, Chowdhury M (1993) Environmental and magnetic field effects on exciplex and twisted charge transfer emission. Chemical Reviews 93:507–535

Bublitz GU, Boxer SG (1997) Stark spectroscopy: applications in chemistry, biology, and materials science. Annual Review of Physical Chemistry 48:213–242

Bunzli JCG (2006) Benefiting from the unique properties of lanthanide ions. Accounts of Chemical Research 39:53–61

Bunzli JCG, Piguet C (2005) Taking advantage of luminescent lanthanide ions. Chemical Society Reviews 34:1048–1077

Burda C, Chen XB, Narayanan R, El-Sayed MA (2005) Chemistry and properties of nanocrystals of different shapes. Chemical Reviews 105:1025–1102

Buriak JM (2002) Organometallic chemistry on silicon and germanium surfaces. Chemical Reviews 102:1271–1308

Burke TJ, Loniello KR, Beebe JA, Ervin KM (2003) Development and application of fluorescence polarization assays in drug discovery. Combinatorial Chemistry & High Throughput Screening 6:183–194

Caminade AM, Padie C, Laurent R, Maraval A, Majoral JP (2006) Uses of dendrimers for DNA microarrays. Sensors 6:901–914

Chao JI, Perevedentseva E, Chung PH, Liu KK, Cheng CY, Chang CC, Cheng CL (2007) Nanometer-sized diamond particle as a probe for biolabeling. Biophysical Journal 93:2199–2208

Charbonniere LJ, Weibel N, Retailleau P, Ziessel R (2007) Relationship between the ligand structure and the luminescent properties of water-soluble lanthanide complexes containing bis(bipyridine) anionic arms. Chemistry-A European Journal 13:346–358

Chudakov DM, Lukyanov S, Lukyanov KA (2005) Fluorescent proteins as a toolkit for in vivo imaging. Trends in Biotechnology 23:605–613

Costa-Fernandez JM, Pereiro R, Sanz-Medel A (2006) The use of luminescent quantum dots for optical sensing. Trac-Trends in Analytical Chemistry 25:207–218

Davydov A (1971) Theory of molecular excitons. Plenum, New York

de Lorimier RM, Smith JJ, Dwyer MA, Looger LL, Sali KM, Paavola CD, Rizk SS, Sadigov S, Conrad DW, Loew L, Hellinga HW (2002) Construction of a fluorescent biosensor family. Protein Science 11:2655–2675

Delehanty JB, Medintz IL, Pons T, Brunel FM, Dawson PE, Mattoussi H (2006) Self-assembled quantum dot-peptide bioconjugates for selective intracellular delivery. Bioconjugate Chemistry 17:920–927

Demchenko AP (1986) Ultraviolet spectroscopy of proteins. Springer, Berlin/Heidelberg/New York

Demchenko AP (2006) Visualization and sensing of intermolecular interactions with two-color fluorescent probes. FEBS Letters 580:2951–2957

Demchenko AP, Klymchenko AS, Pivovarenko VG, Ercelen S, Duportail G, Mely Y (2003) Multiparametric color-changing fluorescence probes. Journal of Fluorescence 13:291–295

Doroshenko AO, Grigorovich AV, Posokhov EA, Pivovarenko VG, Demchenko AP, Sheiko AD (2001) Complex formation between azacrown derivatives of dibenzylidenecyclopentanone and alkali-earth metal ions. Russian Chemical Bulletin, International Edition 50:404–412

Doroshenko AO, Sychevskaya LB, Grygorovych AV, Pivovarenko VG (2002) Fluorescence probing of cell membranes with azacrown substituted ketocyanine dyes. Journal of Fluorescence 12:455–464

Dubertret B, Calame M, Libchaber AJ (2001) Single-mismatch detection using gold-quenched fluorescent oligonucleotides. Nature Biotechnology 19:365–370

Dubertret B, Skourides P, Norris DJ, Noireaux V, Brivanlou AH, Libchaber A (2002) In vivo imaging of quantum dots encapsulated in phospholipid micelles. Science 298:1759–1762

Dubois F, Mahler B, Dubertret B, Doris E, Mioskowski C (2007) A versatile strategy for quantum dot ligand exchange. Journal of the American Chemical Society 129:482–483

Dwarakanath S, Bruno JG, Shastry A, Phillips T, John A, Kumar A, Stephenson LD (2004) Quantum dot-antibody and aptamer conjugates shift fluorescence upon binding bacteria. Biochemical and Biophysical Research Communications 325:739–743

Eggeling C, Widengren J, Rigler R, Seidel CAM (1998) Photobleaching of fluorescent dyes under conditions used for single-molecule detection: evidence of two-step photolysis. Analytical Chemistry 70:2651–2659

Ercelen S, Klymchenko AS, Demchenko AP (2002) Ultrasensitive fluorescent probe for the hydrophobic range of solvent polarities. Analytica Chimica Acta 464:273–287

Eustis S, El-Sayed MA (2006) Why gold nanoparticles are more precious than pretty gold: noble metal surface plasmon resonance and its enhancement of the radiative and nonradiative properties of nanocrystals of different shapes. Chemical Society Reviews 35:209–217

Fan CH, Wang S, Hong JW, Bazan GC, Plaxco KW, Heeger AJ (2003) Beyond superquenching: hyper-efficient energy transfer from conjugated polymers to gold nanoparticles. Proceedings of the National Academy of Sciences of the United States of America 100:6297–6301

Finney NS (2006) Combinatorial discovery of fluorophores and fluorescent probes. Current Opinion in Chemical Biology 10:238–245

Fu AH, Gu WW, Boussert B, Koski K, Gerion D, Manna L, Le Gros M, Larabell CA, Alivisatos AP (2007) Semiconductor quantum rods as single molecule fluorescent biological labels. Nano Letters 7:179–182

Gabriel D, Vernier M, Pfeifer MJ, Dasen B, Tenaillon L, Bouhelal R (2003) High throughput screening technologies for direct cyclic AMP measurement. Assay and Drug Development Technologies 1:291–303

Gao JP, Fu J, Lin CK, Lin J, Han YC, Yu X, Pan CY (2004) Formation and photoluminescence of silver nanoparticles stabilized by a two-armed polymer with a crown ether core. Langmuir 20:9775–9779

Gao XH, Nie SM (2003) Doping mesoporous materials with multicolor quantum dots. Journal of Physical Chemistry B 107:11575–11578

Ge X, Tolosa L, Rao G (2004) Dual-labeled glucose binding protein for ratiometric measurements of glucose. Analytical Chemistry 76:1403–1410

Gerion D, Pinaud F, Williams SC, Parak WJ, Zanchet D, Weiss S, Alivisatos AP (2001) Synthesis and properties of biocompatible water-soluble silica-coated CdSe/ZnS semiconductor quantum dots. Journal of Physical Chemistry B 105:8861–8871

Giepmans BNG, Adams SR, Ellisman MH, Tsien RY (2006) Review - the fluorescent toolbox for assessing protein location and function. Science 312:217–224

Goes M, Lauteslager XY, Verhoeven JW, Holfstraat JW (1988) A blue excitable charge-transfer fluorescent probe and its fluorogenic derivative. European Journal of Organic Chemistry 1998:2373–2377

Goldman ER, Medintz IL, Mattoussi H (2006) Luminescent quantum dots in immunoassays. Analytical and Bioanalytical Chemistry 384:560–563

Goldman ER, Medintz IL, Whitley JL, Hayhurst A, Clapp AR, Uyeda HT, Deschamps JR, Lassman ME, Mattoussi H (2005) A hybrid quantum dot-antibody fragment fluorescence resonance energy transfer-based TNT sensor. Journal of the American Chemical Society 127:6744–6751

Gorman CB, Smith JC (2001) Structure-property relationships in dendritic encapsulation. Accounts of Chemical Research 34:60–71

Grayson SM, Frechet JMJ (2001) Convergent dendrons and dendrimers: from synthesis to applications. Chemical Reviews 101:3819–3867

Grecco HE, Lidke KA, Heintzmann R, Lidke DS, Spagnuolo C, Martinez OE, Jares-Erijman EA, Jovin TM (2004) Ensemble and single particle photophysical properties (two-photon excitation, anisotropy, FRET, lifetime, spectral conversion) of commercial quantum dots in solution and in live cells. Microscopy Research Technique 65:169–179

Gross E, Bedlack RS, Loew LM (1994) Dual-wavelength ratiometric fluorescence measurement of the membrane dipole potential. Biophysical Journal 67:208–216

Guralchuk GY, Sorokin AV, Katrunov IK, Yefimova SL, Lebedenko AN, Malyukin YV, Yarmoluk SM (2007) Specificity of cyanine dye L-21 aggregation in solutions with nucleic acids. Journal of Fluorescence 17:370–376

Han MY, Gao XH, Su JZ, Nie S (2001) Quantum-dot-tagged microbeads for multiplexed optical coding of biomolecules. Nature Biotechnology 19:631–635

Harma H, Keranen AM, Lovgren T (2007) Synthesis and characterization of europium(III) nanoparticles for time-resolved fluoroimmunoassay of prostate-specific antigen. Nanotechnology 18:075604

Haugland RP (1996) Handbook of fluorescent probes and research chemicals. Molecular Probes, Eugene, OR

Haugland RP (2005) The handbook. A guide to fluorescent probes and labeling technologies. Tenth edition. Invitrogen, Eugene, OR

Hemmila I, Laitala V (2005) Progress in lanthanides as luminescent probes. Journal of Fluorescence 15:529–542

Hermant RM, Bakker NAC, Scherer T, Krijnen B, Verhoeven JW (1990) Systematic studies of a series of highly shaped donor-acceptor systems. Journal of the American Chemical Society 112:1214–1221

Hoeben FJ, Jonkheijm P, Meijer EW, Schenning AP (2005) About supramolecular assemblies of pi-conjugated systems. Chemical Reviews 105:1491–1546

Holzapfel V, Musyanovych A, Landfester K, Lorenz MR, Mailander V (2005) Preparation of fluorescent carboxyl and amino functionalized polystyrene particles by miniemulsion polymerization as markers for cells. Macromolecular Chemistry and Physics 206:2440–2449

Hoshino A, Fujioka K, Manabe N, Yamaya S, Goto Y, Yasuhara M, Yamamoto K (2005) Simultaneous multicolor detection system of the single-molecular microbial antigen with total internal reflection fluorescence microscopy. Microbiology and Immunology 49:461–470

Huang B, Wu HK, Bhaya D, Grossman A, Granier S, Kobilka BK, Zare RN (2007) Counting low-copy number proteins in a single cell. Science 315:81–84

Huang XJ, Choi YK (2007) Chemical sensors based on nanostructured materials. Sensors and Actuators B-Chemical 122:659–671

Huhtinen P, Kivela M, Kuronen O, Hagren V, Takalo H, Tenhu H, Lovgren T, Harma H (2005) Synthesis, characterization, and application of Eu(III), Tb(III), Sm(III), and Dy(III) lanthanide chelate nanoparticle labels. Analytical Chemistry 77:2643–2648

Huhtinen P, Vaarno J, Soukka T, Lovgren T, Harma H (2004) Europium(III) nanoparticle-label-based assay for the detection of nucleic acids. Nanotechnology 15:1708–1715

Ishchenko AA, Derevyanko NA, Svidro VA (1992) Effect of polymethine-chain length on fluorescence-spectra of symmetrical cyanine dyes. Optika I Spektroskopiya 72:110–114

Iyer G, Pinaud F, Tsay J, Li JJ, Bentolila LA, Michalet X, Weiss S (2006) Peptide coated quantum dots for biological applications. IEEE Transactions on Nanobioscience 5:231–238

Jameson DM, Croney JC (2003) Fluorescence polarization: past, present and future. Combinatorial Chemistry & High Throughput Screening 6:167–173

Jin T, Fujii F, Yamada E, Nodasaka Y, Kinjo M (2006) Control of the optical properties of quantum dots by surface coating with calix n arene carboxylic acids. Journal of the American Chemical Society 128:9288–9289

Johansson MK, Cook RM (2003) Intramolecular dimers: a new design strategy for fluorescence-quenched probes. Chemistry 9:3466–3471

Juskowiak GL, Stachel SJ, Tivitmahaisoon P, Van Vranken DL (2004) Fluorogenic peptide sequences - transformation of short peptides into fluorophores under ambient photooxidative conditions. Journal of the American Chemical Society 126:550–556

Kessler MA, Wolfbeis OS (1991) New highly fluorescent ketocyanine polarity probes. Spectrochimica Acta Part A-Molecular and Biomolecular Spectroscopy 47:187–192

Khandare J, Kolhe P, Pillai O, Kannan S, Lieh-Lai M, Kannan RM (2005) Synthesis, cellular transport, and activity of polyamidoamine dendrimer-methylprednisolone conjugates. Bioconjugate Chemistry 16:330–337

Kim S, Pudavar HE, Prasad PN (2006) Dye-concentrated organically modified silica nanoparticles as a ratiometric fluorescent pH probe by one- and two-photon excitation. Chemical Communications:2071–2073

Kim Y, Bouffard J, Kooi SE, Swager TM (2005) Highly emissive conjugated polymer excimers. Journal of the American Chemical Society 127:13726–13731

Klymchenko AS, Avilov SV, Demchenko AP (2004) Resolution of Cys and Lys labeling of alpha-crystallin with site-sensitive fluorescent 3-hydroxyflavone dye. Analytical Biochemistry 329:43–57

Klymchenko AS, Demchenko AP (2002) Electrochromic modulation of excited-state intramolecular proton transfer: the new principle in design of fluorescence sensors. Journal of the American Chemical Society 124:12372–12379

Klymchenko AS, Demchenko AP (2003) Multiparametric probing of intermolecular interactions with fluorescent dye exhibiting excited state intramolecular proton transfer. Physical Chemistry Chemical Physics 5:461–468

Klymchenko AS, Demchenko AP (2004) 3-Hydroxychromone dyes exhibiting excited-state intramolecular proton transfer in water with efficient two-band fluorescence. New Journal of Chemistry 28:687–692

Klymchenko AS, Duportail G, Mely Y, Demchenko AP (2003a) Ultrasensitive two-color fluorescence probes for dipole potential in phospholipid membranes. Proceedings of the National Academy of Sciences of the United States of America 100:11219–11224

Klymchenko AS, Duportail G, Ozturk T, Pivovarenko VG, Mely Y, Demchenko AP (2002a) Novel two-band ratiometric fluorescence probes with different location and orientation in phospholipid membranes. Chemistry & Biology 9:1199–1208

Klymchenko AS, Pivovarenko VG, Ozturk T, Demchenko AP (2002b) Elimination of hydrogen bonding effect on solvatochromism of 3-hydroxyflavones. Journal of Physical Chemistry A. 107:4211–4216

Klymchenko AS, Ozturk T, Pivovarenko VG, Demchenko AP (2001a) A new 3-hydroxychromone with dramatically improved fluorescence properties. Tetrahedron Letters 42:7967–7970

Klymchenko AS, Ozturk T, Pivovarenko VG, Demchenko AP (2001b) Synthesis and spectroscopic properties of benzo- and naphthofuryl-3-hydroxychromones. Canadian Journal of Chemistry-Revue Canadienne De Chimie 79:358–363

Klymchenko AS, Pivovarenko VG, Ozturk T, Demchenko AP (2003b) Modulation of the solvent-dependent dual emission in 3-hydroxychromones by substituents. New Journal of Chemistry 27:1336–1343

Kokko L, Lovgren T, Soukka T (2007) Europium(III)-chelates embedded in nanoparticles are protected from interfering compounds present in assay media. Analytica Chimica Acta 585:17–23

Lakowicz JR, Gryczynski I, Gryczynski Z, Tolosa L, Dattelbaum JD, Rao G (1999) Polarization-based sensing with a self-referenced sample. Applied Spectroscopy 53:1149–1157

Landfester K (2006) Synthesis of colloidal particles in miniemulsions. Annual Review of Materials Research 36:231–279

Lian W, Litherland SA, Badrane H, Tan WH, Wu DH, Baker HV, Gulig PA, Lim DV, Jin SG (2004) Ultrasensitive detection of biomolecules with fluorescent dye-doped nanoparticles. Analytical Biochemistry 334:135–144

Liang RQ, Li W, Li Y, Tan CY, Li JX, Jin YX, Ruan KC (2005) An oligonucleotide microarray for microRNA expression analysis based on labeling RNA with quantum dot and nanogold probe. Nucleic Acids Research 33:e17

Liao JH, Swager TM (2007) Quantification of amplified quenching for conjugated polymer microsphere systems. Langmuir 23:112–115

Liu S, Zhang HL, Liu TC, Liu B, Cao YC, Huang ZL, Zhao YD, Luo QM (2007) Optimization of the methods for introduction of amine groups onto the silica nanoparticle surface. Journal of Biomedical Materials Research Part A 80A:752–757

Losytskyy MY, Yashchuk VM, Lukashov SS, Yarmoluk SM (2002) Davydov splitting in spectra of cyanine dye J-aggregates, formed on the polynucleotides. Journal of Fluorescence 12:109–112

Loudet A, Burgess K (2007) BODIPY dyes and their derivatives: syntheses and spectroscopic properties. Chemical Reviews 107:4891–4932

Lu ZK, Lord SJ, Wang H, Moerner WE, Twieg RJ (2006) Long-wavelength analogue of PRODAN: synthesis and properties of Anthradan, a fluorophore with a 2,6-donor-acceptor anthracene structure. Journal of Organic Chemistry 71:9651–9657

M'Baye G, Klymchenko AS, Yushchenko DA, Shvadchak VV, Ozturk T, Mely Y, Duportail G (2007) Fluorescent dyes undergoing intramolecular proton transfer with improved sensitivity to surface charge in lipid bilayers. Photochemical and Photobiological Sciences 6:71–76

Maali A, Cardinal T, Treguer-Delapierre M (2003) Intrinsic fluorescence from individual silver nanoparticles. Physica E-Low-Dimensional Systems & Nanostructures 17:559–560

McQuade DT, Pullen AE, Swager TM (2000) Conjugated polymer-based chemical sensors. Chemical Reviews 100:2537–2574

Meallet-Renault R, Pansu R, Amigoni-Gerbier S, Larpent C (2004) Metal-chelating nanoparticles as selective fluorescent sensor for Cu^{2+}. Chemical Communications:2344–2345

Medintz IL, Trammell SA, Mattoussi H, Mauro JM (2004) Reversible modulation of quantum dot photoluminescence using a protein-bound photochromic fluorescence resonance energy transfer acceptor. Journal of the American Chemical Society 126:30–31

Medintz IL, Uyeda HT, Goldman ER, Mattoussi H (2005) Quantum dot bioconjugates for imaging, labelling and sensing. Nature Materials 4:435–446

Millard AC, Jin L, Wei MD, Wuskell JP, Lewis A, Loew LM (2004) Sensitivity of second harmonic generation from styryl dyes to transmembrane potential. Biophysical Journal 86:1169–1176

Miyawaki A, Nagai T, Mizuno H (2003) Mechanisms of protein fluorophore formation and engineering. Current Opinion in Chemical Biology 7:557–562

Miyawaki A, Nagai T, Mizuno H (2005) Engineering fluorescent proteins. Microscopy techniques. Advances in Biochemical Engineering and Biotechnology 95:1–15

Mizsei J (2007) Gas sensor applications of porous Si layers. Thin Solid Films 515:8310–8315

Moschou EA, Bachas LG, Daunert S, Deo SK (2006) Hinge-motion binding proteins: unraveling their analytical potential. Analytical Chemistry 78:6692–6700

Mujumdar RB, Ernst LA, Mujumdar SR, Lewis CJ, Waggoner AS (1993) Cyanine dye labeling reagents - sulfoindocyanine succinimidyl esters. Bioconjugate Chemistry 4:105–111

Nikiforov TT, Beechem JM (2006) Development of homogeneous binding assays based on fluorescence resonance energy transfer between quantum dots and Alexa Fluor fluorophores. Analytical Biochemistry 357:68–76

O'Riordan TC, Soini AE, Papkovsky DB (2001) Monofunctional derivatives of coproporphyrins for phosphorescent labeling of proteins and binding assays. Analytical Biochemistry 290:366–375

Oesterling I, Mullen K (2007) Multichromophoric polyphenylene dendrimers: toward brilliant light emitters with an increased number of fluorophores. Journal of the American Chemical Society 129:4595–4605

Oh E, Hong MY, Lee D, Nam SH, Yoon HC, Kim HS (2005) Inhibition assay of biomolecules based on fluorescence resonance energy transfer (FRET) between quantum dots and gold nanoparticles. Journal of the American Chemical Society 127:3270–3271

Oueslati I, Ferreira FAS, Carlos LD, Baleizao C, Berberan-Santos MN, de Castro B, Vicens J, Pischel U (2006) Calix 4 azacrowns as novel molecular scaffolds for the generation of visible and near-infrared lanthanide luminescence. Inorganic Chemistry 45:2652–2660

Ow H, Larson DR, Srivastava M, Baird BA, Webb WW, Wiesner U (2005) Bright and stable core-shell fluorescent silica nanoparticles. Nano Letters 5:113–117

Panchuk-Voloshina N, Haugland RP, Bishop-Stewart J, Bhalgat MK, Millard PJ, Mao F, Leung WY, Haugland RP (1999) Alexa dyes, a series of new fluorescent dyes that yield exceptionally bright, photostable conjugates. Journal of Histochemistry & Cytochemistry 47:1179–1188

Pandya S, Yu JH, Parker D (2006) Engineering emissive europium and terbium complexes for molecular imaging and sensing. Dalton Transactions 2006:2757–2766

Papkovsky DB, O'Riordan T, Soini A (2000) Phosphorescent porphyrin probes in biosensors and sensitive bioassays. Biochemical Society Transactions 28:74–77

Papkovsky DB, O'Riordan TC (2005) Emerging applications of phosphorescent metalloporphyrins. Journal of Fluorescence 15:569–584

Papkovsky DB, Ponomarev GV (2001) Spectral-luminescent study of the porphyrin-diketones and their complexes. Spectrochimica Acta Part A-Molecular and Biomolecular Spectroscopy 57:1897–1905

Petrou PS, Kakabakos SE, Christofidis I, Argitis P, Misiakos K (2002) Multi-analyte capillary immunosensor for the determination of hormones in human serum samples. Biosensors & Bioelectronics 17:261–268

Piszczek G (2006) Luminescent metal-ligand complexes as probes of macromolecular interactions and biopolymer dynamics. Archives of Biochemistry and Biophysics 453:54–62

Pivovarenko VG, Klueva AV, Doroshenko AO, Demchenko AP (2000) Bands separation in fluorescence spectra of ketocyanine dyes: evidence for their complex formation with monohydric alcohols. Chemical Physics Letters 325:389–398

Plasek J, Sigler K (1996) Slow fluorescent indicators of membrane potential: a survey of different approaches to probe response analysis. Journal of Photochemistry and Photobiology B-Biology 33:101–124

Qhobosheane M, Santra S, Zhang P, Tan WH (2001) Biochemically functionalized silica nanoparticles. Analyst 126:1274–1278

Rao KVN, Stevens PW, Hall JG, Lyamichev V, Neri BP, Kelso DM (2003) Genotyping single nucleotide polymorphisms directly from genomic DNA by invasive cleavage reaction on microspheres. Nucleic Acids Research 31:e66

Raymo FM, Tomasulo M (2006) Optical processing with photochromic switches. Chemistry-A European Journal 12:3186–3193

Raymo FM, Yildiz I (2007) Luminescent chemosensors based on semiconductor quantum dots. Physical Chemistry Chemical Physics 9:2036–2043

Renikuntla BR, Rose HC, Eldo J, Waggoner AS, Armitage BA (2004) Improved photostability and fluorescence properties through polyfluorination of a cyanine dye. Organic Letters 6:909–912

Riu J, Maroto A, Rius FX (2006) Nanosensors in environmental analysis. Talanta 69:288–301

Roshal AD, Grigorovich AV, Doroshenko AO, Pivovarenko VG, Demchenko AP (1998) Flavonols and crown-flavonols as metal cation chelators. The different nature of Ba^{2+} and Mg^{2+} complexes. Journal of Physical Chemistry A 102:5907–5914

Roshal AD, Grigorovich AV, Doroshenko AO, Pivovarenko VG, Demchenko AP (1999) Flavonols as metal-ion chelators: complex formation with Mg^{2+} and Ba^{2+} cations in the excited state. Journal of Photochemistry and Photobiology A-Chemistry 127:89–100

Runnels LW, Scarlata SF (1995) Theory and application of fluorescence homotransfer to melittin oligomerization. Biophysical Journal 69:1569–1583

Salvioli S, Ardizzoni A, Franceschi C, Cossarizza A (1997) JC-1, but not DiOC(6)(3) or rhodamine 123, is a reliable fluorescent probe to assess Delta Psi changes in intact cells: implications for studies on mitochondrial functionality during apoptosis. FEBS Letters 411:77–82

Sanchez-Barragan I, Costa-Fernandez JM, Valledor M, Campo JC, Sanz-Medel A (2006) Room-temperature phosphorescence (RTP) for optical sensing. Trac-Trends in Analytical Chemistry 25:958–967

Sapsford KE, Medintz IL, Golden JP, Deschamps JR, Uyeda HT, Mattoussi H (2004) Surface-immobilized self-assembled protein-based quantum dot nanoassemblies. Langmuir 20:7720–7728

Sapsford KE, Pons T, Medintz IL, Mattoussi H (2006) Biosensing with luminescent semiconductor quantum dots. Sensors 6:925–953

Scott RW, Wilson OM, Crooks RM (2005) Synthesis, characterization, and applications of dendrimer-encapsulated nanoparticles. Journal of Physical Chemistry B 109:692–704

Shaner NC, Campbell RE, Steinbach PA, Giepmans BNG, Palmer AE, Tsien RY (2004) Improved monomeric red, orange and yellow fluorescent proteins derived from Discosoma sp red fluorescent protein. Nature Biotechnology 22:1567–1572

Shav-Tal Y, Darzacq X, Shenoy SM, Fusco D, Janicki SM, Spector DL, Singer RH (2004) Dynamics of single mRNPs in nuclei of living cells. Science 304:1797–1800

Shcharbin D, Szwedzka M, Bryszewska M (2007) Does fluorescence of ANS reflect its binding to PAMAM dendrimer? Bioorganic Chemistry 35:170–174

Shepard JRE (2006) Polychromatic microarrays: simultaneous multicolor array hybridization of eight samples. Analytical Chemistry 78:2478–2486

Shiohara A, Hoshino A, Hanaki K, Suzuki K, Yamamoto K (2004) On the cyto-toxicity caused by quantum dots. Microbiology and Immunology 48:669–675

Shrestha S, Deo SK (2006) Anthozoa red fluorescent protein in biosensing. Analytical and Bioanalytical Chemistry 386:515–524

Shynkar VV, Klymchenko AS, Duportail G, Demchenko AP, Mely Y (2005) Two-color fluorescent probes for imaging the dipole potential of cell plasma membranes. Biochimica et Biophysica Acta 1712:128–136

Shynkar VV, Klymchenko AS, Kunzelmann C, Duportail G, Muller CD, Demchenko AP, Freyssinet JM, Mely Y (2007) Fluorescent biomembrane probe for ratiometric detection of apoptosis. Journal of the American Chemical Society 129:2187–2193

Shynkar VV, Klymchenko AS, Piemont E, Demchenko AP, Mely Y (2004) Dynamics of intermolecular hydrogen bonds in the excited states of 4′-dialkylamino-3-hydroxyflavones. On the pathway to an ideal fluorescent hydrogen bonding sensor. Journal of Physical Chemistry A 108:8151–8159

Shynkar V, Mely Y, Duportail G, Piemont E, Klymchenko AS, Demchenko AP (2003) Picosecond time-resolved fluorescence studies are consistent with reversible excited-state intramolecular proton transfer in 4′-dialkylamino-3-hydroxyflavones. Journal of Physical Chemistry A. 109: 9522–9529

Soini AE, Seveus L, Meltola NJ, Papkovsky DB, Soini E (2002) Phosphorescent metalloporphyrins as labels in time-resolved luminescence microscopy: effect of mounting on emission intensity. Microscopy Research and Technique 58:125–131

Somers RC, Bawendi MG, Nocera DG (2007) CdSe nanocrystal based chem-/bio-sensors. Chemical Society Reviews 36:579–591

Sun YP, Zhou B, Lin Y, Wang W, Fernando KAS, Pathak P, Meziani MJ, Harruff BA, Wang X, Wang HF, Luo PJG, Yang H, Kose ME, Chen BL, Veca LM, Xie SY (2006) Quantum-sized carbon dots for bright and colorful photoluminescence. Journal of the American Chemical Society 128:7756–7757

Szollosi J, Damjanovich S, Matyus L (1998) Application of fluorescence resonance energy transfer in the clinical laboratory: routine and research. Cytometry 34:159–179

Tan CY, Alas E, Muller JG, Pinto MR, Kleiman VD, Schanze KS (2004) Amplified quenching of a conjugated polyelectrolyte by cyanine dyes. Journal of the American Chemical Society 126:13685–13694

Tauer K, Ahmad H (2003) Study on the preparation and stabilization of pyrene labeled polymer particles in nonpolar media. Polymer Reaction Engineering 11:305–318

Terai T, Kikuchi K, Iwasawa SY, Kawabe T, Hirata Y, Urano Y, Nagano T (2006) Modulation of luminescence intensity of lanthanide complexes by photoinduced electron transfer and its application to a long-lived protease probe. Journal of the American Chemical Society 128:6938–6946

Thomas SW 3rd, Joly GD, Swager TM (2007) Chemical sensors based on amplifying fluorescent conjugated polymers. Chemical Reviews 107:1339–1386

Triulzi RC, Micic M, Giordani S, Serry M, Chiou WA, Leblanc RM (2006) Immunoasssay based on the antibody-conjugated PAMAM-dendrimer-gold quantum dot complex. Chemical Communications :5068–5070

Tsien RY (1998) The green fluorescent protein. Annual Review of Biochemistry 67:509–544

Uchiyama S, Takehira K, Yoshihara T, Tobita S, Ohwada T (2006) Environment-sensitive fluorophore emitting in protic environments. Organic Letters 8:5869–5872

Veinot JGC (2006) Synthesis, surface functionalization, and properties of freestanding silicon nanocrystals. Chemical Communications 40:4160–4168

Vosch T, Antoku Y, Hsiang JC, Richards CI, Gonzalez JI, Dickson RM (2007) Strongly emissive individual DNA-encapsulated Ag nanoclusters as single-molecule fluorophores. Proceedings of the National Academy of Sciences of the United States of America 104:12616–12621

Wang F, Tan WB, Zhang Y, Fan X, Wang M (2006a) Luminescent nanomaterials for biological labelling. Nanotechnology 17:R1–R13

Wang L, Wang KM, Santra S, Zhao XJ, Hilliard LR, Smith JE, Wu JR, Tan WH (2006b) Watching silica nanoparticles glow in the biological world. Analytical Chemistry 78:646–654

Wetzl B, Gruber M, Oswald B, Durkop A, Weidgans B, Probst M, Wolfbeis OS (2003) Set of fluorochromophores in the wavelength range from 450 to 700 nm and suitable for labeling proteins and amino-modified DNA. Journal of Chromatography B Analytical Technologies in the Biomedical and Life Sciences 793:83–92

Woods M, Sherry AD (2003) Synthesis and luminescence studies of aryl substituted tetraamide complexes of europium(III): a new approach to pH responsive luminescent europium probes. Inorganic Chemistry 42:4401–4408

Wosnick JH, Liao JH, Swager TM (2005a) Layer-by-layer poly(phenylene ethynylene) films on silica microspheres for enhanced sensory amplification. Macromolecules 38:9287–9290

Wosnick JH, Mello CM, Swager TM (2005b) Synthesis and application of poly(phenylene ethynylene)s for bioconjugation: a conjugated polymer-based fluorogenic probe for proteases. Journal of the American Chemical Society 127:3400–3405

Wu CF, Szymanski C, McNeill J (2006) Preparation and encapsulation of highly fluorescent conjugated polymer nanoparticles. Langmuir 22:2956–2960

Wu L, Burgess K (2008) Syntheses of highly fluorescent GFP-chromophore analogues. Journal of the American Chemical Society 130:4089–4096

Yao G, Wang L, Wu YR, Smith J, Xu JS, Zhao WJ, Lee EJ, Tan WH (2006) FloDots: luminescent nanoparticles. Analytical and Bioanalytical Chemistry 385:518–524

Yesylevskyy SO, Klymchenko AS, Demchenko AP (2005) Semi-empirical study of two-color fluorescent dyes based on 3-hydroxychromone. Journal of Molecular Structure-Theochem 755:229–239

Yiidiz I, Tomasulo M, Raymo FM (2006) A mechanism to signal receptor-substrate interactions with luminescent quantum dots. Proceedings of the National Academy of Sciences of the United States of America 103:11457–11460

Yuan JL, Wang GL (2006) Lanthanide-based luminescence probes and time-resolved luminescence bioassays. Trac-Trends in Analytical Chemistry 25:490–500

Yushchenko DA, Shvadchak VV, Bilokin MD, Klymchenko AS, Duportail G, Mely Y, Pivovarenko VG (2006) Modulation of dual fluorescence in a 3-hydroxyquinolone dye by perturbation of its intramolecular proton transfer with solvent polarity and basicity. Photochemical and Photobiological Sciences 5:1038–1044

Zhang JK, Zhang WL, Dong SM, Turner APF, Fan QJ, Jia SR (2007) Nano-porous light-emitting silicon chip as a potential biosensor platform. Analytical Letters 40:1549–1555

Zhang L, Feng W (2007) Dendritic conjugated polymers. Progress in Chemistry 19:337–349

Zheng J, Nicovich PR, Dickson RM (2007) Highly fluorescent noble-metal quantum dots. Annual Review of Physical Chemistry 58:409–431

Chapter 5
Recognition Units

At the heart of any chemosensor or biosensor is its recognition unit (receptor). It is constructed for providing selective target binding from a mixture of different and sometimes closely related compounds. The high specificity and affinity of this unit is achieved by its appropriate structures allowing multi-point non-covalent interactions with the target. Such highly selective binding is called *molecular recognition*. In this chapter we discuss different binding units and the principles of their design and construction.

Some targets are small molecules and ions and for their recognition various coordination compounds can be used. Many of the targets, however, are larger molecules such as enzyme substrates, proteins, nucleic acids, macromolecular assemblies or even living cells. Their immense number requires a great variety of means for specific detection. All of these *receptors* or *recognition units* must be transformed into sensors by coupling a dye or nanoparticle to respond to the presence of the target without affecting the binding affinity. Therefore our goal is to achieve optimal binding and efficient labeling of the binder but to still maintain the target binding properties intact while adding the reporter function.

It has to be clearly understood that the sensor detects the *bound* target, whereas information is needed on the *total* concentration of the target present in the system, both bound and unbound. When the equilibrium between the bound and unbound target is established, its distribution between these forms can be achieved based on mass action law. This limits the detection range of any two-state responsive sensor to two orders of magnitude, below and above the *dissociation constant* K_d (Chapter 2). This puts an additional requirement on sensor affinity, which should fit to this rather narrow range of required target concentrations. Too strong binding is as useless as too weak binding. Very important is also the *selectivity* of binding – the ability to discriminate from other species that may be close in structure and properties.

5.1 Recognition Units Built of Small Molecules

The possibilities of organic chemistry to synthesize new molecules with increasing complexity and with more and more specialized functional behavior are tremendous. Organic molecules are able to specifically bind small molecules such as

A.P. Demchenko, *Introduction to Fluorescence Sensing*,
© Springer Science+Business Media B.V. 2009

ions, monosaccharides, amino acids and short peptides. Moreover, some of these compounds have been shown to recognize particular motives at the interfaces and even on the surfaces of protein molecules (Fletcher and Hamilton 2007; Peczuh and Hamilton 2000). The question, whether they can display the properties similar to that of antibodies, is presently under active research.

5.1.1 Crown Ethers, Cryptands, Polyhydroxilic and Boronic Acid Derivatives

Sensing properties can be achieved by a functional modification of the dye itself. In this case the dyes incorporate the target-binding groups (such as *crown ethers* or *cryptands*), which provide a perturbation of the dye electronic structure when the target binds. The targets in these cases are of a small size: the metal ions, such as Ca^{2+}, or small neutral compounds such as glucose.

The literature contains a great number of dyes and their modifications with the incorporation of recognition units (de Silva et al. 1997; Lakowicz 2007; Valeur 2002). Some of them will be discussed in Section 10.3 on the detection of small-molecular targets. Organic dyes and conjugated polymers allow combining, efficiently in one molecule, recognition and reporting properties. Several examples are given below (Fig. 5.1).

Various mechanisms of signal transduction and response can be employed here. They include modulation of reactions, in which the electrons participate in the excited states – the electron, charge and energy transfers (PET, ICT and FRET reactions correspondingly, (see Chapter 6)). The first two of these reactions can exhibit influences by the neighboring charges by electrostatic interactions and therefore these effects are the most applicable for the detection of charged compounds and ions. The modulation of FRET can be used due to the change of the overlap integral on target binding and conformational change leading to the change in the electronic conjugation between fragments of the dye molecule. The mechanisms of these reactions will be discussed in Chapter 6.

As recognition groups, cryptands and crowns are frequently used for binding the ions. Usually their response is based on the PET mechanism and results in quenching. To obtain fluorescence enhancement, more complex constructions have to be developed. Figure 5.1 shows an example of an ion sensor, in which a combination of the PET and FRET mechanisms provides a strong 'OFF-ON' response to ion binding to a cryptand unit.

Fig. 5.1 Examples of fluorescent organic molecules that can bind and detect ions. (**a**) Conjugated polymer with incorporated crown ether groups. The ions, when bound to some of the crown ether groups, produce fluorescence quenching of the whole chain (McQuade et al. 2000). (**b**) The dye containing the azacrown ether group and two carbonyls that provide selective binding of Pb^{2+} ions with a fluorescence enhancement response (Chen and Huang 2002). (**c**) Azacrown ether derivative of 3-hydroxyflavone has two centers of binding bivalent cations, Ma^{2+} and Ba^{2+}. The ions bound at the

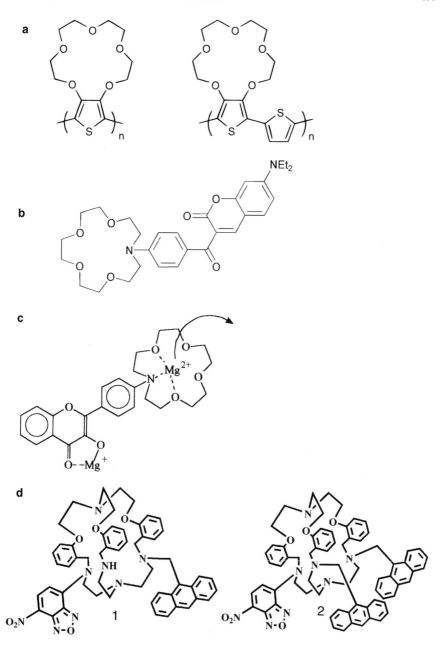

Fig. 5.1 (continued) azacrown site are ejected upon excitation. This leads to different effects of binding in excitation and emission spectra (Roshal et al. 1999). (**d**) A laterally nonsymmetric aza cryptand derivatized with one 7-nitrobenz-2-oxa-1,3-diazole (**1**) and one/two anthracenes (**2**). These compounds give a large enhancement on the binding of Cu^{2+}, Ag^+ and proton. The enhancement is observed in the diazole moiety even when the anthracene fluorophore is excited because of a substantial fluorescence resonance energy transfer from anthracene to the diazole moiety (Sadhu et al. 2007)

5.1.2   *Cyclodextrins*

Cyclodextrins make up a family of *cyclic oligosaccharides* containing six (α-cyclo-dextrin), seven (β-cyclodextrin) and eight (γ-cyclodextrin) α-D-glucose subunits in a ring, creating a conical shape (Fig. 5.2).

The primary hydroxyl groups are directed to the narrow side and the secondary ones are located on the wide side of the torus. With hydroxylic groups facing the outer space, these molecules are highly soluble in water, whereas the inner space can accommodate different low-polar molecules, such as cholesterol. Their binding affinity is determined by cavity size. It increases from α-cyclodextrin (α-CD) to β-cyclodextrin (β-CD) and to γ-cyclodextrin (γ-CD), see Table 5.1.

Inclusion complexes are known as the entities comprising two or more molecules, in which one of them, serving as the '*host*' includes a '*guest*' molecule only by physi-cal forces, i.e., without covalent bonding. Cyclodextrins are typical 'host' molecules that can include a great variety of molecules that have the size conforming to the size of a cavity (Szejtli 1998). Thus, α-cyclodextrin can form an inclusion complex with one, β-cyclodextrin – with two and γ-cyclodextrin – with three pyrene molecules. Cyclodextrins can be produced in large quantities from starch, by a simple and cheap enzymatic conversion, which adds to its popularity as a carrier of drugs, in the cos-metic industry and, of course, in sensing technologies (Szejtli 1998).

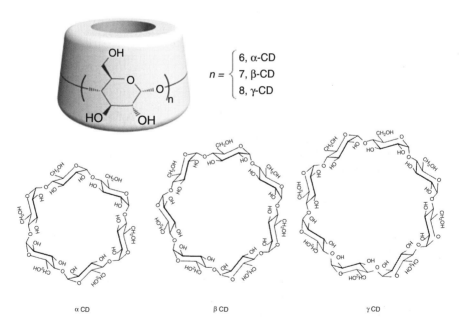

Fig. 5.2 Schematic drawing (above) and the chemical structures (below) of α, β and γ cyclodex-trins (From Wenz et al. 2006; Szejtli 1998)

Table 5.1 Minimum internal diameters d_{min}, cross-sectional areas A_{min} and inner volume V of cyclodextrins

Cyclodextrin	d_{min} (Å)	A_{min} (Å^2)	V (Å^3)
α-CD	4.4	15	174
β-CD	5.8	26	262
γ-CD	7.4	43	427

The presence of a number of exposed hydroxilic groups (18 in α-cyclodextrin, 21 in β-cyclodextrin and 24 in γ-cyclodextrin) allows a practically unlimited number of *chemical modifications* and many of them can modify the selectivity, in the formation of the inclusion complexes, with different molecular guests. The most popular are methylated and hydroxyalkylated derivatives that increase the affinity to low-polar guests. Dimeric and oligomeric forms of cyclodextrins can be easily obtained. They may show an increased affinity to large molecules, even to proteins, incorporating their segments. It was observed that such covalently linked dimeric structures could even disrupt the interactions between protein subunits in multimeric proteins by competition for exposed low-polar residues and thereby inhibit protein aggregation.

Cyclodextrins have received much attention as the very *specific binders* in molecular and supramolecular studies, for providing nano-scale media for chemical reactions (Oshovsky et al. 2007), even for modeling a biomolecular catalysis (Breslow and Dong 1998). It was demonstrated that with proper modification, the sensing with cyclodextrins can be enantioselective, so that a clear discrimination between D- or L-amino acids can be achieved (Pagliari et al. 2004).

Many organic dyes form *inclusion complexes* with cyclodextrins (Al-Hassan and Khanfer 1998). Decrease of polarity, dehydration effects and immobilization effects were detected in the fluorescence response of these dyes. Comparative studies on the binding of different neutral, anionic and cationic dyes (Balabai et al. 1998) demonstrated that the binding is mainly hydrophobic and is only insignificantly affected by the charge of a guest molecule. In all studied cases tight complexes are formed. This is manifested by a decrease of an anisotropy decay, which detects the rotation of the whole complex. Meantime, ultrafast dynamics of guest molecules exist and it has been characterized by time-resolved methods (Douhal 2004).

Inclusion into the cyclodextrin cavity changes the photophysical and spectroscopic properties of organic dye molecules. Thus, for solvatochromic dye Nile Red a remarkable decrease of the non-radiative decay of the ICT state was observed on inclusion into the β-cyclodextrin cavity (Hazra et al. 2004). The influence on the protonation equilibrium (Mohanty et al. 2006) and excited-state proton transfer from pyranine to acetate in gamma-cyclodextrin and hydroxypropyl gamma-cyclodextrin (Mondal et al. 2006) was also reported. Different excited-state reactions, such as ESIPT (Organero et al. 2007), are also affected by binding in nanocavity.

The *fluorescence reporting* signal can be generated in a simple but efficient way by the displacement, by the target molecule, of the dye bound in the cyclodextrin cavity. If the dye is not attached covalently to a cyclodextrin host, we will get a competitor displacement assay (as explained in Chapter 1). If the dye is covalently bound by a flexible link, we can get the direct sensor. The target binding properties can be modulated by chemical substitutions.

An increase of the selectivity and affinity of cyclodextrin to its target is achieved when the sensor is made '*bivalent*' so that the linker connects two cyclodextrin molecules (Yang et al. 2003b). Possessing two beta-cyclodextrin cavities in close vicinity and a functional linker with a good structural variety in a single molecule, these bridged bis(beta-cyclodextrin)s can significantly enhance the original binding ability and molecular selectivity. In this case the possibility of a multipoint interaction with the target increases and recognition of molecules as large as proteins can be achieved by binding the dimer at two exposed sites.

It was shown that biquinolino-modified beta-cyclodextrin dimers (Fig. 5.3) and their metal complexes may serve as efficient fluorescent sensors for the molecular recognition of steroids (Liu et al. 2004). The length of the linker between two monomers is an important variable that can significantly modify steroid binding.

It was shown that a γ-cyclodextrin dimer exhibits a strong molecular recognition ability for bile acids and endocrine disruptors (Makabe et al. 2002). The attached pyrene moieties allowed detecting the binding as the ratio of their monomer and excimer fluorescence intensities. It was indicated that pyrene residues participate in guest binding and serve as hydrophobic cups for the cavities.

Conjugation with peptides expands the possibilities of cyclodextrins for detecting steroid molecules (Hossain et al. 2003). Double labeling with FRET detection can be a method of choice in fluorescence reporting. The construction of more sophisticated conjugates with proteins was also reported, particularly for sensing maltose based on the *E. coli* maltose binding protein (MBP) (Medintz et al. 2003)

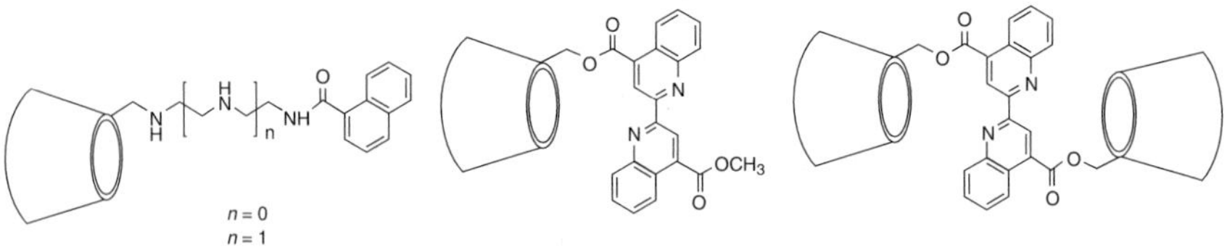

Fig. 5.3 Examples of cyclodextrin conjugates with fluorescence dyes. From left to right: quinoline dye attached through the spacer (n = 1–2); biquinoline dye attached to a monomer; biquinoline dye forming a link between two cyclodextrin molecules (Liu et al. 2004)

5.1.3 Calixarenes

Calixarenes are a class of polycyclic compounds that can allow (by proper chemical substitutions) achieving a high-affinity binding of many small molecules. Their basic structure is presented in Fig. 5.4. Calixarenes are cyclic oligomers obtained by phenol-formaldehyde condensation. They exist in a 'cup-like' shape with a defined upper and lower rim and a central annulus. Their rigid conformation of a skeleton, forming a cavity together with some flexibility of side groups, enables them to act as host molecules. By functionally modifying the upper and/or the lower rims it is possible to prepare various derivatives with differing selectivity for various guest molecules. Mostly they are small molecules and ions but with proper modifications they can target the structural elements of large molecules.

A calix[4]arene structure allows many possibilities for targeting proteins. Their upper rim displays four positions that can be used for covalent attachments. Moreover, the nature and symmetry of such recognition elements may be varied to selectively target the highly irregular surfaces, comprising charged, polar and hydrophobic sites.

A series of synthetic receptors was prepared, in which four peptide loop domains are attached to a central calix[4]arene scaffold (Peczuh and Hamilton 2000). Each peptide loop is based on a cyclic hexapeptide in which two residues have been replaced by a 3-aminomethylbenzoate dipeptide mimetic, which also contains a 5-amino substituent for anchoring the peptide to the scaffold. Through the attachment of various peptide loops, the authors made a series of calix[4]arenes expressing negatively and positively charged regions as well as hydrophobic regions and thus achieved binding to complementary regions of several proteins. They demonstrated the binding to a platelet-derived growth factor (PDGF) and inhibition of its interaction with its cell surface receptor PDGFR. A similar structure interacts specifically with the cytochrome c. The beauty of such structures recognizing particular sites on protein molecules and inhibiting their interactions is represented in Fig. 5.5. The structure resembles a basket of flowers.

Fluorescence reporting on the binding of different targets to calix[4]arene structures is commonly introduced in two ways.

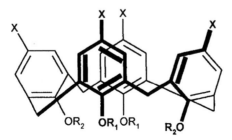

Fig. 5.4 Parent calix[4]arene in a cone conformation. The sites of substitutions with covalent attachment of recognition units and fluorescent dyes in the upper rim are marked X. In the lower rim, the substitution sites are denoted as R_1 and R_2

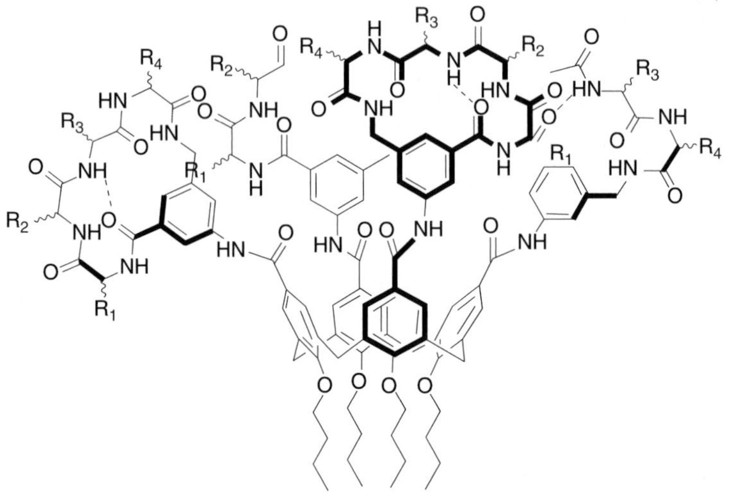

Fig. 5.5 Calix[4]arene derivatives that mimic the properties of antibodies to recognize protein targets. R_1-R_2-R_3-R_4 could be H, CH_2CO_2H; $(CH_2)_4NH_2$. This provides the recognition pattern of negative and positive charges (Reproduced with permission from Peczuh and Hamilton 2000)

(a) Using the property of these molecules to bind different dyes at the site within the cavity, where there occurs the binding of many targets. Then the binding of a target molecule can *displace the fluorescence dye*. When the dye dissociates, it changes its fluorescence properties. The chelating properties of calix[4]arenes towards different dyes have been reported by many researchers (Kubinyi et al. 2005). In all these cases, the formation of the complex leads to dramatic fluorescence quenching and this result suggests a convenient possibility for transforming the receptors into efficient sensors. This suggests using calix[4]arene derivatives in competitor substitution assays. Ample possibilities for the modulation of the target affinities by covalent modifications and a broad choice of responsive dyes with different affinities promise the easy development of simple and efficient assays for many targets.

(b) The *covalent conjugation* of calix[4]arene side groups with fluorescent dyes (Nanduri et al. 2006). In this case, several mechanisms of fluorescence response can be realized. One can be easily applied in ion sensing and is based on the induction/perturbation of a photo-induced intramolecular charge transfer (ICT). The tert-butylcalix[4]arene was synthesized either with one appended fluorophore and three ester groups or with four appended fluorescent reporters (Leray et al. 2001). The dyes were 6-acyl-2-methoxynaphthalene derivatives, which contained an electron-donating substituent (methoxy group) conjugated to an electron-withdrawing substituent (carbonyl group). This is a typical arrangement for fluorescent reporters operating on the basis of the ICT principle (Section 6.1).

It was shown that such molecules responded to binding the ions not only by expected red shifts of the absorption and emission spectra, but also by a drastic enhancement of the fluorescence quantum yield Φ. Thus, the compound with one substituent was synthesized, for which Φ increases from 0.001 to 0.68 upon binding the Ca^{2+} ions. In the case of four substitutions, the emission was strongly depolarized due to the energy transfer (homo-FRET). Regarding the complexation properties, a high selectivity for Na^+ over K^+, Li^+, Ca^{2+} and Mg^{2+} was observed. This selectivity (Na^+/other cations), expressed as the ratio of the stability constants, was found to be more than 400. This result certifies the abilities of calixarenes as useful building platforms in the design of complex systems, in which not only ICT but also other excited-state phenomena (electron and proton transfer, excimer formation and resonance energy transfer) are controlled by ions. The applications mainly concern ion sensing with high selectivity (Valeur and Leray 2007).

Lead and mercury sensing was achieved by calixarene-based sensors bearing two or four dansyl fluorophores (Metivier et al. 2004). These compounds show a high selectivity towards Hg^{2+} and Pb^{2+} over many interfering cations and sensitivity in the 10^{-7} mol l^{-1} concentration range. In this case, the complexation of the mercury ion induces a strong fluorescence quenching due to a well-defined electron transfer process from the fluorophore to the metal center. It was also reported that calix[4]azacrown, having an anthracenyl unit, displayed large chelation-enhanced fluorescence effects with Cs^+, Rb^+ and K^+ over other tested metal cations (Kim et al. 2003).

The sensors based on the calixarene platform with a covalent attachment of a reporter dye can be very efficient. A variety of environment-sensitive dyes were used for attachment, particularly, to an upper rim of their structures. In one of the recent studies, such a modification was made with cyanine dyes (Kachkovskiy et al. 2006), which opens the pathway to extending the sensing techniques to a near-IR range.

In a number of reports the dye was constructed as part of a recognition mechanism and the changes in its interactions with the environment induced by target binding produced the necessary reporter signal. The *enantioselective* (distinguishing stereoisomers) molecular sensing of aromatic amines was achieved using their quenching on the interaction with the chiral host tetra-(S)-di-2-naphthylprolinol calix[4]arene (Jennings and Diamond 2001).

In addition to the role of recognition elements of small molecular sensors, calix[4]arenes play an important role in the construction of fluorescent *functional nanoparticles*. They are used for coating Quantum Dots (Jin et al. 2006), which allows achieving their bright emission and high stability in aqueous solutions. A fluorescent conjugated polymer poly(phenylene ethynylene) containing calix[4]arene-based recognition units displays a sensitivity to be quenched by the N-methylquinolinium ion. This effect is over three times larger than that seen in a control polymer lacking calix[4]arenes (Wosnick and Swager 2004).

The ability of calix[4]arenes to host rare earth cations is very attractive These ions are highly luminescent only in the ligand-bound forms (Section 4.2) and calix[4]arenes may serve as such ligands. The water soluble calix[4]arene derivatives can form luminescent complexes with europium(III) and terbium(III) ions

(Yang et al. 2003a, b). This transforms the whole complex into the site with long-duration emission and allows many new possibilities for sensor developments.

Concluding, we note that combining the remarkable molecular recognition power of calixarenes with proper fluorescence transduction opens new, interesting possibilities. The role of calix[4]arenes in signal transduction mechanisms involving excited-state electron and energy transfers will be discussed in Chapter 6.

5.1.4   Porphyrins

Porphyrin molecule is a heterocyclic macrocycle derived from four pyrole-like subunits interconnected by means of their α carbon atoms via methine bridges (=CH-). This macrocycle is planar and with a high electronic conjugation, it contains 26 π-electrons. Many substitutions can be made at the periphery of the porphyrin ring (Fig. 5.6) and they determine the binding ability to different targets.

Porphyrins can combine *sensing and reporting properties*, since they possess their own light-absorption and fluorescence spectra (Section 4.2). Their fluorescence depends on the presence of a coordinated metal cation in the center of their structure. In most cases, the emission spectra of free base porphyrins and of porphyrins with coordinated cation (metalloporphyrins) are different from each other. Since the two components can be easily distinguished, this suggests the possibility of making sensors for these ions (Purrello et al. 1999). It is hard to do that with natural porphyrins, in which the fluorescence intensity of free base in water is very low and the selectivity of binding to discriminate between different ions in their mixture is not sufficient. Hopefully, the porphyrins scaffold allows for many functional substitutions and complexations that are able to improve, if necessary, their binding properties.

Such improvements can be achieved upon *complex formation with cyclodextrins*. These complexes can be particularly useful for the determination of zinc ion. The zinc binding changes the fluorescence spectrum of porphyrin (Fig. 5.7), which allows ratiometric recording of zinc concentration (Yang et al. 2003b). Upon binding the zinc ion the fluorescence emission of tetraphenylporphyrin at the 656 nm band decreases while that at 606 nm increases, which allows a wavelength-ratiometric detection.

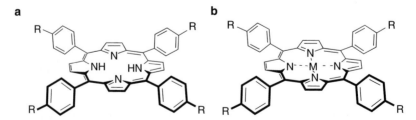

Fig. 5.6 The core structure of synthetic porphyrins. (**a**) Metal-free form. (**b**) Metal-coordinated form. 'R' marks the positions of various substitutions. They can be selected for high-affinity binding to proteins

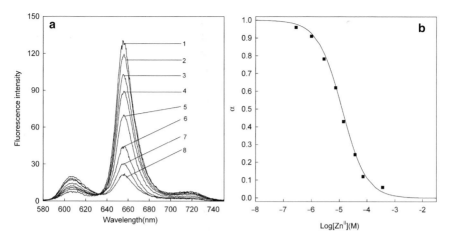

Fig. 5.7 Ratiometric detection of zinc by a fluorescence response in a porphyrins-cyclodextrin complex. (**a**) Effects of the zinc ion concentration on the fluorescence emission spectra (spectra 1–8 correspond to the increasing zinc concentrations from 0 to 7.2×10^5 M). (**b**) Relative fluorescence ratio, α, as a function of $\log[Zn^{2+}]$ at pH 8.0
(Reproduced with permission from Yang et al. 2003b).

The binding of synthetic porphyrins to proteins that contain a heme group in their native form can be very strong. After enhancing the hydrophobicity of the porphyrin core and by increasing the number of peripheral carboxylic acids from 8 to 16, the binding to cytochrome c with a sub-nanomolar dissociation constant was achieved (Ozaki et al. 2006). With $K_d = 0.67 \pm 0.34$ nM this is probably one of the most potent synthetic protein receptors ever designed. The use of porphyrin derivatives as very strong binders has also been demonstrated for other proteins. The strategies for protein surface recognition that offer a new use for porphyrins as molecular scaffolds are discussed in the literature (Tsou et al. 2004).

A novel protein-detecting array based on porphyrins containing peripheral amino acids acting as protein surface receptors has been reported (Zhou et al. 2006). The *array of porphyrin receptors* showed a unique pattern of fluorescence change upon interaction with certain protein samples. Both metal and nonmetal-containing proteins and mixtures of proteins provide distinct patterns, allowing their unambiguous identification.

5.1.5   Dendrimers

Dendrimers (see also Section 4.3) are well-defined, tree-like macromolecules that possess a high degree of order and the possibility to contain selected chemical units in predetermined sites of their structure. With respect to target recognition properties, dendrimers can be considered as unimolecular micelles. Unlike micelles, their structural arrangement does not allow expending their inner volume to solubilize large molecules, nanoparticles or a significant amount of non-miscible co-solvent molecules. But due to the presence of some degree of conformational freedom of chain segments, the dendrimers possess some structural flexibility. They can swell and shrink in response to the change of solvent and can trap and hold small molecules in a structure-dependent manner. Because of their morphology, a high degree of synthetic flexibility and a multitude of identical reaction sites, the dendrimers are highly potential materials for chemical sensing.

A combination of this ability with fluorescence emission properties makes dendrimers really unique as fluorescence sensors. The fact that *conjugated polymers can be made dendritic* is very important (Zhang and Feng 2007). The extent of their conjugation can be made variable. This would allow different functional applications including providing specific target binding and reporting. There can be fully-conjugated dendritic polymers, partially-conjugated dendritic polymers and dendritized conjugated polymers. All of them can maintain the important structural properties of dendrimers.

The *host-guest chemistry* of dendrimers is well described in the literature (Antoni et al. 2007). Their structure allows encapsulating ions or neutral molecules in their internal dynamically formed cavities. It was reported that the binding sites can be created within the cross-linked dendrimers by a mixed-covalent-noncovalent imprinting process (Zimmerman et al. 2003). These binding sites can accommodate different guest molecules, including porphyrins.

The dendrimer-based fluorescent sensors can provide *enantioselective detection*, as demonstrated for the recognition of chiral amino alcohols (Pugh et al. 2000). A strong interest in dendrimers has also emerged from the side of gene and drug delivery, in view of their potential as hosts (Al-Jamal et al. 2006). It is expected that in the near future we will see the broad-scale application of these smart molecules in molecular sensing.

5.1.6 Prospects

The recognition units discussed in this section are designed for the specific binding of three classes of molecules: ions, small molecules and proteins. Organic heterocyclic compounds are better suited for the detection of ions, because the ions can be easily coordinated into heterocycles and because the ion binding can be electronically coupled with the dye response. The well-developed surface of calixarenes and dendrimers, in addition to specific ion binding, allows one to provide the binding of small neutral molecules. Porphyrins can suggest the response in their own emission. Regarding the sensing of small hydrophobic molecules, such as steroids, cyclodextrins must be of great preference. This is because cyclodextrin cavities fit to their sizes, the binding can be specific and involve target penetration into the cavity and formation of strong inclusion complexes. We also observe that in the present conditions, when compared to the great progress in nucleic acid recognition, protein recognition is not well developed, these compounds can offer themselves as promising protein binders. Much like the antibodies, such binders should recognize particular patterns on protein molecular surfaces – the 'hot spots' (see Section 10.5).

5.2 Antibodies and Their Recombinant Fragments

Antibodies (Abs) are powerful recognition tools used in many sensing technologies. They can be developed to bind target compounds (*antigens*) of any size starting from steroids, oligosaccharides and oligopeptides. Usually, in sensing applications, antibodies of immunoglobulin G (IgG) type are used. They are globular glycoproteins with a molecular mass of 150,000–160,000 Da formed of two light (L) chains and two heavy (H) chains joined to form a Y-shaped molecule. This molecule is composed of β-structured domains stabilized by disulfide bonds (see Fig. 1.3). Their *antigen* (Ag) *binding sites* are formed of six loops contributed by two different so-called variable domains (called VL and VH). Because antibodies contain two VL-VH pairs, they are bivalent: they can react with two antigenic binding sites simultaneously. Specificity to any target is determined by the amino acid composition of these domains. The variability of this composition is responsible for the diversity of the antibody repertoire.

Antibodies are used extensively as diagnostic tools in a wide array of different analyses. Because of their unique diversity, they provide a never ending source of molecules with unlimited possibilities for target detection. Antibodies are very efficient in detecting proteins. They recognize regions of the protein surface called *antigenic determinants or epitopes*. The contact area between the protein and bound antibody is relatively large (600–950 Å^2). The contact surfaces exhibit a high level of shape and physico-chemical complementarity.

5.2.1 The Types of Antibodies Used in Sensing

Several types of antibodies and antibody fragments can be used in sensing technologies (Goodchild et al. 2006). *Polyclonal antibodies* are purified from the sera of immunized animals. The serum contains antibodies with varying specificities and affinities. To obtain high affinity antibodies specific for a given target, purification and fractionation steps are needed.

Structurally homogeneous *monoclonal antibodies* are produced by hybridoma cell line clones. Here the problem of achieving high affinity and selectivity is solved by exposing the animals (usually rabbits or mice) to antigens. Then the Ab-producing cells that are raised in their bodies in response to the treatment are selected. Individual Ab-producing animal cells are known to synthesize antibodies of a uniform structure, so when they are fused with myeloma cells, which are capable of continuous propagation, they synthesize identical antibody molecules. These hybrids can be grown for months and years. Their cultivation is already a part of an industry, allowing large-scale antibody production.

Recombinant antibody fragments are attractive for sensing technologies because they allow one to decrease the size of a sensor molecule and also to include substitutions in their primary structures that can facilitate their attachment to surfaces or carriers (including nanoparticles). Groups that can be used for the coupling of fluorescence reporters (particularly, -SH groups) can also be introduced. Whereas antibodies specific for most compounds can be obtained relatively rapidly from the serum of immunized laboratory animals, their site-specific labeling requires the production of recombinant antibody fragments by molecular biology techniques, a time-consuming and technically demanding procedure. Libraries of binding fragments have been created. This allows the *in vitro* selection of a recombinant protein specific for a given target, by phage-display and other library screening techniques (Benhar 2007) without the need for animal immunization.

The most widely used recombinant antibody fragments contain the VL and VH domains linked by a peptide chain (scFv). Antibody fragments with desired binding specificity can be selected from libraries constructed from repertoires of antibody V genes, bypassing hybridoma technology and even immunization. The V gene repertoires are harvested from populations of lymphocytes, or are assembled *in vitro*. They are cloned to display the fragment on the surface of filamentous bacteriophage. Phages that carry scFvs with the desired specificity are selected from the repertoire by panning on antigen; soluble antibody fragments are expressed from infected bacteria and the affinity of the binding of selected antibodies is improved by mutation. The process mimics immune selection and antibody fragments with many different binding specificities have been isolated from the same phage repertoire. The probability of identifying a high-affinity binder increases with library size (Hust and Dubel 2004), which for phage-displayed libraries ranges between 10^7 and 10^{10} clones (Azzazy and Highsmith 2002).

Thus, human antibody fragments have been isolated with specificities against both foreign and self antigens, carbohydrates, secreted and cell surface proteins, viral coat proteins and intracellular antigens from the lumen of the endoplasmic reticulum and the nucleus. Such antibodies have potential as reagents for research and in therapy (Jespers et al. 2004).

Heavy-chain fragments, "*domain antibodies*". Antibody fragments made of heavy chains have become available only recently. Occurring naturally in 'heavy chain' antibodies from camels and now produced in fully humanized form, domain antibodies (dAbs) are the smallest known antigen-binding fragments of antibodies, ranging from 11 to 15 kDa. When they are expressed in a microbial cell culture, they show favorable biophysical properties including solubility and temperature stability. They are well suited for selection and affinity maturation by *in vitro* selection systems such as phage display. dAbs are the monomers and, owing to their small size and inherent stability, they can be formatted into larger molecules to create drugs with prolonged serum half-lives or other pharmacological activities. They bind to targets with high affinity and specificity (Muyldermans 2001).

5.2.2 The Assay Formats Used for Immunoassays

The term '*immunosensor*' has been applied to various direct or indirect detection assays involving antibodies (immunoassays). The outstanding recognition properties of these antibodies allow the realization of these assays in several formats.

Sandwich immunoassays (Fig. 1.11) are the most popular and convenient types of assays that are based on the heterogeneous assay principle. The immobilization of receptor molecules (antibodies or their *antigens*) on solid support, application of the sample and washing steps proceed before target detection. The antigen-antibody complex is routinely detected by using a secondary antibody specific for the antigen or for the constant domain of antibodies. In an *enzyme-linked immunosorbent assay* (ELISA), the secondary antibody is coupled to an enzyme, allowing the achievement of a high detection sensitivity due to enzymatic amplification.

Fluorescence polarization immunoassays. Detection in the change of the polarization (or anisotropy) of the fluorescence emission (see Section 3.2) is one of the most frequently used procedures in sensing with antibodies and is the basis of many commercially available test systems. Rotations of antibodies themselves occur on a time scale of 100 ns or longer, so they are too slow to be detected by common dyes with nanosecond decay. Therefore, this method is mostly used for the detection of relatively small molecules. If the target molecules are not fluorescent, then a competitive assay format (see Section 1.3) is applied. The target molecules compete for the Ab binding sites with target analogs coupled with fluorescent dyes. The test is based on the interplay in the response of a free analog exhibiting low polarization and immobilized on binding to an antibody for which the polarization of the emission is high. The testing can be done when the antibody is free in solution but it can also be immobilized on solid support.

Time-resolved immunoassays. These assays and sensors commonly use the labeled antibodies or antigen analogs (for competitive assays) with chelates of europium(III) or terbium(III) ions possessing lifetimes typically from 0.01 to 1 s. As pointed out in Section 4.2, this allows increasing the sensitivity dramatically, suppressing background emission and avoiding the problems with light-scattering.

FRET immunoassays. They are commonly performed when the antibody is labeled with a FRET donor and the target or competitor is the FRET acceptor (quencher).

It may often be attractive to construct a *direct sensor* by obtaining the response signal from a primary interaction of the antibody and the target antigen in a way that avoids double labeling as in FRET or complicated instrumentation as in polarization or time-resolved immunoassays. However, this is not easy considering that the antigen binding is not associated with large conformational changes, hence 'remote' sensing as described in Chapter 1 may not be efficient. Therefore, contact sensing using environmentally-sensitive covalently attached reporters located in proximity to the antigen-binding site must be used (Fig. 1.3b). This can be achieved by introducing at the desired Ab location a reactive group, such as –SH groups of Cys residues, for coupling the reporter dye. This can be accomplished by recombinant techniques.

The difficulty lies in the choice of a coupling site. In order to find the best solution for this problem, scientists follow two general strategies.

(a) The reporter dye should be located at the *periphery of the binding site*. It should not interfere with the interaction between the receptor and the target but its micro-environment should change in response to Ab-target complex formation (Renard et al., 2002).

(b) The reporter group should be located *within the binding site* in such way that it becomes one of the key participants in antigen binding (Jespers et al. 2004). In the latter case, the reporter molecule can be chemically linked to a hypervariable loop of an antibody repertoire displayed on the phage and this repertoire can be selected for antigen binding. The fluorescence of the probe has to respond quantitatively to antigen binding.

Because of the lack of sufficient data, it is too early to provide a comparative analysis of these two strategies. In experiments, when the dye was coupled to an antibody fragment (single chain variable fragment, scFv), 17 sites were selected for the coupling of the responsive dye (Renard et al., 2002). Being located at the periphery of the binding site, only a few of them provided a fluorescence response to the presence of antigen (lysozyme). A much stronger response, as a substantial decrease of fluorescence intensity, was observed by applying the second strategy (Jespers et al. 2004). Thus, the possibilities for designing semi-synthetic antibody-based sensors with fully integrated reporter dyes should be explored in future research.

5.2.3   *Prospects for Antibody Technologies*

By analyzing the application of antibodies as recognition elements of sensors we obtained very important lessons taught to us by Nature. The generation and synthesis in the living body of these high-affinity binders occurs via selection-amplification steps, starting from a genetically available potentially innumerous library. The random mutation-selection principle can be applied *in vitro* for the creation of libraries of engineered antibodies and their fragments. This allows raising recognition units to non-immunogenic or toxic molecules. Despite significant technical difficulties, academic and commercial laboratories have now developed large (more than 10^{10} variants) phage display antibody libraries, from which diverse sub-nanomolar affinity antibody clones can be isolated. Recently, one such library yielding over 1,000 distinct antibodies against a given antigen was described (Edwards et al. 2003).

Possessing the tools for the selection and amplified synthesis of optimal binders, one can solve the problem of coupling the binding with fluorescence reporting. Despite several promising attempts, direct sensors with the properly attached environment-sensitive dyes still did not find their proper application. With a solution for this problem, antibodies will be able to establish themselves as preferred recognition units in all biosensor technologies, including highly needed microarrays for the detection of carbohydrates and proteins, probably on a proteome scale. A great potential for development in cellular research is expected from a presently almost unexplored possibility of fusing antibodies to different marker proteins (Casadei et al. 1990). Obtaining such molecular hybrids may allow combining two important functions – targeted binding at particular sites and sensing different analytes at these sites.

5.3 Ligand-Binding Proteins and Protein-Based Display Scaffolds

We now address a broad class of proteins possessing important functions of binding and the transport of different substances along their metabolic routes. Their structures are very diverse. Some of them are almost ready biosensors and only the fluorescence response units are needed for incorporation into them. Some others can be used as the scaffolds for the generation of sensors with new functions. An increased understanding of the structure and function of natural ligand-binding proteins together with the advances in protein engineering has also triggered the exploration of various alternative protein architectures.

With these tools, valuable protein-binding molecular scaffolds have been obtained. They represent promising alternatives to antibodies for biotechnological and, potentially, clinical applications (Binz and Pluckthun 2005). Regarding sensor applications, their strong competition with antibodies is expected in the very near future.

5.3.1 Engineering the Binding Sites by Mutations

There are several techniques that can be efficiently applied for the generation of new proteins with desired ligand-binding properties. The most efficient of them start from well-known and almost-optimal protein structures that can be taken as scaffolds. Then sequential steps of random mutations and product selection are taken. Mutations are induced in specific regions of protein structure, usually at the ligand-binding pockets. The frequently applied procedure involves displaying the mutated proteins on the surface of filamentous bacteriophage, the virus that can only infect the bacteria (Uchiyama et al. 2005). Mutations can be induced by standard molecular biology procedures. The phage display technique allows for *in vitro* mutagenesis-selection. The multiple steps of growth of phage-infected bacteria are used to amplify the promising binders and to provide their further selection. This process is often described as '*in vitro* evolution'. It can be especially good for finding the affinity binders for relatively small molecules of metabolites and drugs.

These modern methods of molecular biology allow creating huge combinatorial libraries containing millions of structurally diverse species. Usually these libraries are constructed so that they consist of an underlying constant scaffold and rand-omized variable regions that differ from each other. Why then not synthesize the library of binding proteins *de novo*, since there are all the necessary means for that? The reason is that fully synthetic proteins usually do not fold to compact native-like structures and, if they fold, they possess insufficiently low stability. Therefore, it is reasonable to follow the practical trend – to select the structure of a known protein as a scaffold and to provide targeted or random mutations only at particular sites (Glasner et al. 2007).

The scaffold is usually based on the most rigid elements of protein structure formed by α-helices and β-sheets (Color Plate 5). According to (Hosse et al. 2006), scaffold proteins can be assigned to one of three groups based on the architecture of their backbone:

1. Scaffolds consisting of α-helices (images A–D)
2. Small scaffolds with few secondary structures or an irregular architecture of α-helices and β-sheets (image E) and
3. Predominantly β-sheet scaffolds, representing the majority of proteins used for library display (images F–I)

The scaffolds differ in the presence or the absence of stabilizing disulfide bonds linking spatially separated strands of the protein, a distinction that has conse-quences for the choice of expression system (Hosse et al. 2006). Various reactive groups extending to the surface can be selected for inducing or modifying a particu-lar function, e.g., targeting a particular ligand or providing the fluorescence reporter signal. Because of this enormous variety, achieved by either rational or combinato-rial protein engineering, it is possible to isolate library members binding strongly and specifically virtually to any target.

The scaffold principle was in fact borrowed from Nature. Nature implemented it in the construction of *antibodies* (see Section 5.2 above). The bodies of these molecules are well-defined and highly homologous, except the so-called variable domains. The latter contain 6 hypervariable loops each and it is their variability that is responsible for a population of about 10^8 antibodies of different specificity circulat-ing normally in the human body. Therefore, popular among researchers are the β-sandwich and β-barrel scaffolds, which resemble the antigen-binding variable domains of antibodies (Binz et al. 2005; Binz and Pluckthun 2005).

In contrast, the *ligand-binding proteins* are very conservative by themselves and for the modulation of their binding properties, genetic manipulations are required from the researcher. Protein engineering allows the insertion of structural elements such as folds or loops and, even more, manipulation of whole domains. Thus, proteins with new ligand-binding functions can be engineered through a combinatorial process called random domain insertion (Guntas and Ostermeier 2004). The gene coding domain of one protein can be randomly inserted into the gene sequence of another protein and this hybrid can show not only novel ligand-binding but also allosteric properties.

The design of sensor molecules by *random mutations* at particular sites has shown its efficiency when applied to proteins with natural binding and transport functions.

The results can be compared and even coupled with those obtained by site-directed mutagenesis – amino acid substitutions at pre-defined positions. The latter is the common way of inserting Cys residues that are often needed for labeling with fluorescence reporter dyes that are reactive with –SH groups. The design of such molecular constructs could take lots of skills, especially working with mutant proteins, in which the protein scaffold is stabilized by disulfide bonds. The combination of designed and random-selected protein structures opens wide prospects.

5.3.2   Bacterial Periplasmic Binding Protein (PBP) Scaffolds

In line with their function, transport proteins are commonly highly specific towards their ligands, allowing applications for sensing these ligands as the targets. The *periplasmic binding proteins* of bacteria are the leaders in this sensing strategy. These structurally diverse transport proteins demonstrate high affinity in binding their specific ligands, such as maltose, glucose, glutamine, histidine, phosphate, etc. High-resolution crystallographic structures of the ligand-free and ligand-bound forms showed that PBPs are formed of two domains linked by a hinge and that a hinge-bending motion occurs upon ligand binding (Quiocho and Ledvina 1996). The diversity of biological function, ligand binding, conformational changes and structural adaptability of these proteins have been exploited to engineer biosensors, allosteric control elements, biologically active receptors and enzymes using a combination of techniques, including computational design. PBPs and their ligands have been used as model systems to develop fluorescent sensors based on various transduction principles (Dwyer and Hellinga 2004).

Since the ligand binding site and the fluorescence reporter group may be located in spatially distant areas, these proteins seem appropriate for designing sensors with new ligand binding specificities but sharing similar reporting functions (de Lorimier et al. 2002; Marvin and Hellinga 2001b). Computational methods for redesigning ligand binding specificities of proteins develop actively (Looger et al. 2003). Based on computational findings the mutant binding proteins can be constructed by protein engineering methods. With an environmentally sensitive fluorophore inserted in the hinge between the two domains, they demonstrate experimentally a specific response to the bound targets.

The *maltose binding protein* of E. coli is so far the most efficiently used protein of this family (Medintz and Deschamps 2006). It allows combining two important biosensor properties: specificity of recognition and conformational change. The protein molecule of ~3 × 4 × 6.5 nm in dimension consists in two domains of almost equal size. In open form the binding pocket is exposed to the solvent. Upon binding the maltose, this pocket closes by rotation of domains by ~35° and a lateral twist by ~8° relative to each other. This brings the amino- and carboxy-termini closer by ~0.7 nm. The intra-domain conformation also exhibits some changes (see Color Plate 8). This allows exploring both the contact and remote response to target binding (Dwyer and Hellinga 2004).

5.3.3   Engineering PBPs Binding Sites and the Response of Environment-Sensitive Dyes

There were many successful attempts to transform PBPs into sensors by binding the environmentally sensitive dyes. Initially, in the maltose binding protein a position close to the cleft, where the ligand binds but not involved in the binding itself, was visually selected (Gilardi et al. 1994). It was shown that on binding the maltose, the fluorescence intensity of acrylodan and IANBD dyes increased dramatically. These changes were accompanied by blue shifts of the emission spectra. They can be explained by re-location of the dye from highly polar and exposed to solvent water environment to an environment that is low-polar and screened from contact with the water.

A similar effect was observed in the *phosphate binding protein,* which was suggested as a sensor for inorganic phosphate (P_i) (Brune et al. 1994). Upon P_i binding, the fluorescence spectrum of this label shifts to the blue and undergoes a 5.2-fold increase of emission intensity. The response is very fast, on the time scale of 50 ms.

The possibility to locate the environment-sensitive fluorescence reporter in a position *remote from the ligand binding site* was thoroughly exploited. Such locations that show the largest structural differences in the ligand-free and ligand-bound forms were identified by comparing inter-atomic distances in the two forms (Marvin et al. 1997). Spatial separation of the binding site and reporter groups allows their intrinsic properties to be manipulated independently. In the cited research, three different dyes were coupled at six positions, yielding 18 different constructs. Three out of the 18 constructs showed a larger than two-fold increase in fluorescence intensity. Provided that allosteric linkage is maintained, the ligand binding can therefore be altered without affecting fluorescence reporting. To demonstrate its applicability to biosensor technology, the authors introduced a series of point mutations in the maltose-binding site that lowers the affinity of the protein for its ligand. These mutant proteins were combined in a composite biosensor capable of measuring a target concentration within a 5% accuracy over a concentration range spanning five orders of magnitude.

A successful attempt to modulate the *binding affinity* of a maltose binding protein that does not involve the binding site was demonstrated. This can be done by introducing mutations located at some distance from the ligand binding pocket, which sterically affect the equilibrium between an open, apo-state and a closed, ligand-bound state (Marvin and Hellinga 2001b). The possibility to radically change the specificity of this protein was demonstrated by converting it into a zinc sensor using a targeted design approach (Marvin and Hellinga 2001a). In this new molecular sensor, zinc binding is detected in the form of a fluorescence signal by use of an engineered conformational coupling mechanism linking ligand binding to a reporter group response.

Glucose binding protein has also been the subject of extensive studies. The 'non-allosteric' and 'allosteric' locations of amino acid residues were selected based on the known protein structures in the open and closed forms. These sites were substituted with cysteins for the attachment of environment-sensitive acrylodan and NBD dyes. Allosterically located dyes behaved efficiently, demonstrating a several-fold change of the intensity of the fluorescence signal (Marvin and Hellinga 1998). Strong changes in the fluorescence intensity of 'allosterically' located acrylodan dye were also observed in the glutamate binding protein (Tolosa et al. 2003).

In an extended study, de Lorimier et al. (2002) have conjugated different environmentally sensitive fluorophores at various positions of eleven members of the PBP family. They selected positions that either directly contact with the ligand, are located near the ligand-binding site, or are located away from the binding site, in a region that changes conformation on ligand binding. Approximately a quarter of the 320 conjugates gave a satisfactory sensor response. Binding affinities were mostly affected in the group of constructs where the fluorophores directly contacted the ligand. Wavelength shifts together with the changes of intensity were observed for the environment sensitive dyes, such as acrylodan and NBD. Meanwhile, for some of the labeled PBPs these changes were very little or even undetected, showing that the dye response is strongly position-dependent.

5.3.4   Scaffolds Based on Proteins of the Lipocalin Family

Another class of protein ligand binders, the '*anticalins*', are the artificial products constructed by introducing structural diversity into the binding site of lipocalins, a family of small monomeric ligand-transporting proteins (Weiss and Lowman 2000). Lipocalins are the molecules of 160–180 amino acid residues that are involved in the storage of hydrophobic and/or chemically sensitive organic compounds (Flower et al. 2000). They consist in a β-barrel formed of eight anti-parallel strands, which is the central folding unit forming a conical cavity. The cavity is relatively deep and largely nonpolar. The entrance to this cavity is formed by four loops, which can be randomly mutated for generating molecular pockets with a diversity of shapes and providing the binding sites to different ligands.

The first antiacalins were derived from insect origin. They allowed obtaining sensors for small molecules such as steroids (Korndorfer et al. 2003). In addition to low molecular weight compounds, they can be targeted towards proteins. This was shown for a member of the lipocalin family of human origin, apolipoprotein D. Its function is to transport arachidonic acid and progesterone in various body fluids. After the randomization of 24 amino acids located within the loop region, a mutant was selected that started to bind hemoglobin (Vogt and Skerra 2004).

Anticalins as binders, offer some advantages over traditional antibodies. They are especially important in molecular recognition between receptors and small molecule ligands (Weiss and Lowman 2000). Still, as molecular sensors, their application presently does not go beyond the cases, where the ligands are fluorescent, such as retinol. The proper large-scale applications in fluorescence sensing technologies are still in prospect. These technologies should be adapted to the small size of these proteins, to the rather rigid conformation of their binding sites and to the absence of global conformational changes upon ligand binding.

5.3.5   Other Protein Scaffolds

In addition to scaffolds based on fragments of antibodies, periplasmic ligand-binding proteins and lipocalins, other protein scaffolds have been suggested to create libraries of different binding proteins (Hosse et al. 2006). They were developed with different purposes and focused on mostly pharmaceutical applications. Some of them may be attractive as potential sensors, however. It can be observed that if the scaffold is based on the protein that binds small ligands, then it is easier to achieve specific binding of such ligands. In other cases, the scaffolds are better adapted to large ligands, such as proteins (Hosse et al. 2006). The topology of the interacting surfaces and their complementarity in these cases can be realized more easily. If the three-dimensional structure of a target and a scaffold are known, the mutation sites can be assigned based on a computational design (Wiederstein and Sippl 2005). A variety of protein scaffolds can be analyzed *in silico*, and the results of this analysis fit the experimental data satisfactorily.

Among the prospective sensors are '*affibodies*'. Derived from bacterial cell surface receptors, they represent an engineered version (Z domain) of one of the five stable three-α-helix bundle domains from the antibody-binding region of staphylococcal protein A (Eklund et al. 2002; Renberg et al. 2005). They are small (6 kDa and 58 residues only), highly soluble, do not contain S-S bonds and can be fused on a genetic level with other proteins (Ronnmark et al. 2003). Their small size and the property of spontaneous folding allowed providing their complete chemical synthesis and assembly on an automated peptide synthesizer. Moreover, during this synthetic process the fluorescent dyes and reactive groups for attaching protein to the surface can be incorporated in the desired positions (Engfeldt et al. 2005). A biotin moiety can also be introduced in the same way. This allows providing not only the reporting function but also the binding to a required site. Affibodies have found application in constructing protein microarrays (Renberg et al. 2007) but the presently suggested technologies still rely on target labeling or on a sandwich format.

Thus, we observe that the concept of '*minimal*' protein scaffolds together with the idea of artificial target recognition sites led to many successful developments. The members of several protein families represent promising model systems in this respect. Other examples of 'minimal' protein scaffolds include mutated forms of *cytochrome* b_{562}, *ankyrin repeat domains*, *leucine-rich repeat proteins*, insect *defensin* A, *protease inhibitors* and *scorpion toxins* (Binz and Pluckthun 2005; Hosse et al. 2006). The small size of these proteins allows the efficient use of computer design and the point mutation approach to improve stability and functionality. For scaffolds derived from defensin A and containing 29 amino acids, a library of such mutations has been produced (Yang et al. 2003b).

In order to find scaffolds for new sensors, scientists have started active studies of the proteins that provide immune response in a primitive organism (Binz et al. 2005). It is known that not all adaptive immune systems use the immunoglobulin fold as the basis for specific recognition molecules. Sea lampreys, for example, have evolved an adaptive immune system that is based on *leucine-rich repeat proteins*. Many other proteins, not necessarily involved in adaptive immunity, mediate specific high-affinity interactions. Their transformation into operative fluorescence sensors is a task for future research.

5.3.6 Prospects

Understanding the fact that different protein families, unrelated to immunoglobulins, can provide the basis for specific recognition molecules and can compete with immunoglobulins in many valuable properties, has stimulated rapid research. Such proteins were found primarily among natural ligand-binding proteins. For adaptation to a particular target and an improvement of these properties, an approach based on using the protein structure as a scaffold and applying point and random mutations to provide the changes at particular sites. The examples presented above show that the unlimited possibilities offered by the techniques of random and directed mutations are not enough for making optimal binders. This field needs the knowledge of molecular interactions and of their dynamics. A rigid scaffold as a starting point has proved to be an efficient and prospective approach for this. It should provide an extraordinarily stable protein architecture tolerating multiple substitutions or insertions at the primary structural level. These substitutions should determine affinity and specificity in target binding.

Scaffold-based affinity sensors offer good prospects in biosensor technologies but only on condition that the problem of coupling with the fluorescence reporter response is properly addressed. Protein-ligand binding events usually involve, in different proportions, the combination of binding pocket solvent exclusion (Liu et al. 2005) and conformational change (Flores et al. 2006). A recent study (Yesylevskyy et al. 2006) shows that the number of proteins with global changes of conformation and its dynamics on ligand binding is very limited. Therefore, coupling the binding events with conformational changes cannot be the general mechanism of operation of these sensors. Their prospects will depend strongly on other mechanisms of signal transduction involving those based on direct contact of the target with the reporter dye.

5.4 Designed and Randomly Synthesized Peptides

In the previous section, we discussed the possibilities of making fluorescence sensors based on functional ligand-binding proteins. We observed successful results in attempts to transform them into sensors by producing minor modifications of their structure that mainly involved target binding sites and the sites of reporter binding. Here we discuss and evaluate the possibilities arising from the use of synthetic peptides. They allow *artificial selection of scaffolds* as either folded domains or flexible peptide chains. The peptides possess the necessary properties as recognition units: flexibility to adapt sterically to any target, rigidity in the secondary structure to conserve the binding aptitude and the propensity to form different kinds of non-covalent bonds with a potential target.

To be successful in the generation of designed binding molecules, this approach has to rely strongly on the ability to create and operate with large libraries together with powerful library selection technologies. The advances in protein engineering, selection and evolution technologies can be actively used. But the main advantage of this approach is the possibility of obtaining sensing molecules of the *minimum possible size*. This allows a large-scale application of standard solid-phase techniques of peptide synthesis, which is a rather simple standard technology.

Operating with synthetic peptides has many technical advantages. Peptides offer functional robustness superior to that of most proteins and are well suited for long-term storage in dry, dissolved or immobilized forms. Importantly, they allow full control of the labeling process including a complete labeling at a single site. Oriented immobilization in protein chip formats is readily obtained.

5.4.1 Randomly Synthesized Peptides, Why They Do Not Fold?

Synthetic peptides with a random location of amino acid residues are commonly unfolded. The formation of their native-like structures is a low-probability event that can be observed for very rare sequences. This is because of a strong connection between the amino acid sequence and the three-dimensional organization, known as the *folding code* (Demchenko and Chinarov 1999). The folding code is highly *degenerate* and includes both steric and energetic factors. This means that some substitutions (even many of them) still lead to correct folding. Even larger substitutions are allowed if they are compensated in a structural or thermodynamic sense by other substitutions. The folding code has a *distributed* character, which means that it is not reduced to interactions between particular residues but involves extended elements of structure. In addition, kinetic variables on different hierarchical levels may not only determine the pathway of folding but also the resultant folded structure (Demchenko 2001a; Yesylevskyy et al. 2005).

Because of that, the prediction of folded structures is difficult. Moreover, the folding can be strongly influenced by intermolecular interactions (Demchenko 2001b). All these factors complicate the design of the protein and peptide 3D structures *ab initio*. No less difficult is to design the interaction with the target molecule even if the structure is well known. Therefore, two strategies in the production of peptide binders have been suggested and used:

(a) *Rational design based on already known folded motifs*. This means that if this motif is known, being observed in native proteins, then it is highly probable that it will fold in the same way when produced as a structure of minimal size by synthetic means. By selecting this fold motif as a scaffold structure, one can introduce different modifications inducing or modifying the target recognition and fluorescence response properties. This strategy is commonly known as the *template-based approach* (Singh et al. 2006).

(b) *Selection from a large combinatorial library*. In this case, making a combinatorial peptide library and establishing the selection criteria are needed. The binding to the target could be the major selection criterion, so that this selection could be realized, for instance, by affinity chromatography.

5.4.2   Template-Based Approach

The well-defined secondary and tertiary structures of peptides and proteins are induced and stabilized by interactions in the main chain and by contacts between side groups. Despite a huge conformational space that allows an astronomical number of peptide conformations, the number of folding motifs is very limited (Schulz and Schirmer 1979). The selection of a topological peptide template can be made based on the known 3D structures of proteins. The recent advances in the chemistry of coupling reagents, protecting groups and solid-phase synthesis have made the chemical synthesis of peptides with conformationally controlled and complex structures feasible (Singh et al. 2006).

These peptide templates can be used to construct novel structures with tailor-made functions. Such a peptide-template-based approach demonstrates the utility in achieving a molecular recognition of different targets. A statistical picture of amino acids found at protein-protein interaction sites indicates that proteins recognize and interact with one another mostly through the restricted set of specialized interface amino acid residues Pro, Ile, Tyr, Trp, Asp and Arg (Sillerud and Larson 2005). They represent the three classes of amino acids: hydrophobic, aromatic and charged (one anionic and one cationic). They can be used for constructing the peptide recognition sites.

The automatic *solid phase peptide synthesis* is the key technology of producing the template-designed peptide sensors. The standard procedures allow obtaining peptides as long as 20–30 residues, which can be sufficient for making the designed folds. These techniques allow *parallel synthesis* on the solid support. Combining the surface chemistry with the recent technology of microelectronic semiconductor fabrication, the spatially addressable peptide microarrays can be obtained (Hamada et al. 2005). One can choose between two synthesis methodologies: pre-synthesized peptide immobilization onto a glass or membrane substrate and peptide synthesis *in situ*.

5.4.3   The Exploration of the 'Mini-Protein' Concept

It was interesting to note that a peptide as small as 20 residues can possess a cooperatively folded tertiary structure (Gellman and Woolfson 2002). Such mini-proteins could serve as a fruitful platform for protein design. They serve as leads for sensor applications by positioning all the amino acids necessary for bio-molecular recognition on a compact protein structure. Particularly, they can display a high DNA binding affinity and specificity (Pflum 2004). A 38-amino acid peptide was described as an α-helical hairpin stabilized by two disulfide bridges and presented as a scaffold for future sensor developments (Barthe et al. 2004).

It was proposed that a synthetic 42-residue helix-loop-helix polypeptide that dimerizes to form four-helix bundles could form a scaffold for molecular sensors (Enander et al. 2002). Different functional groups can be incorporated covalently into this scaffold in a site-selective manner. The incorporation of a ligand interacting with the binding site of a protein together with an environment-sensitive dye, allows obtaining a sensor (Fig. 5.8).

This fully designed sensor with an inhibitor of carbonic anhydrase as a recognition unit and a reporting fluorescent dye can detect the binding of carbonic anhydrase (Enander et al. 2002). This binding results in a significant increase of fluorescence intensity, which arises from a disruption of the homodimer in the presence of the target (Enander et al. 2004a). This concept can be applicable to numerous targets, for which the ligand is a small compound that can be grafted on the peptide scaffold. The potential of this approach for microarray applications has been demonstrated by designing an affinity array (Enander et al. 2004b).

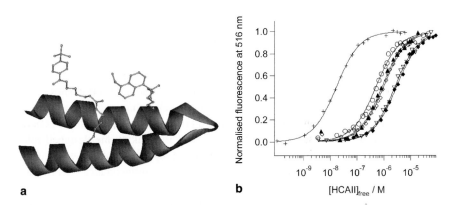

Fig. 5.8 Schematic representation of the scaffold of the helix-loop-helix polypeptide forming a four-helix-bundle when dimerized and proposed as a versatile sensing platform. (**a**) The scaffold is shown with a dansyl probe and benzenesulfonamide attached at amino acid positions 15 and 34, respectively. Attachment of a bifunctional spacer, here aminohexanoic acid, allows one to cova-lently link the benzenesulfonamide residue to a lysine residue side chain in the polypeptide scaf-fold. This construct was used for the detection of human carbonic anhydrase II (HCAII). (**b**) The graph represents a fluorescence response to target binding as a function of its concentration at different lengths of a spacer used for benzenesulfonamide attachment (Reproduced with permis-sion from Enander et al. 2004b).

5.4.4 Molecular Display Including Phage Display

Phage display is a powerful method for the discovery of peptide ligands that are used for analytical tools, drug discovery and target validations (Uchiyama et al. 2005). Phage display technology can produce a great variety of peptides and generate novel peptide ligands. A number of other display platforms include bacterial and yeast display, ribosome display and mRNA display (Levin and Weiss 2006).

Ribosome display (Zahnd et al. 2007) is an *in vitro* selection and evolution technology for proteins and peptides from large libraries. The diversity of the library is not limited by the transformation efficiency of bacterial cells. The random mutations can be introduced easily after each selection round. This allows the facile *directed evolution* of binding proteins over several generations.

The affinity selection of peptides displayed on phage particles can be used for mapping molecular contacts between small molecule ligands and their protein targets (Rodi et al. 2001). Important observations were made in these studies – the binding properties of peptides displayed on the surface of phage particles could mimic the binding properties of peptide segments in naturally occurring proteins. The conformation of these segments can be relatively unimportant for determining the binding properties of these disordered peptides because they adopt 'induced' conformation upon binding to the target. Such *'induced fitting'* to the target (Demchenko 2001b) can be the basis for its molecular recognition (Fig. 5.9).

For the selection of peptide binders from the library, one does not need to know the 3D structures of interacting partners. Moreover, these structures can be flexible. This allows a rapid large-scale identification of potential ligand binding sites and also the easy introduction of a fluorescence reporter responsive to target binding in a number of parameters (wavelength ratiometry, lifetime, anisotropy). The dynamic formation of optimal non-covalent bonds compensates the entropy loss associated with the loss of conformation mobility.

The peptide based sensors are potentially useful for the diagnosis of viral, bacterial, parasitic and autoimmune diseases by detecting the correspondent antibodies (Gomara and Haro 2007). Synthetic peptides can provide uniform, chemically well-defined antigens for antibody analysis, reducing inter- and intra-assay variation. The success of this approach depends on the extent to which synthetic peptides are able to mimic the antigenic determinants (Timmerman et al. 2005).

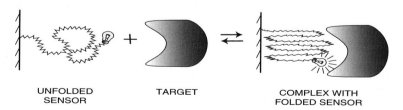

UNFOLDED   TARGET   COMPLEX WITH
SENSOR   FOLDED SENSOR

Fig. 5.9 Illustration of 'induced fitting' in sensor-target interaction. It allows a facile response from a fluorescence reporting dye due to the change of its mobility, hydration or polarity of environment

If *in vitro* selection techniques could produce short polypeptides that tightly and specifically bind to any of a wide range of macromolecular targets, the possibilities for sensor developments would be immense. Some applications have already been demonstrated. One of them is to construct '*peptide beacons*' (Oh et al. 2007). This sensing architecture is based on the increase of rigidity when complexed to a macromolecular target. It is similar to DNA beacons (Section 6.3) and is based on the change of the relative distance between the fluorescent dye and the quencher on target binding. Based on this principle, a robust optical sensor was developed for anti-HIV antibodies (Oh et al. 2007) and the antibodies to p53 protein that are important biomarkers for cancer (Neuweiler et al. 2002).

With peptide sensors, the simplest fluorescence detection is the most difficult in realization. Obtaining a fluorescence response from a single label responding in a wavelength-ratiometric manner requires a new generation of dyes with a two-band emission. Recently, the design, synthesis and functional evaluation of a prototype of a peptide-based sensor for protein analytes was reported (Enander et al., 2008). The high-affinity interaction between the 18-amino acid peptide derived from the antigenic part of a tobacco mosaic virus protein and a recombinant antibody fragment was detected as a substantial change of the intensity ratio of two bands (I_N^*/I_T^*) in emission spectra (Color Plate 6).

The peptide was modified by binding in two different positions with 3-HC dye 6-bromomethyl-2-(2-furanyl)-3-hydroxychromone, an environmentally sensitive fluorophore with a two-band emission. When the peptide was labeled in the C terminus, the I_N^*/I_T^* ratio changed by 40% upon an analyte binding, while labeling close to the residues that are the most important for binding resulted in a construct that completely lacked ratiometric biosensor ability. This result shows that the response of a fluorescence dye in such constructs is strongly position-dependent and that rational design should be complemented by comparative studies of systems with different dye locations.

5.4.5 Antimicrobial Peptides and Their Analogs

Some natural peptides possess functionally important recognition properties and it is straightforward to use them, as well as their analogs, as the binders. Numerous bacteria, plants and higher organisms produce antimicrobial peptides as part of their innate immune system, providing a chemical defense mechanism against microbial invasion. Many of these peptides exert their antimicrobial activity by binding to components of the microbe's surface and disrupting the membrane. They can be incorporated into screening assays for the detection of a pathogenic species (Ngundi et al. 2006). It was shown that the surface-immobilized peptides, such as polymyxins B and E, can be used to detect pathogenic bacteria in two assay formats: sandwich and direct.

Different mutant forms of short antimicrobial peptides have been obtained and selected. They are promising as remedies to fight against antibiotic-resistant microbes. Using *antimicrobial peptides* as recognition elements in an array biosensor detection of these microbes can be achieved (Kulagina et al. 2006).

5.4.6   Advantages of Peptide Technologies and Prospects for Their Development

Peptides as chemical products possess many advantages over protein sensors based on ligand-binding proteins, enzymes or antibodies. They are readily obtained in large amounts by chemical synthesis and the introduction of a fluorescent dye can be a part of this synthesis but not just as a modification at targeted sites, as in proteins.

We also observe that two methods of functional peptide design and production, chemical synthesis and phage display, enrich each other (Uchiyama et al. 2005). The best binders selected from combinatorial libraries can be produced by chemical synthesis on a large-scale level. The synthesis can be robotic. It can be provided directly on arrays to yield a product with highly reproducible properties. In comparison, the large-scale production of mutant recombinant proteins and their subsequent modification is tedious and very expensive. With peptides, all the problems of poor expression levels with the mutated proteins, deleterious effect on the binding affinity of the mutation or of fluorophore coupling or the low yield of this coupling, are avoided.

Synthetic peptides possess a better thermal and chemical stability. They can be easier integrated into nanoparticles, porous materials and polymer gels, or deposited in an array format for the simultaneous detection of many targets (Kodadek 2002). These technologies are expected to combine low cost, speed and convenience, with a wide range of applications in diagnosis and environmental protection.

5.5   Nucleic Acid Aptamers

Aptamers are single-stranded DNA or RNA oligonucleotide sequences that possess the ability to recognize various molecular targets. The targets include peptides and proteins and their recognition occurs with high affinity and specificity (Hamula et al. 2006; Tombelli et al. 2005). The development of aptamers was based on the observation that nucleic acids have a strong potency of not only mutual recognition between the bases forming double-helical structures. They can also fold to form recognition elements for a great number of different targets. Commonly, these properties of oligonucleotides are not observed or are very weak but after the appearance of *selection-amplification techniques* operating with large libraries, the number of aptamers with very strong and extremely selective binding abilities increased tremendously. Much like in peptide selection, to find the best binders one has to create a large pool of similar compounds, a *combinatorial library* and screen this library.

The polyanionic nature of nucleic acids does not prevent them from forming stable three-dimensional structures. These structures are stabilized by an intramolecular formation of short double-helical segments composed of complementary nucleic acid bases. Such structures may exhibit an ideal combination of rigidity and flexibility in intermolecular interactions.

5.5.1   Selection and Production of Aptamers

Aptamers can be identified in oligonucleotide pools by an *in vitro* selection process known as SELEX (*systematic evolution of ligands by exponential enrichment*) (Hamula et al. 2006; Zheng et al. 2006). First, a library of oligonucleotides with a random location of bases has to be obtained. It may contain as much as 10^{15} different molecules and SELEX involves an iterative process of search for the best binders and their selection (Gopinath 2007). SELEX involves repetitive rounds of two processes: (a) partitioning of aptamers from non-aptamers by the affinity method and (b) amplification of aptamers by the *polymerase chain reaction* (PCR).

The sizes of aptamers could vary from tens to thousands of nucleotides. Typically, it is smaller than 200 bases and this is sufficient for their optimal performance. Being single-chain molecules they can fold to make segments of a double-helical structure separated by loops (O'Sullivan et al. 2002). The development of *in vitro* selection and amplification techniques has allowed the identification of specific aptamers, which bind to the target molecules with high affinities. These affinities are frequently comparable with those of monoclonal antibodies, so that their dissociation constants can be observed in the nanomolar to picomolar range. Compared to antibodies, the cross-reactivity of aptamers in binding protein targets is typically minimal (Hicke et al. 2001). Thus, they can discriminate protein targets containing only several amino acid substitutions.

It is important that since aptamers are selected wholly *in vitro*, their specificities can be crafted by the addition of negative selection steps. Presently the high-throughput selection of aptamers can be accomplished by robots. Due to the fact that they can also be potent pharmacological agents (their specific binding reduces protein activity) the amount of selected and identified aptamers has grown tremendously. Aptamer databases were created to help their systematization (Thodima et al. 2006). One of them can be found at http//mfgn.usm.edu/ebl/riboapt.

5.5.2   Attachment of Fluorescence Reporter, Before or After Aptamer Selection?

Aptamers can be labeled with fluorescence reporters after the SELEX procedure selects the optimal binders or even before the SELEX procedure (Fig. 5.10). All classical methods of post-translational modification of nucleic acids can also be applied for labeling aptamers as well as the incorporation of naturally emitting 2-aminopurine (Katilius et al. 2006) and base-substituting dyes. Their fluorescence quantum yield strongly depends on the base stacking interactions when incorporated into double or single stranded DNA. This property can be used to generate a binding-specific fluorescence signal if the aptamers possess the modified fluorescent nucleotide analogues in positions that undergo conformational changes (as shown in Fig. 5.10a). Aptamers that combine recognition and reporting functions are called *signaling aptamers*. The labeling of nucleic acids has been well described in the literature (Cox and Singer 2004).

Similarly to synthetic peptides, for DNA and RNA aptamers the problem of introducing fluorescence reporters can be solved in a most simple and elegant way. The dyes can be introduced covalently into their structures before the selection process (Fig. 5.10b). There is no restriction regarding such modifications and not only single labeling with the environment-sensitive dye but also double labeling with the emitter and quencher or with FRET donor and acceptor can be performed easily. Such a procedure has been described for ATP detecting aptamers (Jhaveri et al. 2000). Nucleotides bearing fluorescent labels can constitute a random pool of sequences, from which the ligand-binding species are selected. The binding species are then screened for aptamers that signal the presence of cognate ligands.

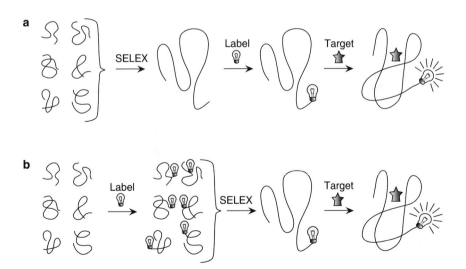

Fig. 5.10 Design of signaling aptamers. (**a**) Labeling an aptamer after the SELEX procedure. The site of labeling is selected with the account of expected strong conformation changes that could change the fluorescence property of the attached dye. (**b**) Signaling aptamers are generated by SELEX using a random-sequence library in which each DNA or RNA molecule is labeled with one or a few dye molecules

With these remarkable properties, aptamers allow all the possibilities for fluorescence reporting, including multiple dye labeling with the observation of excimer formation, PET and FRET. The selection of already labeled aptamers allows minimizing the possible negative effects of post-selection modifications on the aptamer binding to the target (Hesselberth et al. 2000). However, the most attractive idea could be to make a reporter dye that can be involved directly in the process of target recognition. This could allow recording direct changes in the parameters of their fluorescence.

5.5.3   Obtaining a Fluorescence Response and Integration into Sensor Devices

Due to the fact that aptamers are relatively small and flexible molecules and that they may change their conformation significantly on target binding, all the arsenals of fluorescence reporting methods described in Chapter 3 can be applied to aptamer sensing. The results of the application of some of them have been reviewed recently (Nutiu and Li 2005a).

Generating a fluorescence response in aptamers is facilitated by the fact that very often they are largely unstructured in solutions. They fold into well-defined three-dimensional structures upon binding their target molecules (Hermann and Patel 2000; Stojanovic and Kolpashchikov 2004). This allows many possibilities to introduce the reporting signal (Fig. 5.11).

Such a response can be provided by the spectral shifts or in a two-band ratiometric manner. In addition, the retardation of the dye rotation can be observed by the increase of anisotropy. Moreover, the quenching effects can be obtained in lifetime domains. It is surprising in this respect that many researchers prefer less informative intensity sensing, trying to provide a quantitative analysis and facing the problem of proper response calibration. The double labeling with the emitting dye and the quencher can be applied for this.

In fact, *double labeling* is not a great problem if the assay is made in solution and both the 5′-end and 3′-end are available for labeling. To provide the necessary response, the fluorophore and the quencher in one of the forms (with unbound or bound target) should be close together. Luckily, this happens (or can be designed) in some aptamers that structurally resemble 'beacons' and an example of this is the sensor for thrombin (Li et al. 2002). The folding of the DNA chain around the target molecule brings the 5′-end and the 3′-end together, which results in quenching (Fig. 5.12). The same authors showed that the incorporation of an emissive FRET pair at the same sites allows obtaining the wavelength-ratiometric response to target binding.

Structure-switching signaling aptamers are often designed so that they use duplex-complex transition in the conditions of a homogeneous assay (Nutiu and Li 2004, 2005a). The duplex in this case is represented by a double-helical structure labeled with fluorescent dye and a complementary sequence labeled by a quencher. In this structure, the fluorescence is quenched and unquenching occurs when the conformation change occurs and a new complex with the target is formed.

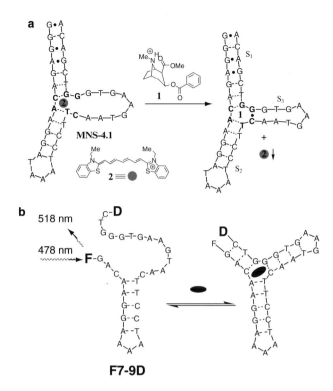

Fig. 5.11 The examples of two sensors responding to the binding of cocaine. (a) Competitive assay based on displacement by cocaine (**1**) of a cyanine dye (**2**). The dye changes its absorbance and aggregates. (b) The assay is based on inducing by cocaine binding of conformational change that approximates the 5' and 3' ends. This results in quenching the fluorescence of fluorescein (**F**) by dabcyl dye (**D**) (Reproduced with permission from Stojanovic et al. 2001; Stojanovic and Landry 2002).

The double labeling of an aptamer with pyrene allows obtaining the two-wavelength ratiometric response on formation of pyrene excimers (Yang et al. 2005). Such a light-switching aptamer was developed for the rapid and sensitive detection of a biomarker protein, platelet-derived growth factor (PDGF). Labeled with one pyrene at each end, the aptamer switches its fluorescence emission from an approximate 400 nm (pyrene monomer) to 485 nm (pyrene excimer) upon PDGF binding. This change of fluorescence spectrum is a result of the conformation rearrangement induced by target binding. The excimer probe is able to effectively detect the picomolar PDGF in a homogeneous assay. Because the excimer has a much longer fluorescence lifetime (approximately 40 ns) than that of the background (approximately 5 ns), the time-resolved measurements can be efficiently used to eliminate the biological background.

Followers of the 'sandwich' methodology also contribute to the field of aptamers sensing. In a recent study (Heyduk and Heyduk 2005) a sensor was developed

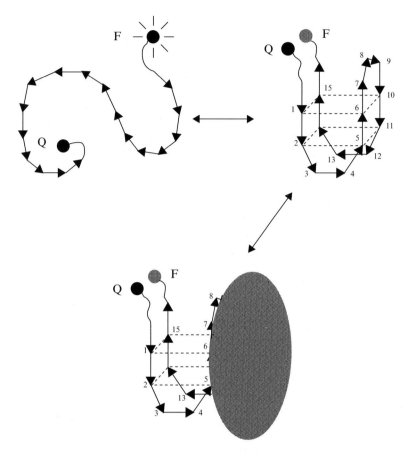

Fig. 5.12 Schematic representation of the conformational change in the thrombin-binding aptamer reported by fluorophore-quencher interaction. At equilibrium, the aptamer switches between two conformations, a nonstructured random coil and a compact intramolecular quadruplex stabilized with internal hydrogen bonds. The addition of thrombin shifts the equilibrium to favor the quadruplex structure, drawing the fluorophore (F) and quencher (Q) closer together and causing fluorescence quenching (Reproduced with permission from Li et al. 2002)

that involves protein-induced co-association of two aptamers recognizing two distinct epitopes of the protein. The aptamers contain short dye-labeled complementary 'signaling' oligonucleotides attached to the aptamer by a non-DNA linker. Co-association of two aptamers with the target protein results in bringing the two oligonucleotides into proximity, producing a large change of FRET signal between the fluorophores.

In cases when there is a change of the double-helical structure on target binding, the response can be obtained with *intercalated dyes*, particularly, with rubidium chelating complexes with response in phosphorescence (Jiang et al. 2004). These complexes have no luminescence in an aqueous solution without DNA but the binding to double-stranded segments of DNA results in their intense emission.

A high affinity ($\sim 10^6$ M^{-1}) to these segments and their change on target binding eliminates the necessity of covalent labeling of aptamers.

An interesting possibility to obtain a fluorescence response on target binding is to apply the dyes by *replacing DNA bases*. These dyes should report on the rupture-formation of the double helix. The fluorescence of 2-aminopurine is sensitive to conformational changes of DNA. Highly fluorescent donor-acceptor *purines* were synthesized recently (Butler et al. 2007). Their absorption spectra located at 320–340 nm appear weakly solvent dependent, while the emission maxima at 360–400 nm show significant long-wavelength shifts upon increases in polarity. They show nearly 100% fluorescence quantum yields and lifetimes up to 3 ns in organic solutions with only a moderate decrease in highly polar solvents, which may be unique for such simple nucleobases. Their application in sensing is expected to follow.

In contrast to purines, the synthesis of responsive isosteric fluorescent *pyrimidine* analogues was not so successful but progress has also been made. In a recent study (Srivatsan and Tor 2007), appending five-membered aromatic heterocycles at the 5-position on a pyrimidine core, a family of responsive fluorescent nucleoside analogues was reported. They contain a furan group as a strong electron donor, which allows exhibiting emission in the visible range. So constructed, ribonucleoside triphosphates are accepted by T7 RNA polymerase as substrates for *in vitro* transcription reactions and are very efficiently incorporated into RNA oligonucleotides, with the generation of fluorescent constructs.

It is certain that fluorescent base replacements of DNA and RNA will continue to find useful applications in sensing techniques that involve specific binding of nucleic acids. They are able to mimic natural nucleobases and, at the same time, report on their interactions with other bases and on binding with proteins.

5.5.4 Aptamer Applications

Small molecule sensing is performed more efficiently with aptamers than with peptides. This can be illustrated by successful isolation from a population of random RNA sequences of subpopulations that bind specifically to a variety of organic dyes. According to rough estimates, one in 10^{10} RNA molecules folds in such a way as to create a specific binding site for small ligands, such as amino acid derivatives, cocaine or ATP (Nutiu and Li 2005b).

Aptamers have shown much promise towards the detection of a variety of *protein targets*, including cytokines. Specifically, for the determination of cytokines and growth factors (Guthrie et al. 2006), several assays making use of aptamers have been developed, including aptamer-based analogs of ELISA, antibody-linked oligonucleotide assay and fluorescence assays based on anisotropy and FRET (Hamula et al. 2006).

The specific detection and quantitation of *cancer-associated proteins* (inosine monophosphate dehydrogenase II, vascular endothelial growth factor and basic fibroblast growth factor) in the context of human serum and in cellular extracts has been realized with aptamers. It is expected that this technology will improve the diagnosis of cancer by enabling the direct detection of the expression and modification of proteins closely correlated with the disease (McCauley et al. 2003).

Aptamers are useful in the studies of *protein-protein interactions*, where a competitive assay format can be applied, in which the aptamers are displaced from protein-protein contact areas.

Aptamer microarrays are becoming one of the most efficient sensing technologies for the multiplex analysis of numerous proteins in parallel, furthering the notion that such arrays may be useful in *proteomics*. Their fabrication is now developed to a great detail (Collett et al. 2005). Fluorescence polarization anisotropy can be used for the measurements of target protein binding both in solutions and on solid support (McCauley et al. 2003).

The solid-phase aptamer-protein interactions are similar to binding interactions seen in solution. Usually when the aptamers are used in heterogeneous sensing platforms, they can be the nucleic acid sequences of different lengths, with one of the ends (either the 3' or 5'-end) being normally used for the binding of the aptamer to the solid support. The other end can be used for carrying a dye. Biotin can be attached to one of the ends, so the aptamers can be spotted on streptavidin-coated slides benefiting from self-assembly based on a very strong streptavidin-biotin interaction (Collett et al. 2005).

In addition to sensor technologies, selected oligonucleotides can be used as '*aptazymes*', the species that possess biocatalytic properties and allow the direct transduction of molecular recognition into catalysis. Together with aptamers they can be used in different bioassays for the detection and quantitation of a wide range of molecular targets (Hesselberth et al. 2000)

5.5.5 Comparison with Other Binders: Prospects

No more than 15 years have passed between the introduction of aptamers technology and its development into one of the most successful fields of molecular sensing. There is no restriction in developing the aptamers as the *sensors to any target*. Selected aptamers can bind to their targets with high affinity and discriminate between closely related target molecules. Aptamers can thus be considered as a valid alternative to antibodies as well as to any peptide or protein bio-mimetic receptors. The unique binding properties of nucleic acids, which are amendable to various modifications, make aptamers perfectly suitable for not only sensing but also for other uses in biotechnology, including pharmaceutical applications (Proske et al. 2005). Such a combination of already realized and potential applications as analytical, diagnostic and therapeutic tools (Tombelli et al. 2005) still further stimulates their development.

Out of all macromolecular binders, only the antibodies can presently compete with library-selected aptamers and peptides in the versatility of their adaptation to any particular target. Regarding other aspects, such as production, storage and application, the aptamers possess many potential *advantages* compared to antibodies. They are technologically more attractive because of their much smaller size. Their large-scale production is expected to be more reproducible and much cheaper. They possess a higher stability to environmental factors, which allows long-term storage without loss of functional properties. Co-synthetic structural modifications, including the introduction of fluorescent groups, can be achieved much more easily. They show superiority to antibodies in specific protein detection (Stadtherr et al. 2005). Moreover, being selected for tripeptide sequences, they can recognize these sequences in a large protein, which makes simultaneous multi-site protein recognition possible (Niu et al. 2007).

Aptamers share with synthetic peptides the benefits regarding manipulation and incorporation into sensor constructs and the main difference between them is on the selection step. The advantage of nucleic acid libraries over protein or peptide libraries is their large size (up to 10^{15} different sequences) and the easy identification of selected binders by an enzymatic amplification with polymerases. The *larger library* offers more molecular diversity and a higher probability of finding the optimal binders. Automated selection procedures now allow the rapid identification of DNA and RNA sequences that can target a broad range of extra- and intracellular proteins with nanomolar affinities and high specificities.

All these advantages are not always clearly seen if the aptamers are accommodated into traditional techniques developed for antibodies, such as ELISA. New techniques exploring their advantages in full should be developed. They will be highly competitive and demonstrate a full potential for the successful realization of extremely valuable properties of aptamers.

5.6   Peptide Nucleic Acids

All current techniques for quantifying specific DNA or RNA exploit the base pair complementarity between the target polynucleotide and a complementary nucleic acid sequence serving as the sensor recognition unit. The major effort of researchers is now directed towards developing the techniques that could dramatically increase the sensitivity of response to this primary interaction. This search was extended to synthetic polymers. It was found that the neutral polymer called peptide nucleic acid demonstrates superior to DNA or RNA binding properties.

5.6.1   Structure and Properties

Peptide nucleic acid (PNA) is an artificial polymer that contains the same side groups (bases) as DNA and RNA and is capable of forming the highly stable complexes with complementary sequences of nucleic acids. Peptide nucleic acid is not an acid! It is an oligomer, in which the entire charged backbone is replaced by an

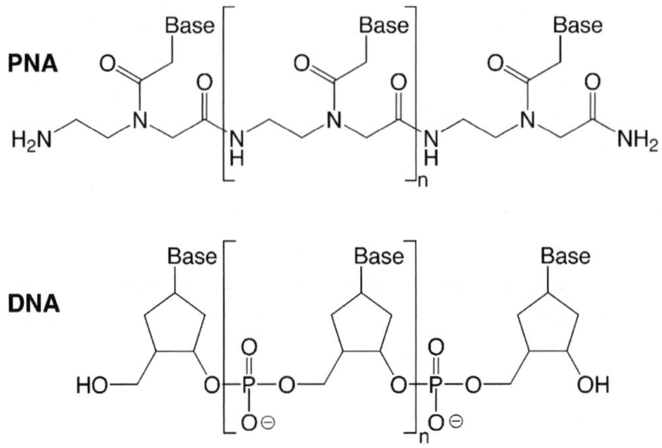

Fig. 5.13 Comparison of PNA and DNA structures showing the structural similarity between them. The similar size of repeat units allows hybridization between the chains in the case of complementarity of side groups (nucleic acid bases). Because of the absence of Coulombic repulsion, PNA binds to DNA more strongly than a complementary DNA chain

uncharged N-(2-aminoethyl) glycine scaffold, to which nucleotide bases are attached via a methylene carbonyl linker. Therefore, PNA can be regarded as DNA with a *neutral peptide backbone* instead of a negatively charged sugar-phosphate backbone of natural nucleic acids (Shakeel et al. 2006). The size and shape of this polymer allows ideal hybridization based on the base-base interaction with DNA and RNA (Fig. 5.13).

Because of the absence of a negative charge, peptide nucleic acids exhibit superior *hybridization properties* than the natural nucleic acids. PNA is chemically stable and resistant to enzymatic cleavage, which provides its stability towards degradation in a living cell. PNA is capable of recognizing specific sequences of DNA and RNA obeying the Watson-Crick hydrogen bonding scheme and the hybrid complexes exhibit extraordinary thermal stability and unique ionic strength effects (Shakeel et al. 2006).

5.6.2   DNA Recognition with Peptide Nucleic Acids

The great interest in PNA as a recognition functionality for DNA or RNA detection appeared after it was demonstrated that *both ssDNA and DNA-PNA hybrids bind cationic conjugated polymer* (polythiophene derivative) with a dramatic change of its fluorescence. A combination of the complementary functions of specific base-to-base recognition of PNA with the extraordinary reporting function of the conjugated polymer suggests a revolutionary change in DNA detection methodology (see Section 10.4).

A combination of surface-bound PNA probes and soluble fluorescent cationic conjugated polymers was suggested for detecting and identifying unlabelled target nucleic acid on microarrays (Raymond et al. 2005). Application of PNA instead of DNA on supported arrays has the advantage of the absence of interaction between PNA and the cationic polymer and only in the presence of target DNA such interaction appears in the form of 'triplex' (Fig. 10.8).

In addition to an active role in DNA recognition and a passive role in reporting, PNA can play an active role when labeled with the dye that can serve as the FRET acceptor. Then, in combination with the conjugated polymer that serves as an antenna and an energy donor, a superquenching effect (Section 4.6) can be achieved with a substantial increase of the sensitivity of the response (Liu and Bazan 2005).

Interaction of PNA with *double-stranded DNA* is the prospective trend in the development of DNA sensing technologies. PNA is able to form specific higher-order (i.e., three- and four-stranded) complexes with DNA. This makes it an ideal structural probe for designing strand-specific dsDNA biosensors (Baker et al. 2006). The formation of higher-order complexes can be detected by dye-labeling of PNA. With the addition to this system of a cationic conjugated polymer, it allows the amplification of the reporting effect.

This finding has important consequences. Commonly, DNA hybridization assays require thermal denaturation producing the dissociation of the two strands, which is necessary for subsequent hybridization. The introduction into a sensing technology of a detection method for dsDNA that eliminates the need for thermal denaturing steps must have good prospects.

The *single nucleotide polymorphism* (SNP) is the major type of variation in the human genome that can characterize the identity of a person on a genetic level. The strategy employing a combination of PNA recognition units, optically amplifying conjugated polymer detectors and S1 nuclease enzymes is capable of detecting SNPs in a simple, rapid and sensitive manner (Baker et al. 2006). The recognition is accomplished by sequence-specific hybridization between the uncharged, fluorescein-labeled PNA probe and the DNA sequence of interest. After subsequent treatment with the S1 nuclease, the cationic polymer associates with the remaining anionic PNA/DNA complex, leading to a sensitized emission of the dye-labeled PNA probe via FRET mechanism. An improvement of this assay can be provided by the additional application of a nonionic detergent (Al Attar et al. 2008).

The techniques addressing the detection of single mismatches in hybridization assays can benefit from the possibility of using PNA segments in which one base is *substituted with a fluorescent dye*. This dye serves as a replacement of a canonical nucleobase with the ability of performing a reporting function (Socher et al. 2008). If a complementary DNA molecule (except at the site of the dye) hybridizes to the probe, the dye exhibits intense fluorescence emission because the stacking in the duplexes enforces its coplanar arrangement. However, a base mismatch at either position, immediately adjacent to the dye, dramatically decreases the fluorescence, presumably because the dye becomes allowed to undergo torsional motions that lead to the rapid depletion of the excited state. The number of dyes tested for this functional response (called *'forced intercallation'*) increases (Bethge et al. 2008).

5.7   Molecularly Imprinted Polymers

The multi-point noncovalent binding that is necessary for the selective detection of an analyte in a mixture with structurally close molecules can be achieved not only with the binding sites formed by natural or synthetic molecules. The recognition element of a sensor can also be a *hole accommodating the target molecule* formed in the structure of a synthetic polymer. If such accommodation can be made by providing a number of sterically fitting noncovalent interactions with the polymer matrix, this may allow realizing sufficient affinity for binding the target and selectivity against non-target molecules (Alexander et al. 2006; Sellergren and Andersson 2000).

In a strict sense, *imprinted polymers* are not the 'molecular sensors', they are the organized macroscopic bodies in which many recognition sites for the same target can be formed by one or several polymer molecules. If the primary events of the target binding are recorded, they can constitute the platform for the development of direct and reagent-independent sensors that combine binding and response. This simple idea turned-out to be very profitable. It is frequently used in both chromatographic separation technology and in drug delivery. In sensor technologies, it offers very interesting prospects.

5.7.1   The Principle of the Formation of an Imprinted Polymer

Imprinted polymers are produced by assembling and co-polymerizing of synthetic monomers and oligomers to form a polymeric network in the presence of a target molecule (the analyte itself or a molecule with a similar structure). This process is often called a *'template-directed polymerization'*. The target serves as a molecular template by creating a *cavity* in a polymer matrix. This cavity is complementary to the template. In addition to sterical fitting, it allows the formation of many noncovalent interactions with the target that became fixed during the polymerization step (Haupt and Mosbach 2000). After the polymer matrix is formed, the target is washed-out leaving 'imprinted' binding sites that are complementary in size and shape to the analyte.

Such a *molecular memory*, when it is introduced into the polymer, allows selective rebinding of the target, i.e., molecular recognition (see Fig. 5.14). The recognition units produced in this way in polymers challenge their natural counterparts that were discussed above, such as ligand-binding proteins and antibodies.

The progress in polymerization techniques, in the formation of molecular imprints and in techniques for trapping/detrapping analytes is described in many publications (Haupt and Mosbach 2000; Mosbach and Haupt 1998). Imprinted polymers can be used in different formats, not only as films but also as microspheres and nanoparticles (Yoshimatsu et al. 2007).

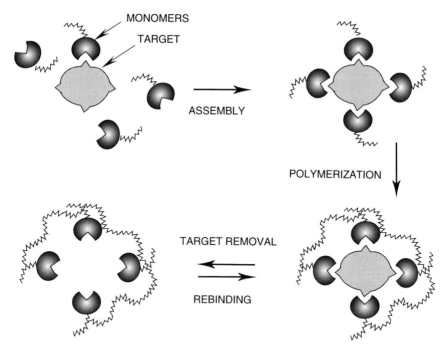

Fig. 5.14 The steps of the formation of recognition sites in an imprinted polymer. In the first step, the monomeric or oligomeric units are incubated with the target so that their noncovalent bonds are formed. Then, polymerization fixes this arrangement and the target is washed-out, leaving a hole that is capable of specific target binding

5.7.2 The Coupling with Reporting Functionality

For the operation of an imprinted polymer as a sensor for the detection of nonfluorescent molecules, the *fluorescence reporter should be incorporated* into a polymer matrix (Rathbone and Bains 2005; Stephenson and Shimizu 2007). This has to be done in a way that allows providing the response to target binding by a detectable change of fluorescence properties. If the target itself contributes to a reporter function (being a fluorescent dye, quencher or FRET acceptor), the sensor performance is easy. For instance, the traces of cancerogenic benzo[a]pyrene in water can be detected by polymer binding, since this binding results in an enhancement of intrinsic benzo[a]pyrene phosphorescence, mediated by a heavy atom incorporated into the polymer (Traviesa-Alvarez et al. 2007). On a general scale, reporting on target binding to polymer sites is a very difficult task.

A number of methods have been suggested for achieving the transduction of the binding event into a *fluorescence signal* (Alexander et al. 2006) but no satisfactory solution to this problem has appeared as yet (Stephenson and Shimizu 2007). It is hard, for instance, to explore the anisotropy sensing, because no free segmental rotation could be allowed in the polymer. Meantime, in view of limitations in other possibilities, such attempts have been made (Hunt and Ansell 2006). Much simpler is the realization of the competitor displacement assay with a fluorescent competitor (Navarro-Villoslada et al. 2007).

Many researchers have attempted the incorporation of fluorescent dyes on the step of the polymerization process. Without a transduction mechanism, this does not resolve the problem of fluorescence reporting. Most of these attempts concern intensity sensing and the fluorescent dye is included into the polymer as a monomeric unit or as a substituent to induce quenching/enhancement on target binding.

5.7.3 Applications

It is amazing that there is virtually no limit to the size and chemical nature of analyzed compounds that can be detected with molecularly imprinted polymers. They have been developed for small organic molecules, such as steroids, amino acids, sugars, drugs, pesticides and also for proteins and even cells (Haupt and Mosbach 1999). In contrast to macromolecules, these binders are resistant to adverse environmental conditions, such as heat and extremes of pH. Their affinity and selectivity can approach that observed for biospecific recognition (Hillberg et al. 2005).

Imprinted polymers are extensively used in many *separation techniques,* such as thin-layer chromatography, high-performance liquid chromatography and solid-phase extraction. Their more intensive application in fluorescence sensing must follow the development of the proper methods of signal transduction and fluorescence detection.

Among the fluorescence-based assays that are attractive for the application we can mention a method for the rapid and sensitive analysis of penicillin-type beta-lactam antibiotics (Urraca et al. 2007). It is based on a competition of tested antibiotics and their labeled analogs.

Substantial progress in sensing with imprinted polymers is expected from the combination of the imprinting methodology with the responsive properties of conjugated polymers. The prototype of this material has been described (Li et al. 2007). It was tested for the detection of 2,4,6-trinitrotoluene and related nitroaromatic compounds. The described polymeric sensor shows remarkable air stability and photostability, high fluorescence quantum yield and reversible analyte binding. Displaying their intrinsic signal amplification capability, fluorescent conjugated polymers are an attractive basis for the design of low-detection–limit sensing devices.

Sensing and Thinking 5: Selecting the Tool for Optimal Target Recognition

The ultimate goal in fluorescence sensing technologies is **to possess a generic methodology allowing sensing of any target and any combination of targets simultaneously by using a simple, unified detection technique**. It would be ideal to select a family of molecular sensors, for which the binding properties of a large number of potential analytes could be induced but which could use the same unified signal-transduction mechanism.

The analysis presented above clearly shows that this goal, in the foreseeable future, is not achievable. Synthetic coordination compounds show optimal performance for detecting only small molecules and ions with some prospects of binding to 'hot spots' of larger molecules. Antibodies and ligand-binding proteins recognize these larger molecules more easily. The biopolymer molecules of larger complexity can be recognized by ligand-binding proteins or antibodies but only based on their short-scale molecular features such as antigenic determinants. Molecular recognition should extend to macromolecules with large contact surfaces, to well-specified molecular patterns on cell surfaces and to extended segments of nucleic acid sequence.

In this respect, biopolymers such as proteins, peptides and nucleic acids often demonstrate the necessary flexibility for forming the interaction sites with a huge number of potential targets. The applications of two principles, *rational design* and *combinatorial library selection*, complement each other. The combination of rigid scaffolds with 'rationally' located flexible recognition units and fluorescence reporters allows achieving highly selective binding and efficient reporting together with the possibility of fine-tuning the affinity to the desired range of target concentrations.

Questions and Problems

1. Why are the small-molecular sensors with a direct fluorescence response better fitted to sensing ions than neutral molecules? What intrinsic signal transduction mechanisms are employed in these sensors? (The reader may refer to Sections 6.1 and 6.2.)
2. Why can cyclodextrins bind cholesterol and its derivatives and extract cholesterol from biomembranes? What are the requirements for such binding?
3. What is the difference in target binding to cyclodextrin monomers and dimers in terms of selectivity and affinity?
4. How do calixarenes and porphyrins interact with each other? Based on these molecules, smart receptors for protein targets can be designed. What principles are implied in this design?
5. What principles are behind molecular recognition by dendrimers?
6. Explain the structure of a typical IgG antibody using Fig. 1.3. How do the flexibility and rigidity of its structure co-participate in realizing its function.?
7. How is the principle of combined flexibility and rigidity realized in ligand-binding proteins?
8. What properties of the maltose binding protein are especially attractive for sensing? Explain how the recognition step can be provided and how it can be coupled with fluorescence response. Compare different possibilities offered by single-labeling and double-labeling to provide the fluorescence response to target binding by this protein.

9. What are the advantages of synthetic peptides over proteins in sensing technologies? How do we obtain them as high-affinity receptors?

10. Compare the affinities to a hypothetical rigid target of a flexible tetrapeptide Leu-Leu-Leu-Asp and of the same sequence incorporated into a rigid structure. Use formulas from Chapter 2 and the rough estimates of a free energy change of 2–3 kJ/mol for the formation of a salt bridge, 10 kJ/mol for hydrophobic interaction between amino acids and 3–4 kJ/mol as an entropy penalty for the suppression of rotation around a single bond.

11. Explain the realization of different technologies to provide the fluorescence response in the cases: (a) aptamers in solution; (b) aptamers attached by one of the terminals to a solid support.

12. What is the mechanism behind the use of intercalating dyes to obtain the response to target binding to aptamers? What are the disadvantages of this approach?

13. The melting of the PNA-DNA double helix (duplex) – will it occur at higher or lower temperatures than the DNA duplex (dsDNA)?

14. Explain the mechanism of the generation of the fluorescence response in the system PNA-ssDNA-conjugated polymer in the cases: (a) PNA is unlabeled; (b) PNA is labeled with a fluorescent dye.

15. For detecting the single nucleotide substitutions, which possibility would you choose to use: short or long receptor sequences? High or low binding affinity?

16. In the case of an imprinted polymer, what are the possibilities to provide a fluorescence response in the following cases: (a) if the polymer does not contain a fluorescence reporter (b) if it contains a reporter that is not in contact with the bound target (c) if the fluorescence reporter is located within the target-binding cavity?

17. Suggest the possibilities to locate the reporter in the imprinted polymer cavities available for direct contact with the target. What detection techniques would you suggest in this case?

References

Al Attar HA, Norden J, O'Brien S, Monkman AP (2008) Improved single nucleotide polymorphisms detection using conjugated polymer/surfactant system and peptide nucleic acid. Biosensors & Bioelectronics 23:1466–1472

Alexander C, Andersson HS, Andersson LI, Ansell RJ, Kirsch N, Nicholls IA, O'Mahony J, Whitcombe MJ (2006) Molecular imprinting science and technology: a survey of the literature for the years up to and including 2003. Journal of Molecular Recognition 19:106–180

Al-Hassan KA, Khanfer MF (1998) Fluorescence probes for cyclodextrin interiors. Journal of Fluorescence 8:139–152

Al-Jamal KT, Ruenraroengsak P, Hartell N, Florence AT (2006) An intrinsically fluorescent dendrimer as a nanoprobe of cell transport. Journal of Drug Targeting 14:405–412

Antoni P, Nystrom D, Hawker CJ, Hult A, Malkoch M (2007) A chemoselective approach for the accelerated synthesis of well-defined dendritic architectures. Chemical Communications:2249–2251

Azzazy HM, Highsmith WE, Jr. (2002) Phage display technology: clinical applications and recent innovations. Clinical Biochemistry 35:425–445

Baker ES, Hong JW, Gaylord BS, Bazan GC, Bowers MT (2006) PNA/dsDNA complexes: site specific binding and dsDNA biosensor applications. Journal of the American Chemical Society 128:8484–8492

Balabai N, Linton B, Napper A, Priyadarshy S, Sukharevsky AP, Waldeck DH (1998) Orientational dynamics of beta-cyclodextrin inclusion complexes. Journal of Physical Chemistry B 102:9617–9624

Barthe P, Cohen-Gonsaud M, Aldrian-Herrada G, Chavanieu A, Labesse G, Roumestand C (2004) Design of an amphipatic alpha-helical hairpin peptide. Comptes Rendus Chimie 7:249–252

Benhar I (2007) Design of synthetic antibody libraries. Expert Opinion on Biological Therapy 7:763–779

Bethge L, Jarikote DV, Seitz O (2008) New cyanine dyes as base surrogates in PNA: forced intercalation probes (FIT-probes) for homogeneous SNP detection. Bioorganic & Medicinal Chemistry 16:114–125

Binz HK, Pluckthun A (2005) Engineered proteins as specific binding reagents. Current Opinion in Biotechnology 16:459–469

Binz HK, Amstutz P, Pluckthun A (2005) Engineering novel binding proteins from nonimmunoglobulin domains. Nature Biotechnology 23:1257–1268

Breslow R, Dong SD (1998) Biomimetic reactions catalyzed by cyclodextrins and their derivatives. Chemical Reviews 98:1997–2011

Brune M, Hunter JL, Corrie JET, Webb MR (1994) Direct, real-time measurement of rapid inorganic phosphate release using a novel fluorescent probe and its application to actomyosin subfragment 1 ATPase. Biochemistry 33:8262–8271

Butler RS, Myers AK, Bellarmine P, Abboud KA, Castellano RK (2007) Highly fluorescent donor-acceptor purines. Journal of Materials Chemistry 17:1863–1865

Casadei J, Powell MJ, Kenten JH (1990) Expression and secretion of aequorin as a chimeric antibody by means of a mammalian expression vector. Proceedings of the National Academy of Sciences of the United States of America 87:2047–2051

Chen CT, Huang WP (2002) A highly selective fluorescent chemosensor for lead ions. Journal of the American Chemical Society 124:6246–6247

Collett JR, Cho EJ, Ellington AD (2005) Production and processing of aptamer microarrays. Methods 37:4–15

Cox WG, Singer VL (2004) Fluorescent DNA hybridization probe preparation using amine modification and reactive dye coupling. Biotechniques 36:114–122

de Lorimier RM, Smith JJ, Dwyer MA, Looger LL, Sali KM, Paavola CD, Rizk SS, Sadigov S, Conrad DW, Loew L, Hellinga HW (2002) Construction of a fluorescent biosensor family. Protein Science 11:2655–2675

Demchenko AP (2001a) Concepts and misconcepts in the analysis of simple kinetics of protein folding. Current Protein & Peptide Science 2:73–98

Demchenko AP (2001b) Recognition between flexible protein molecules: induced and assisted folding. Journal of Molecular Recognition 14:42–61

Demchenko AP, Chinarov VA (1999) Tolerance of protein structures to the changes of amino acid sequences and their interactions. The nature of the folding code. Protein and Peptide Letters 6:115–129

de Silva AP, Gunaratne HQN, Gunnaugsson T, Huxley AJM, McRoy CP, Rademacher JT, Rice TE (1997) Signaling recognition events with fluorescent sensors and switches. Chemical Reviews 97:1515–1566

Douhal A (2004) Ultrafast guest dynamics in cyclodextrin nanocavities. Chemical Reviews 104:1955–1976

Dwyer MA, Hellinga HW (2004) Periplasmic binding proteins: a versatile superfamily for protein engineering. Current Opinion in Structural Biology 14:495–504

Edwards BM, Barash SC, Main SH, Choi GH, Minter R, Ullrich S, Williams E, Du Fou L, Wilton J, Albert VR, Ruben SM, Vaughan TJ (2003) The remarkable flexibility of the human antibody repertoire; isolation of over one thousand different antibodies to a single protein, BLyS. Journal of Molecular Biology 334:103–118

Eklund M, Axelsson L, Uhlen M, Nygren PA (2002) Anti-idiotypic protein domains selected from protein A-based affibody libraries. Proteins-Structure Function and Genetics 48:454–462

Enander K, Dolphin GT, Andersson LK, Liedberg B, Lundstrom I, Baltzer L (2002) Designed, folded polypeptide scaffolds that combine key biosensing events of recognition and reporting. Journal of Organic Chemistry 67:3120–3123

Enander K, Dolphin GT, Baltzer L (2004a) Designed, functionalized helix-loop-helix motifs that bind human carbonic anhydrase II: a new class of synthetic receptor molecules. Journal of the American Chemical Society 126:4464–4465

Enander K, Dolphin GT, Liedberg B, Lundstrom I, Baltzer L (2004b) A versatile polypeptide platform for integrated recognition and reporting: affinity arrays for protein-ligand interaction analysis. Chemistry-A European Journal 10:2375–2385

Enander K, Choulier L, Olsson AL, Yushchenko DA, Kanmert D, Klymchenko AS, Demchenko AP, Mély Y, Altschuh DA (2008) Peptide-Based, Ratiometric Biosensor Construct for Direct Fluorescence Detection of a Protein Analyte. Bioconjugate Chemistry 19:1864–1870

Engfeldt T, Renberg B, Brumer H, Nygren PA, Karlstrom AE (2005) Chemical synthesis of triple-labelled three-helix bundle binding proteins for specific fluorescent detection of unlabeled protein. Chembiochem 6:1043–1050

Fletcher S, Hamilton AD (2007) Protein-protein interaction inhibitors: small molecules from screening techniques. Current Topics in Medicinal Chemistry 7:922–927

Flores S, Echols N, Milburn D, Hespenheide B, Keating K, Lu J, Wells S, Yu EZ, Thorpe M, Gerstein M (2006) The database of macromolecular motions: new features added at the decade mark. Nucleic Acids Research 34:D296–D301

Flower DR, North ACT, Sansom CE (2000) The lipocalin protein family: structural and sequence overview. Biochimica Et Biophysica Acta-Protein Structure and Molecular Enzymology 1482:9–24

Gellman SH, Woolfson DN (2002) Mini-proteins Trp the light fantastic. Nature Structural Biology 9:408–410

Gilardi G, Zhou LQ, Hibbert L, Cass AEG (1994) Engineering the maltose-binding protein for reagentless fluorescence sensing. Analytical Chemistry 66:3840–3847

Glasner ME, Gerlt JA, Babbitt PC (2007) Mechanisms of protein evolution and their application to protein engineering. Advances in Enzymology and Related Areas in Molecular Biology 75:193–239, xii–xiii

Gomara MJ, Haro I (2007) Synthetic peptides for the immunodiagnosis of human diseases. Current Medicinal Chemistry 14:531–546

Goodchild S, Love T, Hopkins N, Mayers C (2006) Engineering antibodies for biosensor technologies. Advances in Applied Microbiology 58:185–226

Gopinath SCB (2007) Methods developed for SELEX. Analytical and Bioanalytical Chemistry 387:171–182

Guntas G, Ostermeier M (2004) Creation of an allosteric enzyme by domain insertion. Journal of Molecular Biology 336:263–273

Guthrie JW, Hamula CLA, Zhang HQ, Le XC (2006) Assays for cytokines using aptamers. Methods 38:324–330

Hamada H, Kameshima N, Szymanska A, Wegner K, Lankiewicz L, Shinohara H, Taki M, Sisido M (2005) Position-specific incorporation of a highly photodurable and blue-laser excitable fluorescent amino acid into proteins for fluorescence sensing. Bioorganic & Medicinal Chemistry 13:3379–3384

Hamula CLA, Guthrie JW, Zhang HQ, Li XF, Le XC (2006) Selection and analytical applications of aptamers. Trac-Trends in Analytical Chemistry 25:681–691

Haupt K, Mosbach K (1999) Molecularly imprinted polymers in chemical and biological sensing. Biochemical Society Transactions 27:344–350

Haupt K, Mosbach K (2000) Molecularly imprinted polymers and their use in biomimetic sensors. Chemical Reviews 100:2495–2504

Hazra P, Chakrabarty D, Chakraborty A, Sarkar N (2004) Intramolecular charge transfer and solvation dynamics of Nile Red in the nanocavity of cyclodextrins. Chemical Physics Letters 388:150–157

Hermann T, Patel DJ (2000) Biochemistry - adaptive recognition by nucleic acid aptamers. Science 287:820–825

Hesselberth JR, Miller D, Robertus J, Ellington AD (2000) In vitro selection of RNA molecules that inhibit the activity of ricin A-chain. Journal of Biological Chemistry 275:4937–4942

Heyduk E, Heyduk T (2005) Nucleic acid-based fluorescence sensors for detecting proteins. Analytical Chemistry 77:1147–1156

Hicke BJ, Marion C, Chang YF, Gould T, Lynott CK, Parma D, Schmidt PG, Warren S (2001) Tenascin-C aptamers are generated using tumor cells and purified protein. Journal of Biological Chemistry 276:48644–48654

Hillberg AL, Brain KR, Allender CJ (2005) Molecular imprinted polymer sensors: implications for therapeutics. Advanced Drug Delivery Reviews 57:1875–1889

Hossain MA, Mihara H, Ueno A (2003) Fluorescence resonance energy transfer in a novel cyclodextrin-peptide conjugate for detecting steroid molecules. Bioorganic & Medicinal Chemistry Letters 13:4305–4308

Hosse RJ, Rothe A, Power BE (2006) A new generation of protein display scaffolds for molecular recognition. Protein Science 15:14–27

Hunt CE, Ansell RJ (2006) Use of fluorescence shift and fluorescence anisotropy to evaluate the re-binding of template to (S)-propranolol imprinted polymers. Analyst 131:678–683

Hust M, Dubel S (2004) Mating antibody phage display with proteomics. Trends in Biotechnology 22:8–14

Jennings K, Diamond D (2001) Enantioselective molecular sensing of aromatic amines using tetra-(S)-di-2-naphthylprolinol calix[4]arene. Analyst 126:1063–1067

Jespers L, Bonnert TP, Winter G (2004) Selection of optical biosensors from chemisynthetic antibody libraries. Protein Engineering Design & Selection 17:709–713

Jhaveri S, Rajendran M, Ellington AD (2000) In vitro selection of signaling aptamers. Nature Biotechnology 18:1293–1297

Jiang Y, Fang X, Bai C (2004) Signaling aptamer/protein binding by a molecular light switch complex. Analytical Chemistry 76:5230–5235

Jin T, Fujii F, Yamada E, Nodasaka Y, Kinjo M (2006) Control of the optical properties of quantum dots by surface coating with calix n arene carboxylic acids. Journal of the American Chemical Society 128:9288–9289

Kachkovskiy GO, Shandura MP, Drapaylo AB, Slominskii JL, Tolmachev OI, Kalchenko VI (2006) New calix[4]arene based hydroxystyryl cyanine dyes. Journal of Inclusion Phenomena and Macrocyclic Chemistry 56:315–321

Katilius E, Katiliene Z, Woodbury NW (2006) Signaling aptamers created using fluorescent nucleotide analogues. Analytical Chemistry 78:6484–6489

Kim JS, Noh KH, Lee SH, Kim SK, Kim SK, Yoon JY (2003) Molecular taekwondo. 2. A new calix[4]azacrown bearing two different binding sites as a new fluorescent ionophore. Journal of Organic Chemistry 68:597–600

Kodadek T (2002) Development of protein-detecting microarrays and related devices. Trends in Biochemical Sciences 27:295–300

Korndorfer IP, Schlehuber S, Skerra A (2003) Structural mechanism of specific ligand recognition by a lipocalin tailored for the complexation of digoxigenin. Journal of Molecular Biology 330:385–396

Kubinyi M, Vidoczy T, Varga O, Nagy K, Bitter I (2005) Absorption and fluorescence spectroscopic study on complexation of oxazine 1 dye by calix 8 arenesulfonate. Applied Spectroscopy 59:134–139

Kulagina NV, Shaffer KM, Anderson GP, Ligler FS, Taitt CR (2006) Antimicrobial peptide-based array for Escherichia coli and Salmonella screening. Analytica Chimica Acta 575:9–15

Lakowicz JR (2007) Principles of fluorescence spectroscopy. Springer, New York

Leray I, Lefevre JP, Delouis JF, Delaire J, Valeur B (2001) Synthesis and photophysical and cation-binding properties of mono- and tetranaphthylcalix 4 arenes as highly sensitive and selective fluorescent sensors for sodium. Chemistry-A European Journal 7:4590–4598

Levin AM, Weiss GA (2006) Optimizing the affinity and specificity of proteins with molecular display. Molecular Biosystems 2:49–57

Li J, Kendig CE, Nesterov EE (2007) Chemosensory performance of molecularly imprinted fluorescent conjugated polymer materials. Journal of the American Chemical Society 129:15911–15918

Li JJ, Fang X, Tan W (2002) Molecular aptamer beacons for real-time protein recognition. Biochemical and Biophysical Research Communications 292:31–40

Liu B, Bazan GC (2005) Methods for strand-specific DNA detection with cationic conjugated polymers suitable for incorporation into DNA chips and microarrays. Proceedings of the National Academy of Sciences of the United States of America 102:589–593

Liu Y, Song Y, Chen Y, Li XQ, Ding F, Zhong RQ (2004) Biquinolino-modified beta-cyclodextrin dimers and their metal complexes as efficient fluorescent sensors for the molecular recognition of steroids. Chemistry 10:3685–3696

Liu Y, Liang P, Chen Y, Zhao YL, Ding F, Yu A (2005) Spectrophotometric study of fluorescence sensing and selective binding of biochemical substrates by 2,2′-bridged biso(beta-cyclodextrin) and its water-soluble fullerene conjugate. Journal of Physical Chemistry B 109:23739–23744

Looger LL, Dwyer MA, Smith JJ, Hellinga HW (2003) Computational design of receptor and sensor proteins with novel functions. Nature 423:185–190

Makabe A, Kinoshita K, Narita M, Hamada F (2002) Guest-responsive fluorescence variations of gamma-cyclodextrins labeled with hetero-functionalized pyrene and tosyl moieties. Analytical Sciences 18:119–124

Marvin JS, Hellinga HW (1998) Engineering biosensors by introducing fluorescent allosteric signal transducers: construction of a novel glucose sensor. Journal of the American Chemical Society 120:7–11

Marvin JS, Hellinga HW (2001a) Conversion of a maltose receptor into a zinc biosensor by computational design. Proceedings of the National Academy of Sciences of the United States of America 98:4955–4960

Marvin JS, Hellinga HW (2001b) Manipulation of ligand binding affinity by exploitation of conformational coupling. Nature Structural Biology 8:795–798

Marvin JS, Corcoran EE, Hattangadi NA, Zhang JV, Gere SA, Hellinga HW (1997) The rational design of allosteric interactions in a monomeric protein and its applications to the construction of biosensors. Proceedings of the National Academy of Sciences of the United States of America 94:4366–4371

McCauley TG, Hamaguchi N, Stanton M (2003) Aptamer-based biosensor arrays for detection and quantification of biological macromolecules. Analytical Biochemistry 319:244–250

McQuade DT, Pullen AE, Swager TM (2000) Conjugated polymer-based chemical sensors. Chemical Reviews 100:2537–2574

Medintz IL, Deschamps JR (2006) Maltose-binding protein: a versatile platform for prototyping biosensing. Current Opinion in Biotechnology 17:17–27

Medintz IL, Goldman ER, Lassman ME, Mauro JM (2003) A fluorescence resonance energy transfer sensor based on maltose binding protein. Bioconjugate Chemistry 14:909–918

Metivier R, Leray I, Valeur B (2004) Lead and mercury sensing by calixarene-based fluoroionophores bearing two or four dansyl fluorophores. Chemistry-A European Journal 10:4480–4490

Mohanty J, Bhasikuttan AC, Nau WM, Pal H (2006) Host-guest complexation of neutral red with macrocyclic host molecules: contrasting pK(a) shifts and binding affinities for cucurbit 7 uril and beta-cyclodextrin. Journal of Physical Chemistry B 110:5132–5138

Mondal SK, Sahu K, Ghosh S, Sen P, Bhattacharyya K (2006) Excited-state proton transfer from pyranine to acetate in gamma-cyclodextrin and hydroxypropyl gamma-cyclodextrin. Journal of Physical Chemistry A 110:13646–13652

Mosbach K, Haupt K (1998) Some new developments and challenges in non-covalent molecular imprinting technology. Journal of Molecular Recognition 11:62–68

Muyldermans S (2001) Single domain camel antibodies: current status. Journal of Biotechnology 74:277–302

Nanduri V, Kim G, Morgan MT, Ess D, Hahm BK, Kothapalli A, Valadez A, Geng T, Bhunia AK (2006) Antibody immobilization on waveguides using a flow-through system shows improved Listeria monocytogenes detection in an automated fiber optic biosensor: RAPTOR (TM). Sensors 6:808–822

Navarro-Villoslada F, Urraca JL, Moreno-Bondi MC, Orellana G (2007) Zearalenone sensing with molecularly imprinted polymers and tailored fluorescent probes. Sensors and Actuators B-Chemical 121:67–73

Neuweiler H, Schulz A, Vaiana AC, Smith JC, Kaul S, Wolfrum J, Sauer M (2002) Detection of individual p53-autoantibodies by using quenched peptide-based molecular probes. Angewandte Chemie-International Edition in English 41:4769–4773

Ngundi MM, Kulagina NV, Anderson GP, Taitt CR (2006) Nonantibody-based recognition: alternative molecules for detection of pathogens. Expert Review of Proteomics 3:511–524

Niu WZ, Jiang N, Hu YH (2007) Detection of proteins based on amino acid sequences by multiple aptamers against tripeptides. Analytical Biochemistry 362:126–135

Nutiu R, Li YF (2004) Structure-switching signaling aptamers: transducing molecular recognition into fluorescence signaling. Chemistry-A European Journal 10:1868–1876

Nutiu R, Li YF (2005a) Aptamers with fluorescence-signaling properties. Methods 37:16–25

Nutiu R, Li YF (2005b) In vitro selection of structure-switching signaling aptamers. Angewandte Chemie-International Edition 44:1061–1065

Oh KJ, Cash KJ, Hugenberg V, Plaxco KW (2007) Peptide beacons: a new design for polypeptide-based optical biosensors. Bioconjugate Chemistry 18:607–609

Organero JA, Tormo L, Sanz M, Roshal A, Douhal A (2007) Complexation effect of gamma-cyclodextrin on a hydroxyflavone derivative: formation of excluded and included anions. Journal of Photochemistry and Photobiology A-Chemistry 188:74–82

Oshovsky GV, Reinhoudt DN, Verboom W (2007) Supramolecular chemistry in water. Angewandte Chemie-International Edition 46:2366–2393

O'Sullivan PJ, Burke M, Soini AE, Papkovsky DB (2002) Synthesis and evaluation of phosphorescent oligonucleotide probes for hybridisation assays. Nucleic Acids Research 30:e114

Ozaki H, Nishihira A, Wakabayashi M, Kuwahara M, Sawai H (2006) Biomolecular sensor based on fluorescence-labeled aptamer. Bioorganic & Medicinal Chemistry Letters 16:4381–4384

Pagliari S, Corradini R, Galaverna G, Sforza S, Dossena A, Montalti M, Prodi L, Zaccheroni N, Marchelli R (2004) Enantioselective fluorescence sensing of amino acids by modified cyclodextrins: role of the cavity and sensing mechanism. Chemistry-A European Journal 10:2749–2758

Peczuh MW, Hamilton AD (2000) Peptide and protein recognition by designed molecules. Chemical Reviews 100:2479–2493

Pflum MKH (2004) Grafting miniature DNA binding proteins. Chemistry & Biology 11:3–4

Proske D, Blank M, Buhmann R, Resch A (2005) Aptamers - basic research, drug development, and clinical applications. Applied Microbiology and Biotechnology 69:367–374

Pugh VJ, Hu QS, Pu L (2000) The first dendrimer-based enantioselective fluorescent sensor for the recognition of chiral amino alcohols. Angewandte Chemie-International Edition 39:3638–3641

Purrello R, Gurrieri S, Lauceri R (1999) Porphyrin assemblies as chemical sensors. Coordination Chemistry Reviews 192:683–706

Quiocho FA, Ledvina PS (1996) Atomic structure and specificity of bacterial periplasmic receptors for active transport and chemotaxis: variation of common themes. Molecular Microbiology 20:17–25

Rathbone DL, Bains A (2005) Tools for fluorescent molecularly imprinted polymers. Biosensors & Bioelectronics 20:1438–1442

Raymond FR, Ho HA, Peytavi R, Bissonnette L, Boissinot M, Picard FJ, Leclerc M, Bergeron MG (2005) Detection of target DNA using fluorescent cationic polymer and peptide nucleic acid probes on solid support. BMC Biotechnology 5:10

Renard M, Belkadi L, Hugo N, England P, Altschuh D, Bedouelle H (2002) Knowledge-based design of reagentless fluorescent biosensors from recombinant antibodies. Journal of Molecular Biology 318:429–442

Renberg B, Shiroyama I, Engfeldt T, Nygren PA, Karlstrom AE (2005) Affibody protein capture microarrays: synthesis and evaluation of random and directed immobilization of affibody molecules. Analytical Biochemistry 341:334–343

Renberg B, Nordin J, Merca A, Uhlen M, Feldwisch J, Nygren PA, Karlstrom AE (2007) Affibody molecules in protein capture microarrays: evaluation of multidomain ligands and different detection formats. Journal of Proteome Research 6:171–179

Rodi DJ, Agoston GE, Manon R, Lapcevich R, Green SJ, Makowski L (2001) Identification of small molecule binding sites within proteins using phage display technology. Combinatorial Chemistry & High Throughput Screening 4:553–572

Ronnmark J, Kampf C, Asplund A, Hoiden-Guthenberg I, Wester K, Ponten F, Uhlen M, Nygren PA (2003) Affibody-beta-galactosidase immunoconjugates produced as soluble fusion proteins in the Escherichia coli cytosol. Journal of Immunological Methods 281:149–160

Roshal AD, Grigorovich AV, Doroshenko AO, Pivovarenko VG, Demchenko AP (1999) Flavonols as metal-ion chelators: complex formation with Mg2 + and Ba2 + cations in the excited state. Journal of Photochemistry and Photobiology A-Chemistry 127:89–100

Sadhu KK, Bag B, Bharadwaj PK (2007) A multi-receptor fluorescence signaling system exhibiting enhancement selectively in presence of Na(I) and Tl(I) ions. Journal of Photochemistry and Photobiology A-Chemistry 185:231–238

Schulz GE, Schirmer RH (1979) Principles of protein structure. Springer, New York

Sellergren B, Andersson LI (2000) Application of imprinted synthetic polymers in binding assay development. Methods 22:92–106

Shakeel S, Karim S, Ali A (2006) Peptide nucleic acid (PNA) - a review. Journal of Chemical Technology and Biotechnology 81:892–899

Sillerud LO, Larson RS (2005) Design and structure of peptide and peptidomimetic antagonists of protein-protein interaction. Current Protein & Peptide Science 6:151–169

Singh Y, Dolphin GT, Razkin J, Dumy P (2006) Synthetic peptide templates for molecular recognition: recent advances and applications. Chembiochem 7:1298–1314

Socher E, Jarikote DV, Knoll A, Roglin L, Burmeister J, Seitz O (2008) FIT probes: peptide nucleic acid probes with a fluorescent base surrogate enable real-time DNA quantification and single nucleotide polymorphism discovery. Analytical Biochemistry 375:318–330

Srivatsan SG, Tor Y (2007) Fluorescent pyrimidine ribonucleotide: synthesis, enzymatic incorporation, and utilization. Journal of the American Chemical Society 129:2044–2053

Stadtherr K, Wolf H, Lindner P (2005) An aptamer-based protein biochip. Analytical Chemistry 77:3437–3443

Stephenson CJ, Shimizu KD (2007) Colorimetric and fluorometric molecularly imprinted polymer sensors and binding assays. Polymer International 56:482–488

Stojanovic MN, Kolpashchikov DM (2004) Modular aptameric sensors. Journal of the American Chemical Society 126:9266–9270

Stojanovic MN, Landry DW (2002) Aptamer-based colorimetric probe for cocaine. Journal of the American Chemical Society 124:9678–9679

Stojanovic MN, de Prada P, Landry DW (2001) Aptamer-based folding fluorescent sensor for cocaine. Journal of the American Chemical Society 123:4928–4931

Szejtli J (1998) Introduction and general overview of cyclodextrin chemistry. Chemical Reviews 98:1743–1753

Thodima V, Pirooznia M, Deng YP (2006) RiboaptDB: a comprehensive database of ribozymes and aptamers. BMC Bioinformatics 7

Timmerman P, Beld J, Puijk WC, Meloen RH (2005) Rapid and quantitative cyclization of multiple peptide loops onto synthetic scaffolds for structural mimicry of protein surfaces. Chembiochem 6:821–824

Tolosa L, Ge XD, Rao G (2003) Reagentless optical sensing of glutamine using a dual-emitting glutamine-binding protein. Analytical Biochemistry 314:199–205

Tombelli S, Minunni A, Mascini A (2005) Analytical applications of aptamers. Biosensors & Bioelectronics 20:2424–2434

Traviesa-Alvarez JM, Sanchez-Barragan I, Costa-Fernandez JM, Pereiro R, Sanz-Medel A (2007) Room temperature phosphorescence optosensing of benzo a pyrene in water using halogenated molecularly imprinted polymers. Analyst 132:218–223

Tsou LK, Jain RK, Hamilton AD (2004) Protein surface recognition by porphyrin-based receptors. Journal of Porphyrins and Phthalocyanines 8:141–147

Uchiyama F, Tanaka Y, Minari Y, Toku N (2005) Designing scaffolds of peptides for phage display libraries. Journal of Bioscience and Bioengineering 99:448–456

Urraca JL, Moreno-Bondi MC, Orellana G, Sellergren B, Hall AJ (2007) Molecularly imprinted polymers as antibody mimics in automated on-line fluorescent competitive assays. Analytical Chemistry 79:4915–4923

Valeur B (2002) Molecular fluorescence. Wiley-VCH, Weinheim

Valeur B, Leray I (2007) Ion-responsive supramolecular fluorescent systems based on multi-chromophoric calixarenes: a review. Inorganica Chimica Acta 360:765–774

Vogt M, Skerra A (2004) Construction of an artificial receptor protein ("anticalin") based on the human apolipoprotein D. Chembiochem 5:191–199

Weiss GA, Lowman HB (2000) Anticalins versus antibodies: made-to-order binding proteins for small molecules. Chemistry & Biology 7:R177-R184

Wenz G, Han BH, Muller A (2006) Cyclodextrin rotaxanes and polyrotaxanes. Chemical Reviews 106:782–817

Wiederstein M, Sippl MJ (2005) Protein sequence randomization: efficient estimation of protein stability using knowledge-based potentials. Journal of Molecular Biology 345:1199–1212

Wosnick JH, Swager TM (2004) Enhanced fluorescence quenching in receptor-containing conjugated polymers: a calix 4 arene-containing poly(phenylene ethynylene). Chemical Communications:2744–2745

Yang CJ, Jockusch S, Vicens M, Turro NJ, Tan W (2005) Light-switching excimer probes for rapid protein monitoring in complex biological fluids. Proceedings of the National Academy of Sciences of the United States of America 102:17278–17283

Yang RH, Chan WH, Lee AWM, Xia PF, Zhang HK, Li KA (2003a) A ratiometric fluorescent sensor for Ag-1 with high selectivity and sensitivity. Journal of the American Chemical Society 125:2884–2885

Yang RH, Li KA, Wang KM, Zhao FL, Li N, Liu F (2003b) Porphyrin assembly on beta-cyclodextrin for selective sensing and detection of a zinc ion based on the dual emission fluorescence ratio. Analytical Chemistry 75:612–621

Yesylevskyy SO, Klymchenko AS, Demchenko AP (2005) Semi-empirical study of two-color fluorescent dyes based on 3-hydroxychromone. Journal of Molecular Structure-Theochem 755:229–239

Yesylevskyy SO, Kharkyanen VN, Demchenko AP (2006) The change of protein intradomain mobility on ligand binding: is it a commonly observed phenomenon? Biophysical Journal 91:3002–3013

Yoshimatsu K, Reimhult K, Krozer A, Mosbach K, Sode K, Ye L (2007) Uniform molecularly imprinted microspheres and nanoparticles prepared by precipitation polymerization: the control of particle size suitable for different analytical applications. Analytica Chimica Acta 584:112–121

Zahnd C, Amstutz P, Pluckthun A (2007) Ribosome display: selecting and evolving proteins in vitro that specifically bind to a target. Nature Methods 4:269–279

Zhang L, Feng W (2007) Dendritic conjugated polymers. Progress in Chemistry 19:337–349

Zheng GX, Shao Y, Xu B (2006) Synthesis and characterization of polyaniline coated gold nano-particle and its primary application. Acta Chimica Sinica 64:733–737

Zhou H, Baldini L, Hong J, Wilson AJ, Hamilton AD (2006) Pattern recognition of proteins based on an array of functionalized porphyrins. Journal of the American Chemical Society 128:2421–2425

Zimmerman SC, Zharov I, Wendland MS, Rakow NA, Suslick KS (2003) Molecular imprinting inside dendrimers. Journal of the American Chemical Society 125:13504–13518

Chapter 6
Mechanisms of Signal Transduction

In previous chapters we discussed two essential components of every fluorescence sensor – the binding-recognition units (receptors) and the fluorescence (luminescence) reporters. It is essential that the process of molecular recognition occurs between the groups of atoms of interacting molecules or interfaces in the *ground electronic states*. In contrast, fluorescence reporting requires *excited states* that involve profound changes in electronic structures and their energies. Thus, coupling mechanisms should exist that connect the two types of events, one molecular and ground-state and the other – electronic and excited-state. The realization of this coupling we will call *transduction* and the molecular or supramolecular structure responsible for this – the *transducer*. According to a more general definition of a transducer, it is a device that is activated by a signal from one system and provides this signal (often in another form) to a different system. Therefore, the collecting and recording of a fluorescent emission by the measurement system and transforming it into an electrical signal is also a signal transduction (Fig. 6.1). Here, we will concentrate on the mechanisms of transduction within the sensor unit on molecular and supramolecular scales and the signal transduction on an instrument level will be discussed in Chapter 9.

In many cases the transduction on a molecular level results in the perturbation of electronic structures of fluorescence reporters leading to changes in the parameters of their emission, as described in Chapter 3. The mechanisms of such a coupling with primary events of target binding are variable and include intra- and intramolecular processes of transfer of the electronic charge. We also observed that the effect of reporting can also be achieved by mechanisms that do not involve perturbation of the electronic structure of a reporter, e.g., by modulation of its rotational mobility. All this provides researchers with a variety of possibilities for optimizing all three functions of a fluorescence sensor: recognition, transduction and reporting.

An additional possibility for signal transduction is the coupling of conformational variables and the nanoscale association phenomena with individual and collective responses of fluorescence emitters. The latter allows the activation of the energy transfer in different formats. This approach is frequently used and therefore deserves a detailed analysis. Finally, we will show that the relay mechanisms

A.P. Demchenko, *Introduction to Fluorescence Sensing*,
© Springer Science + Business Media B.V. 2009

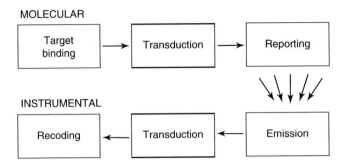

Fig. 6.1 Two levels of signal transduction in fluorescence sensing (molecular and instrumental)

of signal transduction allow not only providing a simple connection of binding and reporting events but also performing more complicated logical operations. Such systems may become elements of future molecular computers.

6.1 Basic Photophysical Signal Transduction Mechanisms

Fluorescent, or in general, luminescent states of molecules or nanoparticles are the *electronic excited states*. Electrons in these states are very reactive and this allows reactions that commonly do not happen in the ground states to proceed. The most fundamental of these reactions are the electron, charge and proton transfers. These reactions are reversible in the sense that in a cycle [reactant excitation → excited-state reaction → relaxation to product ground state → transition to reactant ground state] the same initial ground state species are restored. The interest in these reactions, from the side of sensing technologies, is due to the fact that they allow a one-step deactivation of the excited state (quenching) and, in some cases, generation of new emission bands belonging to the products of these reactions. These reactions can be modulated by target binding to a sensor because of their dependence on non-covalent interactions of reaction sites with the groups of atoms, molecules and surfaces that behave as enhancers or quenchers of their emission. This allows the realization of many different mechanisms for signal transduction.

6.1.1 Photoinduced Electron Transfer (PET)

Electronic excitation is the gain of energy by an electronic structure, which allows its participation in different reactions. Because of this, the excited species can be considered as redox sites that can donate or capture electrons from other sites, being either oxidants or reductants. *Photoinduced electron transfer* (PET) is the process by which an electron moves from one excited site to another site. Important requirements for this reaction are the closeness in space of the two sites and the matching of their oxidation-reduction potentials. The distances for this reaction are much shorter than for FRET, since it needs the crossing of electronic wave functions of initially excited and product states. PET can be facilitated by covalent bonding between donor and acceptor via a short linkage (spacer).

In PET systems only one of the states (reactant or product) is commonly light-emissive. Consequently, this reaction *leads to quenching* of the fluorescence emission of a reactant state. The electron exchange is not a 'profit-free' exchange between electronic systems. It is reversible but when it starts from a singlet (fluorescent) excited state the return is back to the ground state with a non-emissive loss of energy. This reaction can occur between the excited dye and a redox active molecule or group of atoms within the same molecular structure or between two structures. The solvent molecules can also participate in this process (Section 3.1). If the electron donor has a high HOMO (*highest occupied molecular orbital*) level, its transfer is efficient to acceptors with LUMO (*lowest unoccupied molecular orbitals*) of the acceptor.

Known *redox potentials* of interacting partners can be used to choose components so that the PET receptor will transfer an electron to/from the excited lumo/fluorophore, preventing emission. The transduction effect on the binding of the target species can be in the form of a change of the redox potential of the receptor that suppresses the PET reaction, giving rise to emission of the donor. In signal transduction systems based on PET, the relay is in fact a spacer that separates the electronic systems of the donor and acceptor. Both donor and acceptor can be fluorescent dyes that may serve as reporters; in these cases the excited-state energy will also be lost as a result of PET. On excitation, the electronic potential of any dye changes dramatically and it may become a better electron donor or acceptor than its closely located PET partner. This could lead to PET occurring usually much faster than the emission; this is why the fluorescence can be totally quenched.

This process can be exploited in sensor designs of different complexity (de Silva et al. 2001; Valeur and Leray 2000). Consider the simple case in which the fluorescent reporter segment is an electron acceptor and the donor is the group, which is a part of the binding site. Let the target be a metal cation bearing a positive charge (Fig. 6.2). When the target binding site is not occupied, the absorption of the light quantum by the reporter makes it a strong electron acceptor, the electron jumps to it from the receptor site resulting in quenching. This state can be called the OFF state. When the ion is bound, the situation becomes different. The ion attracts the oppositely charged electron from the receptor and the PET to the excited reporter becomes energetically unfavorable. In the absence of PET, the reporter dye remains a bright emitter. This is the ON state. The relative fluorescence intensity can be calibrated as a function of ion concentration.

Such a simple realization of PET has found application mostly in the *sensing of ions* (Section 10.3). In a typical ion sensor, two electronic systems, one participating in ion recognition and the other used for reporting, are connected by a short spacer (Valeur and Leray 2000). Cation binding produces the strong attraction of the electron with the suppression of PET.

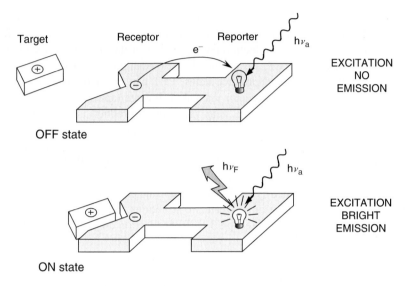

Fig. 6.2 Signal transduction in the simplest PET-based sensor. In a free sensor molecule the fluorescence is quenched due to an efficient electron transfer from donor to acceptor fragments separated by a spacer. Cation binding produces strong electrostatic attraction of the electron from the donor fragment preventing the electron transfer. Fluorescence is enhanced

Sensing of *neutral molecules* in this way has also been made, though with limited success (Granda-Valdes et al. 2000). The schemes that are more complicated allow excluding the target analyte binding from direct influence on PET. For instance, this can be accomplished by influencing the ionization of an attached group (Koner et al. 2007; Tomasulo et al. 2006) and thus extending this transduction mechanism to a much broader range of analytes.

PET can occur not only through bonds but also *through space* and for this process, the most important aspect is the close distance between donor and acceptor. Therefore, in a sensing event the donor and acceptor groups can be put together or moved apart, due to a conformational change in the sensor (Section 6.3).

One more possibility offered by the PET mechanism is the *association or dissociation* of nanoparticles: on association they can play a role of PET donor or acceptor (Section 6.4). Noble metal nanoparticles have proved to be the most efficient electron acceptors from different types of fluorescent donors, including semiconductor Quantum Dots. The latter can be reduced or oxidized at relatively moderate potentials, which tend to vary slightly with the physical dimensions of the nanoparticles (Burda et al. 2005). Therefore, they can be used as efficient electron donors and this possibility makes them efficient fluorescence reporters. Transition metal cations (such as Cu^{+2}) and free radicals quench fluorescence according to the PET mechanism (Section 3.1). This property can be used for their detection.

The PET quenching in *conjugated polymers* deserves special attention. Extremely low amounts of cationic electron acceptors quench the fluorescence of a polyanionic conjugated polymer on a time scale shorter than a picosecond, so that a greater than million-fold sensitivity to the quenching is achieved in comparison with the 'molecular excited states' (Chen et al. 1999). Novel materials such as Quantum Dots suggest new strategies for optimal exploration of the PET phenomenon in sensing. Below we will discuss in more detail the three important cases of PET that are frequently used in sensing.

(a) *Electron transfer between molecular fragments of the same dye molecule.* If a single dye molecule contains two localized electronic systems, the one-electron oxidation-reduction in the excited state results in quenching. Proper construction of the sensor units allows one to achieve complete "on-off" sensor behavior (de Silva et al. 2001). The PET process can be modulated by the target binding at the electron donor or acceptor sites by changing the donor or acceptor strength. Therefore, it is well understood why this mechanism can be applied most efficiently to ion sensing if the ion binds by the ion-chelating group coupled with the π–electronic system. Its application in biomolecular sensing is more difficult, since this needs a rather strong electrostatic perturbation of the electronic properties.

(b) *Intermolecular electron transfer.* At short distances, an electron can be captured by an excited-state fluorophore acting as an electron acceptor from a closely located ground-state electron donor, which also results in quenching. The application of a fluorescence response based on this principle to DNA beacon technology (Knemeyer et al. 2000) allowed achieving a direct homogeneous assay without the double labeling of molecular beacons. The role of the fluorescence quencher is played by a properly located guanosine residue. In sensor proteins, the conformational changes that form or disrupt the PET donor-acceptor pair may produce a dramatic change in the fluorescence response. The attached organic dye can serve as a PET acceptor and intrinsic aromatic amino acids, Trp and Tyr, can be used as the ground-state electron donors (Marme et al. 2003). In contrast to FRET, in PET quenching the emitting donor and acceptor should be located at a much closer mutual distance, usually forming a contact (Heinlein et al. 2003). PET quenching may also occur on the formation of dye dimers and aggregates. In this case it occurs between the unexcited and excited monomers and this effect can also be used in sensing (Johansson and Cook 2003).

(c) *Quenching by spin labels.* Stable nitroxide radicals are known as strong fluorescence quenchers. Their basic mechanism of quenching is also electron transfer and its realization also requires a close proximity of quencher to fluorescent dye. In this case however, the fluorophore always serves as an electron donor and the nitroxide as an acceptor. Such a quenching effect has been known for a long time (Blough and Simpson 1988) and the covalent spin-labeling of proteins and polynucleotides has been described. Whereas the spin-labeled lipids are frequently used for determining the location of proteins and other fluorescent compounds in the membranes (Chattopadhyay and London 1987), the applications of nitroxide radicals in sensing technologies are still very rare. Meanwhile, regarding the double-labeling methods, they offer some advantages mainly because of the much smaller size of the nitroxide group than that of common fluorescent dyes and due to the total "on-off" effect of quenching. In addition to sensing based on proximity, they offer an interesting possibility of sensing based on the variation of the redox potential of their surrounding (Blough and Simpson 1988), since the reduction of nitroxide removes its quenching ability. Based on this property, a method for the determination of ascorbic acid has been suggested (Lozinsky et al. 1999). Double (spin and fluorescence) sensing molecules afford various utilization possibilities (Bognar et al. 2006).

6.1.2 Intramolecular Charge Transfer (ICT)

Intramolecular charge transfer (ICT) is, in principle, also an electron transfer. The difference is that in the latter case, this process occurs within the same electronic system or between systems with a high level of *electronic conjugation* between the partners and it has its own characteristic features. The electronic states achieved in this reaction are not 'charge-separated' but '*charge-polarized*' states. Still, they are the localized states with a distinct energy minima.

The two, PET and ICT, states are easily distinguished by their absorption and emission spectra. In PET, strong quenching occurs without spectral shifts. In contrast, the ICT states are often fluorescent but exhibit changes of intensities. In addition, their excitation and emission spectra may exhibit significant shifts that depend on the environment. This allows wavelength-ratiometric recording (Section 3.6). In some cases a switching of intensity between *two emission bands*, normal (often called locally excited, LE) and achieved in ICT reaction, can occur. This is even more attractive for ratiometric measurement (Section 3.7).

Commonly, the ICT states are observed when the organic dye contains an *electron donating group* (often a dialkylamino group) and an *electron-withdrawing group* (often, carbonyl). If these groups are located at opposite sides of a molecule, electronic polarization is induced. Because in the excited state an electron donor becomes a stronger donor and an acceptor a stronger acceptor, electronic polarization can be substantially increased in the excited-state. A created large dipole moment interacts with medium dipoles, resulting in strong Stokes shifts.

The change in the medium conditions can produce the *LE-ICT switching*. Fluorescence from the normal LE state is commonly observed in low-polar solvents and in cryogenic conditions, where the spectra may contain residuals of vibrational structure. At an increase of polarity and temperature the ICT fluorescence appears, and has a broad and structureless long-wavelength shifted fluorescence band. In polar solvents, these shifts become larger because the solute-solvent dipole interactions are stronger. This is why the so-called 'polarity probes' (Section 4.1) are in fact the ICT dyes. In addition, the ICT states are very sensitive to electric field effects and therefore, to the presence of nearby charges.

From this discussion it is clear that the realization of ICT offers many possibilities. However, the application of this effect cannot be as broad as that of PET and only organic dyes (and not all of them) can generate an efficient ICT emission.

Nevertheless, within this family of dyes one may find many possibilities for producing a fluorescence reporter signal in sensing. They are:

1. *Switching between LE and ICT emissions by direct influence of the target charge* (Fig. 6.3). Particularly, this switching can be produced by binding an ion to either an electron donor or acceptor site. The processes occurring on interaction of the ion-chelation group with a cation have been well described (Valeur 2002). When an electron-donor group is attached to an ion-chelating group, the cation reduces the electron-donating character of this group. Owing to the resulting reduction of the conjugation, a blue shift occurs in the absorption spectra together with a decrease in the molar absorbance. Conversely, a cation interacting with the acceptor group enhances the electron-withdrawing character of this group; the absorption spectrum is thus red-shifted and the molar absorbance is increased. The fluorescence spectra are in principle shifted in the same direction as those of the absorption spectra. An anion produces the opposite effect. As a recent example, a dye exhibiting an ICT reaction was suggested as a sensor to fluoride (Yuan et al. 2007). Fluoride binds to an electron-acceptor group and switches its emission from 500 to 380 nm.

The excited-state charge transfer makes the electron-donor site strongly positively charged. If a cation is bound at this site, it exhibits a strong coulombic

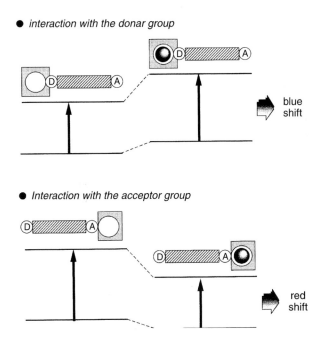

Fig. 6.3 Spectral shifts of ICT sensors resulting from interaction of a bound cation with an electron-donating or electron-withdrawing group (Reproduced with permission from Valeur and Leray 2000)

repulsion up to its dissociation. If this happens on a time scale faster than the emission, no binding will be detected in the fluorescence spectra. Therefore, it will be more efficient to locate the cation binding site close to the electron acceptor group where the binding in the excited state will become stronger. Following the same logic, the sites for anion recognition should be located at the electron-donor sites. A variety of chelator groups have been suggested as recognition units for ions, such as crown ethers, cryptands, coronands and calix[4]arenes (de Silva et al. 1997; Valeur and Leray 2000). The electron donor or acceptor groups can be incorporated into these structures.

2. Coupling of ICT emission with dynamic variables. The ICT states can be stabilized and modulated by the dynamics of surrounding molecules and groups of atoms. The ICT state possesses a strongly increased dipole moment and, in the absence of strong interactions with surrounding dipoles, its energy may lie higher than that of the LE state. Thus, the LE emission at shorter wavelengths will be observed in low-polar or highly viscous environments, where the strong dipole interactions are absent or cannot form by dipole rotations at the time of emission (Fig. 6.4). In contrast, in highly polar environments the surrounding dipoles tend to reorganize their equilibrium configuration, stabilizing the ICT state. This allows the ICT emission to be energetically favorable and therefore intensive. Because of the stronger interaction, a decrease of excited-state energy occurs and the emission is shifted to longer wavelengths.

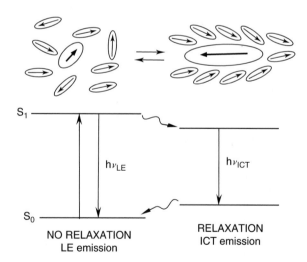

Fig. 6.4   The simplified energy diagram showing the influence of molecular relaxations (with lifetime τ_R) on the energies of LE and ICT states. The ICT states can be strongly stabilized in polar media by the orientation of surrounding dipoles resulting in substantial shifts of fluorescence spectra to lower energies (longer wavelengths)

Thus, an efficient transduction with the detected change of the reporter spectrum can be provided by a change in the dynamics of the molecules and groups of atoms at its location.

3. *Increased environment sensitivity of ICT emission.* The position of the ICT energy level strongly depends on the strength of the dipole-dipole interactions, which is the major effect determining solvent polarity. Exploration of this property allows extending sensing possibilities to detecting nonionic targets. If the target is a neutral compound, its binding can influence the local polarity in the vicinity of the reporter dye. This can be easily done if the testing has to be made in water. Since water is frequently used as a medium for sensing different analytes, it is worth mentioning that it is not only a highly polar liquid but also a medium with very rapid (~1 ps) dielectric relaxations. For sensing in water, different coupled processes can be used, such as providing conformational change, changing the environment of the reporter dye or screening the dye from direct contact with water.

4. *Quenching of intramolecular charge transfer (ICT) states.* The intramolecular charge transfer excited states are characteristic for fluorophores that possess strong electron-donating and electron-accepting groups and a strong quenching of their IST emission can often be observed in polar media and especially in water (Bhattacharyya and Chowdhury 1993; Yatsuhashi et al. 1998). Different factors, such as transition of the dye to a low-polar environment, can induce fluorescence enhancement in these systems.

5. *Modulation of electronic conjugation within dyes exhibiting ICT emission.* This can be accomplished by changing the planarity between electron-donor and acceptor fragments, since the planar structure produces the best π-π electronic coupling. Such disruption of the planarity with the appearance of localized IST emission is well studied in many systems, and correspondent states received the special name *twisted intramolecular charge-transfer* (TICT) states (Grabowski et al. 2003). Planarity can be modulated by covalent modifications of the dye and changing their configurations, which also suggests many possibilities to generate the sensor response. The switching to the TICT state can be used in sensing.

The above-explained general principles of efficient usage of the ICT mechanism in sensing is best demonstrated by the example of designing the Zn^{+2} ion sensor of a new generation (Sumalekshmy et al. 2007). The two compounds, whose formulas are presented below (Fig. 6.5), one containing two hydrogen atoms (**1**) and the other one substituted by fluorine atoms (**2**). These compounds generate strong shifts in the absorption and fluorescence spectra on the binding of zinc ions, which allows precise ratiometric recording. Compound **2** demonstrates a 1:1 ligand-Zn^{+2} binding mode and yields a dissociation constant of $K_d = 2.4\,\mu M$. This compound is strongly asymmetric, which allows an increase of ICT upon excitation, generating a strong dipole moment in the excited state. In this configuration, the metal ion binds to the acceptor rather than to the donor binding site, so that the ion binding increases but does not decrease the charge-transfer character of the excited state.

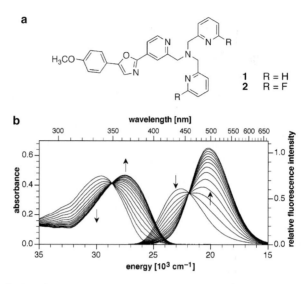

Fig. 6.5 The ICT-based sensor for zinc ion. **(a)** The basic sensor molecule (**1**) and its fluorinated derivative (**2**). **(b)** Absorption (left) and fluorescence emission (right) spectra for the titration of compound (**2**) with zinc ions in micromolar concentration range in methanol. The addition of zinc ions results in the absorption band shifting from 338 to 362 nm and the fluorescence spectra from 441 to 497 nm. Quantum yield grows from 35% to 71% (Reproduced with permission from Sumalekshmy et al. 2007)

This produces substantial shifts of spectra to longer wavelengths. Such an increase, instead of the commonly observed quenching in polar media, increases the brightness. This makes the Zn-bound state even more intensive in emission, making the ratiometric measurements very convenient. This example demonstrates that the general principles work and can be applicable to a broad range of donor-acceptor fluorophores modified with tailored chelating sites for the selective sensing of charged targets.

The transition from LE to ICT emission is not always seen as the appearance of a new band. Meantime, in some aromatic dyes the ICT band is clearly distinguished showing this excited-state reaction as the transition between two localized states (El-Kemary and Rettig 2003; Rurack 2001; Yoshihara et al. 2003). This can be seen in aromatic dyes possessing strong electron-donor and electron acceptor substituents. The ability of these dyes to generate a *two-band emission* is an extremely valuable property for the design of wavelength-ratiometric reporters. An example could be the dimethylamino analogue of boron-dipyrromethene dye, exhibiting well-resolvable LE and ICT emissions (Fig. 6.6). Its dual emission is well resolved, though the quantum yield is low. Substitution of the dimethylamino group into azacrown transforms it into an ion sensor. Coordination of the cation to the nitrogen donor atom of the crown inhibits the charge-transfer process, leading to a cation-dependent enhancement of the LE emission (Kollmannsberger et al. 1998).

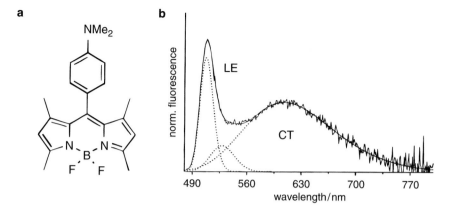

Fig. 6.6 The prototype of a wavelength-ratiometric ion sensor based on intramolecular charge transfer. (**a**) The structure of dimethylamino-substituted boron-dipyrromethene dye. (**b**) Its fluorescence spectrum in diethyl ether and its deconvolution into LE and CT (ICT) contributions (Reproduced with permission from Kollmannsberger et al. 1998)

6.1.3 Excited-State Proton Transfer

The excited states of dye molecules differ from correspondent ground states by the ability to donate protons to the medium or to accept protons. The proton is a particle with a positive charge, so the protonation or deprotonation of a neutral in the ground state molecule makes a charged one in the excited state. Molecules that increase their acidity and more easily lose their protons in the excited states are called '*photo-acids*' and those that increase their basicity are called '*photo-bases*'. These reactions occur with the participation of solvent and, in aqueous solutions, they exhibit variations as a function of pH, which results in the pH-dependent variations of spectra. In many organic molecules both proton-bound and proton-dissociated forms are strongly fluorescent and demonstrate different positions of their spectra (Arnaut and Formosinho 1993), so if these effects are coupled with the sensing events, they are easily recorded.

The equilibrium between two excited-state forms is established on a very short (picosecond) time scale and, in an aqueous solution, it can be shifted not only by changing the solvent pH but also by the solvent's access to the dye and by the dye's local environment. Coupling target binding to these changes can generate a sensor response.

The response when the same molecule contains both proton-donor and proton-acceptor groups in a close proximity could be very efficient. In this case, the proton transfer could be modulated by the solvent but could not depend on it directly, since a proton released by an acidic site can be transferred directly to the basic site. Such a reaction is called the *excited-state intramolecular proton transfer* (ESIPT). It does not require a protic environment and may occur in any solvent, in a solid matrix, even in a vacuum. The strict requirement for this is only on the structure of the molecule exhibiting ESIPT (Formosinho and Arnaut 1993). A proton-donor group is almost invariably a hydroxyl and the basic proton acceptor must be either a heterocyclic nitrogen atom or oxygen in the form of carbonyl. The two groups are usually hydrogen-bonded in the ground state and this bond in the excited state becomes a pathway for proton transfer. The excitation leads to a dramatic redistribution of electronic density, so that proton donors become stronger donors and acceptors – stronger acceptors. Hence, the driving force for ESIPT appears.

In many systems, ESIPT is very fast and irreversible on a fluorescence lifetime scale and therefore, it produces a single ESIPT band in the spectrum, which lacks the switchability needed for sensing. But there are a number of dyes with the very attractive property of producing two bands in emission, one belonging to an initially excited LE state and the other – to the product of an ESIPT reaction. Such dyes have been found among hydroxyphenyl benzothiazole, benzoxazole and benzimidazole derivatives. Their two-band switchability, dependent on the pH and binding of ions, can potentially be explored in sensing (Henary et al. 2007).

Switching between normal and ESIPT emissions can be achieved by changing the aggregation state of the dye. For instance, 2(2′-hydroxyphenyl)benzoxazole (HBO), a typical dye of this group, forms dispersed aggregates in water, exhibiting green ESIPT emission, whereas the dye solution demonstrates a blue normal emission (Huang et al. 2006). It must be noted that on a general scale, these dyes did not find many applications, mainly because of the presence of several ground-state and excited-state forms and the strong quenching of some of these forms.

In Section 4.1 we described the family of 3-hydroxychromone (3HC) dyes that possess the property to observe normal and ESIPT emissions as two well-resolvable and intensive fluorescence bands. The great interest in these dyes from the side of sensing technologies is due to the fact of easy switching between two forms in response to variations of their weak intermolecular interactions with the environment. Now, after describing the properties of the ICT state, we can explain that in accordance with the predictions of Kasha (Kasha 1986), the observed 'normal' emission is in fact the emission from the ICT state that is separated from the ESIPT state by a small energy barrier. The power of these dyes is in the fact that directly and without any side reactions, they combine ICT and ESIPT mechanisms to report on the change of weak interactions with their environment by intensity variations of two bands with different colors in their emission. Azacrown-substituted analogs of these dyes were tested as ion sensors and the two-wavelength ratiometric response was observed both in the excitation and emission spectra (Roshal et al. 1998, 1999).

6.1.4   Prospects

A photoinduced electron transfer can be considered as the most general photophysical mechanism of signal transduction in fluorescence sensing. It can be realized with all fluorescence and luminescence emitters, including all kinds of emitting nanoparticles and in this sense, can provide the optimal way of communication between inorganic and organic worlds in the form of exchange of electrons. Usually, this exchange results in complete quenching and different configurations of participating PET donors and acceptors can find application as the response elements in sensing.

Being so easily achieved and so broadly observed, the PET effect is well suited only to intensity sensing, since the quenched states cannot provide a reference for time-resolved anisotropy or wavelength-shifting measurements. There is an active search for PET sensors with dual (LE and PET) emissions and the first promising results have been reported (Rurack et al. 2002). This raises expectations for the success of this trend.

In contrast, ICT has the basic ability to generate new bands in emission or produce strong spectral shifts but this mechanism is available only to organic dyes. Moreover, these dyes should possess certain very necessary elements of structure, such as electron donor and acceptor groups separated in space to provide a high dipole moment. It is known that fluorescence in such systems can be strongly quenched in highly polar media, so the design of the best fluorescence reporters based on ICT (or, probably, on TICT) remains for the future.

Protonation-deprotonation of fluorescent dyes has found straightforward application in pH sensing and can be used on a broader scale when it is coupled with other reactions induced by target binding in water. Signal transduction with the aid of ESIPT is free from this limitation; the informative signal can be obtained in different solvents and upon modulation of the solvent/sensor environment. This prospective trend needs to be further developed.

The success of signal transduction depends strongly on the recognition unit and on its coupling to the reporter. Promising in this respect is the ability of calixarenes as useful building platforms in the design of multichromophoric systems in which basic photophysical phenomena (electron, charge and proton transfers), as well as excimer formation and resonance energy transfer, are controlled by ions (Valeur and Leray 2007). More elaborate recognition units and transducers are needed for sensing various neutral targets.

6.2   Signal Transduction via Excited-State Energy Transfer

The mechanism of long-range communication between dye molecules and also with and between light absorbing/emitting particles, is frequently discussed throughout this book. In Section 3.5 the elements of the general theory of *Förster resonance energy transfer* (FRET) are presented and in Section 4.1 the reader may find a list of organic dyes that are frequently used as the FRET donors and acceptors. In Chapter 4 it was also indicated that not only the organic dyes but also different kinds of metal chelating complexes and nanoparticles could be both donors and acceptors of FRET and some applications of this effect were indicated.

It is important to point out that FRET is *mechanistically reversible* and dyes with similar excitation and emission spectra may exchange their excited-state energies.

In Section 4.3, in relation to confinement and concentration effects exhibited by the dyes in polymeric matrices, we discussed this process occurring between dyes of the same structure, homo-FRET. We also mentioned the possibility of 'cascade' FRET between dyes possessing different excitation and emission spectra that could modulate emission properties.

There are many other possibilities to observe and apply the effects arising from electronic interaction between the dye molecules or nanoparticles and transfer of their energies. Here, we will discuss the possibilities of the FRET mechanism to be used for signal transduction in the chain of events between initial target binding and the generation of an efficient reporting signal.

In addition to the commonly referred mechanism of the excited-state energy transfer, which is the *long-range inductive resonance* (Förster mechanism), there is an *exchange resonance* mechanism of transfer (Dexter mechanism) operating at short distances, up to $1.5–1.8$ nm only. The latter is governed by specific orbital interactions and decays exponentially with the distance. Strictly speaking, the term 'FRET' *should refer only to a long-range transfer* extending to $5–7$ nm but, since the two mechanisms operate in the same direction, many authors do not distinguish them. Both these mechanisms can be involved, for instance, in interactions between nanoparticles and conjugated polymers (Hunyadi and Murphy 2006).

6.2.1   Directed Excited-State Energy Transfer in Multi-fluorophore Systems

We begin this sub-section with the formulation of three rules that are important for the explanation of the *collective effects* in the excited-state energy transfer in a system composed of different types of dyes participating in the transfer process.

1. In a system composed of structurally similar dyes located in identical environments and possessing identical absorption and emission spectra and also quantum yields and lifetimes, FRET (in this case homo-FRET) is just the *exchange of energies between the dyes undetected by spectroscopic or lifetime measurements*. It can only be revealed by the measurements of anisotropy and only in the case if the dyes occupy different orientations in space and these orientations do not change rapidly on the scale of fluorescence decay (solid or highly viscous dye environments). This rule follows from the identity of all parameters determining the FRET efficiency (Section 3.5) and from the kinetic scheme, in which the forward and reverse transfer rates are equal to and higher than the rates of non-emissive and emissive transfers to the ground state.

2. In a system composed of structurally similar dyes possessing identical fluorescence spectra but exhibiting different effects of quenching, the *effect of quenching will be observed for the whole system*. The quenching efficiency will be determined by the dye with the shortest lifetime τ. This follows from the kinetic

scheme in which for this dye the rate of non-radiative deactivation of the excited state is comparable to or faster than the rate of transfer. If the transfer is efficient and this dye is fully quenched, we will then observe quenching in the whole ensemble.

3. In a system composed of structurally dissimilar dyes possessing different excitation and emission spectra, *the transfer will be directional* from the dyes possessing a short-wavelength to those possessing long-wavelength absorption and emission spectra (taking account of rule 2). This is because of the overlap of integral *J* for the transfer being larger if the donor emission is at shorter and the acceptor absorption is at longer wavelengths than in the opposite case (Fig. 3.9). Thus, the energy flows from blue (short-wavelength) to red (long wavelength) emitters.

The latter case is the *hetero-FRET*. If the transfer is efficient, the quenching of the final acceptor quenches the whole ensemble. In a general case, the transfer and the quenching efficiencies are determined by the interplay of rate constants of all transformations of excited states, starting from that of an initially excited donor (Varnavski et al. 2002).

Based on these rules one can easily describe the effect of 'superquenching', which can be observed in dye-doped polymeric particles, in conjugated polymers and dendrimers (Chapter 4). Moreover, manipulating with a single trigger dye, one can provide an efficient collective sensor response by switching on and off the whole ensemble of fluorescence emitters.

By manipulating with hetero-FRET, one can design an efficient '*wavelength converter*'. One can choose the primary donor with the excitation spectrum that ideally fits the used light source and the final acceptor with a position of emission maximum at the desired wavelength. This wavelength can be chosen throughout all the visible range, down to near-infrared. 'Intermediate' dyes serving as both donors and acceptors may be needed to fill in the gap in the energy transfer chain.

The transfer, in which several different dyes provide the chain of transfer events to achieve a very significant shift in emission wavelength (as shown in Fig. 6.7), is called a '*cascade energy transfer*' (Serin et al. 2002). If the goal is obtaining bright emission of the final acceptor, it is important to avoid the self-quenching interactions of the dyes on this pathway, the 'light traps'.

The direct application of such wavelength converters is found in *multiplex assays* (simultaneous analysis of many targets). In homogeneous assays the possibility for distinguishing several types of sensor-target interactions will appear if we are able to label every type of sensor molecule with a certain '*barcode*' that could be recognizable in chromatography or flow cytometry. Dyes (and nanoparticles) emitting at different wavelengths may serve as these barcodes (see Section 9.4.1).

Another interesting possibility that can be used in sensing is '*FRET-gating*', when adding or removing the intermediate in the multi-step FRET (for instance, dye

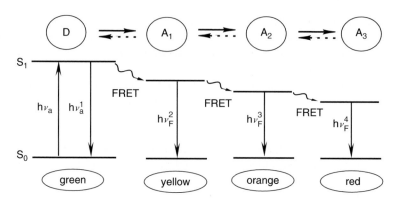

Fig. 6.7 Cascade systems with directed excited-state energy transfer. With each step, the emission spectrum shifts to longer wavelengths

A_2 in Fig. 6.7) modulates the fluorescence response. The following example illustrates this possibility (Wang et al. 2004). It was observed that the excitation of a cationic water-soluble conjugated polymer results in inefficient FRET to ethidium bromide dye intercalated within the bases of a double-stranded DNA molecule. When a fluorescein label is attached to one terminus of the DNA, an efficient FRET starts to be observed from the polymer through fluorescein to ethidium bromide. These experiments show that the proximity and conformational freedom of fluorescein provide a FRET gate to dyes intercalated within DNA, which are optically amplified by the properties of the conjugated polymer donor. The overall process provides a substantial improvement over previous homogeneous conjugated polymer-based DNA sensors. It is also interesting to note that the multistep energy transfer of intrinsically excited DNA occurs between its bases and this is one of its important mechanisms of protection against photodamage (Yashchuk et al. 2006).

6.2.2 Light-Harvesting (Antenna) Effect

Directed excited-state energy transfer can be organized in such a way that a large number of strong light-absorbing donors, when excited, transfer their energy to a much smaller number of acceptors. The latter may not be good light absorbers but because they are excited via efficient energy transfer from many donors, their fluorescence may be increased dramatically. This principle of *light-harvesting* is used in natural systems of photosynthesis that collect an enormous amount of solar energy by exciting the so-called antenna pigments and re-directing it to reaction centers, where it is converted into redox chemical energy. The primary absorbers serve as 'antennas' to provide the most efficient collection and transfer of energy. Such '*antenna effects*' are also in the background of the dramatic increase in luminescence of weakly absorbing lanthanide ions when they form coordination complexes with strongly absorbing ligands (Section 4.2). This principle is very general and it may be used for optimizing fluorescence (and, in general, luminescence) properties of many molecular and supramolecular systems, from dimers of organic dyes to complex nano-composites.

In the conditions of a proper arrangement, the fluorescence intensity of the acceptor will depend upon the *light absorption efficiency of the donor*. This suggests the selection of donors with a high molar absorbance and providing an increased number of them in the system. An example of such a solution can be found in the work on blended conjugated polymer nanoparticles (Huang et al. 2007). A 90% conversion of blue light into green, yellow and red emissions was achieved by the addition of less than 1% of polymer emitting at these wavelengths. It was stated that because of such composition the fluorescence brightness of the blended polymer nanoparticles can be much higher than that of inorganic quantum dots and dye-loaded silica particles of similar dimensions.

The antenna effect is illustrated in Fig. 6.8. The necessary conditions for its efficient implementation are the *high molar absorbance* of antenna dyes, *efficient energy transfer* to the acceptor dye and *high quantum yield* of emission of the latter. Such a system of signal transduction allows providing the response in a sensor operation on several steps: energy exchange (homo-FRET) and quenching in the antenna system, energy transfer to the acceptor and quenching of the acceptor emission. But probably the most important is the possibility to manipulate with reporter properties of the acceptor. It may be an environment-sensitive two-band or wavelength-shifting dye or a dye forming exciplexes (Chapter 3). Its spectroscopic

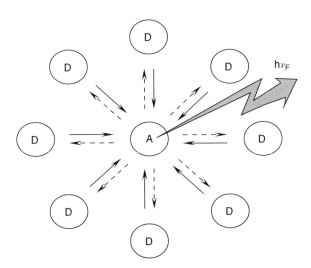

Fig. 6.8 Illustration of the antenna effect. Great number of donors (D) absorb light and transfer the excitation energy to a small number of acceptors (A). Excitation of acceptors via FRET increases their emission intensity and lifetime in comparison with the direct excitation. Homo-FRET between the donors is not shown

properties should not be different from those observed on its direct excitation but can be seen on a much brighter level.

A dramatic increase of the fluorescence emission intensity of the acceptor dye due to the 'focusing of antenna' compared to direct excitation is called an 'amplification' by some authors. It should be stressed that there is no amplification on the level of the acceptor dye, since its quantum yield does not increase. All that happens is that, via the antenna, it becomes excited much more frequently than on direct excitation. This is the reason for its highly increased apparent brightness. Photodegradation of such a dye also occurs at a much faster rate, in accordance with the frequency of the acts of excitation (Forde and Hanley 2005).

6.2.3 Peculiarities of FRET with and Between Nanoparticles

The long-range interactions leading to FRET are not limited to fluorescence and are characteristic of any emissive excited state, whatever its origin. All these phenomena can be adequately described by the *Förster-type inductive dipole resonance* mechanism. In their description we use the term 'dye' in its general sense, which includes organic dye molecules and different kinds of molecular complexes and nanoparticles. But there are definitely peculiarities of energy transfer with and between the particles, the size of which is comparable with the Förster radius R_0 (Eq. 3.12) since particles of such size cannot be approximated by point dipoles and therefore evaluation of the distance between them is not very definitive. The involvement of dipole orientation factor, κ^2, may also lead to confusion.

Moreover, the *distance dependence* of energy transfer may be different from that described by the $1/R^6$ function. This is because of the different mechanism of this interaction. In the inductive-resonance mechanism of FRET developed by Förster, two interacting point dipoles are considered, whereas in the case of conducting particles, the donor dipole interacts with a 'Fermi gas' of nearly free conduction electrons. Based on such modeling it was suggested that on the interaction of a small organic dye and a large conducting metal nanoparticle, the $1/R^4$ function should be observed. This suggestion was confirmed in an experiment (Yun et al. 2005) in which a gold nanoparticle was appended to the 5′ end of one DNA strand as the energy acceptor and a fluorescein dye to the 5′ end of the complementary strand as the energy donor. Analysis of the energy transfer as a function of DNA lengths (15, 20, 30, 60 base pairs) allowed testing the distance dependence in the range of 5–25 nm. The result showed conformity with the $1/R^4$ function. This is one of the results showing that for signal transduction in fluorescence sensing, the application of nanoparticles may allow the extension of transduction to longer distances, up to 15–20 nm.

An extended distance dependence of FRET *between nanoparticles* is also suggested based on quantum chemical calculations (Wong et al. 2004). The authors came to the conclusion that the FRET rate at large separations exhibits several features not anticipated in the conventional theory: the actual rate shows a much weaker short-range distance dependence, in this version of theory, closer to $1/R^2$ in contrast to the $1/R^6$ value. The Förster expression overestimates the energy transfer rate by more than two orders of magnitude at short separation (<1 nm). The distance at which the Förster rate is recovered is observed to be rather large (similar to 10 nm). Thus, the Förster expression seems to be inappropriate for condensed-phase systems where the donors and acceptors are large particles that can be closely packed, as, for example, in thin films. This and other results show that in principle, the particles possess more possibilities to manipulate with FRET effects and to optimize them than the conventional dyes (Jares-Erijman and Jovin 2003).

6.2.4   The Optimal Choice of FRET Donors

In Chapter 4 we described different types of fluorescence reporters that offer a tremendous arsenal of possibilities. Here we will try to outline the major trends leading to the improvement of their properties via energy transfer. Many different types of reporters are applied as donors and acceptors in FRET because each of them has its advantages and drawbacks. Here the idea is to try to use FRET as the mechanism that could allow strengthening the strong and suppressing or eliminating the weak points in fluorescence reporting by coupling fluorescence emitters and allowing the excited-state energy to transfer between them.

We start with optimizing FRET donors and try to formulate our requirements to them. Our primary aim is to achieve the *highest possible brightness*, which could determine the absolute sensitivity of the sensor response. It is determined primarily by high molar absorbance (Section 4.1). Here, we have a choice between dye-doped nanoparticles, Quantum Dots and conjugated polymers (Chapter 4). Organic dyes and luminescent metal conjugates are less attractive for this role.

The other very attractive property of a FRET donor could be the *long excited-state lifetime*. This means the long lifetime of the illumination of the acceptor via FRET. In this case, an acceptor with even a short-living emission will attain the property of a long-lifetime emission. In this respect, lanthanide luminescent complexes (Charbonniere et al. 2006) and metal chelates producing phosphorescence emission (Maliwal et al. 2001) are on the front line. Quantum Dots with their lifetimes of tens of nanoseconds but still not microseconds are second and organic dye emitters represented also by dye-doped nanoparticles, conjugated polymers and fluorescent proteins, are behind them.

6.2.5   Lanthanides as FRET Donors

The spectroscopic and temporal emission properties of *lanthanide chelates* make them very attractive as donors in detection techniques based on FRET (Selvin 2002). This is mainly because of their long emission lifetimes. This allows increasing the duration of the transfer and enhancing, in this way, the FRET efficiency. Due to the convenient positions of lanthanide emission bands for overlapping the absorption spectra of many organic dyes these dyes can be used as FRET acceptors. Even more attractive as such acceptors are semiconductor Quantum Dots (QDs), in view of their much higher brightness.

The exploration of these benefits has already started and it follows two lines. Donor (lanthanide chelate) and acceptor (organic dye or Quantum Dot) can be combined in the same hybrid construct, in which the spectral properties will be optimized. In one of the recent studies (Charbonniere et al. 2006) it was amply demonstrated that the long lifetime of the donor allows transferring the energy to the acceptor *during all this lifetime*. Efficient FRET from metal chelate donors to the QD acceptors was evidenced both by the sensitization of QD emission and by a thousand-fold increase of their luminescence decay time, reaching some hundreds of microseconds. This opens the way for many new applications.

The other possibility is to use the advantages of lanthanide chelates as the FRET donors when the donor-acceptor pair is formed directly in the test system. One of such applications could be a homogeneous immunoassay (Mathis 1995), which uses two specific antibodies that bind the same antigen at two different sites; one is labeled by the FRET donor and the other by an acceptor. Other coupling mechanisms such as avidin-biotin binding (Charbonniere et al. 2006) can be used for different types of assays. To achieve increased stability, the lanthanide chelates can be incorporated into polymeric or silica nanoparticles (Hai et al. 2004).

Thus, based on the unique properties of lanthanides, hybrid fluorescence emitters that display long emission wavelengths, long decay times and high quantum yields can be constructed based on the application of the FRET mechanism. The skill in constructing these hybrids is to amplify favorable and suppress unfavorable properties of participating components. If the donor emits light with a long lifetime and transfers its energy to an unquenched acceptor, then the emission of the latter should also become of long duration, like that of the donor. This makes the lanthanide complexes very attractive FRET donors. They can be covalently linked with a long-wavelength emitting acceptor (Selvin 2002). Importantly, the quantum yield of the formed hybrid can be high. It can approach that of the higher quantum yield acceptor, rather than the lower quantum yield typical of donor metal-ligand complexes.

Composite luminescent particles with these useful properties, combining long-wavelength emission and long lifetimes, must have numerous applications in biophysics, clinical diagnostics, DNA analysis and drug discovery. Their sensing properties can be induced based on different effects: variation of FRET donor-acceptor distances and of quantum yields and lifetimes of donors and acceptors.

The advantages of using long-lifetime donors in fluorescence sensing technologies based on FRET were discussed above with regard to lanthanide complexes. The same reasoning applies to the chelating complexes of transition metals. As shown (Maliwal et al. 2001), the donor in this case could be a ruthenium metal-ligand complex and the acceptor a common dye, such as Texas Red. It was shown that due to FRET, the long-lifetime donor results in a long-lived component in the acceptor decay.

There are good examples in the literature on the illustration and application of this principle. It was reported that the energy transfer from a phosphorescent chelating complex of iridium to a semiconductor QD increases the lifetime of the latter from 40 to 400 ns (Anikeeva et al. 2006). The dramatic (by two or more orders of magnitude) increase of lifetimes of organic dyes can also be achieved by an energy transfer from the excited europeum luminescent and ruthenium phosphorescent complexes.

Combining the two optimization parameters (brightness and lifetime) one may infer that the optimal FRET donors could be nanoparticles doped in high density with luminescent lanthanide chelates. Such particles have been described in the literature (Casanova et al. 2006; Monat et al. 2007) and their efficient FRET to organic dyes (Casanova et al. 2006) and Quantum Dots (Charbonniere et al. 2006) have been demonstrated.

6.2.6   Quantum Dots as FRET Donors

In addition to already mentioned advantages as FRET donors (such as the possibility to excite fluorescence in a very broad wavelength range, narrow and strongly size-dependent emission bands and high brightness (Section 4.4)), Quantum Dots possess other useful properties. They are reasonably resistant to photobleaching and their high brightness is also displayed on two-photon excitation. In addition, their emission is almost insensitive to medium conditions. Such independence, however, causes difficulties on their incorporation into sensing techniques, in which they have to be responsive. Therefore, a good combination could be a pair consisting of a Quantum Dot and a fluorescent dye or fluorescent protein connected by the FRET mechanism. In these constructs, the Quantum Dots serve as the energy donors and the dyes as acceptors (Jares-Erijman and Jovin 2003).

Color Plate 7 illustrates some of these possibilities. With variation of the properties of FRET receptors, many possibilities appear for inducing the QD conjugates to become environment-sensitive and pH-sensitive. They can detect proteases, nucleases or other hydrolytic enzymes and allow observing DNA synthesis by incorporation into its chain of fluorescent bases.

6.2.7   The Optimal Choice of FRET Acceptors

If the best FRET acceptors are the best quenchers, then one may consider for this purpose the *noble metal nanoparticles*. It has been shown (Stojanovic and Stefanovic 2003) that gold nanoparticles quench the fluorescence of cationic polyfluorene conjugated polymer nanoparticles with the Stern-Volmer constants approaching 10^{11} M^{-1}, which is several orders of magnitude higher than any previously reported conjugated polymer-quencher pair and by 9–10 orders of magnitude higher than the organic dye-quencher pairs. Three factors may account for this extraordinary efficiency: amplification of the quenching via rapid internal energy or electron transfer, electrostatic interactions between the cationic polymer and anionic nanoparticles and the ability of gold nanoparticles to quench via efficient energy transfer. Inhibition assay of biomolecules based on FRET between Quantum Dots (donors) and gold nanoparticles (acceptors) has also been reported (Oh et al. 2005).

In most known applications to date, organic dye molecules dominate as FRET acceptors from various kinds of donors. This is because of their two important properties: they are small molecules that allow easy and versatile conjugation with nanoparticles and they may be highly responsive as reporters. Their excitation via FRET increases the sensitivity of their response but may not change the mechanism of this response based on its interactions and dynamics.

The donor-acceptor pairs can easily be selected from QDs of different sizes and the series of dyes possessing also gradual shifts of their fluorescence spectra, such as Alexa dyes. The FRET between quantum dots emitting at 565, 605 and 655 nm as energy donors and these dyes with absorbance maxima at 594, 633, 647 and 680 nm as energy acceptors were recently tested (Nikiforov and Beechem 2006). As a first step, the covalent conjugates between all three types of QDs were prepared as well as each of the Alexa labels. The FRET can be easily detected and the binding quantified in a homogeneous system on the formation of complexes avidin-biotin and antigen-antibody. Competitive binding assays can also be developed in such systems.

The approach, based on QD donors and dye acceptors, has found many applications: DNA and RNA hybridization assays, immunoassays, construction of biosensors based on ligand-binding proteins and enzymes. Usually, the following configuration is used: QDs are conjugated with the receptor molecules and organic dyes are used either for target labeling or for labeling the target analogs in competitor displacement assays.

In the previous sub-section, we observed that the combination of Quantum Dots as FRET donors and organic dyes as acceptors is very efficient. It can be expected that the reverse combination of dyes as donors and QDs as acceptors will not be efficient at all! These expectations were confirmed in experiments (Clapp et al. 2005).

6.2.8 Prospects

The *Förster resonance energy transfer* has often been considered as an important tool to generate fluorescence sensors based on the distance dependence of the transfer and the possibility to change it during the sensing event. We now observe an active development of research methodology that uses this phenomenon for a related but different purpose – for dramatic *improvement of the fluorescent reporting* unit, facilitating signal transduction and reporting and increasing its sensitivity. This became possible due to the introduction as fluorescent emitters of different kinds of nanoparticles that are much superior to organic dyes in photophysical behavior but are inferior in responsive abilities. Their conjugation and use of FRET to improve the response properties opens up an amazingly broad field for research and development.

The FRET mechanism allows operating all the parameters of fluorescence emission within required ranges: positions of band maxima, anisotropies and lifetimes. An efficient switch of the emission to near-infrared can be achieved in this way, which opens new possibilities in sensing. Particularly attractive is the prospect of *whole body sensing*, since the soft human tissues are relatively transparent in this wavelength range (800–1,000 nm) and long lifetimes allow one to eliminate light-scattering interference. They are offered by oligometalic complexes in which the coordinated metal ions are connected by an efficient directed excited-state energy transfer (Bunzli and Piguet 2005).

6.3   Signal Transduction via Conformational Changes

There are many possibilities to incorporate conformational changes triggered by target binding into the process of signal transduction. Many such changes in flexible molecular structures occur with low activation energies, so the switch between two conformations can be coupled to target binding. Those that cannot proceed in the ground state can sometimes be observed in the excited state, where the activation barriers for rotations around certain bonds are dramatically reduced. Thus, responding sensors involving both large-scale and small-scale conformational changes can be constructed. The electron transfer and energy transfer on a molecular level are distance- and orientation-dependent. Therefore, the change in the distances and orientations between the partners in these reactions can be used as tools in signal transduction.

6.3.1   Excited-State Isomerism in the Reporter Dyes and Small Molecules

There are several classes of dye molecules that are highly stable in the ground state but *easily isomerize in the excited state*. One of the classical examples for this is stilbene. In a trans-form it is strongly fluorescent but in solutions that allow its rapid isomerization into cys-stilbene, its fluorescence is very weak. Based on this fact, one can make a sensor in which the target binding will influence the stilbene isomerization. Such a possibility can be suggested based on results obtained with anti-trans-stilbene antibodies (Simeonov et al. 2000). In a complex with antibodies trans-stibene exhibits a bright emission, indicating that its isomerization is suppressed. For our purposes, the antibody may serve as a prototype of a target to show the efficient modulation of the response on its binding. The other feasible idea is to modify stilbene with target-recognition moiety and use such a stilbene-antibody associate as a sensor. Following this reasoning, a sensor for mercury ions was constructed (Matsushita et al. 2005). Stilbene fluorescence is bright but becomes strongly quenched by these ions.

In some cases, the effect of binding can be coupled with the redistribution of *rotamers* (Henary et al. 2007) and on the conformational change in the sensor unit bringing two distant pyrene groups together with the formation of an excimer (Schazmann et al. 2006). The designed molecules exhibiting conformational change can be labeled with two dyes to provide a response to ion binding by *intramolecular FRET* (Petitjean and Lehn 2007).

6.3.2 Conformational Changes in Conjugated Polymers

In Section 4.6 we reviewed the properties of conjugated polymers as fluorescence reporters and argued that their response can be based on two unique features, 'superquenching' and the dependence of fluorescence emission on *chain conformation*. The modulation of chain conformation on target binding has found many applications, especially in the detection of nucleic acids (Section 10.4). Here, we must stress that the chain conformation produces a dramatic impact on the properties of a π-conjugated system that influences the fluorescence, since it determines the conjugation length.

The conformation-dependent fluorescence of conjugated polymers has been shown by many researchers. It is important that these polymeric molecules allow many different modifications with the attachment of side groups and these groups may determine the conformational state of the backbone and thus, the fluorescence signal. Of special interest are the modifications with the charged side groups that transform low-polar polymers into polyelectrolytes and allow target binding based on electrostatic interactions.

The example presented below demonstrates this (Fig. 6.9). It was shown that electrostatic interactions with negatively and positively charged peptides altered the fluorescence spectrum of poly{3-[(*S*)-5-amino-5-carboxyl-3-oxapentyl]-2,5-thiophenylene hydrochloride}shown in Fig. 6.9, (Nilsson et al. 2003). The addition of a positively charged peptide in a random-coil conformation forces the polymer to adopt a nonplanar conformation, which leads to an increased and blue-shifted intensity of the emitted light. After the addition of a negatively charged peptide with a random-coil conformation, the backbone of the polymer adopts a planar conformation and an aggregation of the polymer chains occurs, seen as a red shift and a decrease in the intensity of the emitted light. A calcium ion sensor was suggested based on a different conformation of this polymer in the unbound form and on its binding to Ca^{+2}-free and Ca^{+2}-bound forms of calmodulin.

With the proper modification of side groups, these polymers can specifically recognize charged molecules, such as ATP (Li et al. 2005). But the most significant effect is recorded on binding to regular fibrillar protein structures that appear in some pathologies, such as Alzheimer's disease. These pathological protein aggregates exhibit a β-sheet structure and the polymer binding to them produces remarkable changes in their spectra (Herland et al. 2008).

A family of glycopolythiophenes containing sialic acid or mannose ligands was prepared and evaluated for its ability to bind lectins, viruses and bacteria (Baek et al.

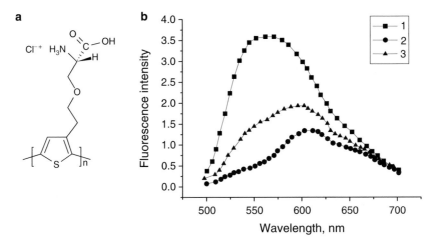

Fig. 6.9 The conjugated polyelectrolyte poly{3-[(S)-5-amino-5-carboxyl-3-oxapentyl]-2,5-thio-phenylene hydrochloride} (**a**) and its fluorescence spectrum (**b**). The fluorescence spectrum in unbound form (1) changes upon binding to calmodulin. In the bound form, the fluorescence spectrum responds to the presence of Ca^{+2} ions as shown in comparison of Ca^{+2}-free (2) and Ca^{+2}-bound (3) forms (Redrawn from Nilsson and Inganas 2004)

2000). The binding interactions resulted in an unusual red-shift in the visible absorption of the polymer backbone, suggesting a lengthening of the effective conjugated length upon interaction of the ligand with its congnate receptor.

The binding of the conjugated polymer with cationic substituents to DNA is based on electrostatic interaction and depends on the DNA conformation. On binding to DNA, the polymer has to change its conformation and thus generate the sensing signal (Fig. 4.28). This methodology found application in DNA hybridization assays, in which a conjugated polymer started to play the role of 'hybridization transducer' (Dore et al. 2006). It was successfully applied to nucleic acid aptamer sensors (Ho and Leclerc 2004). Since single-stranded DNA (ss-DNA) can specifically bind to various targets, including complementary ss-DNA, ions, proteins, drugs and so forth and on this binding the oligonucleotide probe often undergoes a conformational transition, the range of the application of this methodology can be very broad (De Schryver et al. 2005).

6.3.3 Conformational Changes in Peptide Sensors and Aptamers

There is nothing artificial in the mechanism of recognition between flexible peptides and nucleotides and their well-organized targets. A great number of 'natively unfolded' proteins have been discovered in recent decades, which even required revising the paradigm on interactions of proteins with their substrates and ligands. It became clear that the '*lock-and-key*' and '*induced fit*' mechanisms should be complemented by the mechanism of '*induced and assisted folding*' (Demchenko 2001). Many cases were found in which the natively unfolded protein on its specific interaction with another protein or nucleic acid exhibits the unfolding → folding conformational change to form a stable and well organized complex (Fig. 6.10a).

There is a special class of proteins called molecular chaperones, which, interacting with unfolded proteins, provide their transition into the folded state. Moreover, examples were found of interacting partners, so that both of them are unfolded and upon interaction they find a common pathway to the folded states (Fig. 6.10b). In both these cases, the interactions between the partners can be very specific, starting from their initial contacts. The whole process follows strict thermodynamic and kinetic rules and involves millions of decisions of the chains for abortion or for re-enforcement of contact. Therefore, the formed complex is also very specific, satisfying the requirement for molecular recognition.

The strategy to make virtually any protein into a sensor with signal transduction based on *target-induced folding* has been developed (Kohn and Plaxco 2005). Rational genetic engineering can destabilize the protein structure in such a way that it becomes naturally unfolded. Binding the target should make the folded state thermodynamically more stable and induce the unfolding-folding transition. These results make a fair background for understanding how the unfolded peptide selected from the library without a regular structure and in the absence of a formed binding

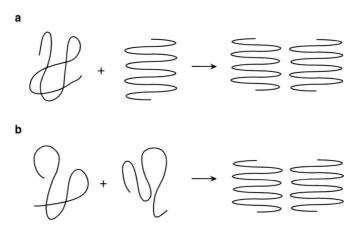

Fig. 6.10 Illustration of the mechanisms of 'induced and assisted folding'. (**a**) Unfolded protein folds upon interaction with the other folded protein used as a template. (**b**) Two unfolded proteins assist each other to attain folded conformations. In both cases, specific recognition occurs on the level of groups of atoms. It is sequentially enforced in the course of folding

site, can serve as a very specific binder. The specificity of interaction appears if the design allows a strong coupling of the chain folding and recognition.

Such a profound change in the conformation of the recognition unit provides many possibilities for signal transduction in sensing. Many fluorescence reporting mechanisms can be applied to such systems. Meanwhile, one precaution has to be taken. Modification with fluorescent dye can be a strong intervention into the process of folding. Therefore, such modification should preferably to be made in a site that has a minimal involvement in the process of folding. Alternatively, the modification should be made during the step of selecting the prospective receptors from a combinatorial library, so that the selection has to be made taking into account the folding and recognition properties of the peptide-dye conjugate.

The *zinc finger* presents a classical example of transforming the designed peptide into a sensor. It is derived from a small (25–30 residue) protein domain that selectively binds the zinc ion with dissociation constants in the picomolar range. In the absence of zinc, the peptide is unfolded and it folds to coordinate the Zn ion (Fig. 6.11). The primary coordination sphere involves two His and two Cys residues. The chemical synthesis of this peptide was easily performed with the incorporation of fluorescence reporters into this structure. In this case, the signal transduction does not involve the interaction of a reporter with coordinated zinc and can be provided by the dyes that interact between themselves and a report based on FRET or PET mechanisms. Therefore, many modifications are possible to improve the performance of this sensor (Walkup and Imperiali 1997).

The conformational adaptation of the receptor in forming the complex with a target is present in sensing *aptamers*. In Section 5.5 we described in detail the selection, synthesis and operation of nucleic acid aptamers. Here we will focus on the mediation

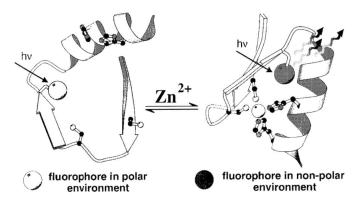

Fig. 6.11 The zinc-induced folding of zinc finger peptide. The conformational change in peptide is detected with the environment-sensitive dye. In unfolded state the dye is exposed to water and indicates its position in a polar environment. Its environment changes to low-polar on the folding of peptide triggered by zinc ion (Reproduced with permission from Walkup and Imperiali 1996)

and generation of the fluorescence signal coupled with conformational change in these sensor molecules. Many aptamers (especially with a short sequence) are structureless in free states in solutions and some others possess only the elements of a double-helical structure that are flexible and easy to unfold. A number of studies have demonstrated that the target binding is coupled with their ordering into a well-defined three-dimensional structure (Hermann and Patel 2000; Stojanovic and Kolpashchikov 2004). This suggests many possibilities for providing a fluorescence response with the realization of the principle of direct sensing.

An interesting possibility for signal transduction can be found with aptamers. Since they reorganize and in many cases form the double-helical structure on target binding, this process can be detected by the dyes that can *specifically intercalate* into this structure. Particularly, such a response can be obtained with intercalating rubidium chelating complexes that can provide a dramatic increase in phosphorescence (Puleo et al. 2006). This eliminates the necessity of the covalent labeling of aptamers. The response to conformational change in aptamers can also be detected with the aid of a cationic conjugated polymer (Ho and Leclerc 2004).

6.3.4   Molecular Beacons

Molecular beacons (other name 'molecular hairpins') are the direct molecular sensors designed for the detection of specific nucleic acid sequences based on the formation of complementary double helix structures. Nucleic acids are widely known for their ability to form Watson-Crick double helical structures that are stabilized by hydrogen bonds between complementary base pairs. The formation of these structures is strongly sequence-specific. The double helix can be formed between two different chains with complementary structures but also in a single chain when it folds on itself to form a double helix. This allows making sensors out of single-chain nucleic acids that fold into a specially designed structure containing two structural components, a loop and a stem (Fig. 6.12). They can be considered as a special case of aptamers, with specificity to recognize sequences in nucleic acids.

The loop is the recognition element containing a complementary nucleic acid sequence to the target sequence that has to be detected in a pool of many different oligonucleotides. The stem is formed of two complementary sequences that fold on themselves to flank the loop. Hybridization of the loop part with the target DNA or RNA sequence generates a conformational change in this structure, its stretching-out. As a result, the chain segments forming a stem disconnect and move apart. This motion is detected, being transformed into fluorescence signals. Since the nucleic acid sequence of the loop can be made complementary to that of any nucleic acid target, molecular beacons can be used as a generic technology for the recognition of specific nucleic acid sequences (Tyagi and Kramer 1996).

The distance between the 3′ and 5′ ends, being very short in a folded beacon, dramatically increases on target binding. Thus, the fluorescence reporter signal can be generated based on recording the disconnection between the stem segments. The

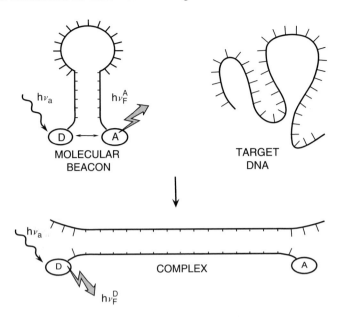

Fig. 6.12 Schematic representation of the generation of a fluorescence signal by the hybridization of nucleic acid sequences with molecular beacons. A molecular beacon is a hairpin-shaped nucleic acid with the sequence of the loop complementary to the sequence in the target nucleic acid. In the absence of the target, the two dyes, donor (D) and acceptor (A) that can be a fluorophore and a quencher are located closely and interact. When the loop hybridizes with a target sequence, the stem is disrupted and the two labels move apart generating a fluorescence response of the donor

dye can be attached to one of these chain terminals or to both of them. There are several possibilities to generate the reporter signal.

1. The *increase of distance* between two beacon ends can be used in detection by FRET to a non-fluorescent acceptor (quencher). The introduction of two labels, donor and acceptor, is needed in this case. In the sensing event (hybridization with complementary sequence) the separation between two dyes increases, causing fluorescence enhancement. This approach has been explored in a number of works (Marras 2006; Marras et al. 2006).

2. More advantageous is the detection technique in which both the *FRET donor and acceptor are fluorescent*. This possibility was realized recently with the use of cyanine dyes Cy3 and Cy5 (Ueberfeld and Walt 2004). When the donor and acceptor dyes are in close proximity, an emission of the acceptor is observed. On hybridization, this emission disappears and the emission of the donor is enhanced. It was found that the ratio of donor-acceptor intensities is independent of the amount of the probe and provides a quantitative measure of the free target concentration.

3. The quenching of the fluorescence signal is achieved by *photoinduced electron transfer* (PET) quenching (see Section 6.1). In PET, quenching the emitting dye and the quencher (electron donor or acceptor) should be present in the stem structure at a very close distance from each other (Heinlein et al. 2003), in contrast to a much larger distance (of several nanometers) that is needed in FRET sensing. Since in the folded state of the molecular beacon the distance between the two chain terminals is very short, complete quenching can be achieved, so that fluorescence enhancement appears on hybridization. The advantage of this technology is in the possibility to avoid a second labeling. The nucleic acid guanosine base can play the role of PET quencher of complementary dye in beacon construction (Knemeyer et al. 2000).

4. One of the hairpin ends can be *immobilized on the metal surface* (Dudley et al. 2002). If the other terminal contains the dye, the fluorescence of this dye will be quenched due to contact with the surface (also due to the PET mechanism). Fluorescence enhancement will be observed on hybridization.

Various hybridization techniques based on the stem-loop structure have been developed (Broude 2002). One of the suggestions was on providing the signal amplification by coupling the sensing event with a *biocatalytic reaction*. Nucleic acids that behave as enzymes (named ribozymes and deoxyribozymes) were used to construct catalytic molecular beacons. For example, the specific oligonucleotide recognition property of the stem-loop structure was coupled with catalytic activity, such as the RNase activity of the hammerhead deoxyribozyme (Stojanovic et al. 2001).

The sensing DNA or RNA oligonucleotide can successfully be substituted by artificial oligomers – *peptide nucleic acids* (Section 5.6). In their structures the negatively charged ribose or deoxy ribose phosphate is replaced by an uncharged N-(2-aminoethyl)-glycine scaffold, to which the nucleobases are attached via a methylene carbonyl linker (Brandt and Hoheisel 2004). The target recognition by these beacons is less sensitive to ionic strength and their fluorescence is not affected by DNA-binding proteins.

The principle of the molecular beacon acquires more and more general importance. A successful result on its application in detecting proteins has been reported recently (Thurley et al. 2007). The *peptide beacon* (Fig. 6.13) uses the ability of a designed peptide to change its conformation between target-bound and target-free forms with fluorescence reporting based on double labeling.

6.3.5 *Proteins Exhibiting Conformational Changes*

There are several classes of *ligand-binding proteins* that exhibit global conformational changes, such as the motions of their domains (Gerstein et al. 1994). The *hinge-bending motions of protein domains* provide the most pronounced example of large-amplitude slow motions in a number of proteins, which are coupled with their functioning. These proteins posses two well-separated domains, which

a

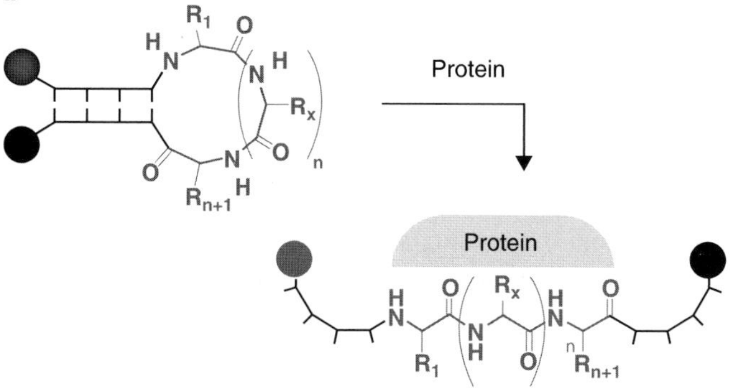

b

Hairpin Peptide Beacons:

Ac-Lys-t t g g-Glu-Pro-Gln-pTyr-Glu-Ile-Pro-c c a a -Lys-Gly ^{CONH_2} **1**
Pyb Pyb

Ac-Lys-t t g g-Pro-His-Pro-Phe-His-Stat-Ile -c c a a-Lys-Gly ^{CONH_2} **2**
NIR Dabcyl

Fig. 6.13 The prototype of the peptide molecular beacon for detecting proteins. The same principle as in DNA beacons is applied here. Whereas the recognition part is peptide, the stem part is made of complementary peptide nucleic acid (PNA) sequences. **(a)** The principle of beacon formation. R_i are the side groups of amino acids. **(b)** The sequences of peptide molecular beacons. Lower case letters indicate PNA nucleotides. Pyb is pyrene butanoic acid, NIR is the near-IR dye and Dabcyl is the quencher (Reprinted with permission from Thurley et al. 2007)

undergo a large hinge-bending motion upon ligand binding. In the open ligand-free state the domains are far from each other, while in the closed ligand-bound state the domain becomes tightly packed. The ligand is often captured in the cleft between the domains, which both form the binding pocket (Color Plate 8).

Many attempts have been made to apply the ligand-binding proteins as molecular sensors for the detection of their ligands (de Lorimier et al. 2002). The major problem here is to indicate the act of ligand binding by generating the measurable response signal. If one puts the fluorescent reporter group into the ligand-binding site it may interfere with the ligand binding. Therefore, it is preferable to locate the reporter group outside the ligand-binding site but allow for some mechanism of signal transduction to this group, which could be a conformational change. It is interesting to note that *calmodulin* and several periplasmic *ligand-binding proteins* are the only proteins in which this principle was realized in practice and led to the design of direct molecular biosensors for the detection of correspondent ligands with a fluorescent group at a remote site (Dattelbaum et al. 2005; Mizoue and Chazin 2002). In these cases the ligand binding changes the *intramolecular dynamics* of the protein and induces conformational change that allows transduction of the signal to reporter groups without its direct contact with the ligand (Mizoue and

Chazin 2002). The small number of these successful cases is an indication of the difficulties in extending this principle of sensing to other ligand-binding proteins.

Analysis of the intra- and inter-domain dynamics of 157 ligand-binding proteins (Yesylevskyy et al. 2006) allows formulation of the necessary conditions, which should be satisfied in order to design the molecular biosensor with the responsive groups not being in direct contact with the ligand. The *intramolecular dynamics* of such a biosensor should be sensitive to the ligand binding events in order to transfer the signal from the binding site to the reporter group. This can only be achieved if (a) the protein undergoes significant conformational change upon the ligand binding; (b) the intra-domain conformational changes are small in comparison with the relative motion of domains and (c) the strength of the inter-domain contacts is significant in the ligand-bound conformation and very small in the ligand-free conformation.

The technique developed in (Yesylevskyy et al. 2006) can be applied to the candidate proteins to see if these conditions are fulfilled for them. This will allow minimizing the possible design errors caused by the choice of the protein, which is not suitable for the design of sensors, where the reporter group is not in contact with the ligand. An algorithm was developed, which allows physically consistent simulations of slow large-scale protein dynamics, based on subdividing the protein into the hierarchy of the rigid-body-like clusters (Yesylevskyy et al. 2007). This technique allows predicting a closed conformation, staring from the open one in agreement with their known crystallographic structures. Such techniques are promising in deciphering the character of protein motions involved in the signal transduction from the ligand binding site to the reporter group.

It was shown in (Yesylevskyy et al. 2006) that the periplasmic *ligand-binding proteins* and *calmodulin* (Color Plate 8) are the most probable targets for the biosensor design. One of such proteins, the *maltose binding protein* (MBP) is the most successful sensor protein, which is extensively studied and widely used. Many genetic mutants of MBP were produced to modulate specificity and affinity in target recognition and different spectroscopic techniques of signal transduction and reporting (Gilardi et al. 1994; Medintz and Deschamps 2006). In particular, a competitor displacement assay was suggested for maltose binding, in which the competitor was the labeled β-cyclodextrin derivative (Medintz et al. 2003). The use of a Quantum Dot – organic dye pair allowed the use of distance-dependent interactions between them, resulting in PET quenching or FRET (Medintz and Deschamps 2006). Located distantly in open conformation, they become close neighbors in a closed conformation. A different construction using conformation-dependent FRET efficiency from the ruthenium complex to Quantum Dot was also found to be efficient (Raymo and Yildiz 2007).

The successful application of sensors based on protein molecules exhibiting conformational changes allows avoiding direct contact of a fluorescence reporter with the target but depends critically on a position of the right reporter at the right place. An example of a successful application of a two-color ratiometric dye is the observation of the response of this dye to a conformational change of enzyme inhibitor α_1-antitrypsin with elastase (Fig. 6.14). The ratiometric response of the dye is the result of the change of its interaction with the molecular environment.

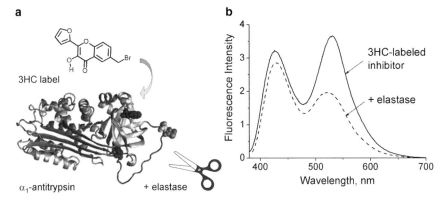

Fig. 6.14 An example of the application of environment-sensitive dye for the detection of conformational change in protein (Courtesy of Y. Mely et al., to be published). **(a)** Three-dimensional representation of the native human α_1-antitrypsin. The 3-hydroxychromone (3HC) label was attached by Cys-232 residue of the protein. Elastase enzyme (presented as the scissors) interacts with α_1-antitrypsin and cleaves the peptide bond in the inhibitor's loop resulting in the conformation change. **(b)** Steady-state fluorescence emission spectra of 10 μM of the labeled α_1-antitrypsin in the absence (—) and in the presence (----) of elastase

6.3.6   Prospects

Conformational changes in the receptor molecular unit together with the implementation of the reporting mechanism based on electron, proton or energy transfer provided by fluorescence reporters allow realizing the principle of direct sensing (Section 1.6). Based on these changes we are able to obtain a direct response to target binding in the form of changing the parameters of emission. Dependent upon the system design, it is possible to select conformational isomerizations in the reporting dye or the changes in the conformation of conjugated polymers with their strong effects on *electronic conjugation* between the fragments of dye molecule or the monomers in a polymeric structure.

The folding reactions in polypeptides and oligonucleotides selected as the best binders from large libraries offer dramatic progress in sensing technologies. These flexible sensors should be selected for being committed to folding only in complexes with the targets. Though they are structureless or possess only a small amount of three-dimensional structure, they can be selected to contain all geometrical and bonding features for correct target recognition. Sensing by 'induced folding' must find many applications as a simple, universal and economical way to address many targets.

Molecular beacons are a specialized version of aptamers that realize transition from an intramolecular (hairpin) to an intermolecular (complex) folded state of oligonucleotide. Detecting responses to this change is easy and can be provided in a number of ways. Finally, regarding the ligand binding proteins and their genetic versions, their conformational changes are very specific, as specific as their recognition. Time will show whether a generic methodology will be developed on their basis.

6.4 Signal Transduction via Association and Aggregation Phenomena

Since it is well documented that the reduction of size and dimensionality of different types of nanoparticles results in quantum confinement effects that dramatically influence their spectroscopic properties, it is natural to suggest that the aggregation of these particles will provide the necessary signal for sensing. In clustered assemblies of the particles, new collective properties can be observed, such as the *coupled plasmon absorbance*, inter-particle electron and energy transfer and the appearance of electrical conductivity (Barbara et al. 2005). Thus, possessing unique responsive properties, nanoparticles can change them in the course of aggregation coupled with target binding.

The strong color that can be detected visually on illumination of the nanoparticles is due to a combination of two optical phenomena: light *absorption* and *scattering* (Yguerabide and Yguerabide 1998). Due to this fact, the correct term could be 'extinction' instead of 'absorption', since it describes both absorption and scattering components to the observed colors.

6.4.1 Association of Nanoparticles on Binding a Polyvalent Target

One of the easiest possibilities to detect a bivalent or polyvalent target is to provide a sensing system in which this target (or a competitor that can be substituted by the target) can simultaneously react with the binding sites on several nanoparticles. This reaction may induce aggregation or at least cluster formation, in analogy to a well-known immunoprecipitation method, where a bivalent antibody reacting with a polyvalent antigen induces the formation of precipitate. With the introduction of nanoparticles into the sensor development this principle has obtained a new stimulus for development. An easily detectable change in the optical properties of gold and silver nanoparticles occurs on their association due to *plasmon-plasmon interactions* between adjacent particles. The characteristic visually detected red color of gold colloid changes to a bluish-purple upon colloid aggregation (Fig. 6.15).

This property has found application in *polynucleotide detection* (Elghanian et al. 1997). Upon introduction into the test system of the target polynucleotide, the gold nanoparticles labeled with nucleotides possessing attached complementary sequences start to associate. This induces a dramatic change in the visually detected color from red to blue.

Because of the extremely strong light absorption of gold nanoparticles, the straightforward application of this colorimetric method allows a detection limit of ~10 femtomoles of oligonucleotide. This is already 50 times more sensitive than

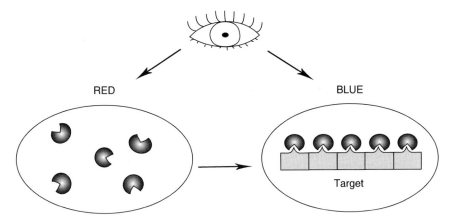

Fig. 6.15 The change of visually detected reflected/scattered color of gold nanoparticles on their association caused by their binding by the polyvalent target

sandwich hybridization detection methods based on fluorescence detection. In order to further increase this sensitivity, the intensity and color of scattered light was suggested to be recorded (Storhoff et al. 2004). This scatter-based method is so sensitive that it allows detection of zeptomole quantities of nucleic acid targets.

There are many other detection methods suggested with the use of the color change of noble metal nanoparticles. The color change based on the analyte-induced aggregation/deaggregation test was suggested for the detection of low concentrations of *heavy metal ions* (Pb^{2+}, Cd^{2+}, Hg^{2+}) in aqueous solutions. The ensemble in this case consists of gold nanoparticles functionalized with alkanethiol chains carrying carboxylate groups at the distal terminal ends. The aggregation of the particles upon the addition of metal ions occurs due to ion coordination by these groups belonging to two neighbor particles. Dextran-coated gold nanoparticles aggregated with concanavalin A were used for competitive *glucose sensing* in an extended millimolar-micromolar range (Aslan et al. 2004).

6.4.2   Association-Induced FRET and Quenching

The change in physical separation between nanoparticles induces changes in energy- and electron transfer efficiency. One of the typical applications of using *association between Quantum Dots* is the sensing of *potassium ions* by observing FRET with the change of color between small (green) and large (red) CdSe/ZnS nanocrystals (Chen et al. 2006). This assay showed excellent sensitivity.

The association of *conjugated polymers* in solutions commonly leads to quenching. Since such an association can be modulated by target binding (Jiang et al. 2007), this provides an additional possibility for fluorescence sensing with an efficient response.

Quantum Dots that emit at 525 and 585 nm are used to encode aptamer-linked nanostructures sensitive to adenosine and cocaine, respectively. In addition to Quantum Dots, the nanostructures also contain gold nanoparticles that serve as quenchers. The addition of target analytes disassembles the nanostructures and results in increased emission from Quantum Dots. Simultaneous colorimetric and fluorescent detection and quantification of both molecules in one pot has been demonstrated (Liu et al. 2007).

6.5   Integration of Molecular and Digital Worlds

Sensing and computing are always connected and both exhibit dramatic advancements in miniaturization leading them to the molecular level. In information technology, miniaturization is responsible for the outstanding progress seen in recent decades. However, it still remains to be based on macroscopic concepts and their realization in silicon-based devices, the reduction in size of which is approaching its limit. The next and definitely ultimate limit of miniaturization is that of molecules, since molecules are the smallest bodies in which information transfer can be achieved in the form of directed translocation of charge. Information inputs and outputs on a molecular level can be provided by optical means and this makes an additional strong connection between sensing and computing. Moreover, we will see that molecular-level systems analogous to those that are discussed in this book are capable of transferring, processing and storing information in digital form.

6.5.1   The Direct Recording of Digital Information from Molecular Sensors

The sensitivity of fluorescence sensors rapidly approaches the level of counting the number of molecules. In such systems, statistical laws that allow ensemble-averaging of studied properties may fail to be valid and the fluctuations of these properties may be great. Meantime, the provided response is commonly based on statistical averaging and obtaining the analog output (Fig. 6.16).

Fundamentally, a *digital signal* can contain more information and can be processed, transmitted or stored more efficiently and with less noise than an analog signal (Najmabadi et al. 2007). Because of this, digital processing has become the primary choice in the development of electronic systems. Digital systems operate with discrete numbers that are usually presented in binary code (base 2), so that numeric values are expressed by two symbols, 1 for '*true*' and 0 for '*false*'. The elementary process of the interaction of single molecules can in principle be optimally expressed in digital form. Instead, we commonly collect an analog output signal, which is continuous in both time and amplitude but noisy, and we then convert it into a digital form for processing and analysis. The smaller the system, the greater may be the problem of noise.

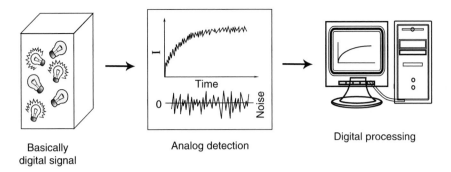

Fig. 6.16 Typical sequence of signal processing in molecular sensing. The primary response to target binding is digital (YES or NO). Detection involves ensemble-averaging of elementary signals to provide analog output. Then the signal is digitized for further processing and analysis

Noise, or random variations in a signal, can also appear during different steps of processing or transmitting the data and an analog signal is not resistant to it. It can be reduced by different kinds of frequency filters but cannot be eliminated completely. In contrast, in digital signals this type of noise cannot distort the quality of the information, as any signal close to a particular value (0 or 1 in binary systems) will be interpreted as that value (Najmabadi et al. 2007). The value of any analog signal can be converted into digital form by representing it in binary format by the combination of (0)s and (1)s, which is often done with the sensor output signals. This does not eliminate the noise, however.

There are experiments that clearly show the digital (YES or NO) nature of sensor-target interaction. *Fluorescence correlation spectroscopy* is able to extract useful information from the noise (Thompson et al. 2002). This method requires the creation of an illuminated, very small, (sub-femtoliter) open-volume element by a focused laser beam. Fluorescence intensity fluctuations will appear due to spikes of emission intensity of single labeled molecules when these molecules enter and leave this volume. The autocorrelation function provides information on the rates of diffusion of these molecules and on the changes of these rates if they bind other molecules and move more slowly. The binding constants can be determined in this way, in solution with the labeling of one of the interacting partners. Meantime, this method does not allow a single-molecular digital response and this problem is still waiting for a solution.

6.5.2 Hybrid Molecular-Digital Systems

The new concept of a hybrid system comprises molecular analysis through direct *interaction between molecules and digits*. In such systems, the sensor recognition events are not coupled with a response provided by an optical or any other physical phenomenon. Rather, a true interaction between digits and molecules is achieved, since here a molecular event is detected by its ability to manipulate a digital sequence (Najmabadi et al. 2007).

The *compact disk* (CD) format for the fabrication of microarrays could be implemented for molecular detection in a method analogous to data reading off a conventional compact disk. Using the same principle as reading from common CD disks, the digital information is recorded. In one of the versions of this approach, molecules do not create a digital signal but alter an existing digital stream. Positive molecular interaction events on the surface of the CD create an interference within the digital steam as the CD is read by the player. Sensor molecules are deposited on a chemically modified CD surface after a specific data pattern is written onto the internal digital layer of a CD. Molecular screening is performed by reading the data on this CD before and after the addition of a tested sample (La Clair and Burkart 2006). These systems are called *hybrid* because here the analytical device and the computing device are not the two separate entities, as they are in conventional techniques. The analytical system becomes an integral part of the digital communication and it can benefit from sophisticated logic and networking structures found in contemporary computing. It remains difficult to predict the future of this and related technologies but it is a fact that molecular and digital worlds started their communication on a technical level.

6.5.3 Logical Operations with Fluorescent Dyes

In previous chapters we always implicitly assumed that by letting the sensor response function be of any shape, the transduction function (that connects the number of formed complexes with the response that they produce) will always be linear (or weighted linear as explained in Section 2.5). This means the existence of *strict proportionality* between the amount of *bound* target and the detected signal. However, the transduction mechanism can be more complicated. For instance, there may be a need to obtain a response on the amount of one analyte that conditionally depends on the presence or absence of a second analyte. Speaking in the language of information technology, such analysis can include *logic gates* operations. Recently, great effort has been devoted to design and investigate chemical sensors with a non-linear and modulated performance. Some of the molecular and nano-scale devices can perform logic operations of remarkable complexity.

These *logic operations* are formally the same as those performed in digital computing. They are based on Boolean algebra, which is in the background of the operation of modern computers dealing with 0 and 1 binary code. With single input (e.g., analyte binding) and single output (e.g., switch-on of fluorescence intensity) only two operations are possible, YES when input is 0 or 1 and the output is the same and NOT, which is the opposite: when input is 0 or 1, the output is 1 or 0. With two inputs (such as two different ions bound to two different sites) more complicated logical operations are possible. When both inputs are 1 and 1 the AND operation gives output 1, OR operation gives 1 and NOR operation gives 0. When both inputs are 0 and 0 the AND operation gives 0, OR operation gives 0 and NOR operation gives 1. If the first input is 1 and the second input is 0 then the AND operation gives 0, OR operation gives 1 and NOR operation gives 0. In the INHIBIT logic gate, if both inputs are 0 or 1, the output will be 0 and if they are different, the output will be 1. This gives us examples of how the sensors can perform operations similar to those in modern computers.

The introduction of the language of information science into sensing technology, if it is not a mental exercise, should have some important reason. The reason is that by using this language we can design and classify the sensors with *nonlinear and multiparametric operation*. For instance, if in a given system we need to obtain the response to the presence of one analyte on the background of the presence of a second analyte and the absence of a third analyte, in the traditional way we have to apply three separate sensors and then analyze the three output signals. But we are, in principle, able to devise a molecular or supramolecular construction that will do this itself and provide us with an already processed single output (Fig. 6.17).

There are many results along the line of this development and many attempts to re-consider the previously obtained data in line with new concepts (Tomasulo et al. 2006). Guides to these results can be found in recently published reviews (Magri et al. 2007) and only a few of them that are closely related to the development of sensor technologies will be mentioned here. Mainly, the suggested constructs are based on the products of synthetic organic chemistry. They are made by assembling several ion recognition sites with two or more organic heterocycles, which provide the signal transduction and response. This transduction is based on the basic mechanisms that were discussed above, PET, ICT and FRET. It was shown that efficient signal processing can be achieved with one or several fluorophores and electron donors or acceptors coupled with ion binding sites (Straight et al. 2007). Even the combination

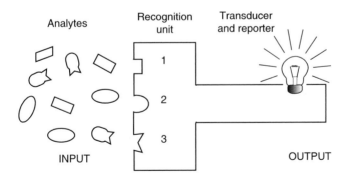

Fig. 6.17 Schematic presentation of molecular logic gate. The sensor consists of several (shown 3) types of analyte binding sites. The sensor performs a logical operation to provide an output signal at a designed combination of occupations of these sites

of the ion chelating group and the pH-titrated group may be surprisingly efficient for providing two inputs (concentrations of ions and protons).

In the construction of logic gates, special attention is being paid to *photochromes* (Gust et al. 2006). They are organic molecules that are isomerized by light between two stable forms. They can be covalently linked to other dyes, so that the changes in their properties resulting from photoisomerization can be used to switch the electron and energy transfer on or off. With the aid of photochromes the problem of not only the transformation but also of the storage of information on a molecular level can be resolved.

Other developments of *molecular logic gates* include designing constructs involving DNA (Okamoto et al. 2004), deoxyribozymes (Stojanovic and Stefanovic 2003), peptide nucleic acid (PNA) (Cheah et al. 2005), DNA aptamers (Yoshida and Yokobayashi 2007) and cationic conjugated polymer/DNA assemblies with ethidium bromide as an intercalating reporter dye (Zhou et al. 2005).

For connecting the elements performing logic functions, different types of *'molecular wires'* are being developed. They may include organic molecular wires such as oligo(2,5-thiophene ethynylene)s, oligo(1,4-phenylene ethynylene)s, oligo(1,4-phenylene vinylene)s, aromatic ladder oligomers, oligophenylenes, polyphenylenes, acetylene oligomers, carbon nanotubes and organometallic constructs. For providing connection by electron conductance, silver ion clusters have been suggested (Barbara et al. 2005) With discrete states, the several-atom silver clusters exhibit molecule-like behavior, with strong visible fluorescence and robust optical properties. They can be electrically contacted in thin films to produce the first electroluminescent single molecules.

Molecular wires can also be developed based on an excitonic conduction mechanism that can be realized in conducting polymers (Yashchuk et al. 2000). In connecting the functional elements, these neutral (electron + hole) quasi-particles can display a one-way conductivity.

One of the major problems in *'molecular computing'* is the development of multi-channel outputs that requires a fluorescent reporter to be present in several recognizable emissive states (Perez-Inestrosa et al. 2007). This problem, in principle, can be resolved with color-changing fluorescent dyes.

Interestingly, smart molecules allow not only computing but also *protection of the information* on the molecular scale (Margulies et al. 2007). By exploiting the principles of molecular Boolean logic, a molecular device can be developed that mimics the operation of an electronic keypad lock, e.g., a common security circuit used for numerous applications in which access to an object or data is restricted to a limited number of persons. The 'password' should include not only the proper combination of inputs but also the correct order by which they are introduced. It was shown that different 'passwords' can be coded by a combination of two chemical and one optical input signal, which can activate, separately, blue or green fluorescence output channels from pyrene or fluorescein fluorophores. The information in each channel is a single-bit light output signal that can be used to authorize a user, to verify authentication of a product or to initiate a higher-level process.

6.5.4 Prospects

Organic molecules and heterocycles of different origin allow not only the transduction of a signal into an emissive form but also its non-linear transformation, resembling the operation of elementary digital computational devices. The molecular level and *discrete nature of intermolecular interactions* that are in the background of sensing suggest the idea of obtaining the output information already in digital form and we observe an active search for its realization. There is a possibility to go deeper and request designed molecules to perform smart logical operations. Many of these attempts are successful and presently we observe newly synthesized molecules that are able to perform rather complicated operations of logic gates.

With all these successful developments, the future of post-silicone digital molecular electronics is still obscure. This field needs new conceptual solutions and we observe the presentation of many schemes suggesting logical operations on a molecular level. Such attempts often look like logical exercises that do not attempt to address any real application in computing. This is because most of these transformations have a *time limit of reagent addition* and its diffusion, therefore they cannot be conducted within a time frame comparable to the operation of electrical circuits. Probably, only reactions with elementary steps of femtosecond duration will be useful for these applications. Fortunately, the reactions of excited-state-electron, charge, proton and energy transfers – offer such possibility. The fact that these reactions are triggered by light and allow optical output (Syromyatnikov et al. 1996) may also solve another important problem, which is the integration of elementary functions and providing interconnections required for computation. Finding practical solutions on this trend are probably possible, though in the unforeseeable future.

In strong contrast, systems with *slow diffusion-controlled chemical inputs* have every chance of development into smart sensors of the next generation. They will allow, instead of a linear response, complicated multiparametric logical operations, resulting in an already processed output signal. To achieve this the sensors should be able to exist in several states with programmed switching realized between them. No doubt, digital logics will be a common language in these developments.

Sensing and Thinking 6: Coupling Recognition and Reporting Functionalities

Summing up, we list the basic mechanisms of signal transduction that operate in fluorescence sensors coupling the target binding and fluorescence (luminescence) response.

The shortest on the scale of distances is the *intramolecular charge transfer* (ICT). It can operate within a π-conjugated electronic system and is able to directly transform the weak interaction energy of target binding into spectral shifts. The elementary act in intramolecular proton transfer (ESIPT) occurs at even shorter distances (the proton is much heavier than the electron and its wavefunction is shorter) but in organic dyes this transfer is coupled with conformational isomerizations, making-breaking of hydrogen bonds and ICT. The 3-hydroxychromone family of dyes is unique among well-described systems, in which ESIPT is coupled only with ICT and this allows obtaining a strong two-band ratiometric response.

Photoinduced proton transfer (PET) can operate at greater distances between localized electronic systems. It can be both intramolecular and intermolecular. Intramolecular transfer requires separation between donor and acceptor fragments by only one or two single bond bridges and intermolecular transfer needs direct contact between them. All these mechanisms are basic and universal but for their realization they need proper molecular structures that can be the most efficiently provided by organic dyes.

The *Förster resonance energy transfer* (FRET) can operate at greater distances, up to 8–10 nm. It is much more universal and can be realized with all kinds of emissive molecules and particles as the donors and all kinds of light-absorbing acceptors if the conditions regarding distance, orientation and spectral overlap are satisfied. Collective effects in FRET allow efficient light-harvesting and wavelength-shifting, which allows broad-range manipulations with the wavelength and sensitivity of response.

Conformational isomerizations add a new dimension to signal transduction. They offer many possibilities for coupling the target binding with the response. They are especially important in cases of large and neutral target molecules, the binding of which does not produce the perturbation sufficient for the direct modulation of IST, PET or ESIPT. Target-induced *aggregation of nanoparticles* is an additional possibility for detecting large-sized targets.

The ability of molecular sensors to display the information in *digital form* and to transform it in order to perform elementary logical operations, opens new horizons for exploration and promises new advanced products.

Questions and Problems

1. Explain the relation between photoinduced electron transfer (PET) and fluorescence quenching. Does PET always lead to quenching? Is quenching always the result of PET?
2. What is the distance dependence for PET? What other factors determine PET efficiency? Suggest several examples for using these factors in the construction of sensors.
3. Based on what criteria would you differentiate ICT and PET? Suggest the sensor designs that can be realized with PET but cannot with ICT and *vice versa*. What methods of fluorescence detection shown in Fig. 3.1 are applicable for the detection of a response based on PET and what on ICT?

4. Identify electron-donating and electron-withdrawing groups in the structure of ketocyanine dye described in Section 4.1. Suggest how the dipole moment is oriented in this molecule.

5. How do LE and ICT emissions depend on solvent polarity? Can the switching between them be induced by changing the temperature? If yes, in what direction is this change and what is its mechanism?

6. What are the differences in requirements for observation and reaction mechanisms between intermolecular and intramolecular proton transfers?

7. List and explain *collective* effects in FRET. What will be the spectroscopic effects in the following cases: (a) dye molecules of only one type are present in the system; (b) molecules of one type with an added quencher; (c) molecules of one type (donors) with an added fluorescence acceptor?

8. Explain the antenna effect. Is the energy transfer between the donors (homo-FRET) important for manifestation and the application of the antenna effect?

9. Evaluate the properties of different dyes and nanoparticles as FRET donors and acceptors. Select the optimal donor-acceptor pairs for (a) measurements of intensity; (b) two-band ratiometric measurements; (c) measurements of lifetimes.

10. What is the induced folding of macromolecules? How can it be used in sensing?

11. Can molecular beacons be immobilized on a surface? What transduction mechanisms can be used when a molecular beacon is in solution and when it is immobilized?

12. Try to suggest alternative structures to molecular beacons performing the same function – hybridization with single strand DNA with the reporting based on conformational change.

13. What are the advantages and disadvantages of maltose binding protein as a scaffold for the design of various sensors? What are the optimal detection methods to be used with this protein as the sensor: (a) with single labeling? (b) with double labeling?

14. Why do the aggregates of noble metal nanoparticles change their color?

15. What is gained and what is lost if a sensor with its operation based on the principle of logical gate allows three inputs and a single output?

References

Anikeeva PO, Madigan CF, Coe-Sullivan SA, Steckel JS, Bawendi MG, Bulovic V (2006) Photoluminescence of CdSe/ZnS core/shell quantum dots enhanced by energy transfer from a phosphorescent donor. Chemical Physics Letters 424:120–125

Arnaut LG, Formosinho SJ (1993) Excited-state proton-transfer reactions .1. Fundamentals and intermolecular reactions. Journal of Photochemistry and Photobiology A-Chemistry 75:1–20

Aslan K, Lakowicz JR, Geddes CD (2004) Tunable plasmonic glucose sensing based on the dissociation of Con A-aggregated dextran-coated gold colloids. Analytica Chimica Acta 517:139–144

Baek MG, Stevens RC, Charych DH (2000) Design and synthesis of novel glycopolythiophene assemblies for colorimetric detection of influenza virus and E. coli. Bioconjugate Chemistry 11:777–788

Barbara PF, Gesquiere AJ, Park SJ, Lee YJ (2005) Single-molecule spectroscopy of conjugated polymers. Accounts of Chemical Research 38:602–610

Bhattacharyya K, Chowdhury M (1993) Environmental and magnetic field effects on exciplex and twisted charge transfer emission. Chemical Reviews 93:507–535

Blough NV, Simpson DJ (1988) Chemically mediated fluorescence yield switching in nitroxide-fluorophore adducts: optical sensors of radical/redox reactions. Journal of the American Chemical Society 110:1915–1917

Bognar B, Osz E, Hideg K, Kalai T (2006) Synthesis of new double (spin and fluorescence) sensor reagents and labels. Journal of Heterocyclic Chemistry 43:81–86

Brandt O, Hoheisel JD (2004) Peptide nucleic acids on microarrays and other biosensors. Trends in Biotechnology 22:617–622

Broude NE (2002) Stem-loop oligonucleotides: a robust tool for molecular biology and biotechnology. Trends in Biotechnology 20:249–256

Bunzli JCG, Piguet C (2005) Taking advantage of luminescent lanthanide ions. Chemical Society Reviews 34:1048–1077

Burda C, Chen XB, Narayanan R, El-Sayed MA (2005) Chemistry and properties of nanocrystals of different shapes. Chemical Reviews 105:1025–1102

Casanova D, Giaume D, Gacoin T, Boilot JP, Alexandrou A (2006) Single lanthanide-doped oxide nanoparticles as donors in fluorescence resonance energy transfer experiments. Journal of Physical Chemistry B 110:19264–19270

Charbonniere LJ, Hildebrandt N, Ziessel RF, Loehmannsroeben HG (2006) Lanthanides to quantum dots resonance energy transfer in time-resolved fluoro-immunoassays and luminescence microscopy. Journal of the American Chemical Society 128:12800–12809

Chattopadhyay A, London E (1987) Parallax method for direct measurement of membrane penetration depth utilizing fluorescence quenching by spin-labeled phospholipids. Biochemistry 26:39–45

Cheah IK, Langford SJ, Latter MJ (2005) Concept transfer - from genetic instruction to molecular logic. Supramolecular Chemistry 17:121–128

Chen CY, Cheng CT, Lai CW, Wu PW, Wu KC, Chou PT, Chou YH, Chiu HT (2006) Potassium ion recognition by 15-crown-5 functionalized CdSe/ZnS quantum dots in H_2O. Chemical Communications:263–265

Chen LH, McBranch DW, Wang HL, Helgeson R, Wudl F, Whitten DG (1999) Highly sensitive biological and chemical sensors based on reversible fluorescence quenching in a conjugated polymer. Proceedings of the National Academy of Sciences of the United States of America 96:12287–12292

Clapp AR, Medintz IL, Fisher BR, Anderson GP, Mattoussi H (2005) Can luminescent quantum dots be efficient energy acceptors with organic dye donors? Journal of the American Chemical Society 127:1242–1250

Dattelbaum JD, Looger LL, Benson DE, Sali KM, Thompson RB, Hellinga HW (2005) Analysis of allosteric signal transduction mechanisms in an engineered fluorescent maltose biosensor. Protein Science 14:284–291

de Lorimier RM, Smith JJ, Dwyer MA, Looger LL, Sali KM, Paavola CD, Rizk SS, Sadigov S, Conrad DW, Loew L, Hellinga HW (2002) Construction of a fluorescent biosensor family. Protein Science 11:2655–2675

Demchenko AP (2001) Recognition between flexible protein molecules: induced and assisted folding. Journal of Molecular Recognition 14:42–61

De Schryver FC, Vosch T, Cotlet M, Van der Auweraer M, Mullen K, Hofkens J (2005) Energy dissipation in multichromophoric single dendrimers. Accounts of Chemical Research 38:514–522

de Silva AP, Gunaratne HQN, Gunnaugsson T, Huxley AJM, McRoy CP, Rademacher JT, Rice TE (1997) Signaling recognition events with fluorescent sensors and switches. Chemical Reviews 97:1515–1566

de Silva AP, Fox DB, Moody TS, Weir SM (2001) The development of molecular fluorescent switches. Trends in Biotechnology 19:29–34

Dore K, Leclerc M, Boudreau D (2006) Investigation of a fluorescence signal amplification mechanism used for the direct molecular detection of nucleic acids. Journal of Fluorescence 16:259–265

Dudley AM, Aach J, Steffen MA, Church GM (2002) Measuring absolute expression with microarrays with a calibrated reference sample and an extended signal intensity range. Proceedings of the National Academy of Sciences of the United States of America 99:7554–7559

Elghanian R, Storhoff JJ, Mucic RC, Letsinger RL, Mirkin CA (1997) Selective colorimetric detection of polynucleotides based on the distance-dependent optical properties of gold nanoparticles. Science 277:1078–1081

El-Kemary M, Rettig W (2003) Multiple emission in coumarins with heterocyclic substituents. Physical Chemistry Chemical Physics 5:5221–5228

Forde TS, Hanley QS (2005) Following FRET through five energy transfer steps: spectroscopic photobleaching, recovery of spectra, and a sequential mechanism of FRET. Photochemical & Photobiological Sciences 4:609–616

Formosinho SJ, Arnaut LG (1993) Excited-state proton-transfer reactions.2. Intramolecular reactions. Journal of Photochemistry and Photobiology A-Chemistry 75:21–48

Gerstein M, Lesk AM, Chothia C (1994) Structural mechanisms for domain movements in proteins. Biochemistry 33:6739–6749

Gilardi G, Zhou LQ, Hibbert L, Cass AE (1994) Engineering the maltose binding protein for reagentless fluorescence sensing. Analytical Chemistry 66:3840–3847

Grabowski ZR, Rotkiewicz K, Rettig W (2003) Structural changes accompanying intramolecular charge transfer: focus on twisted intramolecular charge transfer states and structures. Chemical Reviews 103:3899–4031

Granda-Valdes M, Badia R, Pina-Luis G, Diaz-Garcia ME (2000) Photoinduced electron transfer systems and their analytical application in chemical sensing. Quimica Analitica 19:38–53

Gust D, Moore TA, Moore AL (2006) Molecular switches controlled by light. Chemical Communications:1169–1178

Hai XD, Tan MQ, Wang G, Ye ZQ, Yuan JL, Matsumoto K (2004) Preparation and a time-resolved fluoroimmunoassay application of new europium fluorescent nanoparticles. Analytical Sciences 20:245–246

Heinlein T, Knemeyer J-P, Piestert O, Sauer M (2003) Photoinduced electron transfer between fluorescent dyes and guanosine residues in DNA-hairpins. Journal of Physical Chemistry B 107:7957–7964

Henary MM, Wu YG, Cody J, Sumalekshmy S, Li J, Mandal S, Fahrni CJ (2007) Excited-state intramolecular proton transfer in 2-(2′-arylsulfonamidophenyl)benzimidazole derivatives: the effect of donor and acceptor substituents. Journal of Organic Chemistry 72:4784–4797

Herland A, Thomsson D, Mirzov O (2008) Decoration of amyloid fibrils with luminescent conjugated polymers. Journal of Material Chemistry 18:126–132

Hermann T, Patel DJ (2000) Biochemistry - adaptive recognition by nucleic acid aptamers. Science 287:820–825

Ho HA, Leclerc M (2004) Optical sensors based on hybrid aptamer/conjugated polymer complexes. Journal of the American Chemical Society 126:1384–1387

Huang B, Wu HK, Bhaya D, Grossman A, Granier S, Kobilka BK, Zare RN (2007) Counting low-copy number proteins in a single cell. Science 315:81–84

Huang J, Peng AD, Fu HB, Ma Y, Zhai TY, Yao JN (2006) Temperature-dependent ratiometric fluorescence from an organic aggregates system. Journal of Physical Chemistry A 110:9079–9083

Hunyadi SE, Murphy CJ (2006) Tunable one-dimensional silver-silica nanopeapod architectures. Journal of Physical Chemistry B 110:7226–7231

Jares-Erijman EA, Jovin TM (2003) FRET imaging. Nature Biotechnology 21:1387–1395

Jiang H, Zhao XY, Schanze KS (2007) Effects of polymer aggregation and quencher size on amplified fluorescence quenching of conjugated polyelectrolytes. Langmuir 23:9481–9486

Johansson MK, Cook RM (2003) Intramolecular dimers: a new design strategy for fluorescence-quenched probes. Chemistry 9:3466–3471

Kasha M (1986) Proton-transfer spectroscopy - perturbation of the tautomerization potential. Journal of the Chemical Society-Faraday Transactions II 82:2379–2392

Knemeyer JP, Marme N, Sauer M (2000) Probes for detection of specific DNA sequences at the single-molecule level. Analytical Chemistry 72:3717–3724

Kohn JE, Plaxco KW (2005) Engineering a signal transduction mechanism for protein-based biosensors. Proceedings of the National Academy of Sciences of the United States of America 102:10841–10845

Kollmannsberger M, Rurack K, Resch-Genger U, Daub J (1998) Ultrafast charge transfer in amino-substituted boron dipyrromethene dyes and its inhibition by cation complexation: a new design concept for highly sensitive fluorescent probes. Journal of Physical Chemistry A 102:10211–10220

Koner AL, Schatz J, Nau WM, Pischel U (2007) Selective sensing of citrate by a supramolecular 1,8-naphthalimide/calix[4]arene assembly via complexation-modulated pK(a) shifts in a ternary complex. Journal of Organic Chemistry 72:3889–3895

La Clair JJ, Burkart MD (2006) Geometry in digital molecular arrays. Organic & Biomolecular Chemistry 4:3052–3055

Li C, Numata M, Takeuchi M, Shinkai S (2005) A sensitive colorimetric and fluorescent probe based on a polythiophene derivative for the detection of ATP. Angewandte Chemie-International Edition in English 44:6371–6374

Liu J, Lee JH, Lu Y (2007) Quantum dot encoding of aptamer-linked nanostructures for one-pot simultaneous detection of multiple analytes. Analytical Chemistry 79:4120–4125

Lozinsky E, Martin VV, Berezina TA, Shames AI, Weis AL, Likhtenshtein GI (1999) Dual fluorophore-nitroxide probes for analysis of vitamin C in biological liquids. Journal of Biochemical and Biophysical Methods 38:29–42

Magri DC, Vance TP, de Silva AP (2007) From complexation to computation: recent progress in molecular logic. Inorganica Chimica Acta 360:751–764

Maliwal BP, Gryczynski Z, Lakowicz JR (2001) Long-wavelength long-lifetime luminophores. Analytical Chemistry 73:4277–4285

Margulies D, Felder CE, Melman G, Shanzer A (2007) A molecular keypad lock: a photochemical device capable of authorizing password entries. Journal of the American Chemical Society 129:347–354

Marme N, Knemeyer JP, Sauer M, Wolfrum J (2003) Inter- and intramolecular fluorescence quenching of organic dyes by tryptophan. Bioconjugate Chemistry 14:1133–1139

Marras SA (2006) Selection of fluorophore and quencher pairs for fluorescent nucleic acid hybridization probes. Methods in Molecular Biology 335:3–16

Marras SA, Tyagi S, Kramer FR (2006) Real-time assays with molecular beacons and other fluorescent nucleic acid hybridization probes. Clinica Chimica Acta 363:48–60

Mathis G (1995) Probing molecular-interactions with homogeneous techniques based on rareearth cryptates and fluorescence energy-transfer. Clinical Chemistry 41:1391–1397

Matsushita M, Meijler MM, Wirsching P, Lerner RA, Janda KD (2005) A blue fluorescent antibody-cofactor sensor for mercury. Organic Letters 7:4943–4946

Medintz IL, Deschamps JR (2006) Maltose-binding protein: a versatile platform for prototyping biosensing. Current Opinion in Biotechnology 17:17–27

Medintz IL, Goldman ER, Lassman ME, Mauro JM (2003) A fluorescence resonance energy transfer sensor based on maltose binding protein. Bioconjugate Chemistry 14:909–918

Mizoue LS, Chazin WJ (2002) Engineering and design of ligand-induced conformational change in proteins. Current Opinion in Structural Biology 12:459–463

Monat C, Grillet C, Domachuk R, Smith C, Magi E, Moss DJ, Nguyen HC, Tomljenovic-Hanic S, Cronin-Golomb M, Eggleton BJ, Freeman D, Madden S, Luther-Davies B, Mutzenich S, Rosengarten G, Mitchell A (2007) Frontiers in microphotonics: tunability and all-optical control. Laser Physics Letters 4:177–186

Najmabadi P, La Clair JJ, Burkart MD (2007) A systems perspective to digital structures in molecular analysis. Organic & Biomolecular Chemistry 5:214–222

Nikiforov TT, Beechem JM (2006) Development of homogeneous binding assays based on fluorescence resonance energy transfer between quantum dots and Alexa Fluor fluorophores. Analytical Biochemistry 357:68–76

Nilsson KP, Rydberg J, Baltzer L, Inganas O (2003) Self-assembly of synthetic peptides control conformation and optical properties of a zwitterionic polythiophene derivative. Proceedings of the National Academy of Sciences of the United States of America 100:10170–10174

Nilsson KPR, Inganas O (2004) Optical emission of conjugated polyuelectrolite. Calcium-induced conformational changes in calmodulin and calmodulin-calcineurin interactions. Macromolecules 37:9109–9113

Oh E, Hong MY, Lee D, Nam SH, Yoon HC, Kim HS (2005) Inhibition assay of biomolecules based on fluorescence resonance energy transfer (FRET) between quantum dots and gold nanoparticles. Journal of the American Chemical Society 127:3270–3271

Okamoto A, Tanaka K, Saito I (2004) DNA logic gates. Journal of the American Chemical Society 126:9458–9463

Perez-Inestrosa E, Montenegro JM, Collado D, Suau R, Casado J (2007) Molecules with multiple light-emissive electronic excited states as a strategy toward molecular reversible logic gates. Journal of Physical Chemistry C 111:6904–6909

Petitjean A, Lehn JM (2007) Conformational switching of the pyridine-pyrimidine-pyridine scaffold for ion-controlled FRET. Inorganica Chimica Acta 360:849–856

Puleo CM, Liu K, Wang TH (2006) Pushing miRNA quantification to the limits: high-throughput miRNA gene expression analysis using single-molecule detection. Nanomedicine 1:123–127

Raymo FM, Yildiz I (2007) Luminescent chemosensors based on semiconductor quantum dots. Physical Chemistry Chemical Physics 9:2036–2043

Roshal AD, Grigorovich AV, Doroshenko AO, Pivovarenko VG, Demchenko AP (1998) Flavonols and crown-flavonols as metal cation chelators. The different nature of Ba2+ and Mg2+ complexes. Journal of Physical Chemistry A 102:5907–5914

Roshal AD, Grigorovich AV, Doroshenko AO, Pivovarenko VG, Demchenko AP (1999) Flavonols as metal-ion chelators: complex formation with Mg2+ and Ba2+ cations in the excited state. Journal of Photochemistry and Photobiology A-Chemistry 127:89–100

Rurack K (2001) Flipping the light switch 'on' - the design of sensor molecules that show cation-induced fluorescence enhancement with heavy and transition metal ions. Spectrochimica Acta A 57:2161–2195

Rurack K, Danel A, Rotkiewicz K, Grabka D, Spieles M, Rettig W (2002) 1,3-Diphenyl-1H-pyra-zolo[3,4-b]quinoline: a versatile fluorophore for the design of brightly emissive molecular sensors. Organic Letters 4:4647–4650

Schazmann B, Alhashimy N, Diamond D (2006) Chloride selective calix[4]arene optical sensor combining urea functionality with pyrene excimer transduction. Journal of the American Chemical Society 128:8607–8614

Selvin PR (2002) Principles and biophysical applications of lanthanide-based probes. Annual Review of Biophysics and Biomolecular Structure 31:275–302

Serin JM, Brousmiche DW, Frechet JMJ (2002) Cascade energy transfer in a conformationally mobile multichromophoric dendrimer. Chemical Communications 2002:2605–2607

Simeonov A, Matsushita M, Juban EA, Thompson EHZ, Hoffman TZ, Beuscher AE, Taylor MJ, Wirsching P, Rettig W, McCusker JK, Stevens RC, Millar DP, Schultz PG, Lerner RA, Janda KD (2000) Blue-fluorescent antibodies. Science 290:307–313

Stojanovic MN, Kolpashchikov DM (2004) Modular aptameric sensors. Journal of the American Chemical Society 126:9266–9270

Stojanovic MN, Stefanovic D (2003) Deoxyribozyme-based half-adder. Journal of the American Chemical Society 125:6673–6676

Stojanovic MN, de Prada P, Landry DW (2001) Catalytic molecular beacons. Chembiochem 2:411–415

Storhoff JJ, Lucas AD, Garimella V, Bao YP, Muller UR (2004) Homogeneous detection of unamplified genomic DNA sequences based on colorimetric scatter of gold nanoparticle probes. Nature Biotechnology 22:883–887

Straight SD, Liddell PA, Terazono Y, Moore TA, Moore AL, Gust D (2007) All-photonic molecular XOR and NOR logic gates based on photochemical control of fluorescence in a fulgimide-porphyrin-dithienylethene triad. Advanced Functional Materials 17:777–785

Sumalekshmy S, Henary MM, Siegel N, Lawson PV, Wu Y, Schmidt K, Bredas JL, Perry JW, Fahrni CJ (2007) Design of emission ratiometric metal-ion sensors with enhanced two-photon cross section and brightness. Journal of the American Chemical Society 129:11888–1889

Syromyatnikov VG, Yashchuk VN, Ogulchansky TY, Savchenko IA, Kolendo AY (1996) Some light-sensitive imide molecular systems with the determined functional properties. Molecular Crystals and Liquid Crystals Science and Technology Section A-Molecular Crystals and Liquid Crystals 283:293–298

Thompson NL, Lieto AM, Allen NW (2002) Recent advances in fluorescence correlation spectroscopy. Current Opinion in Structural Biology 12:634–641

Thurley S, Roglin L, Seitz O (2007) Hairpin peptide beacon: dual-labeled PNA-peptide-hybrids for protein detection. Journal of the American Chemical Society 129:12693

Tomasulo M, Yildiz I, Kaanumalle SL, Raymo FM (2006) pH-sensitive ligand for luminescent quantum dots. Langmuir 22:10284–10290

Tyagi S, Kramer FR (1996) Molecular beacons: probes that fluoresce upon hybridization. Nature Biotechnology 14:303–308

Ueberfeld J, Walt DR (2004) Reversible ratiometric probe for quantitative DNA measurements. Analytical Chemistry 76:947–952

Valeur B (2002) Molecular fluorescence. Wiley-VCH, Weinheim

Valeur B, Leray I (2000) Design principles of fluorescent molecular sensors for cation recognition. Coordination Chemistry Reviews 205:3–40

Valeur B, Leray I (2007) Ion-responsive supramolecular fluorescent systems based on multi-chromophoric calixarenes: a review. Inorganica Chimica Acta 360:765–774

Varnavski OP, Ostrowski JC, Sukhomlinova L, Twieg RJ, Bazan GC, Goodson T (2002) Coherent effects in energy transport in model dendritic structures investigated by ultrafast fluorescence anisotropy spectroscopy. Journal of the American Chemical Society 124:1736–1743

Walkup GK, Imperiali B (1996) Design and evaluation of a peptidyl fluorescent chemosensor for divalent zinc. Journal of the American Chemical Society 118:3053–3054

Walkup GK, Imperiali B (1997) Fluorescent chemosensors for divalent zinc based on zinc finger domains. Enhanced oxidative stability, metal binding affinity, and structural and functional characterization. Journal of the American Chemical Society 119:3443–3450

Wang S, Gaylord BS, Bazan GC (2004) Fluorescein provides a resonance gate for FRET from conjugated polymers to DNA intercalated dyes. Journal of the American Chemical Society 126:5446–5451

Wong KF, Bagchi B, Rossky PJ (2004) Distance and orientation dependence of excitation transfer rates in conjugated systems: beyond the Forster theory. Journal of Physical Chemistry A 108:5752–5763

Yashchuk V, Kudrya V, Losytskyy M, Suga H, Ohul'chanskyy T (2006) The nature of the electronic excitations capturing centres in the DNA. Journal of Molecular Liquids 127:79–83

Yashchuk VMN, Syromyatnikov VG, Ogul'chansky TY, Kolendo AY, Prot T, Blazejowski J, Kudrya VY (2000) Multifunctional macromolecules and structures as one-way exciton conductors. Molecular Crystals and Liquid Crystals 353:287–300

Yatsuhashi T, Nakajima Y, Shimada T, Tachibana H, Inoue H (1998) Molecular mechanism for the radationless deactivation of the interamolecular charge-transfer excited singlet state of aminofluorenones through hydrogen bonds with alcohols. Journal of Physical Chemistry A 102:8657–8663

Yesylevskyy SO, Kharkyanen VN, Demchenko AP (2006) The change of protein intradomain mobility on ligand binding: is it a commonly observed phenomenon? Biophysics Journal 91:3002–3013

Yesylevskyy SO, Kharkyanen VN, Demchenko AP (2007) The blind search for the closed states of hinge-bending proteins. Proteins 71:831–843

Yguerabide J, Yguerabide EE (1998) Light-scattering submicroscopic particles as highly fluorescent analogs and their use as tracer labels in clinical and biological applications - II. Experimental characterization. Analytical Biochemistry 262:157–176

Yoshida W, Yokobayashi Y (2007) Photonic boolean logic gates based on DNA aptamers. Chemical Communications:195–197

Yoshihara T, Galievsky VA, Druzhinin SI, Saha S, Zachariasse KA (2003) Singlet excited state dipole moments of dual fluorescent N-phenylpyrroles and 4-(dimethylamino)benzonitrile from solvatochromic and thermochromic spectral shifts. Photochemical and Photobiological Sciences 2:342–353

Yuan MS, Liu ZQ, Fang Q (2007) Donor-and-acceptor substituted truxenes as multifunctional fluorescent probes. Journal of Organic Chemistry 72:7915–7922

Yun CS, Javier A, Jennings T, Fisher M, Hira S, Peterson S, Hopkins B, Reich NO, Strouse GF (2005) Nanometal surface energy transfer in optical rulers, breaking the FRET barrier. Journal of the American Chemical Society 127:3115–3119

Zhou YC, Zhang DQ, Zhang YZ, Tang YL, Zhu DB (2005) Tuning the CD spectrum and optical rotation value of a new binaphthalene molecule with two spiropyran units: mimicking the function of a molecular "AND" logic gate and a new chiral molecular switch. Journal of Organic Chemistry 70:6164–6170

Chapter 7
Supramolecular Structures and Interfaces for Sensing

An important step towards the perfection of fluorescent sensors can be made by their self-assembly into supramolecular structures and by their attachment to flat or porous surfaces and to different particles. During this step, a substantial improvement in performance can be attained and new important functionalities added. Scaling up from small single molecules to larger molecular ensembles aims at the achievement of two important goals: *improvement of molecular recognition*, especially with the targets of large size and complexity and *exploration of collective properties* of fluorescence reporters that offer enhanced sensitivity. These technological breakthroughs are a part of '*bottom-up*' approach to technological products with advanced functions (Balzani et al. 2003). In this chapter we will see how these functions can be realized, starting from passive support materials up to self-assembled matrices for very specific recognition of large-molecular-mass targets.

7.1 Building Blocks for Supramolecular Sensors

Supramolecular structures, with the function of fluorescence sensing, are seen as self-organized complexes between or with designed partners, in which recognition and reporting can be coupled in a most efficient way. In an effort to develop better sensors, an important role is given to nanoscale support materials. They include different types of molecules and particles representing inorganic, organic and biological worlds (Oshovsky et al. 2007).

7.1.1 Carbon Nanotubes

Tubular structures are probably mechanically the most rigid and functionally versatile modules that can be used both as nanoscale support materials and as transducers. Their inherent well-defined structural features, such as cylindrical dimensionality,

A.P. Demchenko, *Introduction to Fluorescence Sensing,*
© Springer Science+Business Media B.V. 2009

forming precisely established inner and outer volumes, are very useful for this. Especially attractive are the nanotubular structures that possess their own optical properties useful in sensing.

These are *carbon nanotubes*, which are very promising materials as molecular wires in electronics and also as the building blocks of various nanocomposites. These tubes have a tunable near-infrared emission (at 1,200–1,400 nm) that responds to changes in the local dielectric function (Choi and Strano 2007) but remains stable to permanent photobleaching. This emission can be quenched by appropriate redox-active dyes and with proper construction this effect can be used in sensing (Satishkumar et al. 2007).

Single-stranded DNA (ssDNA) interacts with carbon nanotubes noncovalently, wrapping around them by means of π-stacking interactions between the nucleotide bases and the nanotube sidewalls. In contrast, such interaction of double-stranded DNA (dsDNA) is significantly weaker. This difference in the binding interactions of carbon nanotubes with ssDNA and dsDNA can be used for DNA detection but the problem is in providing the reporter signal. It can be based on the absorption and near-infrared fluorescence of these materials and also on their ability to quench the fluorescence of organic dyes. The labeled testing ssDNA wraps around the nanotube, providing fluorescence quenching of the label and its hybridization with the complementary DNA strand results in the disruption of the complex, with the nanotube resulting in fluorescence enhancement (Yang et al. 2008).

The *absorption and emission spectra* of carbon nanotubes correspond to the window of transparency of human tissues, which, together with relatively low light-scattering in this range, is promising for the design of *in-vivo* sensors. The chemistry of the surface modification of these tubes develops rapidly. The fact that the attachment of fluorescent dyes commonly results in quenching should not discourage researchers, since many possibilities still have not been tried. These problems can be resolved together with the integration of nanotubes into various devices and designing the integrated sensing elements (Mahar et al. 2007). This makes for good prospects for the development of smart sensors, in which the signal transduction is provided by fluorescence quenching and charge transfer. The first attempts along this line have been successful (Barone et al. 2005).

7.1.2   Core-Shell Compositions

Nanosensors are often designed based on a core-shell principle. This allows combining maximum functionality with facile production. The core part commonly carries the supporting and reporting functions (for instance, it can be magnetic), whereas the shell is responsible for target recognition and signal transduction, together with such necessary properties as solubility or adhesion to a particular surface.

Silica nanoparticles or *polystyrene beads* can be selected as the cores and different compositions, based on their coating, can be used, for instance, with a thin layer of gold (Cao et al. 2006). In this case, the optical properties can be modulated by the thickness of the metal layer. Compositions based on silver nanoparticles as cores and silica shells can also be formed. They are used in techniques based on metal-enhanced fluorescence (Section 8.6). Many other possibilities exist for such core-shell combinations, including cores formed by Quantum Dots (Medintz and Deschamps 2006).

7.1.3 Polynucleotide Scaffolds

Let us for the moment forget about a very important functional role of DNA and RNA and consider oligo- and polynucleotides of different lengths as scaffolds for the assembly of nanocomposite sensors. Their role may not be limited to nucleic acid hybridization but may allow an extremely broad range of applications in sensors operating on a supramolecular level. Their great advantage is the possibility of a *designed location* in the sequence of nucleic acid bases of different building blocks that can provide programmable affinity to the target.

As an example, we can refer to sensors based on DNA, in which deoxyuridine bases are modified with the attachment of porphyrins (Fendt et al. 2007). It has been shown that such constructions can serve as versatile building blocks for generating porphyrin arrays on the nanometer scale. The porphyrins can be introduced site-specifically into the DNA strands and the electronic properties of the formed array can thus be tuned. Single-stranded and double-stranded DNA provide different possibilities for stacking the porphyrin units that influence dramatically the spectroscopic properties. These porphyrin-DNA hybrids offer a practically unlimited number of possibilities for further functionalization.

The ability of DNA to be the scaffolds for assembling different molecules on the nanometer scale should find many applications in creating complex molecular assemblies. These systems allow the positioning of peptides, proteins or other supramolecular sensor components in distinct patterns with *precise spacing*. The well-defined structure and spacing of DNA are properties that are required for making the templates for secondary components in a bottom-up approach toward self-assembly.

7.1.4 Peptide Scaffolds

The ability of peptide chain to form three-dimensional structures of various complexity and functional use will stimulate researchers for many years to come. Innumerable possibilities for structure formation based on variation in a sequence formed by 20 amino acids can be used not only for the construction of sensor units (as we observed in Section

5.4) but also for making the scaffolds for supramolecular sensors. Here, we present an example of the formation of scaffolds for such sensors by association of peptides.

Peptides with the properties of forming *self-assembled tubular structures* were first suggested as anti-microbial agents. Later, it was realized that such peptide structures can serve as ideal scaffolds for supramolecular sensors (Gao and Matsui 2005). Surprisingly, it was found that even peptides as short as Phe-Phe dipeptide can form stable nanotubes. Via their molecular-recognition functions, the self-assembled peptide nanostructures can be further organized to form nanowires, nanoparticles and even 'nanoforests' (on solid support). It has been demonstrated that cyclic peptides, formed as planar rings by alternating L- and D- amino acids, self-assemble via hydrogen bonding to tubular open-ended and hollow structures (Brea et al. 2007). The application of such molecular self-assemblies as the sensor building blocks use simple production methods, it is robust, practical and affordable. It is also beneficial that smart functionalities can be added at desired positions in peptide nanotubes through well-established chemical and peptide syntheses. They may include both recognition and reporting units with the exploration of fluorescence-based technologies.

The other types of peptide-based scaffolds are their associates based on the principle of *amphiphilicity* (the separation of polar and unpolar sites). Their polar sites can be represented by one to two charged amino acids and a hydrophobic part – by four or more sequential hydrophobic amino acids. They form ordered nanostructures similar to those formed by lipids (Jiang et al. 2007). There are many opportunities for the hierarchical formation of supramolecular assemblies of a larger complexity based on these peptide assemblies.

The principle of the formation of the third type of these scaffolds is borrowed from the known examples of appearance *in vivo* of *amyloid fibrils,* which are associated with a large number of human diseases. Application of peptides forming amyloid fibrils as scaffolds was suggested based on the fact that they form *very rigid, β-sheeted structures* (del Mercato et al. 2007).

Peptides assembled based on noncovalent intermolecular interactions allow the realization of three of the most important types of these interactions: (a) electrostatic interactions based on the complementarity of charged groups, (b) hydrogen bonding that involves complementarity of the proton donor and acceptor groups and (c) hydrophobic interactions that provide stabilization of contacts between low-polar groups in polar media. Individually, these bonds are weak but their collective action provides significant stabilization of intermolecular complexes.

7.2   Self-Assembled Supramolecular Systems

With the increase of size and complexity, the preparation of molecular conjugates by covalent synthesis becomes a very difficult, time-consuming and even unrealizable choice. In contrast, supramolecular chemistry based on self-assembly may offer the creation of nanoscale systems with a broad range of sensor applications. The self-assembly of composite sensor units is based on the same principles as the sensing itself – on complementarity between interacting partners and the formation of non-covalent interactions between them, such as hydrogen bonds, salt bridges, solvation forces, π-π stacking and coordination of metal ions. Due to the collective character of these bonds, the interactions between the partners can compete in strength with covalent bonds.

Supramolecular chemistry has developed into a strong interdisciplinary field of research. One of its rules is the *requirement for topological saturation*. Saturation by non-covalent interactions should provide a collective effect that should bring to the system a necessary structural stability. Additional stability may be achieved with covalent cross-links formed after the assembly. A 'hot spot' should remain only for the recognition site and, if necessary, for providing binding to a surface or location in a heterogeneous structure, such as the living cell.

7.2.1   Affinity Coupling

In general, there are two principles of formation of supramolecular structures: affinity coupling and self-assembly. Examples of *affinity coupling* are the interactions antigen-antibody and biotin-avidin (or streptavidin) that are the most frequently used for the creation of supramolecular structures. The advantage of antibodies is their versatility, since antigenic determinants can be inherently present or artificially incorporated into any structure of protein, nucleic acid or carbohydrate origin. They can be obtained as binding sites for many artificial molecules, including organic dyes. The possibilities of manipulating with them have grown with the introduction of monovalent single-chain *antibody fragments* (Piervincenzi et al. 1998). Thus, for efficient highly selective high-affinity coupling, one of the assembling units should contain an antibody or its fragment and the other – an antigenic determinant recognized by it. With a broad variation of affinities ($K_d \approx 10^{-8}$–10^{-13} M) the antigen-antibody interactions may not in all cases be sufficiently strong for the formation of stable supramolecular structures.

The binding of *avidin* or *streptavidin* with biotin is different. Protein avidin can be found in the white of chicken eggs. It binds a rather small molecule *biotin* (244.3 Da) with a uniquely high affinity ($K_d \approx 10^{-14}$–10^{-15} M). Since each avidin molecule can bind four biotins or biotin-labeled units, complex supramolecular constructions can be designed. Streptavidin is a bacterial analog of avidin; it has similar properties to avidin but can be produced in bacteria (Color Plate 9).

Streptavidin is a globular 60 kDa protein with a high (compared to other proteins) thermal and pH-stability. Therefore, if streptavidin is part of one component of the structure and biotin is attached to its another component, these components find each other in solution and associate irreversibly, with the formation of a series of noncovalent bonds between complementary groups of interacting surfaces that are formed together and are, therefore, as strong as a covalent bond. The streptavidin-biotin pair has now become a standard tool for providing a strong and specific assembling of supramolecular structures. In contrast to antibodies, their complementary surfaces do not allow much structural variation but there are many possibilities for reacting or substituting groups that are outside their contact areas.

In the biological world, a number of additional partner pairs can be formed that are characterized by high affinities. Such are the pairs lectin – carbohydrate or enzyme – suicide inhibitor possessing subnanomolar dissociation constants. However, they have not found extensive practical use. Thus, the streptavidin – biotin pair remains the most popular for generating molecular junctions.

It is surprising that by now organic synthetic chemistry could not suggest a proper substitute for the streptavidin-biotin pair that could compete in simplicity, stability and price. Presently the binding properties of biotin but not of avidin or streptavidin can be simulated with designed peptides. The trivalent vancomycin/ D-Ala-D-Ala- complex may probably be suggested as a pair with relatively tight binding of smaller molecular weight units (Rao et al. 1998).

Affinity coupling is an extremely valuable tool in intracellular research. For many of these studies it is necessary that the fluorescently labeled partner is intro- duced from outside into the cell, where it should find its second partner out of thousands of types of other molecules. Unfortunately, neither the streptavidin- biotin nor antigen-antibody pair can be efficiently used for this task. The solutions to this problem and their illustrations will be discussed in Chapter 11.

7.2.2 Self-Assembly

Self-assembly involves the use of topological complementarity and saturation of intermolecular interactions between the partners on a broader scale, with the possible adaptation of conformation of partners to achieve this saturation (Lehn 1995).

Self-assembling and self-organizing methodologies are powerful tools for the "bottom-up" approach for the realization of complex structures with functional properties. Recently, this concept has been extended to the design of fluorescent chemosensors, providing new, exciting potential for the development of innovative sensing systems (Mancin et al. 2006). The idea of this approach is again borrowed from Nature. Living cells do not appear as a result of 'nanofabrication'. All the processes leading to the formation of supramolecular structures, such as chromatin, membranes, microfilaments or viruses, start from the synthesis of *individual molecules* and their *subsequent assembly*. Being the building blocks for such assembly, biomolecules are adapted for assembly by their inner structures, forming the contact surfaces. The same is required from artificial systems.

In such self-assembling systems, the principles of molecular recognition should be actively explored. In addition to steric complementarity there should be maximal intermolecular contact, with maximal saturation by noncovalent bonds. Throughout this chapter the reader can evaluate the relative importance of different types of these interactions for particular structures. Thus, amphiphilic structures favorable for *hydrophobic interactions* are essential for self-assembly by incorporation into micelles and lipid bilayers but they are 'lubricants' that lack the specificity that may be often necessary. *Electrostatic interactions* can provide the necessary recognition pattern, especially when they are assembled into clusters. They govern the layer-by-layer formation of multi-layer scaffolds. Here, we will discuss the third very important type of noncovalent intermolecular interactions, hydrogen bonds.

Hydrogen bonds are short-range, unidirectional and specific. Individual bonds, having an energy of the order of 10–15 kJ/mol, are rather weak but their collective action can result in a much stronger effect. It is enough to mention that these bonds provide stability to the α-helix in proteins and to the double helix in DNA. In self-assembling molecular structures, quadruple bonds of acceptor-acceptor-donor-donor (AADD) type have been suggested as a powerful strategy for creating nanocomposite structures (Corbin et al. 2002). Constructed following this principle, the ureidodeazapterin-based module (Fig. 7.1) reveals an unprecedented stability of dimers based on their AADD self-complementary with the dimerization constant $K_{dimer} > 5 \times 10^8$ M^{-1}. Linking covalently two such units together with a semi-rigid spacer, a number of stable dendritic structures can be obtained.

There is a very strong demand for new modules and strategies that can amplify the relatively weak strength of a hydrogen bond to provide more stable assemblies. The example presented above demonstrates clearly how to achieve this.

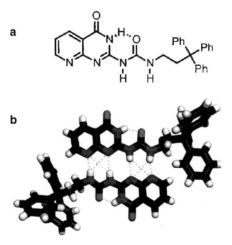

Fig. 7.1 Hetrocyclic structure containing two donors and two acceptors for hydrogen bonding (**a**) and the model of the formation of the hydrogen-bonded dimer (**b**) (Reproduced with permission from Corbin et al. 2002)

Collective effects of rather weak bonds can bring results that are quite satisfactory for the formation of stable and specific supramolecular structures. Experimental evidence exists that such *multivalent interactions* can increase the affinity, compared to univalent interactions, by many orders of magnitude (Baldini et al. 2007).

Many different hydrogen-bonded supramolecular structures comprising tens of hydrogen bonds have been suggested (Vriezema et al. 2005). These structures can be less stable in water and other protic solvents, since these solvents can themselves be good hydrogen bonding partners.

The full advantage of *hydrogen-bonding complementarity* can be explored in precise and programmable self assembly, based on the DNA double helix (Seeman 2003). The partners containing complementary sequences can assemble in a thermally reversible manner over a range of length scales. DNA can be conjugated to other materials (molecules and nanoparticles) or attached to a solid support. It is expected that the organizational capabilities of structural DNA nanotechnology, which are just beginning to be explored, will find broad application.

7.2.3   *Two-Dimensional Self-Assembly of S-Layer Proteins*

Peptides and proteins provide a variety of opportunities for controlled and specific self-assembly. Mechanistically, peptide self-assembly can result from amphiphilic properties and from complementarity of partner structures. This complementarity can pre-exist in components but can also appear in the course of interaction, the process known as '*induced folding*' (Demchenko 2001). Designed peptides forming the dimers can be used for assembling the nanoparticles (Aili et al. 2007).

Proteins that can assemble into *two-dimensional crystals* deserve special attention (Sleytr et al. 2007). Such proteins form the surface layers (S-layers) characteristic for prokaryotic organisms (archaea and bacteria). These proteins vary in molecular mass of subunits (40–200 kDa) and the thickness of the formed layer (from 5–20 up to 70 nm). As S-layers are the periodic structures, they exhibit identical physicochemical properties for each molecular unit down to the subnanometer level and possess pores of identical size and morphology. Protein engineering allows obtaining their truncated forms and also covalently coupled hybrids with other proteins, of which the most interesting is the coupling with streptavidin. The latter allows obtaining complicated constructions based on two-dimensional S-layers as the building blocks.

Many applications of *S-layers* in nanobiotechnology depend on the ability of isolated subunits to re-crystallize into monomolecular lattices in suspension or on suitable surfaces and interfaces, such as liposomes (see Section 7.6). Because of these properties and using the possibility of modifications with various protein reagents, S-layer lattices can be exploited as scaffolding and patterning elements for generating more complex supramolecular assemblies and structures. One such possibility is illustrated in Fig. 7.2.

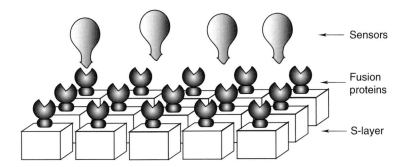

Fig. 7.2 Schematic representation of two-dimensional structures formed of engineered S-layer proteins fused to a functional or structure-forming protein (e.g., streptavidin). The latter can form supramolecular structures by affinity coupling (such as streptavidin with biotin) or by self-assembly. Different possibilities can be realized in these constructions, such as biotinilated nanoparticles or antibodies. This ensures the binding of receptor elements in a defined spacing and orientation

7.2.4   Template-Assisted Assembly

The self-organization of essential sensor components on a proper template is an important approach to the realization of functioning self-assembled structures as sensors. In this case, the receptor and the reporter do not need to interact directly. The communication between them can be provided by the *spatial proximity* of these two units on the template only, so that a signal transduction mechanism can be intermolecular but efficient. Thus, the template can play an important role both in the formation of the functional supramolecular structure and in the maintenance of its integrity. The results discussed in the previous section provide the ideas, what templates can be chosen and how they can be properly used.

The main advantages of template-assisted assembly are the possibility of choosing the template and its immobilization strategy and broad possibilities in the optimization of sensor performance. Depending on the template, *little or no synthetic modifications of the receptor and reporter are needed* and this allows an easy selection of their optimal configuration. The spatial proximity of a large number of sensor units in the assembly leads to collective effects and properties that may contribute to the improvement of the sensors performances. Different types of template have been used to guide the self-organization of sensors, spanning from micellar aggregates to monolayers, to glass surfaces and, more recently, to nanoparticles. Each of these templates has its own peculiar features, which are reflected in the characteristic performance of the resulting sensor.

An additional benefit of template-assisted assembly is the possibility of *micropatterning* – the formation on the surface of molecular clusters with desirable properties. The active areas for accommodating these clusters, with a size of 100 nm or less, can be pre-fabricated on the surface. This approach, with the aid of surface chemistry, merges fabrication and self-assembly and thus combines bottom-up and top-down approaches (Henzie et al. 2006).

7.2.5   Micelles: The Simplest Self-Assembled Sensors

Whereas dendrimers can be viewed as rather rigid micelles with unimolecular properties, the real *detergent micelles* are very flexible structures. The mobility of these detergent molecules in these structures is high and their properties are strongly concentration-dependent. At low concentrations they exist as separate molecules and only when their concentration exceeds a certain critical value (called *critical micelle concentration*, CMC), are the micelles formed. Detergent molecules are *amphiphilic*. They consist of a polar head and a low-polar (usually, hydrocarbon) tail. In highly polar solvents (e.g., water) the micelle tends to expose solvent to its heads and hides the tails by putting them together (Fig. 7.3). In a low-polar solvent (e.g., hexane) the reverse micelle is formed, in which the tails extend to the solvent and the heads are screened from the solvent by assembling at the micelle surface. Despite the flexibility of these structures, they possess a very important property of molecular selectivity, based on polarity that is efficiently used in sensing.

This molecular selectivity results in the ability to solubilize molecules and particles that are *insoluble* or *low soluble* in the main solvent. There are many organic fluorescent dyes that can be very good sensors for ions. In the meantime, the sensing cannot be provided in a simple way since the ions are insoluble in the organic phase. To solve this problem, sensor dyes can be incorporated into the interior of direct micelles. The small polar recognition sites can be extended out of the micelle or the ions can diffuse for a short distance at the polar interface into the

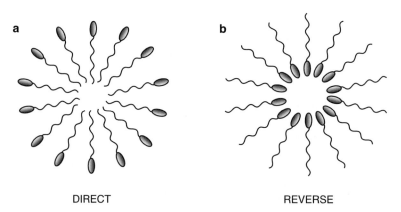

DIRECT   REVERSE

Fig. 7.3 Schematic representation of direct (**a**) and reverse (**b**) micelles. Direct micelles are formed in water and in other highly polar media. Their polar heads stretch out and the assembled hydrophobic tails form the low-polar 'nano-phase', which can solubilize low-polar molecules. On the opposite, in reverse micelles formed in low-polar solvents the low-polar tails stretch-out and polar heads assemble. They can solubilize water and other highly polar molecules

micelle. In both cases, efficient ion recognition may occur. This principle is the most efficiently applied for the determination of Cu^{2+} (Fernandez et al. 2004) and Hg^{2+} (Wang et al. 2007) ions.

Interestingly, the small size of the micelle and, correspondingly, its small inner volume, may allow the incorporation of ion receptor molecules together with a fluorescence reporter *even without their covalent linkage*. Thus, the only common feature, insolubility in water, can induce different sensor components to compose into an efficient sensor within the micelle interior, as shown in Fig. 7.4. Such an approach, though allowing only homogeneous assays, demonstrates many attractive advantages. It is simple and instead of complicated chemistry it requires just a simple mixing of components. Moreover, it provides a simple way of optimization of sensor system performance by variation of the composition of the system.

It is known that many, even low-polar, organic solvents contain traces of water that are very difficult to determine. Equally difficult is to determine *the traces of low polar organic solvents in water*. Micelles allow quantitative determination of different compounds based on the distribution in their concentration between bulk and micellar phases. Thus, the small concentrations of water or other polar impurities in low-polar solvents, such as hexane, can be determined by their dramatically increased concentrations within reverse micelles. To demonstrate this we made reverse micelles that incorporated our two-color ratiometric 3HC dyes (Klymchenko and Demchenko 2002a). An "empty" reverse micelle in hexane binds water molecules with high affinity. This results in some changes in the micelle structure, which influences the dye environment. As a result of this we observe the dramatic change of the ratio of intensities of the two fluorescence bands when single water molecules (with respect to molecules of

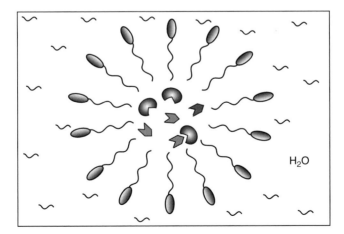

Fig. 7.4 Illustration of the principle of micellar sensing. The low solubility of both the sensor and the target in the main solvent results in their solubilization inside the micelle interior. This increases manifold their local concentrations and allows optimization of the sensor sensitivity

detergent) are solubilized in the micelle (Fig. 7.5). Nanomolar amounts of water molecules in low-polar solvents can be detected in this way. Note that in this case the water and sensor molecules do not interact directly and the spectroscopic effect is exactly the opposite to that expected in the case of solvent polarity increase.

Accordingly, micelles formed in water allow solubilizing molecules that are water low-soluble, thus facilitating their determination. We commonly observe that when water-soluble fluorescent dyes are used as reporters, their incorporation into the micelle may substantially reduce fluorescence quenching by water (discussed in Section 3.1), which is an important benefit.

More complex problems can be solved in a similar way. For instance, a sample of water contaminated with hydrocarbons can be tested for the presence of extremely low levels of harmful *heterocyclic hydrocarbons*. When a micelle-forming detergent is added to this sample together with hydrophobic sensor molecules, then the detection of these heterocyclic compounds can easily be provided. In this system,

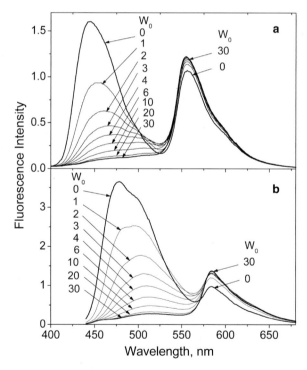

Fig. 7.5 Fluorescence spectra of 3-hydroxychromone dyes FE (**a**) and FA (**b**) in reverse micelles of 0.1 M detergent AOT in hexane recorded at an increase of the amount of solubilized water (the numbers indicate the molar water:AOT ratio, W_0). Reproduced with permission from (Klymchenko and Demchenko 2002a)

the formation of direct micelles is needed. They incorporate hydrocarbons and form within the micelle interior a hydrophobic microphase, in which the concentration of the analyte can be increased a million-fold! Such dramatically increased local concentrations can easily be detected.

The sensor molecules, if they are *proteins*, usually cannot be dissolved in an organic solvent without the loss of their function. But in the interior of reverse micelles, in the presence of a small amount of hydration water, they remain structurally intact and functional. In this system, the target molecules penetrating into the micelle from the main phase can form complexes with the sensors and thus can be analyzed. This sensing principle is analogous to that of micellar biocatalysis. An enzyme (or the sensor in our case) can interact with low-polar substrates or targets dissolved in the main apolar solvent.

7.2.6 Prospects

New, rapidly developing chemistry based on the principles of self-assembly and self-organization tends to complement and to a significant extent, substitute common organic synthetic chemistry in an attempt to produce supramolecular systems of great complexity and a high level of functionality. The new strategy requires the design and synthesis of a limited number of relatively simple building blocks, which are then allowed to self-organize into complex structures. Following this, without a strong synthetic effort, one can assemble the recognition unit together with the signal transduction and reporting unit into one block with superior properties. In this case, there may be no need for covalent links between these units; they only have to be designed according to a productive strategy and the noncovalent interactions between them are quite sufficient for efficient functioning. It could be ideal to develop one or several 'standard' reporting units that could be compatible to self-assemble with the library of receptors raised to detect a huge variety of targets. We will then be able to use a unified technique for their detection. However, this idea is still in its infancy.

Interaction with the template and incorporation into other systems that tend to self-associate, such as micelles, produces additional confinement and has a stabilizing effect. Still, the detergent micelles are often 'too dynamic' for this purpose, since their structures are concentration-dependent and very sensitive to environmental conditions, such as temperature and ionic strength. In this respect, micelles are not strong enough in competition with adsorbed monolayers and phospholipid bilayers, which will be considered below. One property makes them unique, however. They can concentrate the impurities from the solvent volume in their interior, the polarity of which differs from the main solvent. This micelle solubilization effect increases dramatically the sensitivity in the detection of these impurities.

Scientists indicate that despite the fact that there are interesting examples of large self-assembled structures, there is still a general lack of understanding as to how to introduce the standardized procedures for the step-by-step synthesis of nano-sized composites. This direction, in the meantime, is the only way to fill the gap between the miniaturization of macroscopic approaches (known as the 'top-down' principle) and chemical synthesis, which has a limit on increasing the complexity of molecular structures.

7.3   Conjugation, Labeling and Cross-linking

In the development of supramolecular sensors, synthetic chemistry plays a very important role in modifying the molecular and supramolecular structural blocks by covalent attachment of recognition and reporting units and additional stabilization of self-assembled structures with covalent bonds. Several key approaches on this pathway will be outlined below.

7.3.1   Conjugation and Labeling

Amino groups and *SH-groups* are the usual targets for *chemical modifications* and their presence at the sites of desired modification and their absence at other sites may guarantee the necessary precision (Brinkley 1992). SH-targeting is more actively used with peptides and proteins because these groups, present as Cys residues, are less abundant than the amino groups (N-terminal and that of Lys residues). Solid phase synthesis of peptides and site-directed mutagenesis of large proteins are used for the incorporation of SH- groups into desired sites of the sequence (Gilardi et al. 1994; Sloan and Hellinga 1998), for subsequent attachment of fluorescent dyes according to well-known chemistry. Usually, the fluorescence dyes are synthesized together with the reactive group for labeling. The most frequently used reactions for labeling the amino and SH groups are described in Fig. 7.6.

Three-dimensional structures of macromolecules, if they are known, are actively used for the determination of the optimal labeling sites. Comparative studies of sensor proteins in which the *engineered* Cys *locations* were rationally designed, based on the strongest differences of environments of these sites in unbound and bound forms derived from X-ray structures (Marvin et al. 1997), are witness to the usefulness of this approach. The retention of the native protein structure during SH labeling is especially important for proteins that contain significant amounts of S-S bonds that are necessary for supporting this structure and particularly for antibodies. Thus, the skill is to select the location of the introduced SH- group in such a way that allows correct protein folding with correct formation of S-S bonds that are essential for the structure and, simultaneously, to keep the introduced SH-group free and available for chemical modification. It was found (Renard et al. 2002) that a mild reducing treatment is necessary to reactivate mutant Cys residues before coupling with SH-reactive dyes. The reaction of coupling in these conditions allows preserving the essential disulphide bonds of the antibody variable domains. In the

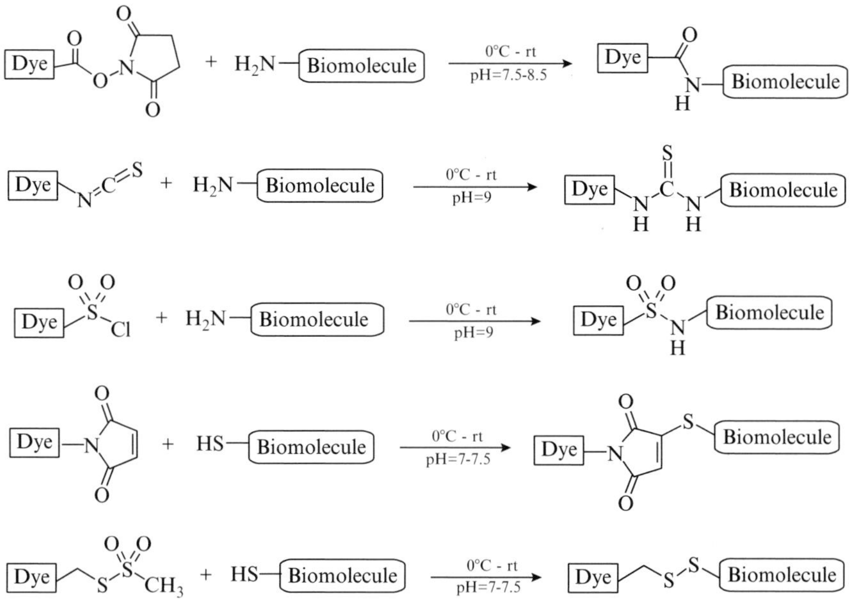

Fig. 7.6 Reactions of labeling amino and sulfhydryl groups that can be used for the attachment of fluorescent dyes

same study, the structural and energetic criteria were applied to locate the sites of the introduction of reactive Cys residues by mutagenesis. The authors selected the target residues as belonging to the topological neighborhood of the antigen in the structure of the complex between antibody and antigen, its absence of functional importance for binding the antigen and its solvent accessibility. The feasibility of this rule has been shown in experiments.

7.3.2 Co-synthetic Modifications

The *co-synthetic incorporation* of fluorescent dyes into polymer and biopolymer structures is less frequently used, though such procedures may be of choice for sensing peptides and also oligonucleotides as aptamers. In Section 5.4 we observed that in some sensing technologies the large globular protein structures are not required and the necessary effect of target recognition can be achieved by smaller peptides. They are more flexible in application and more tolerant to environmental conditions, with a better performance regarding reproducibility and multi-use operation. In this case, peptides can be obtained by chemical solid-phase synthesis and any possible variation of amino acid sequence can easily be achieved. They may include co-synthetic incorporation of fluorescent reporter groups at any position in the sequence, which makes the site-specific modification of SH- or NH$_2$- groups unnecessary (Link et al. 2003).

An alternative possibility, the incorporation of *non-natural amino acids* into a cell-free peptide and protein biosynthesis on ribosome, is also possible. In this case, a non-natural aminoacyl-tRNA is constructed and incorporated into a synthetic process (Katzen et al. 2005).

Regarding covalent labeling of nucleic acids, both 5′ and 3′ terminals are available for this. There are also many new possibilities to synthesize artificial fluorescent nucleic acid bases that can be incorporated co-synthetically.

7.3.3 *Chemical and Photochemical Cross-linking*

Self-assembly, if necessary, can be combined with *chemical cross-linking* and pho-topolymerization (Mancin et al. 2006). This will provide higher stability to the formed structures and allow the integration of molecules with different functions. Such applications are immense and the reader is advised to consult classical mono-graphs (Hermanson 1995) and reviews (Brinkley 1992; Brunner 1993), in which these modifications are described in sufficient detail.

7.4 Supporting and Transducing Surfaces

Surfaces play very special roles in fluorescence sensing. Despite the fact that, unlike potentiometric or cantilever sensors, their presence is not so critical for the reporting event, the sensor assembly on the surface and the sensing based on the interaction with it can allow additional possibilities.

The formation of an organized assembly can be achieved by the self-organization of receptors and fluorescent reporters on a *surface as a template* (Arduini et al. 2007). In this way, one can form a construction in which the target binding and reporting units do not interact directly and the communication between these two units is deter-mined only by their spatial closeness, ensured by the template. Already in this simple construction there are many possibilities for modification and optimization of the sensor by a simple adjustment of the ratio of components. With the combination of covalent binding and self-organizing methodologies, more complicated multifunc-tional sensors can be constructed. In this section we discuss different technologies for the formation of surfaces with an active and passive role in sensing.

7.4.1 *Surfaces with a Passive Role: Covalent Attachments*

The role of the surface as a passive support can be very important in many sensor constructions. When the sensor molecules, nanoparticles or supramolecular structures are immobilized on the surface, *they cannot diffuse* freely in a solution and mix with other sensors. Many attractive analytical advantages can be derived from the sensor attachment to the surface. We indicate several such new possibilities.

(a) The sensors composed in this way allow their usage for many times with simple sample addition-washing protocols or providing continuous monitoring in the flow.
(b) In such many-use systems, reagent additions can easily be applied, for instance, the addition of a fluorescent competitor.
(c) Due to fixed spatial separation between sensor molecules, the sensing of a large number of analytes (multiplex sensing) can be accomplished on the same plate, simultaneously. For such multiplex sensing *microarrays* (sometimes called biochips) can be developed for the simultaneous sensing of hundreds or even thousands of targets (see Section 9.4).

The choice of methods for surface immobilization depends on the chemical structures of the sensor and surface; these methods are well described in original literature and reviews (Sobek et al. 2006). They do not differ much from those methods that are used in other sensor technologies, where surface immobilization is also needed, such as surface plasmon resonance. Meantime, each sensor and each application may demand its own solution; the demands on applied solid support are thereby as manifold as the applications themselves. Immobilization should satisfy many requirements. In particular, it should involve proper orientation of sensors and availability of receptor sites for target binding.

Hydrogel layers allow a high density of sensor immobilization. Hydrogels are the three-dimensional (3D) polymer networks that swell but do not dissolve in water. They can be made optically transparent and fluorescent-free, being formed from different hydrophilic polymers, such as polyacrylamide and polyethylene glycol (Svedhem et al. 2001). Their development into 3D microarrays for DNA hybridization and protein detection (Rubina et al. 2003, 2004) allows substantially increasing the density of sensor locations. One of the proposed procedures (Burnham et al. 2006) is the following. A disulphide-cross-linked derivative of hydrogel is deposited on a surface of quartz or silicon. The application of the reducing agent then provides the generation of reactive SH groups throughout the hydrogel, leading to 'activated hydrogel'. These SH groups can be readily modified with the attachment of proteins or other functional groups, resulting in functional hydrogel.

To provide efficient affinity coupling with the use of a biotin-streptavidin pair, SH-reactive biotin derivative is applied. It yields *biotinilated hydrogel*, which is ready to bind streptavidin. The latter is particularly useful for the immobilization of biotinylated aptamers (Schaferling et al. 2003). Since biotinylated proteins are deposited from a solution, such a system is advantageous for the deposition of proteins that need delicate conditions of treatment. Generally, in addition to an increased binding capacity, immobilization within three-dimensional hydrogels offers many advantages over binding to flat, two-dimensional surfaces, such as a 'wet' and friendly polar environment with the stabilizing effect of the gel matrix. A disadvantage is the slow rate of establishing the target-binding equilibrium, which is limited by target diffusion in the gel.

Different materials that can serve as supporting media, transducers and even imprinted recognition elements, can be made in a *sol-gel process*. Both inorganic and organic/inorganic composite materials can be obtained by this procedure. The colloidal suspensions or 'sols' are formed via hydrolysis of alkoxy metal groups in the precursors with subsequent polycondensation. The obtained transparent, glass-like structures can attain different shapes and may be obtained with high porosity. The pore size distribution can be controlled by chemical composition and by the reaction conditions. Importantly, fluorescent dyes can be incorporated into these structures in several ways: by adsorption within the pores, by incorporation into the reaction mixture (due to mild conditions of synthesis there is no decomposition of dyes) and also by covalent binding to the finally obtained material. Covalent attachment is preferred for obtaining sensors of high stability.

7.4.2 Self-Assembled Monolayers

Self-assembled monolayers (SAMs) are monomolecular ordered structures that can form spontaneously on a chemically active surface (Flink et al. 2000; Ruckenstein and Li 2005). They can be prepared simply by adding a solution of monolayer-forming molecules onto the surface and washing off the excess. The monolayer is formed due to the strong interaction energy of one of the groups (usually, terminal) of a monolayer former with an active surface (Sharpe et al. 2007). It can be additionally stabilized by side interactions in the monolayer.

Gold and silver surfaces form very stable and ordered monolayers with thiol-containing compounds because of the strong binding energy between them (Fig. 7.7a). In the case of alkane thiols deposited on gold, this energy can be as high as 100–150 kJ/mol. SAM of long chain alkane thiol produces a highly packed and ordered surface, which can provide a membrane-like micro-environment, useful for immobilizing biological molecules. In addition to the low molecular mass thiol derivatives, self-assembled monolayers on gold can be formed by SH-substituted peptides, proteins, carbohydrates (Revell et al. 1998), DNA and peptide nucleic acids (Briones and Martin-Gago 2006).

It has been reported that polythiophene monolayers on gold retain their fluorescent properties (Zotti et al. 2008), as well as some modified fluorescent dyes (Kriegisch and Lambert 2005). Meantime, a strong *quenching of fluorescence* of closely located dyes by PET and FRET mechanisms is a well-known property of gold surfaces. This property can be useful (Perez-Luna et al. 2002). One such possibility is the following. By immobilizing the analyte of interest (or its structural analogue) to a metal surface and exposing it to a labeled receptor (e.g., antibody), the fluorescence of the labeled receptor, being in close proximity to the metal,

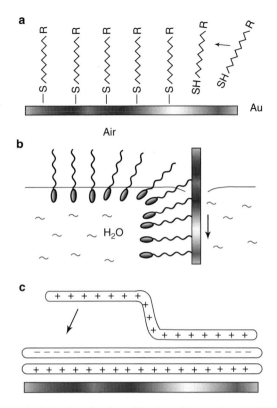

Fig. 7.7 Different methods for functional modification of the surface (**a**) Self-assembled mono-layers. They are frequently used for spontaneous modifications of gold and silver surfaces with compounds containing sulfur. R denotes the active group needed for further modifications. (**b**) Langmuir-Blodgette films. The films are first formed on a liquid surface and then transferred to a solid-liquid interface. (**c**) Films obtained by the layer-by-layer procedure, in which the layering is provided based on electrostatic interaction with the previously formed layer.

becomes quenched. If a free analyte is present, the labeled receptor dissociates from the metal surface with an increase of fluorescence intensity.

In many cases, quenching by a metal surface has to be avoided and another type of support has to be used. *Alkyl silane* molecules (e.g., octadecyltrichlorosilane) form self-assembled structures on silicon oxide surfaces and also, a number of different surfaces can be modified with alkyl carboxylates. Recent developments in the chemistry of SAMs on glass suggest new technologies for fluorescent sensor design. Trialkoxysilanes or halogenosilanes reacting with hydroxylated surfaces with the formation of SAMs offer such possibilities (Sullivan and Huck 2003).

Alkylsiloxanes are formed on the surface of silicon or glass. The reactive siloxane groups can form a cross-linked network on a surface by condensing with hydroxyl groups on the surface, water and neighboring siloxanes. This binding is more stable than that of alkanethiolates on gold but there is a stronger limitation on the chemistry of the second functional group that will extend from the formed surface and will be needed for further modifications.

Designed polypeptides with variable repeat length containing N-terminal dicysteine have been suggested as scaffolds for the surface immobilization of Quantum Dots (Medintz et al. 2006). Many other methods have been developed for modification of the surfaces of these particles, which attain increasing importance in sensing.

Hydrophobic surfaces can be covered with *phospholipid monolayers*, which can be formed spontaneously and the stability of such structures can be increased by thermal treatment (Stine et al. 2005). Such a coating may help to solubilize the hydrophobic particles in polar media and also incorporate target-binding functionalities into them.

The dense and highly regular two-dimensional distribution of exposed reactive functional groups of SAMs offers many possibilities for further attachment to the solid support of the whole sensors or their *assembly on the surface* from receptor and reporter components. SAMs offer a simple and straightforward method to generate modified surfaces with a transducing function. This property is of extreme value for making functional microfluidic devices (Section 9.5), in which, due to the narrow size of the produced channels, a high surface-to-volume ratio can be generated. Even in these microscopic structures, the deposition of SAMs is simple, it requires only the flow of a solution or a gas stream of adsorbate molecules through the channels.

Thus, one of the actively used functions of SAMs is the *hosting of biosensor molecules and cells*. This includes the confining and aligning of biological macromolecules and the immobilization of cells in a way that prevents their adhesion. The other important problem is the immobilization of fluorescence reporters and providing a signal transduction function to these reporters. Thus, the general role of SAMs is to form an interface between the surface and the molecular sensors or nanosensors. Such a possibility can be realized both on flat surfaces and on particles of different sizes.

Supramolecular interactions, such as of the π-π type, have recently been exploited to make sensitive glass surfaces. There have been reports on the functionalization of glass surfaces (quartz, glass slides and silica particles) with 2,7-diazapyrene derivatives for the detection of catecholamine neurotransmitters, such as dopamine (Cejas and Raymo 2005) (Fig. 7.8).

Thus, the most attractive feature of self-assembled monolayers is their ability to easily change the surface properties of the layer, including its binding properties, by tuning the chemistry of the terminal group that appears on its surface without significant influence on the formation of these layers.

7.4.3   Langmuir-Blodgett Films

The formation of *Langmuir-Blodgett films* (Fig. 7.7b) is an alternative to the self-assembly method of producing regular monolayers on a solid support surface (Davis

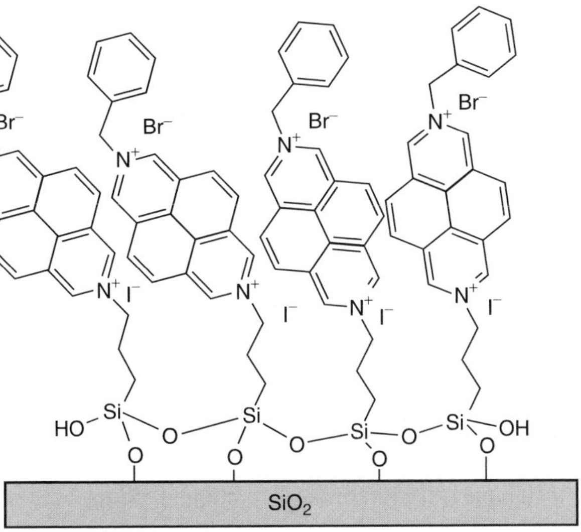

Fig. 7.8 Schematic structure of 2,7-diazapyrenium monolayers formed on silicon substrates that are able of selective binding of dopamine (Cejas and Raymo 2005)

and Higson 2005). This method is less demanding on the strength of the interaction of the monolayer-forming molecules with the support but requires them to be *amphiphilic*, i.e., to be composed of polar and apolar parts separated in space. The monolayer is primarily formed at the air-liquid interface and is then transferred to the surface of the solid plate by immersing it in this liquid. By repeating the operation of immersion and emersion several times, several regular monolayers can be deposited. When primarily formed at the air-water interface and then transferred to the hydrophobic support, these films expose polar groups. This is very useful for further modifications, such as attaching various molecular and nano-composite sensors operating in aqueous media. Langmuir-Blodgett films with all their modifications made in solvent can be used in air, which makes them applicable to the sensing of vapors.

There are many possibilities of forming Langmuir-Blodgett monolayers in a mixed composition with the inclusion of fluorescent dyes (Hussain et al. 2006). In particular, amphiphilic analogs of polarity-sensitive dyes can be mixed with fatty acids to form stable monolayers (Alekseeva et al. 2005). The formation of stable films by luminescent complexes of europium has also been described (Xiang et al. 2006). Conjugated polymers can be used as monolayer formers (Mattu et al. 2006). All this suggests many attractive possibilities for fluorescence sensing but, surprisingly, this methodology is rarely used.

It should be indicated that Langmuir-Blodgett films when formed do not allow easy incorporation of compounds dissolved in an external solution. Such incorporation can be better provided during the step of their formation; the incorporated molecules should be amphiphilic and possess some similarity with the major component of the film. There is also a possibility to make a second layer on top of the film.

If this layer is formed by phospholipids, many possibilities appear for the incorporation of a broad variety of different molecules, including biomembrane proteins. There are results (Davis and Higson 2005) showing that single-strand DNA molecules can be incorporated into the films formed by octadecylamine and they can be hybridized with complementary strands.

7.4.4 Layer-by-Layer Approach

The *layer-by-layer technique* (Fig. 7.7c) is a version of self-assembly that is available to polyelectrolites and uses the property of a molecular layer formed by a charged polymer to absorb a molecular layer of a polymer with an opposite charge (Ariga et al. 2006). Such layer-by-layer technique allows obtaining multilayers with precise thickness. Since many natural and synthetic polymers contain the charges, this technique is useful for the formation of the surfaces with desired properties. In addition, the distance from the surface monolayer to solid support can be precisely controlled. A smaller-size charged molecule including fluorescent dyes (Egawa et al. 2006) can participate in the formation of the layers.

The layer-by-layer structures are not limited to flat surfaces but can be assembled onto different *charged particles*. In comparison with Langmuir-Blodgett films, they do not show a very high degree of order. These films are highly stable, due to the multiple interactions providing the charge neutralization. Meanwhile, the order in the polymer layers themselves cannot be very high, which makes the structures rather amorphous. In aqueous solutions, they can be affected by the pH and ionic strength, which is not always a favorable property.

The layers formed by *conjugated polymers* deserve special attention. The polyelectrolitic nature of many of them gives very good prospects for making 'responsive' layers with their inclusion. It is known that their fluorescence superquenching effect in thin films is much greater compared to that in isolated conjugated polymer molecules in solution. This is because the efficient exciton length is larger in two dimensions than in a one-dimensional polymer chain. The precise deposition of the quenchers that could be a part of the reporting mechanism is possible in this system.

The layer-by-layer approach can be extended to *nanoparticles*. The stepwise construction of a novel kind of self-assembled organic/inorganic multilayers based on multivalent supramolecular interactions between guest-functionalized dendrimers and host-modified gold nanoparticles has been reported (Crespo-Biel et al. 2005). Such supramolecular layer-by-layer assembly yields the growth of the plasmon absorption band in proportion to the number of layers. The interactions between optically responsive monolayers can be studied as a function of distance between them, which is regulated by a number of intermediate inert layers. It was found (Ianoul and Bergeron 2006) that in order to minimize the quenching of the fluorescence signal, twenty polyelectrolyte monolayers are necessary between the nanoparticles and the dye, deposited by a layer-by-layer technique producing a 15–20 nm separation cushion.

The combination of the layer-by-layer method of multilayer formation with other fabrication techniques, such as spin-coating, spraying and photolithography, offers many new possibilities for the creation of sensor arrays (Ariga et al. 2007).

7.4.5   Prospects

The design of fluorescence sensors using functionalized surfaces has many technical advantages. It also extends the possibilities of sensing, for instance, to the gas phase. With surface immobilization, it is much easier to form sensors that allow multiple uses, continuous application of the tested medium and target monitoring, washing-out the non-target species or reagent additions. The surface deposition allows the spatial separation of sensors to different targets and therefore provides multiplex sensing.

These are the 'trivial' and most easily understood advantages of 'passive' surfaces. Even for them, there is a lot of potential for improvement by making the right choice from different possibilities. A surface covered with polymer hydrogel is very attractive, since it allows the 'three-dimensional' incorporation of sensor units into the gel matrix. However, this can be done only for small-sized sensor molecules or particles and the response of these sensors is slow because of retarded diffusion. In contrast, being attached to SAMs, the sensor recognition sites are equally exposed and their response is faster but the limited number of receptors inherent to a planar surface restricts the sensor density.

Surfaces can play an active role by participating in signal transduction and such participation is versatile. Gold and silver surfaces are very good as SAM formers and also as fluorescence quenchers and this favorable combination can be used in the design of many sensors exploring the variation of distance between the surface and fluorescent dye (or nanoparticles). These and other, more inert, surfaces can be modified with the introduction of SAM, made of fluorescent molecules or of a layer formed by a fluorescent conjugated polymer. This offers many additional possibilities. A variety of amphiphilic, 'fatty-acid-type' molecules can form Langmuir-Blodgett monolayers and multilayers with the incorporation of fluorescent dyes and this possibility should not be overlooked.

Finally, in addition, the surface can be extensively used as a platform for the solid-phase synthesis of peptides and oligonucleotides with analyte receptor functions. Affinity coupling and self-assembly as important steps for making sensors of high complexity can be efficiently provided on the surface. The thin film technology is also one of the most powerful strategies for the immobilization of sensor units on artificial devices, such as optodes.

7.5 Functional Lipid Bilayers

The membrane of a living cell is probably one of the most distinct examples of how Nature uses the self-assembly of relatively simple building blocks to create highly-organized structures. The membrane can be viewed as a structure formed of lipids with different integral proteins and glycolipids incorporated into it. Lipids alone can form artificial self-assembled structures, which in many properties resemble the cell membranes, the basic element of which is the *lipid bilayer*.

Lipid bilayers (Fig. 7.9) can form structures of different curvatures, from planar membranes to small nanoparticles (the smallest size is ~50 nm) with a closed volume. The latter are called the *lipid vesicles* or *liposomes*. Their stability allows filling their inner volume to use them as drug carriers. Bilayers themselves allow the incorporation of membrane proteins, making them convenient models for biological membranes.

Common phospholipids forming these structures are molecules with polar heads and two long (16–18 carbon atoms) hydrocarbon tails. The bilayer consists of two monolayers, in which the apolar tails come together and the polar heads stretch out to form two outer surfaces, extending to an aqueous solvent. When the vesicle is formed, some of the heads appear facing the inner volume and the others extend to the outer volume. Bilayer membranes, especially in the form of vesicles, offer very interesting possibilities for fluorescence sensing.

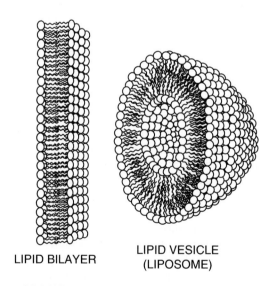

LIPID BILAYER

LIPID VESICLE
(LIPOSOME)

Fig. 7.9 Schematic view of lipid bilayer and vesicle with closed volume formed by this bilayer (liposome). The stability of these structures is determined by the amphiphilic properties of the constituting molecules. Polar heads tend to be exposed to an aqueous solvent, whereas hydrophobic tails tend to be screened from it

7.5.1 Liposomes as Integrated Sensors

Liposomes are self-organized membrane structures, they can spontaneously form in water from the suspension of hydrated lipids (phospholipids or sphyngolipids) by sonication, injection through a porous membrane or just by gentle agitation. They possess very important collective properties. If they are made of a single type of lipid, they exhibit remarkable cooperativity in structural transitions. Thus, as a function of temperature, phospholipid molecules can change the structural arrangement in the bilayer at a fixed temperature within a transition range as narrow as 0.6°C. These transitions between structural forms are extremely sensitive to the incorporation of some other molecules, such as cholesterol.

The structural and dynamic properties of the bilayers can change in a specific manner upon incorporation of many different compounds (and the general anesthetics are well-known examples). Important to note, all these changes involve only the arrangements and dynamics of lipids and do not produce a change in the overall structure and integrity of the liposome. Being very thin (about 4 nm in width), the bilayers possess tremendous depth-dependent gradients of polarity, hydration and electric fields. They are practically impermeable to ions, even to protons.

The latter properties make the bilayers ideal transducers in sensor technologies. From the time of discovery, liposomes have been the objects of active research with various fluorescent dyes serving as the 'probes' for their structure and dynamics. The synthesis of the derivatives of wavelength-ratiometric two-color fluorescent dyes made possible their spontaneous incorporation into the bilayer at the desired depth and orientation (Klymchenko et al. 2002b) (Fig. 7.10). This allowed determining the polarity and hydration in the bilayer as easily distinguishable parameters for

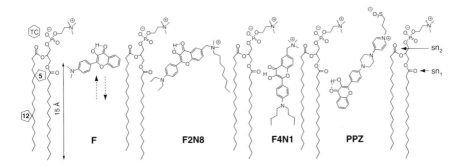

Fig. 7.10 The first-generation 3-hydroxychromone dyes for biomembrane studies. Dye **F** does not occupy a well-determined position (its motion is shown by arrows). Dependent on its position, it may or may not form a hydrogen bond with water. Location of Dye **F2N8** is at the polar interface. Dye **F4N1** in vertical orientation goes deeper into the bilayer, on the level of the lipid sn_1 carbonyls. Dye **PPZ** is in inverted orientation, it goes deeper than the carbonyl groups. Such a location is supported by the results on quenching by nitroxide paramagnetic quenchers, which were covalently attached to the lipids (shown in the left-hand part of the figure) (Reproduced with permission from Klymchenko et al. 2002b)

the first time (Klymchenko et al. 2004a, b) and the characterization of its electro-static potential (Klymchenko et al. 2003).

Thus, we know that a liposome can interact with different molecules and this interaction changes its collective structural and dynamic properties, without changing its integrity. We also know that these interactions can depend strongly on the lipid composition of the membrane and in this sense, can be very specific. We possess novel fluorescence dyes as tools to provide the most sensitive response to these interactions. What remains is to make the sensors. The lipid bilayer itself can serve as the sensor element for the efficient detection and characterization of molecular interactions. Possessing high cooperativity, the *whole bilayer structure* can serve as a composed nanosensor, combining the binding and reporting functions. The trans-duction occurs due to the cooperative properties of the bilayer. Located in the bilayer, fluorescent dyes provide the reporting signal.

This idea was explored for the development of a prototype sensor for *choles-terol*. The changes in lipid membranes upon incorporation of cholesterol are well known. They are especially great for bilayers made of lipid sphyngomyelin, which are able to form very rigid structures with cholesterol, "rafts". We formed vesicles made of sphyngomyelin, incorporating functional 3HC dyes and observed dramatic changes of florescence color upon incorporation of cholesterol (Turkmen et al. 2005). These experiments were reproduced with a number of dyes and one of these results is presented in Fig. 7.11.

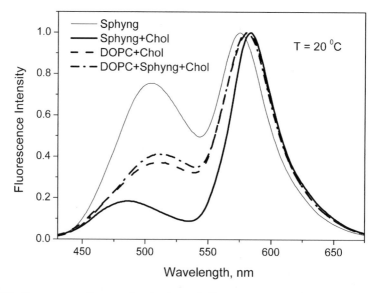

Fig. 7.11 The prototype for a two-band ratiometric fluorescence sensor for cholesterol that uses sphingomyelin (SM) bilayer vesicles with functionalized 3HF dye (Turkmen et al. 2005). The thin, solid line represents the spectrum in the SM bilayer. Upon addition of cholesterol, the relative intensity of the short-wavelength band dramatically (by ~80%) decreases (thick solid line). This does not happen when cholesterol is added to phosphatidylcholine vesicles (compare dashed and dot-dashed lines) (Reproduced with permission from Demchenko 2005)

We observe that upon incorporation of cholesterol, the blue (500 nm) band loses about 80% of its relative integral intensity. This dramatic change of color can easily be detected, even visually and, of course, with the application of a simple fluorescence detection technique. The absence of the effect of cholesterol on the spectra in phosphatidylchoine vesicles demonstrates that a specific raft-forming structure is needed for the generation of spectroscopic changes.

7.5.2 Stabilized Phospholipid Bilayers

The weak points of liposomes are well-known to those who work with them. It is not easy to make them homogeneous in size and their long-term stability is poor. Therefore, the idea emerged to stabilize them by forming them around micro- or nanoparticles, to form *core-shell liposomes*. The cores of these particles may be of noble metals, polymeric or composed of Quantum Dots, with the exploration in full of their signal transduction and fluorescence possibilities. In such a construction the particle serves as the 'core'; it is surrounded by a lipid bilayer, the 'shell'.

The core determines the particle size and shape and protects it from spontaneous decomposition. It can be selected in order to obtain additional valuable properties. For instance, it can be made magnetic (Shinkai 2002), which will facilitate the separation of sensor particles from tested media and their repeated use. Alternatively, being a polymer or silica bead doped with fluorescent dye or Quantum Dot, it can provide fluorescent emission. The procedures for making these composites are simple due to the spontaneous formation of bilayers (Mornet et al. 2005).

The membrane shell can possess both receptor and reporter properties and can be modified for this in many ways. Lipid self-assembly on particles allows receptor insertion and amplification of receptor-target recognition. One such incorporated receptor can be monosialoganglioside GM1, which can specifically bind *cholera toxin* (Carmona-Ribeiro 2001). The utility of self-assembled vesicles, bilayers or monolayers at interfaces is limited only by our own imagination.

One of the ways for improving the stability of phospholipid bilayers sensors is their formation on the surface of polymer latex beads (Fig. 7.12). These composites, called *lipobeads* (Ng et al. 2001), can be transformed into fluorescent sensors by the incorporation into the bilayer of various fluorescent dyes that possess sensing properties or are conjugated with recognition units. Such lipobead nanosensors have been suggested for intracellular measurements of pH, Ca^{2+} and O_2 (Ma and Rosenzweig 2005). There are extensive possibilities in the modification of the surfaces of these particles with bioactive molecules, for the development of biosensors.

The *planar surfaces* of optical waveguides and microfluidic devices can also be modified by depositing lipid bilayers, resulting in their stabilization and allowing the incorporation of different types of recognition units. Among recent developments is the formation of a bilayer membrane that recognizes cholera toxin directly on the surface of a planar optical waveguide (Kelly et al. 2006). This approach can be extended to the development of sensor arrays on a planar support (Yamazaki et al. 2005).

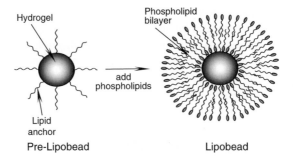

Fig. 7.12 Schematic illustration of the self-assembly of a lipid membrane around a core nanoparticle made of hydrogel or any other core material that is surface-conjugated with lipid for increasing stability of the complex

The incorporation of integral membrane proteins into the arrayed membranes enables the study of ligand/receptor binding, as well as interactions with intact living cells.

7.5.3   Polymersomes

Polymersomes are *synthetic analogs of liposomes* made of amphiphilic vesicle-forming block copolymers. The advantage of polymersomes over liposomes is their increased stability, which contributes to the increased lifetime of these structures. In contrast to common liposomes, which have a high tendency to collapse with time, polymersomes can be rendered eternally stable. This can be achieved by chemical cross-linking (Discher et al. 2002).

For the formation of bilayers, the polymersomes have to be assembled according to the same principle as the lipid vesicles. They consist of two blocks – polar heads and hydrophobic tails. The vast amount of available monomers and the ability to vary the ratio of the two blocks make it possible to tune the properties of the resulting vesicles, for example, vesicle size, polarity, stability, etc. In general, the membranes of block copolymer vesicles possess a higher thickness and less fluidity compared to liposomes (Antonietti and Forster 2003). These properties can vary over a broad range, depending on the polymer composition (Bermudez et al. 2002). Polymersomes are still awaiting application in sensor technologies.

7.5.4   Formation of Protein Layers over Lipid Bilayers

The layers formed by two-dimensional crystallization of *S-layer proteins* (Section 7.3) usually need some support for their formation. It was demonstrated that they can readily be formed over liposomes (Moll et al. 2002). This allows activation of an important line of development in combining the S-layer and lipid membrane technologies (Huber et al. 2006). Although the development of S-layer technologies was focused primarily on liposomal drug delivery systems, an important field of future applications emerges in sensor technologies.

The interaction of S-layer proteins with lipid molecules is noncovalent. Electrostatic interaction between exposed carboxy groups on the inner face of the S-layer lattice and the zwitterionic lipid head groups at the outer face of the liposomes is primarily responsible for the binding and defined orientation of the S-layer subunits. As a result, they can form a closed shell around the liposome and this allows obtaining composites with essentially increased stability. However, these interactions do not substantially reduce lipid mobility in the plane of the bilayer. A formed closed lattice structure is amendable to different modifications. Moreover, since S-layer proteins can be genetically modified and fused with other structure-forming proteins, supramolecular functional constructs can be assembled.

7.5.5   Prospects

There are many very positive features offered by lipid bilayers and their analogs. These structures can be formed spontaneously and they allow the spontaneous incorporation of different functional molecules, both receptors and reporters. They can respond to target binding by changes in their integral and easily measurable properties, such as lipid order and dynamics. When they form particles with closed inner volumes (liposomes), these particles can retain many different molecules or particles with diverse functional possibilities in this volume. Composites containing single particles surrounded by the bilayer are of special importance because they allow strongly increased stability of the bilayer towards aggregation or decomposition and also because of the possibility to provide additional useful properties to the core particles. Synthetic lipid analogs and bilayer-forming polymers expand these possibilities.

Sensing and Thinking 7: Extended Sensing Possibilities with Smart Nano-ensembles

In reading this chapter, we observe a myriad of opportunities for manipulating with interactions of molecules, nanoparticles and surfaces. Since there is no limit regarding the sensor construct, the choice of optimal solution can be made on a very broad scale. This may involve not only molecular complexes and responding nanoparticles but also self-assembled supramolecular structures, composite nanoparticles (that combine recognition and response properties) and various types of surfaces and interfaces. Clearly, this relatively new field of development of fluorescence sensors is far from being mature. Whereas the advance in recognition units and fluorescence reporters, though promising many improvements, follows already well-established trends, the optimization of their performance based on the principles of integration and self-assembly promises new, exciting discoveries that will result in new, useful products.

Synthetic strategies based on covalent linking, affinity coupling and self-assembly offer many advantages. They include the minimization of synthetic work, the ease of modification and optimization of the sensor and the possibility to tune its properties by a simple adjustment of the ratio of components. The applications of these new concepts tend to not only combine the target binding and the response but also improve this response by creating an optimal supramolecular ensemble, allowing the activation of optimal communication and signal transduction between its structural elements. The recognition units can be formed by an assembly of the functional molecules, with the appearance of new recognition properties.

Incorporation onto planar surfaces and nano-sized structures provides the support for functional materials that allows their stabilization and integration into self-assembled nanoscale recognition units. The use of nanoparticles can make these recognition units nano-sized and multi-valent. In addition, the self-assembled systems allow great possibilities for the introduction of fluorescence response functionality, thus providing additional important properties. In the assembly, the receptor and the reporter units may not interact directly and the signal on target binding can be transmitted due to the spatial closeness ensured by the template and even indirectly, via a change in the integral property of the nanoscale system.

Improving the functionality always increases the complexity. To overcome the observed difficulties, libraries of sensor building blocks have to be created. The well-established streptavidin-biotin pair is not optimal for many applications, therefore new pairs for molecular assembly, especially of synthetic origin, are being actively sought. Their libraries are expected to be made available and then easily combined, to produce the most suitable system for the desired use. Being kept close to the template, they can collectively operate and give new properties to the sensor device.

Questions and Problems

1. Evaluate the prospective application of carbon nanotubes. Can they serve as support materials? Or as optical signal transducers? Or as FRET donors or acceptors with organic dyes?
2. Why are nanoscale materials with core–shell composition so attractive? How can they combine recognition and response properties?
3. What is the advantage of a DNA scaffold compared to one made of synthetic polymers? How does the DNA structure allow the incorporation of bulky substituents? Provide the modeling and compare your predicted structures with the results on porphyrins-DNA composites described in ref. (Fendt et al. 2007).

4. Is it possible to design the partner for self-assembly, based on the known three-dimensional structure of the other partner? On what principles should this design be based?

5. How many intermolecular hydrogen bonds should be formed to make the dimeric structure stable?

6. Explain why the support surface formed by a S-layer protein allows obtaining structures in which the sensor elements can face only one side. What factors could determine the sidedness?

7. Why does the operation of sensors based on micellar structures not allow a broad-range variation of micelle-forming detergent concentrations? Explain the 'concentration-of-the-target' effect of micelles.

8. What are the structural and energetic requirements for the formation of self-assembled monolayers?

9. What are the advantages of hydrogel layers? What additional limitations are imposed on the use of this support material?

10. Can Langmuir-Blodgett films be formed on a polar surface? Can they be formed in the presence of a detergent? Explain how to incorporate fluorescent dye into these films.

11. In what medium should the layer-by-layer assembly be made?

12. List the most important properties of phospholipid bilayers and the vesicles (liposomes) made of them. What are the reasons and what are the means to stabilize them? What new properties can be obtained?

13. How could you define the recognition element and signal transduction mechanism in a liposome cholesterol sensor?

14. On what length scales can you estimate the applications of 'top-down' and 'bottom-up' approaches? Does a gap between them exist and if so, how can it be filled?

References

Aili D, Enander K, Baltzer L, Liedberg B (2007) Synthetic de novo designed polypeptides for control of nanoparticle assembly and biosensing. Biochemical Society Transactions 35:532–534

Alekseeva VI, Marinina LE, Savvina LP, Ibrayev NH (2005) Spectral and luminescent properties of nile red dye in Langmuir-Blodgett films. Molecular Crystals and Liquid Crystals 427:471–478

Antonietti M, Forster S (2003) Vesicles and liposomes: a self-assembly principle beyond lipids. Advanced Materials 15:1323–1333

Arduini M, Rampazzo E, Mancin F, Tecilla P, Tonellato U (2007) Template assisted self-organized chemosensors. Inorganica Chimica Acta 360:721–727

Ariga K, Nakanishi T, Michinobu T (2006) Immobilization of biomaterials to nano-assembled films (self-assembled monolayers, Langmuir-Blodgett films, and layer-by-layer assemblies) and their related functions. Journal of Nanoscience and Nanotechnology 6:2278–2301

Ariga K, Hill JP, Ji QM (2007) Layer-by-layer assembly as a versatile bottom-up nanofabrication technique for exploratory research and realistic application. Physical Chemistry Chemical Physics 9:2319–2340

Baldini L, Casnati A, Sansone F, Ungaro R (2007) Calixarene-based multivalent ligands. Chemical Society Reviews 36:254–266

Balzani V, Credi A, Venturi M (2003) Molecular logic circuits. Chemphyschem 4:49–59

Barone PW, Baik S, Heller DA, Strano MS (2005) Near-infrared optical sensors based on single-walled carbon nanotubes. Nature Materials 4:86–92

Bermudez H, Brannan AK, Hammer DA, Bates FS, Discher DE (2002) Molecular weight dependence of polymersome membrane structure, elasticity, and stability. Macromolecules 35:8203–8208

Brea RJ, Vazquez ME, Mosquera M, Castedo L, Granja JR (2007) Controlling multiple fluorescent signal output in cyclic peptide-based supramolecular systems. Journal of the American Chemical Society 129:1653–1657

Brinkley M (1992) A brief survey of methods for preparing protein conjugates with dyes, haptens, and cross-linking reagents. Bioconjugate Chemistry 3:2–13

Briones C, Martin-Gago JA (2006) Nucleic acids and their analogs as nanomaterials for biosensor development. Current Nanoscience 2:257–273

Brunner J (1993) New photolabeling and cross-linking methods. Annual Review of Biochemistry 62:483–514

Burnham MR, Turner JN, Szarowski D, Martin DL (2006) Biological functionalization and surface micropatterning of polyacrylamide hydrogels. Biomaterials 27:5883–5891

Cao YC, Hua XF, Zhu XX, Wang Z, Huang ZL, Zhao YD, Chen H, Liu MX (2006) Preparation of Au coated polystyrene beads and their application in an immunoassay. Journal of Immunological Methods 317:163–170

Carmona-Ribeiro AM (2001) Bilayer vesicles and liposomes as interface agents. Chemical Society Reviews 30:241–247

Cejas MA, Raymo FM (2005) Fluorescent diazapyrenium films and their response to dopamine. Langmuir 21:5795–5802

Choi JH, Strano MS (2007) Solvatochromism in single-walled carbon nanotubes. Applied Physics Letters 90

Corbin PS, Lawless LJ, Li ZT, Ma YG, Witmer MJ, Zimmerman SC (2002) Discrete and polymeric self-assembled dendrimers: hydrogen bond-mediated assembly with high stability and high fidelity. Proceedings of the National Academy of Sciences of the United States of America 99:5099–5104

Crespo-Biel O, Dordi B, Reinhoudt DN, Huskens J (2005) Supramolecular layer-by-layer assembly: alternating adsorptions of guest- and host-functionalized molecules and particles using multivalent supramolecular interactions. Journal of the American Chemical Society 127:7594–7600

Davis F, Higson SPJ (2005) Structured thin films as functional components within biosensors. Biosensors & Bioelectronics 21:1–20

del Mercato LL, Pompa PP, Maruccio G, Della Torre A, Sabella S, Tamburro AM, Cingolani R, Rinaldi R (2007) Charge transport and intrinsic fluorescence in amyloid-like fibrils. Proceedings of the National Academy of Sciences of the United States of America 104:18019–18024

Demchenko AP (2001) Recognition between flexible protein molecules: induced and assisted folding. Journal of Molecular Recognition 14:42–61

Demchenko AP (2005) Optimization of fluorescence response in the design of molecular biosensors. Analytical Biochemistry 343:1–22

Discher BM, Bermudez H, Hammer DA, Discher DE, Won YY, Bates FS (2002) Cross-linked polymersome membranes: vesicles with broadly adjustable properties. Journal of Physical Chemistry B 106:2848–2854

Egawa Y, Hayashida R, Anzai JI (2006) Multilayered assemblies composed of brilliant yellow and poly(allylamine) for an optical pH sensor. Analytical Sciences 22:1117–1119

Fendt LA, Bouamaied I, Thoni S, Amiot N, Stulz E (2007) DNA as supramolecular scaffold for porphyrin arrays on the nanorneter scale. Journal of the American Chemical Society 129:15319–15329

Fernandez YD, Gramatges AP, Amendola V, Foti F, Mangano C, Pallavicini P, Patroni S (2004) Using micelles for a new approach to fluorescent sensors for metal cations. Chemical Communications:1650–1651

Flink S, van Veggel F, Reinhoudt DN (2000) Sensor functionalities in self-assembled monolayers. Advanced Materials 12:1315–1328

Gao XY, Matsui H (2005) Peptide-based nanotubes and their applications in bionanotechnology. Advanced Materials 17:2037–2050

Gilardi G, Zhou LQ, Hibbert L, Cass AE (1994) Engineering the maltose binding protein for reagentless fluorescence sensing. Analytical Chemistry 66:3840–3847

Henzie J, Barton JE, Stender CL, Odom TW (2006) Large-area nanoscale patterning: chemistry meets fabrication. Accounts of Chemical Research 39:249–257

Hermanson GT (1995) Bioconjugation techniques. Academic, San Diego, CA

Huber C, Liu J, Egelseer EM, Moll D, Knoll W, Sleytr UB, Sara M (2006) Heterotetramers formed by an S-layer-streptavidin fusion protein and core-streptavidin as a nanoarrayed template for biochip development. Small 2:142–150

Hussain SA, Paul PK, Bhattacharjee D (2006) Role of microenvironment in the mixed Langmuir-Blodgett films. Journal of Colloid and Interface Science 299:785–790

Ianoul A, Bergeron A (2006) Spatially inhomogeneous enhancement of fluorescence by a monolayer of silver nanoparticles. Langmuir 22:10217–10222

Jiang H, Zhao XY, Schanze KS (2007) Effects of polymer aggregation and quencher size on amplified fluorescence quenching of conjugated polyelectrolytes. Langmuir 23:9481–9486

Katzen F, Chang G, Kudlicki W (2005) The past, present and future of cell-free protein synthesis. Trends in Biotechnology 23:150–156

Kelly TL, Lam MCW, Wolf MO (2006) Carbohydrate-labeled fluorescent microparticles and their binding to lectins. Bioconjugate Chemistry 17:575–578

Klymchenko AS, Demchenko AP (2002a) Probing AOT Reverse Micelles with Two-Color Fluorescence Dyes Based on 3-Hydroxychromone. Langmuir 18:5637–5639.

Klymchenko AS, Duportail G, Ozturk T, Pivovarenko VG, Mely Y, Demchenko AP (2002b) Novel two-band ratiometric fluorescence probes with different location and orientation in phospholipid membranes. Chemistry & Biology 9:1199–1208

Klymchenko AS, Duportail G, Mely Y, Demchenko AP (2003) Ultrasensitive two-color fluorescence probes for dipole potential in phospholipid membranes. Proceedings of the National Academy of Sciences of the United States of America 100:11219–11224

Klymchenko AS, Duportail G, Demchenko AP, Mely Y (2004a) Bimodal distribution and fluorescence response of environment-sensitive probes in lipid bilayers. Biophysical Journal 86: 2929–2941

Klymchenko AS, Mely Y, Demchenko AP, Duportail G (2004b) Simultaneous probing of hydration and polarity of lipid bilayers with 3-hydroxyflavone fluorescent dyes. Biochimica Et Biophysica Acta-Biomembranes 1665:6–19

Kriegisch V, Lambert C (2005) Self-assembled monolayers of chromophores on gold surfaces. Supermolecular Dye Chemistry 258:257–313

Lehn J-M (1995) Supramolecular chemistry. VCH, Weinheim

Link AJ, Mock ML, Tirrell DA (2003) Non-canonical amino acids in protein engineering. Current Opinion in Biotechnology 14:603–609

Ma AH, Rosenzweig Z (2005) Synthesis and analytical properties of micrometric biosensing lipobeads. Analytical and Bioanalytical Chemistry 382:28–36

Mahar B, Laslau C, Yip R, Sun Y (2007) Development of carbon nanotube-based sensors - a review. IEEE Sensors Journal 7:266–284

Mancin F, Rampazzo E, Tecilla P, Tonellato U (2006) Self-assembled fluorescent chemosensors. Chemistry-A European Journal 12:1844–1854

Marvin JS, Corcoran EE, Hattangadi NA, Zhang JV, Gere SA, Hellinga HW (1997) The rational design of allosteric interactions in a monomeric protein and its applications to the construction of biosensors. Proceedings of the National Academy of Sciences of the United States of America 94:4366–4371

Mattu J, Johansson T, Holdcroft S, Leach GW (2006) Highly ordered polymer films of amphiphilic, regioregular polythiophene derivatives. Journal of Physical Chemistry B 110:15328–15337

Medintz IL, Deschamps JR (2006) Maltose-binding protein: a versatile platform for prototyping biosensing. Current Opinion in Biotechnology 17:17–27

Medintz IL, Sapsford KE, Clapp AR, Pons T, Higashiya S, Welch JT, Mattoussi H (2006) Designer variable repeat length polypeptides as scaffolds for surface immobilization of quantum dots. Journal of Physical Chemistry B 110:10683–10690

Moll D, Huber C, Schlegel B, Pum D, Sleytr UB, Sara M (2002) S-layer-streptavidin fusion proteins as template for nanopatterned molecular arrays. Proceedings of the National Academy of Sciences of the United States of America 99:14646–14651

Mornet S, Lambert O, Duguet E, Brisson A (2005) The formation of supported lipid bilayers on silica nanoparticles revealed by cryoelectron microscopy. Nano Letters 5:281–285

Ng CC, Cheng YL, Pennefather PS (2001) One-step synthesis of a fluorescent phospholipid-hydrogel conjugate for driving self-assembly of supported lipid membranes. Macromolecules 34:5759–5765

Oshovsky GV, Reinhoudt DN, Verboom W (2007) Supramolecular chemistry in water. Angewandte Chemie-International Edition 46:2366–2393

Perez-Luna VH, Yang SP, Rabinovich EM, Buranda T, Sklar LA, Hampton PD, Lopez GP (2002) Fluorescence biosensing strategy based on energy transfer between fluorescently labeled receptors and a metallic surface. Biosensors & Bioelectronics 17:71–78

Piervincenzi RT, Reichert WM, Hellinga HW (1998) Genetic engineering of a single-chain antibody fragment for surface immobilization in an optical biosensor. Biosensors & Bioelectronics 13:305–312

Rao JH, Lahiri J, Isaacs L, Weis RM, Whitesides GM (1998) A trivalent system from vancomycin center dot D-Ala-D-Ala with higher affinity than avidin center dot biotin. Science 280:708–711

Renard M, Belkadi L, Hugo N, England P, Altschuh D, Bedouelle H (2002) Knowledge-based design of reagentless fluorescent biosensors from recombinant antibodies. Journal of Molecular Biology 318:429–442

Revell DJ, Knight JR, Blyth DJ, Haines AH, Russell DA (1998) Self-assembled carbohydrate monolayers: formation and surface selective molecular recognition. Langmuir 14:4517–4524

Rubina AY, Dementieva EI, Stomakhin AA, Darii EL, Pan'kov SV, Barsky VE, Ivanov SM, Konovalova EV, Mirzabekov AD (2003) Hydrogel-based protein microchips: manufacturing, properties, and applications. Biotechniques 34:1008–1014, 1016–1020, 1022

Rubina AY, Pan'kov SV, Dementieva EI, Pen'kov DN, Butygin AV, Vasiliskov VA, Chudinov AV, Mikheikin AL, Mikhailovich VM, Mirzabekov AD (2004) Hydrogel drop microchips with immobilized DNA: properties and methods for large-scale production. Analytical Biochemistry 325:92–106

Ruckenstein E, Li ZF (2005) Surface modification and functionalization through the self-assembled monolayer and graft polymerization. Advances in Colloid and Interface Science 113:43–63

Satishkumar BC, Brown LO, Gao Y, Wang CC, Wang HL, Doorn SK (2007) Reversible fluorescence quenching in carbon nanotubes for biomolecular sensing. Nature Nanotechnology 2:560–564

Schaferling M, Riepl M, Pavlickova P, Paul H, Kambhampati D, Liedberg B (2003) Functionalized self-assembled monolayers on gold as binding matrices for the screening of antibody-antigen interactions. Microchimica Acta 142:193–203

Seeman NC (2003) At the crossroads of chemistry, biology, and materials: structural DNA nanotechnology. Chemistry & Biology 10:1151–1159

Sharpe RBA, Burdinski D, Huskens J, Zandvliet HJW, Reinhoudt DN, Poelsema B (2007) Template-directed self-assembly of alkanethiol monolayers: selective growth on preexisting monolayer edges. Langmuir 23:1141–1146

Shinkai M (2002) Functional magnetic particles for medical application. Journal of Bioscience and Bioengineering 94:606–613

Sloan DJ, Hellinga HW (1998) Structure-based engineering of environmentally sensitive fluorophores for monitoring protein-protein interactions. Protein Engineering 11:819–823

Sleytr UB, Egelseer EM, Ilk N, Pum D, Schuster B (2007) S-Layers as a basic building block in a molecular construction kit. FEBS Journal 274:323–334

Sobek J, Bartscherer K, Jacob A, Hoheisel JD, Angenendt P (2006) Microarray technology as a universal tool for high-throughput analysis of biological systems. Combinatorial Chemistry & High Throughput Screening 9:365–380

Stine R, Pishko MV, Hampton JR, Dameron AA, Weiss PS (2005) Heat-stabilized phospholipid films: film characterization and the production of protein-resistant surfaces. Langmuir 21:11352–11356

Sullivan TP, Huck WTS (2003) Reactions on monolayers: organic synthesis in two dimensions. European Journal of Organic Chemistry 1:17–29

Svedhem S, Hollander CA, Shi J, Konradsson P, Liedberg B, Svensson SCT (2001) Synthesis of a series of oligo(ethylene glycol)-terminated alkanethiol amides designed to address structure and stability of biosensing interfaces. Journal of Organic Chemistry 66:4494–4503

Turkmen Z, Klymchenko AS, Oncul S, Duportail G, Topcu G, Demchenko AP (2005) A triterpene oleanolic acid conjugate with 3-hydroxyflavone derivative as a new membrane probe with two-color ratiometric response. Journal of Biochemical and Biophysical Methods 64:1–18

Vriezema DM, Aragones MC, Elemans J, Cornelissen J, Rowan AE, Nolte RJM (2005) Self-assembled nanoreactors. Chemical Reviews 105:1445–1489

Wang JB, Qian XH, Qian JH, Xu YF (2007) Micelle-induced versatile performance of amphiphilic intramolecular charge-transfer fluorescent molecular sensors. Chemistry-A European Journal 13:7543–7552

Xiang XM, Qian DJ, Li FY, Chen HT, Liu HG, Huang W, Feng XS (2006) Fabrication of europium complexes with 4′-(4-methylphenyl)-2,2′: 6′,2″-terpyridine and 4,4′-dinonyl-2,2′-dipyridyl at the air-water interface and their emission properties in Langmuir-Blodgett films. Colloids and Surfaces A-Physicochemical and Engineering Aspects 273:29–34

Yamazaki V, Sirenko O, Schafer RJ, Nguyen L, Gutsmann T, Brade L, Groves JT (2005) Cell membrane array fabrication and assay technology. BMC Biotechnology 5:18

Yang R, Jin J, Chen Y, Shao N, Kang H, Xiao Z, Tang Z, Wu Y, Zhu Z, Tan W (2008) Carbon nanotube-quenched fluorescent oligonucleotides: probes that fluoresce upon hybridization. Journal of the American Chemical Society 130:8351–8358

Zotti G, Vercelli B, Berlin A (2008) Monolayers and Multilayers of Conjugated Polymers as Nanosized Electronic Components. Accounts of Chemical Research 41: ASAP Article, 10.1021/ar8000102

Chapter 8
Non-Conventional Generation and Transformation of Response

Most of our discussions in previous chapters were concentrated on sensors that can be excited by light and emit an informative light signal about the target binding. Meanwhile, this is not the only possibility for obtaining reporting information. Dyes, or *luminophores* in general, can also be excited in a chemical reaction (*chemiluminescence*), in a biochemical transformation (*bioluminescence*) and in a reaction at an electrode (*electroluminescence* and more specifically, *electrochemiluminescence*). The reporting can be provided by deactivation of the excited state, not only in the form of emission but also in the form of electron transfer to the conducting surface. This allows producing an electrical signal directly, avoiding emission and detection of light. Moreover, the emission of a miniaturized semiconductor or polymeric light source can be directly coupled with the sensing event. Such coupling can be provided with the response of a miniaturized detector.

Excitation via the *evanescent field* effect is a powerful tool to introduce spatial resolution into the sensor system; it can be combined with different sensing technologies. If introduced into heterogeneous assays, it allows a direct response to the target binding.

Though the systems exploring these possibilities develop quite rapidly, they cannot easily overcome the problems arising due to the low sensitivity of a produced or detected signal on the background of a high noise level, which is obviously present in miniaturized systems. Therefore, this chapter is complemented by a section describing the mechanisms of amplification and luminescence signals by coupling it with *plasmonic interactions*, which are, in a sense, also non-conventional.

8.1 Chemiluminescence and Electrochemiluminescence

Chemiluminescence is an emission that occurs as a result of a chemical reaction and commonly is observed in solutions, being started by mixing the reagents. In contrast, *electrochemiluminescence* is an emission coupled with a redox reaction at the electrode. Thus, it can usually be observed in a solution in a thin pre-electrode layer. Both of them are chemical reactions, in which the reagents are consumed and in which chemical stoichiometry is observed. Each product of this reaction appears in the

A.P. Demchenko, *Introduction to Fluorescence Sensing*,
© Springer Science+Business Media B.V. 2009

excited state and receives enough energy to emit light but, because of different coupled deactivation events, the quantum yields of this emission rarely exceed 1%.

8.1.1 Chemiluminescence

Chemiluminescence is also known as 'cold-light' emission. It occurs when an intermediate is formed in a chemical reaction that decomposes into a product with its electrons appearing in the excited state. Such species are able to relax to the ground state with the emission of light. A classical example of such a reaction is the reaction of *luminol* (5-amino-2,3-dihydro-1,4-phthalazinedione) and a strong oxidizer, such as hydrogen peroxide, in the presence of sensitizers, such as iron and copper (Fig. 8.1). A strong blue emission is produced in this reaction.

This reaction can be primarily used for the determination of *organic peroxides* (Baj and Krawczyk 2007) but is not limited to them. Any compound that influences the rate of reaction producing the luminescence can be determined and the major problem is the selectivity of this assay. It has been reported on the application of

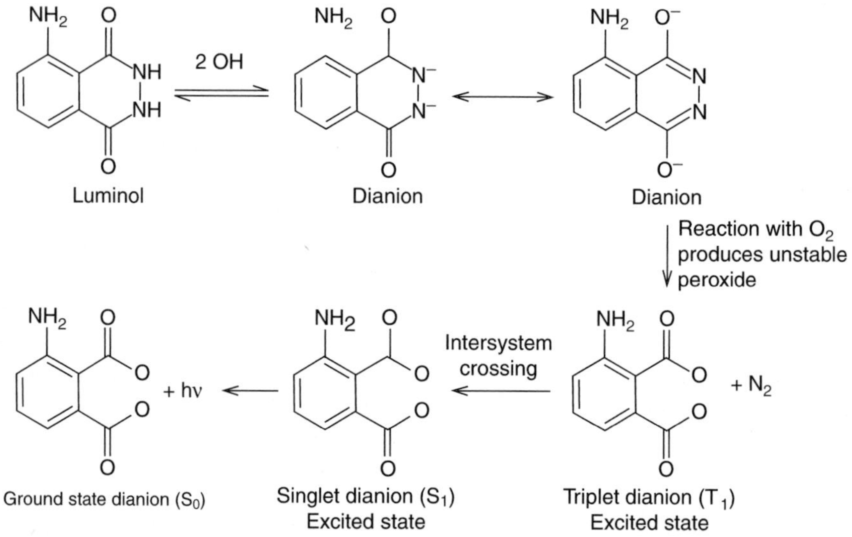

Fig. 8.1 The scheme of luminol transformations leading to chemiluminescence. First, luminol is activated with an oxidant to dianion. Usually, a solution of hydrogen peroxide (H_2O_2) and a hydroxide salt in water is used as the activator. In the presence of a catalyst (such as an iron-containing compound), the hydrogen peroxide is decomposed to form oxygen and water. The oxygen produced in this reaction reacts with luminol dianion to yield organic peroxide. The latter is very unstable and immediately decomposes with a loss of nitrogen to produce a 5-aminophthalic acid with electrons in an excited state. As the excited state relaxes to the ground state, the excess energy is liberated as photons, visible as a blue light (From Wikipedia)

this reaction for the determination of carbon monoxide in blood serum samples (Xie et al. 2007) and of sugars in food (Li and He 2007). It was found that silver, gold and platinum nanoparticles enhance chemiluminescence in a luminol – hydrogen peroxide system. This fact opens many possibilities for determining the compounds that not only inhibit the reaction, but also influence the binding of its substrates to the particle surface. The list of such compounds includes uric acid, ascorbic acid, estrogens and phenols (Xu and Cui 2007).

The method of chemiluminescence progresses by absorbing ideas from other luminescence techniques. Thus, a resonance energy transfer (FRET) between *chemiluminescent donors* and luminescent quantum-dots as acceptors has been suggested as an improvement of this methodology (Huang et al. 2006). Although high resolution in time is not applicable to this method, many attempts have been made to decrease the response time by decreasing the reaction volume and increasing the speed of mixing.

Interestingly, luminol is used with great success by forensic investigators for the chemiluminescent detection of *trace amounts of blood* left at crime scenes. The area under investigation can be sprayed with solutions of luminol and activator. The detection is based on the fact that iron cations are necessary components of the blood protein hemoglobin, which is present in all blood. They catalyze the chemical reaction of luminol. This leads to a blue glow that is easily detected in the dark, revealing the location of even infinitesimal amounts of blood.

8.1.2   Enhanced Chemiluminescence

The use of chemiluminescence in detection technologies is not limited by the direct influence of the target on the reaction producing the emission. The application in sensing can be realized by labeling the sensor unit with the catalyst and adding all the substrates of the reaction into the test medium. *Horseradish peroxidase* is an enzyme that is used in many technologies, including chemiluminescence. It transforms peroxides with the generation of oxygen. When it is coupled to one of the partners interacting in the sensing event and the substrate (the transformation of which changes its absorption or emission) is added, an amplified signal can be generated. When this enzyme is coupled to the receptor molecule, then the presence of this complex can be detected with great sensitivity in a micro-flow format (Yakovleva et al. 2003). In this technique, to produce the chemiluminescent signal the enzyme catalyzes the conversion of the added substrate (usually, a luminol derivative) into a sensitized reagent. Its further oxidation by hydrogen peroxide produces an excited species that emits light, allowing ultrasensitive detection. In detecting protein targets, a femtomolar level of sensitivity can be achieved.

Chemiluminescent detection increases the sensitivity of *immunoassays*. For instance, peroxidase can be attached to an antibody that specifically recognizes the antigen. Thus, in an ELISA assay with peroxidase labeling, the substitution of colorimetric detection by chemiluminescence allows a ten-fold increase in sensitivity (Zhang et al. 2006a).

Chemiluminescence is attractive in many respects but one point is not so pleasant for spectroscopists. On application, the emission intensity is often very low. In such cases with optical excitation, one has to do the obvious thing – increase the intensity of incident light. In chemiluminescence this cannot be done because this intensity is basically zero.

8.1.3   Electrochemiluminescence

Electrogenerated chemiluminescence (ECL), also called electrochemiluminescence, is chemiluminescence produced directly or indirectly as a result of electron transfer between an electrode and some solution species or species bound to the electrode surface (Kulmala and Suomi 2003). These electron transfer reactions generate excited states that emit luminescent light. Thus, this process is essentially the conversion of electrical energy into radiative energy.

Many dyes and metal complexes that produce common luminescence or chemiluminescence can also generate ECL (Richter 2004). One example is tris-(2,2′-bipyridine) chelate, $Ru(bpy)_3^{2+}$, which is already well known to the reader as the luminescent metal chelating complex (Section 4.2). The complexes of ruthenium(III), such as $Ru(bpy)_3^{2+}$, are excited along the one-electron oxidation-reduction pathway. Application of voltage to an electrode in its presence results in light emission that is similar to that observed on photoexcitation and can be detected at very low concentrations (10^{-11} M).

ECL reactions are rather efficient. Their *quantum yield* can be defined as the ratio of the number of emitted quanta to the number of redox events leading to the generation of excited states. The ECL reaction for $Ru(bpy)_3^{2+}$ has a quantum efficiency of 5%. This value is frequently used as a reference for the determination of quantum yields by a comparative approach (Richter 2004). The other remarkable feature of ECL reactions is their *reversibility*. In contrast to substrates of chemiluminescence reactions in solutions, $Ru(bpy)_3^{2+}$ is not consumed during the photochemical cycle, it relaxes to its ground state (Fig. 8.2), which allows a single molecule to participate in many reaction cycles.

The *indirect generation* of ECL is also possible. In this case, a 'co-reactant' is needed, which is a substance that attains strong oxidation or reduction potential in reaction at the electrode. The produced intermediate interacts with ECL luminescent species to generate their excited states. The use of a co-reactant is rather limited in aqueous solvents, since the available potential range in water is too narrow to generate the required energetic precursors (Richter 2004). A useful co-reactant in an aqueous medium is the oxalate anion and also tri-n-propylamine that work together with $Ru(bpy)_3^{2+}$. Usually, such reactions are performed at high and constant concentrations of the co-reactant.

Many inorganic complexes and organic molecules have been tested in ECL systems (Richter 2004), meantime $Ru(bpy)_3^{2+}$ remains the most requested luminophore for different applications. With its use as a single reporter it is not easy to explore multiple labeling or the introduction of an internal standard. By attaching suitable groups to the bipyridine moieties, it can be used to label antibodies for application

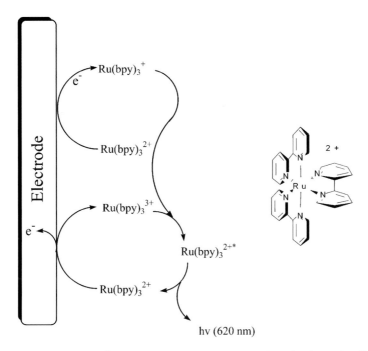

Fig. 8.2 Structure of Ru(bpy)$_3^{2+}$ and proposed mechanism giving rise to its electrochemiluminescence (ECL) emission. Ru(bpy)$_3^{2+*}$ represents an excited molecule that emits light (Reproduced with permission from Richter 2004)

in sandwich assays (Miao and Bard 2003) or for labeling the single-strand DNA to be used in hybridization techniques (Dennany et al. 2003). Introduction into the Ru(bpy)$_3^{2+}$ system of the ion-chelating groups allows obtaining the ion sensors operating similarly to that producing a fluorescence response (Richter 2004).

Many other versions of common fluorescence detection techniques have been adapted for the ECL technique (Fahnrich et al. 2001). ECL demonstrates remarkable flexibility in the direction of miniaturization and imaging. It can be used in the form of a microelectrode and an ultra-microelectrode technique to produce images on electrode surfaces and also as a light source for near-field optical microscopy (Zu et al. 2001).

8.1.4 Cathodic Luminescence

A special version of ECL is cathodic luminescence (Kulmala and Suomi 2003). Its mechanism is basically different from that discussed above. Instead of being based on the electron-transfer of chemiluminescent reactions of electrochemically generated reactants in solutions it involves the direct formation of excited states at the electrode surface. This type of high-voltage cathodic luminescence results from the injection of hot electrons into the aqueous electrolyte solution, with the formation of hydrated electrons (Fig. 8.3).

This technique, though totally dependent on the electrode phenomenon, is closer to common photoluminescence in several aspects. First, it allows spatial resolution and, in particular, the excitation of those luminophores that are *close to the electrode surface*, resembling evanescent wave excitation (see Section 8.5). Here the distance limit is about 50 nm from the tunnel emission electrode surface. This allows obtaining a reporter signal only from those labels that are attached to species adsorbed to the surface, whereas more distant label molecules are not excited. The other labeled species, remaining in solution, are not excited during the lifetime of the hot electrons, so they do not provide a signal. This allows carrying out homogeneous assays, which avoids washing-out the unbound species.

The other attractive feature of this technique is the possibility of a time-resolved study. Similarly to photoexcitation, an electric pulse on an electrode starts the emission that can be then observed as a *time-resolved decay*. This allows the detection of signals from several labels simultaneously and they may be resolved by differences in decay times. If it is necessary to resolve contributions from sensors emitting at

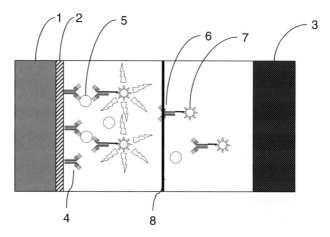

Fig. 8.3 The measuring principles of an immunoassay (sandwich format) employing thin insulating film-coated tunnel emission electrodes. The working electrode consists of a base material (1) and an insulating film (2) coated with antibodies (4) that are specific to an analyte (5). The counter electrode (3) does not participate in generating the response. The second antibody (6) specific to the analyte is attached to the label (7). This results in the formation of immunocomplexes (4—7) on the surface of the insulating film. The analyte level can be quantified after this reaction step by electric pulse excitation of the label molecules (7) involved in the complex formation. Line (8) marks the distance from the electrode, within which excitation is possible (Reproduced with permission from Kulmala and Suomi 2003)

different wavelengths, this possibility also exists. Wavelength and time discrimination can be combined in an efficient separation of signals emerging from different labels.

The special feature of hot electron injection into aqueous solutions is that luminophores having very different *optical and redox properties* can be simultaneously excited (Kulmala and Suomi 2003). This opens many possibilities for their selection. The absence of results from systematic studies does not allow one to draw general conclusions but since it has been reported that this type of electrogenerated emission can be obtained from organic dyes, metal chelates and also Quantum Dots, it can be expected that the range of applicable reporters is rather broad.

The most intensively studied in this application were luminescence metal complexes with aromatic ligands. An injection of hot electrons into an aqueous solution provides the possibility to excite them electrochemically by a *ligand-sensitized mechanism*. As in photoexcitation (Section 4.2), the ligand is first excited and then it transfers the energy to the central ion that finally emits light. The best studied in this respect are Tb(III) phenolic chelates. With a lifetime of 1.7–2.4 ms, these complexes allow the efficient use of time-resolved measuring techniques. It was stated that the time-resolved ECL measurements provide excellent signal-to-noise ratio and about the same sensitivity of detection as the time-resolved photoluminescence of lanthanide(III) chelates (Kulmala and Suomi 2003). This methodology has been successfully applied to immunoassays.

8.1.5   Solid-State Electroluminescence

Both chemiluminescence and electrochemiluminescence reactions occur in solutions. In contrast, electroluminescent materials, which emit light upon the application of electric voltage, need solid conduction connections for the electric current supply and therefore they are limited to solid films. They are used in light-emitting diodes and different screen displays but almost without direct application as receptors-reporters in sensing. This situation may change dramatically after finding the unique properties of several-atom gold (Au_n) and silver (Ag_n) *nanoclusters* (Section 4.5). These clusters behave like molecules, exhibiting discrete electronic states and electronic transitions between these states, giving rise to light emission (Lee et al. 2005).

This new class of single-molecule fluorophores has been created and electrically contacted in thin films to produce the first *electroluminescent single molecules* (Lee and El-Sayed 2006). A direct reporter of nanoscale charge injection and transport through discrete energy levels can result in the bright electroluminescence of silver ions. With such small devices, single-molecule light-emitting diodes (LEDs) and optoelectronic logic gates can be created. These experiments, utilizing the small size and quantum behavior of individual Ag nanoclusters, open the new field of single-molecule optoelectronics, with strong prospects in sensing technologies.

8.1.6   Essentials of the Techniques and Their Prospects

Simple and cheap chemiluminescent techniques have found many followers and are frequently used in clinical diagnostics. Electrogenerated chemiluminescence offers further increased simplicity in making the flow cells with sample injection unnecessary but requiring locating the receptors at the surface of conducting materials. Both of these techniques involve the generation of excited states by species that undergo highly energetic electron transfer reactions. In chemiluminescence, the process leading to the production of light is an essentially irreversible chemical reaction that has to be started by reagent mixing (usually, their injection into a flow cell). The intensity is measured in a steady-state regime and emission decay-associated time-resolved measurements here are not possible. The optical technique in this case is amazingly simple; it needs neither a light excitation unit nor wavelength filtering in the emission beam. The microscopic volume flow-injection cell can be located on top of the photomultiplier detector and this is the most important part of the instrumentation.

Since in electrogenerated chemiluminescence the emission is initiated and controlled by switching the electrode voltage, time-resolved or time-gated luminescence can be easily detected with reporters exhibiting long lifetimes. No separation of the bound and unbound target is needed, since only the target bound in a thin pre-electrode layer can be detected. Such a combination of spatial and temporal selection allows discrimination against background emissions without any wavelength selection. This allows very simple instrumentation.

Regarding the prospects, we have to consider that starting the response reaction 'by syringe' is less attractive than by light or by electrical pulse, especially concerning miniaturization and multiplex applications. Future developments will show if the sensor technologies based on chemically or electrochemically generated luminescence will represent strong competition to the techniques using optical excitation. Simplicity of instrumentation is an attractive feature but it cannot be considered decisive. The limiting factor for this technology at present is the rather small choice of available reagents that can be used as reporters.

8.2   Bioluminescence

Bioluminescence is also chemiluminescence but it takes place in various living organisms, the most known of which are fireflies. It is a rather widespread phenomenon in the living world, involving a broad range of species, from bacteria to fishes. In contrast to common chemiluminescence, bioluminescence reactions are highly efficient, with quantum yields approaching 88%. Their applications in sensing can be as exotic as the reactions themselves.

8.2.1 The Origin of Bioluminescence

Luminescent organisms synthesize specific substrate *luciferin* together with the enzyme *luciferase*. This enzyme catalyzes the formation of the intermediate complex of luciferin with ATP (adenosine tri-phosphate, the ubiquitous intracellular energy source). This complex oxidizes when combined with oxygen, producing a very strong 'cold' emission.

The general scheme of this two-step reaction is the following:

$$\text{luciferin} + \text{ATP} + \text{luciferase} \rightarrow \text{luciferase} \bullet \text{luciferyl adenylate} + PP_i$$
$$\text{luciferase} \bullet \text{luciferyl adenylate} + O_2 \rightarrow \text{luciferase} + \text{oxyluciferin} + \text{AMP} + CO_2 + light$$

The first step is the formation of an enzyme-bound luciferyl-adenylate and the release of the inorganic phosphate, PP_i. The second step is an oxidative decarboxylation of luciferyl adenylate that results in the production of oxyluciferin, AMP, carbon dioxide and emission of light. The chemistry of this transformation for the firefly's *luciferin-luciferase system* is presented in Fig. 8.4.

It should be noted that both luciferin and luciferase are generic terms for a substrate and its associated enzyme that catalyzes a light-producing reaction. Their particular molecular structures are variable among different species. The best studied and the most frequently used firefly luciferase is a monomeric protein with a molecular mass of 62 kDa. It emits yellow-green light with a peak emission at 560 nm.

One of the applications of this phenomenon is the *production of images* inside living objects without the necessity of applying an external light source (Frangioni 2006). The image can be created by spontaneous emission, resulting from a specific biochemical reaction.

8.2.2 Genetic Manipulations with Luciferase

By means of genetic engineering, the reaction giving rise to bioluminescence can be produced in different animals, e.g., in nude mice. Of course, making different animals luminescent can be good fun, resulting in possibly good applications.

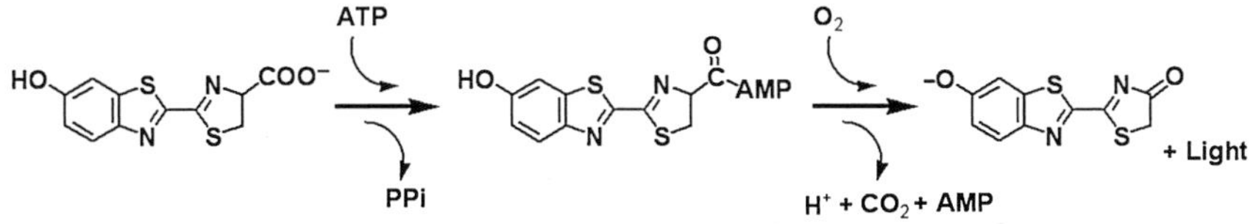

Fig. 8.4 The two-step chemical reaction catalyzed by firefly luciferase resulting in the generation of light emissions

Meanwhile, probably more important during the present step is to make luminescent whole-cell biosensors based on responsive bacteria and yeasts (Section 9.5). This allows achieving several aims.

(a) To provide *analytical detection* of compounds that are present in trace amounts but produce a significant biological effect, such as steroid hormones (Fine et al. 2006).
(b) To obtain an *integral response* to a particular substance on a cellular level. This response will contain information not only on the concentration of the substance but also on its influence on the cell, its efficiency, tolerance, etc.
(c) To provide the *parallel screening* of many different compounds, including those suspected as hazards and those tested as potential drugs, on the level of cellular effects.

The milestone for this trend could be the design of an efficient 'mini-luciferase', a small fragment of this enzyme that could retain the catalytic properties but be more convenient in use. Some success with this trend has already been achieved (Paulmurugan and Gambhir 2007).

8.2.3 Bioluminescence Resonance Energy Transfer

The excited states producing bioluminescent emission can also transfer the energy to other emitters in a mechanism based on *Förster resonance energy transfer* (FRET). This mechanism is of general origin and does not involve any specificity in the generation of excited states, whether it is by light or by chemical reaction. In this regard, the new term, 'bioluminescence resonance energy transfer' (BRET), which was introduced by some authors to describe this type of energy transfer seems to be incorrect. For sensing technologies, this type of energy transfer is interesting, since it can transform many fluorophores that are useful in sensing into a new class of bioluminescent probes.

Particularly, this opens the interesting possibility of activating the sensors via an internal bioluminescent molecular 'light source'. This idea has been efficiently realized with Quantum Dots (So et al. 2006). '*Self-illuminating Quantum Dots*' that represent their hybrids with luciferase have been produced and tested. Since the excitation source in this case is not needed, the Dots incorporated into the cells provide much more efficient sensing than in a common manner of excitation. 'From inside' they are excited with a great selectivity, which allows virtually eliminating the light scattering and autofluorescence (of cellular pigments, porphyrins and flavins).

Another interesting application has been found in research with the aim of determining the *interactome* (the whole pattern of interactions of macromolecules inside the living cell). A bioluminescent luciferase can be genetically fused to one candidate protein and a green fluorescent protein mutant fused to another protein of interest. Interactions between the two labeled partner proteins can bring the luciferase and green fluorescent protein close enough for resonance energy transfer to occur, thus changing the color of the bioluminescent emission. The energy transfer from the bioluminescent donor should be particularly useful for testing the protein interactions within a variety of native cells, especially with integral membrane proteins or proteins targeted to specific organelles (Xu et al. 1999).

The molecular imaging of protein-protein interactions in living cells modulated by binding small molecules raises not only theoretical but also practical interest in clinical diagnostics and the development of new drugs. For *in vivo* monitoring of estrogen-like compounds, the homogeneous assay can use the combination of luciferase with a protein from the green fluorescent protein family. A procedure for evaluating the presence of estrogen-like compounds has been developed and optimized. The estrogen alphareceptor (ERalpha) forms homodimers as a result of binding the estrogen-like compounds, such as 17-beta estradiol. The fusion of the bioluminescent donor and acceptor to monomers allows one to visualize the estrogen binding (Michelini et al. 2004).

8.2.4 Prospects

The attractive feature of bioluminescence is the possibility of providing target detection and imaging inside the living cell in the absence of external excitation (Roda et al. 2004). This requires genetic manipulation of the luciferase enzyme and the presence of all necessary substrates of the catalyzed reaction. The encountered difficulties have been successfully surpassed, so that even 'mini-luciferase' and fragments of this enzyme, which restore the enzyme activity upon assembly, are now available. With the aid of this tool, bacterial, yeast and human cultured cells can be transformed into efficient 'whole-cell' biosensors (Section 9.5). Sensing inside the cell can occur according to the energy-transfer mechanism, which does not require external excitation and very promising results for this trend have already been obtained.

8.3 Two-Photon Excitation, Up-Conversion and Stimulated Emission

Being well-known to physicists, the phenomena observed upon manipulation with optical flux density and with two-step transition to the excited state followed by spontaneous emission from this state have been almost forgotten in sensor development. Now the situation has changed dramatically and the great advantages in the exploration of these phenomena have started to find proper application.

8.3.1 Two-Photon and Multi-Photon Fluorescence

An interesting phenomenon is behind this technique. A molecule can be excited not only by the light quantum with the energy that corresponds to the energy of its electronic transition, but also by *two or more quanta*, the total energy of which corresponds to the energy of this transition (Fig. 8.5). For such summation of their energies these quanta should appear together in the right place at the right time and their simultaneous absorption should occur in a single event, which is very rare. To increase its probability, high photon flux energy is needed. The light beam has to be focused into a very small volume and the high temporal flux density can be achieved by the application of ultra-short pulses. For the present sensing technologies, only excitation with two or three photons of equal energies is practically useful, since this can be provided with one laser. The energy of two-photon excitation has to correspond to one-half of the electronic transition energy, so that to excite the dye absorbing light at 400 nm (25,000 cm^{-1}), a laser emitting at about 800 nm (12,500 cm^{-1}) should be applied (So et al. 2000).

What is the advantage of this technique? In contrast to common single-photon excitation with the emission changing linearly with the intensity of excitation light, the fluorescence emission on two-photon excitation *varies with the square power* of the excitation intensity. This quadratic relationship between excitation and emission allows focusing the excitation power to a very small volume without observing the emission outside this volume. In this way, we achieve a very high spatial resolution. Species located in very small focal volumes (counted in femtoliters!) can be selectively excited, while the rest of the sample remains in the dark. With these developments, new possibilities are opened in three linked areas: fluorescence microscopy (see Section 11.1), detection of single molecules (Section 11.2) and analytical methods that enable small-volume detection.

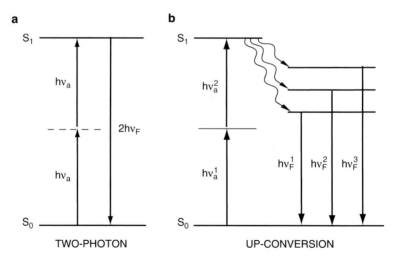

Fig. 8.5 Energy level diagram for two-photon excitation (**a**) and for up-conversion in lanthanide-doped crystals (**b**). Note that in the first case the 'intermediate' energy level is 'virtual', non-existent in reality and only the precise matching of two quanta in time and space may result in the addition of their energies. In the second case, the intermediate state is the real long-living state and electrons excited by low-energy quanta can reside in it while waiting for the second quantum to be excited to a level of higher energy

In addition to these attractive features, the problems with *scattered light* are eliminated. This is because the scattered light appears at the wavelength of excitation, which is commonly in the near-IR; this is quite a different spectral region than that from which fluorescence is observed. A high penetration of light of 800–1,000 nm in human tissues together with the possibility of activating highly localized photochemical reactions, opens new possibilities for non-invasive clinical diagnostics and the photodynamic treatment of diseases (Papkovsky et al. 2000).

To selecting the dye for two-photon spectroscopy or microscopy, one has to know that in these excitation conditions some properties are different from those observed at excitation by single photons. Due to different quantum mechanical selection rules, the two-photon and single-photon excitation spectra are not equivalent and for two photons they are often much broader. The relative molar absorbances of many dyes may exhibit an unexpected variation: high brightness with single photons may transform to a low brightness at two-photon excitation. Examples of opposite behavior are also known. A special effort is being made to synthesize dyes with especially high two-photonic absorbance. Particularly, some styryl dyes were found to be good two-photon emitters for DNA detection and imaging (Yashchuk et al. 2007). Recently, a very detailed study of both symmetric and asymmetric cationic cyanine dyes has been performed and general regularities of their structure-property relations established. Enhancing the two-photon absorbance can be achieved by extending the polymethine chain and also by an increase in the donor strengths of the terminal groups (Fu et al. 2007).

Recently, the study of the two-photonic behavior of various conjugated polymers and nanoparticles has started with very promising results (Drobizhev et al. 2005). A strong enhancement of two-photon absorption can be achieved with fluorescent-conjugated dendrimers (Varnavski et al. 2002). Their application in cellular studies is expected in the future.

Usually, the two-photon fluorescence is excited by *ultra-short laser pulses*. As we have seen above, this is not because the high time resolution is actually needed but because the highest density of photons can be concentrated within the shortest pulses, increasing the probability of two-photon absorption. The ideal, though expensive, possibility for this is to use the basic frequency of a Ti-sapphire laser, which is able to provide pulses as short as 100 fs in a tuning range of 800–1,000 nm. This allows exciting all light-emitters that possess absorption bands in the violet-blue spectral range.

Lasers with picosecond and nanosecond pulse widths can also be used but less efficiently. There is a much broader choice of them, starting from mode-locked Nd:YAG and the dye lasers that can be pumped by Nd:YAG and ending with recently-developed pulsed diode lasers. Lasers with a long pulse duration may not only decrease the peak pulse density but also increase the possibility of up-conversion (excitation from excited state), which may decrease the quantum yield. These effects depend upon the applied dye.

Surprisingly, the sensor technology based on two-photon excitation can be realized in a simple and cost-effective way. A group of Finnish researchers (Hanninen et al. 2000) used a microchip Nd:YAG laser producing sub-nanosecond pulses for the excitation of fluorescence and *latex microparticles* were used as matrices for carrying the binder units to detect target binding. When a microparticle appears in the focal volume of two-photon excitation, the confocally arranged scattering detector monitors its arrival and then the two-photon excited fluorescence measurement is triggered. The signals from the particles are recorded one by one and their statistical analysis performed. Such a procedure allows avoiding any separation or washing steps (Fig. 8.6).

This methodology has been demonstrated for immunoassays, pathogen detection assays (viral assays) and some research assays and its success has initiated many suggestions for its application on a larger scale. The microparticles can be encoded with different luminescent dyes and thus separation-free multi-target analysis becomes possible in a single assay.

8.3.2   Up-Conversion Technique with Nanocrystals Possessing Lanthanine Guests

We now introduce the reader to a methodology that is also able to obtain visible emission with excitation in the near-IR but is based on a mechanism that differs entirely from that of two-photon excitation. Due to this difference, it cannot operate

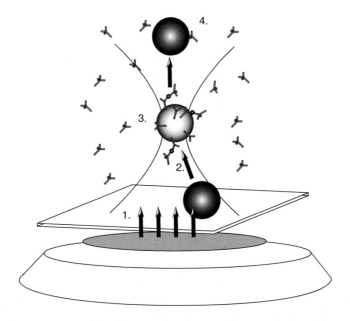

Fig. 8.6 The operation principle of a sensor based on two-photon excitation. A latex particle is pushed by the optical forces of excitation light (1, 2) and moving through a very small (1 fl) detection volume (3) of two-photon excitation, generating the fluorescence response signal. Once measured, the particles are pushed away by the same optical forces (4) (Courtesy of Prof. Pekka Hänninen, University of Turku)

with common organic dyes or nanoparticles and is limited to very special materials. The physical term *'up-conversion'* means excitation from already achieved excited states. Thus, an excited molecule can absorb additional light quantum to be raised to a higher energy level.

This effect is frequently used in physics for measuring the very short lifetimes of these states. In this focus, it may not be interesting for sensing, at least at the present stage of technology development. There are luminescent materials that allow exploring this in a very efficient way, however.

In ceramic crystalline host matrices doped with trivalent lanthanide ions, such as Yb^{3+} and Er^{3+}, two important properties, a *variety of electronic excited states* and the *long lifetimes* of these states uniquely meet. The long-lifetime excited states can operate as metastable states excited from a ground state to be excited again with the gain of energy to reach the emissive states (see Fig. 8.5). Here, the site being excited to a low-energy electronic state can wait until the next quantum brings it to a higher excited state, from which, commonly after relaxation to lower levels, the emission occurs at higher energies. Therefore, this emission is free from the restrictions imposed on two-photon excitation requiring very high irradiation densities and can be excited by conventional light intensities.

Except for such rare earth crystalline materials, no other presently known material exhibits such properties. Thus, on application of the sensing technique to composite and biological systems, even in the presence of many contaminants, there will be *no background emission*. Such emission is simply removed spectrally, without the need for temporal resolution. Due to such sequential absorption, the excitation energies are sufficient to convert near-IR excitation light into emission at visible wavelengths. Thus, the incorrectly named 'anti-Stokes' emission appears to be shifted dramatically (but not necessarily with the doubling of energy) to shorter wavelengths with respect to excitation. For example, this emission can be excited by an inexpensive laser diode at 980 nm and measured at different wavelengths, from green to red. The valuable properties of lanthanide ions, such as their narrow emission spectra and long-lifetime decays, are retained (De la Rosa et al. 2005).

Excitation at near-infrared and emission at red wavelengths are particularly attractive for the analysis of strongly colored and fluorescent samples, which are often of concern in clinical immunoassays and in high-throughput screening (Kuningas et al. 2005). Particularly, this approach allows sensor applications to the *whole blood*, so that blood pigments do not interfere with assays.

The straightforward application of nanoparticles made of up-converting materials is in cell and tissue *imaging*. With their aid, the high penetration depth of near-IR light, together with the absence of light-scattering and fluorescence background, allows obtaining ideal images in the visible (Chatterjee et al. 2008).

No less fascinating are the applications of up-conversion emitters as *FRET donors*. Due to extremely sharp and narrow emission bands characteristic of luminescent lanthanide ions, no donor emission can be detected at the wavelengths of the sensitized emission of the acceptor. Also, no acceptor can be excited directly because the excitation is in the near-IR. This means that the donor and acceptor emission channels do not overlap, which is almost impossible to achieve with organic dyes.

This property can be used for precise ratiometric FRET measurements in the steady state. As an example, this version of FRET was applied to a competitive homogeneous immunoassay for 17 beta-estradiol (E2) in serum, using an up-converting nanocrystal as the donor and a small-molecular dye as the acceptor (Kuningas et al. 2006). The detection limit was at sub-nanomolar concentrations. In general terms, the FRET technique based on up-converting particles as the donors has unique advantages compared to the present homogeneous luminescence-based methods and can enable an attractive assay system platform for clinical diagnostics and for high-throughput screening approaches.

The large size (>100 nm) of the presently available up-converting nanoparticles limits some applications. The limiting factor in the sensitivity of assays is also the size and shape distributions of these crystals, leading to the distribution of the reporter signal. The development of novel methods of obtaining monodispersed nanocrystals (Boyer et al. 2007) is expected to solve this problem.

8.3.3   Sensors as Lasers and Lasers as Sensors

Many fluorescent molecules and nanoparticles exhibit an *amplified stimulated emission* (ASE) in solutions under extensive illumination. This phenomenon is in the background to the operation of dye lasers. Under intensive light (pumping by external laser) the reversal in the population of the ground and excited state can be achieved. The depopulation from a highly populated excited state produces bright emission with a very characteristic spectrum (Fig. 8.7). On the background of the broad and structureless 'normal' emission band there appear very sharp and highly intensive lines, due to the generation of amplified emission. Commonly, it is a single line and its position in comparison to a band maximum of normal emission is shifted towards longer wavelengths. If the responsive dyes or nanoparticles are part of a laser cavity, then their linear single photon response can be replaced with a nonlinear one with a dramatic increase of intensity due to the amplified emission of many photons.

For a long time the ASE phenomenon has not been requested in sensor technologies. In the meantime, it can be applied in a straightforward manner if a system is found in which a sensing signal can be introduced as a switch between spontaneous and stimulated emissions. Since stimulated emission can be produced by two-photon excitation (He et al. 2006), there is in principle a possibility to introduce spatial resolution into this interesting methodology. Unfortunately, many organic dyes, the ASE of which has been studied in most detail, are not attractive for this application because of their low photostability.

The solution of the problem of signal transduction in sensor applications of ASE emitters can be found with CdSe nanoparticles (Somers et al. 2007), as shown in Fig. 8.7b. The appearance of sharp lines in the emission spectra above the minimum pump energy at which ASE can be observed (*the lasing threshold*), can be referenced against the intensity of normal emission at other wavelengths. The transduction mechanism in sensing becomes the influence of the target on the threshold position. The mechanism for increased ASE sensitivity can be mediated via the influence on

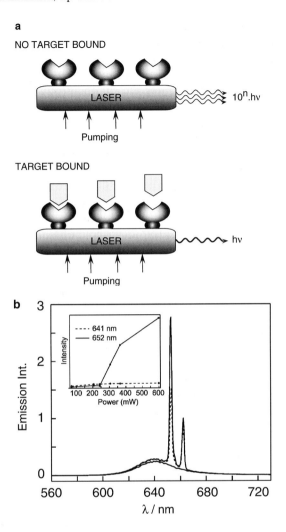

Fig. 8.7 Application of amplified stimulated emission in sensing. (**a**) Design concept to achieve sensing on short-length scales. A single photon emitting center is replaced by a laser, which can be turned off or on by the recognition of a target. The laser cavity is pumped with intensive light to produce the level inversion between ground and excited states. One of the possibilities is to couple the target binding with a strong quenching effect that destroys the level inversion and removes amplified stimulated emission. (**b**) Fluorescence from the CdSe nanocrystals showing normal and amplified stimulated emissions at intensive excitations. The spectrum transforms into discrete sharp lines once the lasing threshold is crossed. The insert shows the non-linear increase in intensity at a threshold pump intensity at 652 nm while the fluorescence at 641 nm shows a linear response. The sensor response could be in the appearance-disappearance of the ASE peak (Reproduced with permission from Somers et al. 2007)

excited-state kinetics. The lasing threshold depends strongly on the material emission lifetime: short lifetimes result in less intensive accumulation of excited-state species and thus in higher lasing thresholds. Longer lifetimes result in lower lasing thresholds due to the higher probability of populating the excited state.

In *conjugated polymers* with a long-lifetime emission, more emissive excitons are generated. The introduction of a quencher decreases the exciton lifetime, thereby resulting in increased lasing threshold. The amplified sensing response can therefore be obtained by pumping the conjugated polymer material close to the intrinsic lasing threshold and observing the switching between spontaneous and stimulated emissions.

These ideas were practically realized in the design of a sensor responsive to vapors of *trinitrotoluene* (TNT), which is a dangerous explosive that can be used in terrorist attacks. In specially designed conjugated polymer films, the ASE observation allowed increasing its sensitivity to TNT manifold (Rose et al. 2005). The adsorption of the trace vapors of these explosives on thin polymer films introduces non-radiative deactivation pathways that compete with populating the excited state and thus with the stimulated emission. As expected, the induced cessation of the lasing action and associated sensitivity enhancement is most pronounced when the films are pumped at intensities near to their lasing threshold.

8.4   Direct Generation of the Electrical Response Signal

From reading the previous sections, we can derive that there are many alternatives to common optical excitation by visible light. Here, we will provide a closer look at the side of detection of the reporter signal. Are there alternatives to the generation of light emission that has to be detected by electronic recording? It has to be acknowledged that electrochemical methods, being strong competitors to fluorescence (luminescence) sensing, offer the most straightforward approach to signal conversion, from target interaction to electrical signal in an electrode-coupled reaction. Is there any possibility to merge these technologies in order to select the optimal solution? These possibilities will be discussed below.

8.4.1   *Light-Addressable Potentiometric Sensors (LAPS)*

In the dark state of semiconductor layers the electric conductivity is absent because of the absence of free electrons. They appear at light illumination and this may induce the photocurrent. If the semiconductor is represented by a thin film exposed to a medium with a variable concentration of protons or ions, due to the polarization of the charges, the electrical conductance can change. The electric current itself and its variation can only be observed if the film is illuminated by light and the energy of the light quanta is sufficient for an electron transfer to a conduction zone. The number of these electrons will depend on the polarization of the semiconductor surface that can be produced by ions in the test sample. Thus, with laser excitation we can obtain a direct electrical response.

Light-addressable potentiometric sensors are built based on this principle (Fig. 8.8). The sensor plate is made of a thin semiconductor layer with an insulating layer on the front surface exposed to a tested medium and an ohmic contact on the back surface.

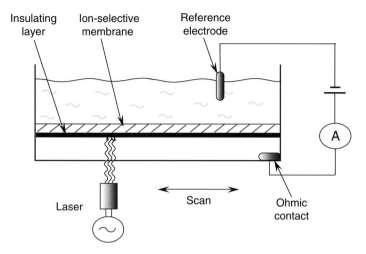

Fig. 8.8 Schematic presentation of a LAPS device. The laser scans the semiconductor thin film which, through an insulating layer, is exposed to a sample that can induce the polarization effect by a variation of pH or concentrations of ions. The laser scans the whole semiconductor surface and only an illuminated site can produce the response

The laser-induced photocurrent is usually modulated with a frequency typically in the range of 10–20 kHz that requires time-integrated measurements. The sensing surface of the LAPS can be modified to implement sensitivity and selectivity to various chemical species. It can be represented by an ion-selective membrane. This membrane can be substituted by an arrayed combination of materials, thus a single sensor plate can be used as a multi-analyte sensor (Yoshinobu et al. 2005). Scanning with a laser will provide the necessary spatial selectivity, producing an image that can achieve a ~1 μm resolution.

If a living cell is deposited on top of an insulating layer, its electrical activity can be detected (Xu et al. 2006). This technique can be used to detect changes of the action potential towards different drugs. By scanning the laser along the LAPS sensor, the surface potentials at any desired position can be detected.

8.4.2   Photocells as Sensors

Photoinduced electron transfer (PET) is an important signal transduction mechanism that is in the background of many sensor technologies (Section 6.1). Being a short-distance effect, it usually involves the electron movement from one molecule or its fragment to the other molecule or fragment and results in quenching. Usually, the reporting signal is optical, it is the ON-OFF switching in emission. We observe that the electron can also be transferred to the *conducting surface*, such as the metal surface. Thus, in principle, an *electric signal* can be generated by the transfer of electrons from an excited dye.

In semiconductors the electrons injected into their conductance bands can generate electric currents. Devices based on this principle have been developed for solar energy conversion (Gratzel 2007). Absorbing the energy of light, different organic dyes, Quantum Dots or luminescent metal chelates serving as the light harvesters provide a heterogeneous electron transfer from their electronically excited states into the wide conduction band of the semiconductor. The electrons then migrate to the collector electrode generating an electrical current. The efficiency of these solar energy transducers has reached the ten percent level and continues to grow. The wide band gap oxide semiconductors of mesoporous or nanocrystalline morphology, usually nanocrystalline TiO_2, have been selected for their optimal use. These photocell devices received the name *Gratzer cells* (Peter 2007).

The efficiency of these cells depends on many different factors and can be modulated by variables that allow coupling with target receptor functionalities to produce the sensors. Directly involved could be the dye-semiconductor distance, the dye orientation with respect to the surface and the medium conditions influencing, via the classical Pekar-Markus factor, the electron transfer rate. In the potentially possible design of efficient sensors these factors could be the points of application of the transduction signal. There are many possibilities to control and amplify this signal. It is quite possible to upgrade this system with FRET donors and modulators, to use the receptor-reporter mechanisms that are already established in sensing.

Summarizing, we have shown only two possibilities for merging photochemistry and electrochemistry in order to substitute the optical response into an electrical one. This field is open for active exploration. It should be noted that the use of an electrode instead of a light detection unit leads to smaller instruments and reduces the cost of their production. Microfabrication is more suited for the construction of electrochemical than of optical detectors. This raises a strong motivation to look for simpler and cheaper detection methods than the detection of light.

8.5   Evanescent-Wave Fluorescence Sensors

Evanescent wave excitation is primarily the possibility to achieve spatial selection in optical excitation, since this technique allows exciting those species that are located close to the interface between the two media. The power of this approach is in its ability to be combined with different sensor technologies, offering their improvement and, often, their dramatic transformation.

8.5.1   Excitation by the Evanescent Field

The phenomenon of the *refraction of light* at the interface between the media of different refractive indices is known from textbooks. The refraction strongly depends on the angle of the incident beam, so that with the increase of this angle

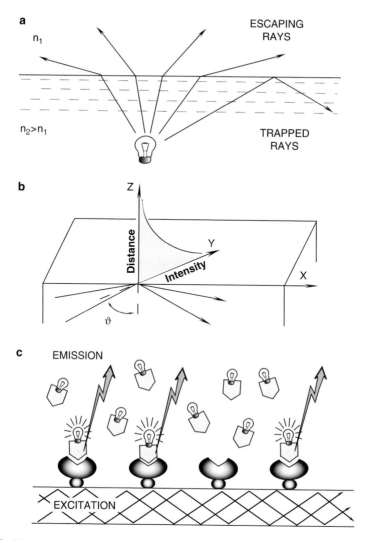

Fig. 8.9 The total internal reflection and its application in sensing using optical waveguides. **(a)** The transmission and reflection of light entering from a medium with a higher refractive index n_2 to a medium with a lower refractive index n_1. The rays incident at angles higher than a critical angle $\theta_c = \sin^{-1}(n_1/n_2)$ are totally reflected. **(b)** The rays internally reflected into the medium with n_2 produce an evanescent wave in the medium with n_1. This wave decays exponentially as a function of distance from the surface, on a nanometer scale. **(c)** The evanescent wave is formed in an optical waveguide by internally reflected light rays propagating inside it. The generated evanescent wave is able to provide optical excitation of dyes located close to the surface. This phenomenon is used in sensing by selective detection of surface-immobilized labeled species

the condition appears when the light is totally reflected (Fig. 8.9a). This occurs when the angle at which the light propagating in a medium of a higher refractive index n_2 encounters a medium of a lower refractive index n_1, is higher than a certain critical angle θ_c. The latter is determined by the equation $\theta_c = \sin^{-1}(n_1/n_2)$. It is interesting to note that the very attractive sparkling properties of finely processed diamonds are due exactly to this phenomenon.

However, the *total internal reflection* is not truly total. A small amount of electromagnetic field propagating with the light frequency in the medium with n_2 still penetrates a short distance into the medium of lower n_1 in the form of a standing wave known as the *evanescent wave* (Taitt et al. 2005). This evanescent wave is identical in frequency to the incident light (Fig. 8.9b).

The evanescent wave decays exponentially with the distance from the surface. In practically realizable situations, it propagates to a distance of 100–200 nm, which depends on refractive indices, the incidence angle and the wavelength. Its energy is sufficient to excite the fluorescence of species located within the range of action of this field.

When special materials and a special configuration of the excitation beam were developed, this phenomenon found important applications in sensing. The technique used for this is called *evanescent field excitation* (by the physical principle) and *total internal reflection* spectroscopy/microscopy (by the method of its realization). This technique can be realized on optical waveguides (see Section 9.2) and in fluorescence microscopy format (Section 11.1).

The basic principle of the operation of evanescent-wave sensors is illustrated in Fig. 8.9c. The light propagates along an optical waveguide without leaving it, exhibiting total internal reflection from its walls. The sensors (that may be composed of receptor units only) are attached to the waveguide surface and extend into the solvent. The difference between the refractive indices of a solvent water medium (n_1 = 1.33–1.37) and fused silica from which the optical waveguides are usually made (n_2 = 1.46) is quite significant for providing this optical trapping. The labeled targets that are bound to the sensor can be detected by their fluorescence emissions. If the labeled species are not bound and are not under the influence of an evanescent field, they are not excited and remain in the dark.

Since only the bound species demonstrate the fluorescence response and those remaining in the solution do not respond for providing or modulating the reporting signal there emerges the possibility to avoid washing-out the unbound species, since being at a sufficient distance, they remain in the dark. With the exploration of this effect, many different sensor configurations based on different principles can be realized. For instance, the labeled species is not necessarily the target but the competitor. Another possibility is to make the sensors attached to waveguides fluorescent and the target or its competitor in this case can be the quencher. Moreover, under specific conditions of illumination, this effect can be used in microscopy (Section 11.1). In this case, only the area of the studied object that is attached to a glass support will selectively produce the response.

8.5.2   Applications in Sensing

Evanescent wave sensing technology cannot be considered separately from other developments in fluorescence sensing. Moreover, it can be used as a platform for the adaptation of different, already developed approaches and allows their substantial improvement. Since all these sensors can measure surface-specific binding events by collecting the emission selectively from the surface, many heterogeneous assays can be transformed into direct sensors. The versatility of this technology is seen from its application by using both fiber-optic and planar array fluorescence sensors (Section 9.2).

The most extensive application that this technology has found is in *immunoassays*. Evanescent wave biosensors utilizing antibody-based sandwich fluorescence assays have been described for the detection of gram-negative and gram-positive bacteria, viruses and different proteins, including toxins. In fact, any compounds to which immunoassays are applicable can be determined by the application of this method. Competitive and displacement assays can be established for the detection of small molecular weight compounds, such as hormones and neurotransmitters. The reader can find many references in a recently published review on this field (Taitt et al. 2005).

One of the most interesting application areas is the *hybridization*-based detection of target DNA/RNA molecules. It has been shown that the molecular beacon technique is also quite applicable in this case (Liu and Tan 1999). Biotinylated molecular beacons have been designed and immobilized on an optical fiber core surface via biotin-avidin or biotin-streptavidin interactions. They can be used to directly detect, in real-time, the target DNA/RNA molecules without using competitive assays. The sensor demonstrates rapid, stable, highly selective and reproducible performance.

Spatial resolution offered by this technology is extremely attractive for assays in which the binding of a labeled target or competitor can be measured with the excitation confined to a narrow pre-surface layer. The other application is in cellular research with the aid of a specialized *evanescent field microscope* (Section 11.1). In general, evanescent-field sensing can be considered as a universal platform that may encompass many sensing technologies targeted to many analytes and provide, with spatially selective excitation, very sharp images.

8.6   Plasmonic Enhancement of Emission Response

Solid supports are often used for the immobilization of sensor molecules and their integration into nano-composites. Should they play only a passive role? Below we will see that the role of the supporting surface can be very active in the generation and modulation of sensor response. This is because of specific *metal plasmonic effects* that can modulate all kinds of excited-state phenomena. For years their role was in shadow because of frequently observed fluorescence quenching of organic dyes at the interface with metal. They occur due to excited-state electron transfer (PET) to the metal conduction band operating at short distances (Section 6.1), with the possible involvement of FRET, the operation of which extends to distances of several nanometers (Section 6.2). Quite recently, it was recognized that plasmonic effects near metal surfaces not only allow the emission to be observed but can produce a strong enhancement. These effects have become a hot topic both in research and in sensor development, promising many new possibilities. They have developed into methods which, by increasing dramatically the total emission per reporter molecule, provide an increased sensitivity and a wider dynamic range for chemical analysis, medical diagnostics and *in vivo* molecular imaging.

8.6.1 *Surface Plasmon-Field Enhanced Fluorescence*

In the previous section, we discussed the effect of fluorescence excitation by evanescent wave. This wave appears at the interface between media of different refractive indices and extends into the medium with a lower refractive index when the incident light beam is directed at an angle higher than critical and all incident light is reflected. It is also known that the evanescent field is dramatically enhanced when a thin (about $50\,\mu m$) layer of noble metal (gold or silver) is deposited on the solid glass surface. Since this effect depends on the refractive index in the contacting medium, a variation of this index can be detected in a thin solvent layer close to the surface. Such detection is used in the *surface plasmon resonance* (SPR) sensing technique.

The enhancement of the evanescent field occurs due to resonant coupling between the incident radiation and the surface plasmon wave. The accumulation of molecules or particles at the metal surface in the course of sensor-target interaction causes the interfacial refractive index to change. This change can be detected by the shift of the resonance angle for surface plasmon excitation, making the SPR method the most popular label-free technique for affinity-based sensing. In conventional SPR, the detection occurs by observing the resonance angle shifts or, in simplified instrument versions, by the change of reflectance monitored at a fixed angle of incidence that is close to the resonance angle. The well-known resonant enhancement of Raman scattering bands is also related to this phenomenon.

With the knowledge of these findings, the reader must come up with the idea of combining the benefits offered by evanescent-wave excitation with the possibility of plasmonic enhancement of this emission. The technical realization of this possibility has led to the establishment of *surface plasmon field-enhanced fluorescence spectroscopy* (SPFS). It exploits the strong SPR-generated evanescent field for the excitation of surface-confined fluorescent dye (Ekgasit et al. 2004). The resonant excitation of an evanescent surface plasmon mode is used here to excite the dyes chemically attached to the target biomolecules and located close to the metal/solution interface. In this area, the dye is exposed to the strong optical fields, giving rise to significantly enhanced emission. The emitted fluorescence photons are then detected, which can be used for determining the concentration of the analyte. The use of fluorescence detection schemes in combination with the resonant excitation of surface plasmons has been shown to considerably increase sensitivity for bioana-lyte detection (Neumann et al. 2002).

Since it is technically possible to monitor the *reflected light* (for obtaining the SPR signal) and *emitted light* (as in evanescent-wave fluorescence sensing) simul-taneously from the same surface, this approach can provide two complementary channels of information. Many benefits can be obtained from such a marriage. For instance, unlabeled targets of large mass, producing a strong change of refractive index in the interfacial region, are active in SPR. In contrast, the small-size fluores-cent dyes produce a strongly amplified effect in fluorescence in the same region, without a detectable change in the SPR signal. The unwanted quenching effect of PET to the metal layer can be made beneficial; it can be transformed into a dis-tance-dependent sensing signal with a strongly enhanced dynamic range of response. In this configuration, the sensitivity of the fluorescence-based response can be much higher than that of SPR. Both of them allow not only equilibrium measurements but also the study of association-dissociation kinetics.

An interesting property of surface-plasmon fluorescence spectroscopy is the ability to generate an *angle-resolved emission*. It has been shown that by micro-scopic observation of Quantum Dot conjugates with DNA on gold films, multicolor images can be obtained. They can be used for the simultaneous qualitative analysis of QD-conjugated analyte DNA strands (Robelek et al. 2004). In a later report (Gryczynski et al. 2005), it was shown that upon excitation of Quantum Dots the surface plasmons emitted a hollow cone of radiation (due to symmetry conditions) into an attached hemispherical glass prism located at a narrow angle. This direc-tional radiation preserves the spectral properties of QD emission and is highly polarized, irrespective of the excitation polarization. The development of these new techniques in combination with the use of highly photostable Quantum Dots addresses the issue of bioassay sensitivity and offers substantial improvements.

8.6.2   Enhancement of Dye Fluorescence Near Metal Nanoparticles

If, instead of a solid layer, the dye interacts with metal nanoparticles, additional important factors can modulate emission in this system. Particularly, the scattering of incident radiation on nanoparticles and the already described (Section 4.5) plas-mon confinement effect, induce new, unusual features to emitted radiation. A com-prehensive understanding of this inherently nanoscale process emerges. It leads to design strategies for optimizing the fluorescence enhancement and its use in many applications.

The basic effect of the enhancement of emission intensity by neighboring metal material is very different from that commonly observed in fluorescence, where the free-space quantum yield, Φ_0 and lifetime, τ_0, are observed to change in the same direction. As described in Section 3.1, the following equations relate Φ_0 and τ_0 to radiative and non-radiative rate constants:

$$\Phi_0 = k_r / (k_r + \Sigma k_{nr} = 1 / [1 + (\Sigma k_{nr} / k_r)] \tag{8.1}$$

$$\tau_0 = 1 / (k_r + \Sigma k_{nr}) \tag{8.2}$$

Thus, in a common free-space case, the ratio of *radiative* (k_r) and the sum of all *non-radiative* (k_{nr}) *rate constants* of excited-state decay change in such a way that, being highly fluorescent, the dye should emit much faster than all other deactivation processes (with the rates Σk_{nr}) occur. In contrast, plasmonic enhancement modulates the *intrinsic decay rates*. It was found that in proximity to metal structures, the changes of Φ and τ (marked with index m) are quite different and the correspondent equation for Φ_m can be written as:

$$\Phi_m = (k_r + k_m) / (k_r + k_m + \Sigma k_{nr}) \tag{8.3}$$

Here k_r is the unmodified radiative decay rate, k_m is the metal-modified radiative decay rate and Σk_{nr} are, as above, the nonradiative rates combining all dynamic quenching effects. Similarly, the metal-modified lifetime, τ_m, of a fluorophore is decreased by an *increased radiative decay rate*:

$$\tau_m = 1 / (k_r + k_m + \Sigma k_{nr}) \tag{8.4}$$

Unusual predictions follow from the theory of plasmon enhancement effects. It can be seen that as the value of k_m increases, the quantum yield, Φ_m also increases, instead of the commonly expected decrease. Nevertheless, this increase results in the decrease of the emission lifetime, τ_m. These predictions have been confirmed in experiments (Aslan et al. 2005a).

Metallic nanoparticles (especially silver, gold and copper) are known to scatter light efficiently and in an angular-dependent fashion and silver particles are several-fold better scatterers than gold (Yguerabide and Yguerabide 1998). Subsequently, light scattering by silver and gold nanoparticles can be detected at concentrations as low as 10^{-16} M. The scattered light by metallic nanoparticles is the highest at observation angles of $0°$ and $180°$ with respect to the incident light, i.e., in the backward and forward directions. The angular dependence of the scattered light depends on the size, shape and composition of the nanoparticles.

It is essential that enhanced fluorescence emission exactly reproduces this angular dependence. In addition, the angular dependence can be changed upon binding of a labeled target protein. It was suggested to use this property for the development of *angle-ratiometric fluorescence sensing* (Aslan et al. 2007b).

In dye-modified silver nanoparticles adsorbed on a solid support, in which these effects have been studied in most detail, there is an interaction between two electronic systems. One is of the dye and the other is the system of confined free electrons of the metal particle, the resonance motion of which is limited by the particle size. In Section 4.5 we described their properties and showed that such confinement effects result in spectroscopic behavior typical of strong light absorbers and weak light emitters. Thus, what produces the highly intensified emission in such systems?

Until recently, these effects were interpreted as surface plasmon enhancement phenomena modulating the dye emission. Thus, the central role was given to the dyes as emitters operating under the influence of a strong evanescent field (see Fig. 8.9). Therefore, it was considered that the dyes were both excited species and sole emitters. The role attributed to the surface plasmons was only that of modifiers of their intrinsic radiative decay rate (Geddes and Lakowicz 2002). This interpretation has shifted to a greater role for metal particles after it was discovered that surface plasmon-coupled emission is directional and can propagate at a unique angle from the back of the film (Aslan et al. 2005b). This allowed suggesting a model that considers nonradiative energy transfer occurring from excited distal fluorophores to the surface plasmon electrons in films formed of adsorbed particles, as depicted in Fig. 8.10 below. The surface plasmons in turn radiate, providing under certain conditions the photophysical characteristics of the coupled dyes. In essence, radiation of the whole system occurs.

The final selection between these models will be made when the influence of many factors affecting the metal enhancement have been studied in detail. Primarily, this phenomenon must depend on the *radiative decay rate* and quantum yield of the fluorescent molecules, providing more possibility for the enhancement of weak fluorescence emitters. Such regularity is observed in experiments and many exam-

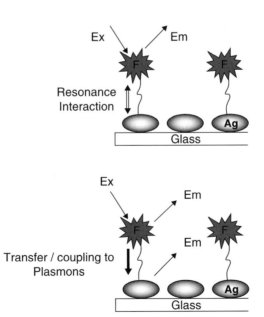

Fig. 8.10 Early interpretation of metal-enhanced fluorescence whereby fluorophores undergo a radiative decay rate modification (top) and current thinking, the fluorescence emission is partially plasmon coupled and the system radiates (bottom) (Reproduced from Aslan et al. 2005b)

ples confirming it can be found in the literature (Aslan et al. 2005a). The fluorescence enhancement of low-quantum yield fluorophores, converting poor emitters into good ones, may prove to be of great practical significance.

This phenomenon should also depend on the size and shape of nanoparticles, since this factor determines the quantum efficiency of the scattering component of their extinction spectrum. Larger particles are better scatterers (Yguerabide and Yguerabide 1998) and stronger enhancement effects have been reported for them (Yagai et al. 2007). The light scattering, being increased with nanoparticle size, should also depend upon the energy of the plasmon excitation, since the optical cross-section of the nanoparticle on resonance is strongly enhanced relative to its off-resonance value.

Since resonance energy transfer is involved, important factors affecting the intensity of the fluorescence emission must be also the orientation of the dye dipole moments relative to the nanoparticles normal surface and the overlap of the absorption and emission bands of the dye with the plasmon band of the metal. There have been a number of observations demonstrating with the same silver nanoparticles the universal character of the enhancement for dyes absorbing and emitting from UV to near-IR (Aslan et al. 2005a). In contrast, some recent data found a strong correlation between the fluorescence intensity of the dye and the degree of spectral overlap with the plasmon resonance of the nanoparticle (Chen et al. 2007), which in combination with properly selected particle size, can provide a strong wavelength-selective enhancement (Yagai et al. 2007). These two properties must be coupled, since the optical cross-section of the nanoparticle on resonance is strongly enhanced relative to its off-resonance value. Presently, the role of the plasmon resonance energy with respect to the excitation and emission energies of a fluorophore is not well understood and needs further experimentation and analysis.

Much effort has been made to establish the *optimum distance* between the dye and the metal particle. Whereas fluorescence from a molecule directly adsorbed onto the surface of a metallic nanoparticle is strongly quenched, at a distance of a few nanometers from the nanoparticles it can be strongly enhanced. However, if the distance is too large, the enhancement effect must again vanish. Experiments focused on such distance optimization were performed with metal particles covered with an inert material of variable thickness, with the dye adsorbed or covalently attached to its surface. Silica shells (Tovmachenko et al. 2006) and inert stearic acid spacer layers (Ray et al. 2007) were used for this purpose. The results presented in different publications exhibit a rather broad variation in the estimation of such optimal distance, from 5–10 nm to more than 20 nm. Probably, the optimal distance has to be evaluated together with the other variables discussed above.

The generality of the observed phenomenon is important. The interaction with metal nanoparticles produces emission enhancement not only of organic dyes but also of Quantum Dots (Gryczynski et al. 2005), conjugated polymers (Park et al. 2007) and metal-chelator luminophores (Zhang et al. 2006b). An immense variety of potentially useful nanocomposites can be constructed exhibiting enhanced emission. An interesting possibility, for example, can be realized when the dye is incorporated into the silica core of beads that are coated by a silver layer to generate porous but continuous metal nanoshells (Zhang et al. 2006b). Being brightly emissive, such particles are simultaneously protected from the quenching effect of the solvent.

Metal enhancement is demonstrated for *two-photon excitation*. Simultaneous absorption of two photons occurs if the density of incident photons is high (Section 8.3), which can be achieved by pulsed lasers with a high peak power and sharp focusing. Metal nanoparticles can provide such an enhancement of excitation rate that is sufficient for generating the two photon emissions near their surfaces (Gryczynski et al. 2002). It may be possible that such an approach will change two-photon microscopy (Section 11.1) by making scanning with ultra-short laser pulses unnecessary; so that, wide-field optics will be sufficient.

Notably, the effect of metal-nanoparticle emission enhancement is not limited to fluorescence only. It has also been discovered for *chemiluminescence* (Chowdhury et al. 2006). This indicates that the surface plasmons can be directly excited from chemically-induced electronic excited states and excludes the possibility that they are created by incident excitation light. This observation should stimulate new developments in chemiluminescence sensing methods (Section 8.1), which are methods that use very cheap instrumentation but suffer from low levels of emission signals.

Phosphorescence emission can also be enhanced upon interaction with metal (Zhang et al. 2007) and this effect is clearly observed with metal nanoparticles (Zhang et al. 2006b). Since the triplet states that give rise to phosphorescence are chemically reactive, the proximity of the metal nanoparticles produces their activation and acceleration, in particular, generating singlet oxygen and superoxide radicals (Zhang et al. 2007). On the other hand, the shorter lifetime of the metal-enhanced fluorescence results in less time for the excited state photoreactions and thus more excitation-emission cycles can occur prior to *photobleaching*. Thus, increased photostability is an important additional benefit.

8.6.3 Application of Metal-Nanoparticle Enhancement

A straightforward use of metal enhancement is to provide *increased sensitivity* to traditional detection techniques. An example of this application is the recent work of (Aslan et al. 2006), in which the application of this approach resulted in the rapid detection of several femtomoles of RNA and without the use of a time- and reagent-consuming PCR procedure. It has a strong potential to be developed into different single-molecular detection methodologies.

The other important advantage in sensing applications is the *localized nature* of the enhancement effect. The enhanced evanescent field mode of excitation localizes the excitation volume in close proximity to the metal nanostructures. As in other evanescent-field excitation techniques, this allows exciting only the bound species and eliminates the need for any washing steps. In contrast to plain surfaces, nanoparticles possess dramatically larger surface areas. Together with their effect as scatterers, this offers broad flexibility in assay development.

In this regard, the approach based on metal particle enhancement has a strong potential for application in *high throughput screening* and other assays dealing with a great number of samples and determining many parameters. Such possibility of *multiplex detection* has been demonstrated for DNA microarrays, in which the surface-bound silver nanoparticles were used as fluorescence enhancers (Sabanayagam and Lakowicz 2007).

8.6.4 *Microwave Acceleration of Metal-Enhanced Emission*

Low-power microwave heating is increasingly used in different techniques of chemical synthesis to accelerate the reaction rates (Tokuyama and Nakamura 2005). Microwaves with a frequency of 0.3–300 GHz lie between the infrared and radio-frequency electromagnetic radiations. They are strongly absorbed by mobile dipoles in the medium, so that polar systems of any size are selectively heated and non-polar ones remain transparent to this radiation. For metals, the attenuation of microwave radiation arises from the motions of extremely mobile conductance electrons. The charge carriers in the metal, which are displaced by the electric field, are subjected to resistance due to collisions with the lattice phonons. They provide energy to the lattice resulting in the local heating. This so-called Ohmic heating of the metal nanoparticles occurs in addition to the heating of any medium, such as water surrounding the nanoparticles.

So the *localized heating* around the metal nanostructures can be achieved without heating the whole volume. This accelerates the local diffusion and specific binding (recognition) between macromolecules close to them. Thus, the speed of receptor-target binding can be increased, which is highly needed for increasing the efficiency of the operation of sensors. Thus, the combination of metal enhancement and localized heating by microwaves makes the sensing both very sensitive and rapid (Aslan and Geddes 2005).

When the time-consuming steps of sample pre-treatment and after-treatment are eliminated or substantially reduced, then the *assay duration* becomes limited by the sensor-target incubation time. The process of recognition between the target and receptor involves many trial-and-error steps on a molecular level, so that many minutes and sometimes hours are often needed to establish the equilibrium between the free and bound targets. Such a slow rate is particularly characteristic for the most popular assays involving macromolecules, in which multi-point specific contacts need to be established, such as DNA hybridization assays and immunoassays. It was recently demonstrated that for both types of assays, with the application of microwaves, the incubation times can be reduced manifold, to seconds.

Thus, it has been shown (Aslan et al. 2006, 2007a) that by combining the effects of silver nanoparticles on dye emission with low-power, localized microwave heating, a model *DNA hybridization assay* to detect 23-mer targets can be kinetically complete within 20 s. This gives an acceleration of up to 600-fold, as compared to an identical hybridization assay run without microwaves. The result is identical, which excludes any destruction or denaturation of interacting partners. The microwave-induced temperature jump within close proximity to the silver nanoparticles was determined to be ca. 1°C from studies using a thermally-responsive fluorophore, which is quite insufficient to produce any structural changes. Simultaneously, by metal enhancement, a greater than ten-fold increase in assay sensitivity was achieved. The speed of the assay, together with high sensitivity, are especially important in the identification of such dangerous bioagents as Anthrax (DNA) or Ebola (RNA) viruses.

The limiting factors in current *immunoassays* are their slow speed and low sensitivity. To increase the sensitivity, additional operations (such as catalytic amplification) are frequently introduced but they further increase the time of test duration. In our discussion on the new horizons in clinical diagnostics (Section 12.3) we consider the possibility of microwave-heated metal-enhancement as an attractive technology for the future. This statement is based on recent findings that this approach allows reducing the incubation times in immunoassays from many minutes to a few seconds, with a dramatic increase in sensitivity (Aslan et al. 2007c). Thus, this new technology provides a method with which ultrafast and ultra-bright immunoassays can be realized.

In both DNA hybridization assays and immunoassays it was noticed that the use of low-power microwaves to kinetically accelerate assays significantly improves molecular recognition; it reduces both the rate and extent of *nonspecific absorption*. This is a significant finding, since the assay sensitivity (analytical detectability) is often limited by the overall extent of the non-specific binding.

In view of these findings, it is highly probable that this new technology will fundamentally change the ways by which the immunoassays and other diagnostic tools are currently employed in clinical medicine, adding speed and sensitivity.

8.6.5   Prospects

Presently, we are at the most romantic stage of exploration of the phenomenon of metal enhancement of fluorescence. This is definitely a new and very prospective tool in sensing technologies that has already showed superiority over traditional detection techniques. Meanwhile, it is still difficult to predict whether a particular sized metal structure will strongly quench or strongly enhance the fluorescence or other type of emission of a definite molecular dye or nanocomposite. There are many factors influencing this enhancement and notably, the interplay between enhancement and quenching. These factors can belong to the emitter (quantum yield, lifetime), to the metal (plain surface or nanoparticles, particle size and shape, resonance absorption energy) or to both of them (orientation and distance).

In the case of fluorescent molecules located at very short distances from a metal surface, quenching takes place. Electromagnetic-field enhancement occurs at longer distances. As a result, there is an optimal dye-to-metal distance for fluorescence enhancement. With good understanding of this phenomenon, switching between enhancement and quenching can provide the broadest dynamic ranges hardly available to any other sensing technique.

It is presently certain that the phenomenon of metal enhancement is of general origin. It appears irrespective of the nature of the excited state (singlet or triplet) and of the manner of excitation (optical or a chemical reaction). This adds many new dimensions to the research and practical use of this phenomenon.

Such a bright perspective stimulates detailed studies in all aspects, including the sizes, shapes and composition of metal nanostructures and also the nature, spectroscopic properties and locality of fluorophores. The role of augmentation of optical intensity incident on the dye molecule through near field enhancement of modification of the radiative decay rate of the emitter and of increasing the coupling efficiency of the fluorescence emission to the far field through nanoparticle scattering, deserve detailed studies.

Sensing and Thinking 8: Eliminating Light Sources and Photodetectors: What Remains?

In this chapter we tried to compare seemingly incomparable techniques. They offer development of fluorescence sensing but in different, sometimes unexpected, directions. The reader is encouraged to analyze them and develop his/her own opinion as to which of them offer substantial progress with a broad-scale practical implementation and which will end up in research laboratories. Different criteria for such comparison can be used in addition to the common detection limit and sensitivity. (a) Speed of analysis. (b) Spatial resolution with extension to microarray techniques. (c) Possibility of extension to microbead assays. The reader may consider different constructions, combining two or several of these non-conventional approaches.

Questions and Problems

1. Why is chemoluminescence restricted to a very narrow range of reactions? What kinds of chemical substances can interfere with these assays? Evaluate the possibilities of this method for sensing multiple samples.
2. Are the mechanisms of quenching of chemiluminescence and bioluminescence emissions similar to that of common luminescence, or different?
3. Why is spectral resolution for the detection of chemiluminescence commonly not needed and the sample cell can be located on top of the detector?
4. Estimate the achievable time resolution with chemiluminescence and electro-chemiluminescence methods.
5. What are the reasons for introducing the luciferin-luciferase system into different living cells by genetic engineering? Evaluate the possible applications of this approach.

6. What are the conditions for obtaining the two-photonic emission? Can these photons be of different energies?
7. Can the two-photon emission be polarized and can it allow application in polarization assays?
8. What are the differences in the excitation of two-photon and up-conversion luminophores? Does up-conversion allow high spatial resolution? What are the special properties of up-converting luminophores as the FRET donors? Can they also serve as FRET acceptors?
9. What are the requirements for the observation of amplified stimulated emission? How can fluorescence reporting be introduced into this process?
10. What are the possibilities to obtain the electrical signal directly instead of generating a reporter emission and recording it with a photodetector?
11. Explain the performance of LAPS. What are the areas of its application? For multiplex applications, do we need to make a separate system containing the semiconductor plate and the electrodes for every analyte?
12. What is the distance dependence for evanescent wave excitation? Compare it with the sizes of macromolecules, nanoparticles and cells.
13. What is special in metal nanoparticles so that they provide the effect of enhancement of the fluorescence? What is its dependence on the particle size and the dye-particle distance?
14. Explain the mechanism by which the enhancement of fluorescence intensity can be coupled with shortening the lifetime.
15. Why can the microwave heating be local and focused only on metal particles in solutions? How can it decrease the duration of the assays?

References

Aslan K, Geddes CD (2005) Microwave-accelerated metal-enhanced fluorescence: Platform technology for ultrafast and ultrabright assays. Analytical Chemistry 77:8057–8067

Aslan K, Gryczynski I, Malicka J, Matveeva E, Lakowicz JR, Geddes CD (2005a) Metal-enhanced fluorescence: an emerging tool in biotechnology. Current Opinion in Biotechnology 16:55–62

Aslan K, Leonenko Z, Lakowicz JR, Geddes CD (2005b) Annealed silver-island films for applications in metal-enhanced fluorescence: interpretation in terms of radiating plasmons. Journal of Fluorescence 15:643–654

Aslan K, Huang J, Wilson GM, Geddes CD (2006) Metal-enhanced fluorescence-based RNA sensing. Journal of the American Chemical Society 128:4206–4207

Aslan K, Malyn SN, Bector G, Geddes CD (2007a) Microwave-accelerated metal-enhanced fluorescence: an ultra-fast and sensitive DNA sensing platform. Analyst 132:1122–1129

Aslan K, Malyn SN, Geddes CD (2007b) Angular-dependent metal-enhanced fluorescence from silver colloid-deposited films: opportunity for angular-ratiometric surface assays. Analyst 132:1112–1121

Aslan K, Malyn SN, Geddes CD (2007c) Microwave-accelerated surface plasmon-coupled directional luminescence: application to fast and sensitive assays in buffer, human serum and whole blood. Journal of Immunological Methods 323:55–64

Baj S, Krawczyk T (2007) Chemiluminescence detection of organic peroxides in a two-phase system. Analytica Chimica Acta 585:147–153

Boyer JC, Cuccia LA, Capobianco JA (2007) Synthesis of colloidal upconverting NaYF4: Er3+/
    Yb3+ and Tm3+/Yb3+ monodisperse nanocrystals. Nano Letters 7:847–852
Chatterjee DK, Rufaihah AJ, Zhang Y (2008) Upconversion fluorescence imaging of cells and
    small animals using lanthanide doped nanocrystals. Biomaterials 29:937–943
Chen Y, Munechika K, Ginger DS (2007) Dependence of fluorescence intensity on the spectral overlap
    between fluorophores and plasmon resonant single silver nanoparticles. Nano Letters 7:690–696
Chowdhury MH, Aslan K, Malyn SN, Lakowicz JR, Geddes CD (2006) Metal-enhanced chemi-
    luminescence: radiating plasmons generated from chemically induced electronic excited
    states. Applied Physics Letters 88:173104
De la Rosa E, Salas P, Diaz-Torres LA, Martinez A, Angeles C (2005) Strong visible cooperative
    up-conversion emission in ZrO2:Yb3+ nanocrystals. Journal of Nanoscience and
    Nanotechnology 5:1480–1486
Dennany L, Forster RJ, Rusling JF (2003) Simultaneous direct electrochemiluminescence and
    catalytic voltammetry detection of DNA in ultrathin films. Journal of the American Chemical
    Society 125:5213–5218
Drobizhev M, Rebane A, Suo Z, Spangler CW (2005) One-, two- and three-photon spectroscopy
    of pi-conjugated dendrimers: cooperative enhancement and coherent domains. Journal of
    Luminescence 111:291–305
Ekgasit S, Thammacharoen C, Yu F, Knoll W (2004) Evanescent field in surface plasmon reso-
    nance and surface plasmon field-enhanced fluorescence spectroscopies. Analytical Chemistry
    76:2210–2219
Fahnrich KA, Pravda M, Guilbault GG (2001) Recent applications of electrogenerated chemilu-
    minescence in chemical analysis. Talanta 54:531–559
Fine T, Leskinen P, Isobe T, Shiraishi H, Morita M, Marks RS, Virta M (2006) Luminescent yeast
    cells entrapped in hydrogels for estrogenic endocrine disrupting chemical biodetection.
    Biosensors & Bioelectronics 21:2263–2269
Frangioni JV (2006) Self-illuminating quantum dots light the way. Nature Biotechnology
    24:326–328
Fu J, Padilha LA, Hagan DJ, Van Stryland EW, Przhonska OV, Bondar MV, Slominsky YL,
    Kachkovski AD (2007) Molecular structure - two-photon absorption property relations in
    polymethine dyes. Journal of the Optical Society of America B-Optical Physics 24:56–66
Geddes CD, Lakowicz JR (2002) Metal-enhanced fluorescence. Journal of Fluorescence
    12:121–129
Gratzel M (2007) Photovoltaic and photoelectrochemical conversion of solar energy. Philosophical
    Transactions of the Royal Society of London Series A-Mathematical Physical and Engineering
    Sciences 365:993–1005
Gryczynski I, Malicka J, Shen YB, Gryczynski Z, Lakowicz JR (2002) Multiphoton excitation of
    fluorescence near metallic particles: enhanced and localized excitation. Journal of Physical
    Chemistry B 106:2191–2195
Gryczynski I, Malicka J, Jiang W, Fischer H, Chan WCW, Gryczynski Z, Grudzinski W, Lakowicz
    JR (2005) Surface-plasmon-coupled emission of quantum dots. Journal of Physical Chemistry
    B 109:1088–1093
Hanninen P, Soini A, Meltola N, Soini J, Soukka J, Soini E (2000) A new microvolume technique
    for bioaffinity assays using two-photon excitation. Nature Biotechnology 18:548–550
He GS, Zheng Q, Prasad PN, Grote JG, Hopkins FK (2006) Infrared two-photon-excited visible
    lasing from a DNA-surfactant-chromophore complex. Optics Letters 31:359–361
Huang XY, Li L, Qian HF, Dong CQ, Ren JC (2006) A resonance energy transfer between chemi-
    luminescent donors and luminescent quantum-dots as acceptors (CRET). Angewandte
    Chemie-International Edition 45:5140–5143
Kulmala S, Suomi J (2003) Current status of modern analytical luminescence methods. Analytica
    Chimica Acta 500:21–69
Kuningas K, Rantanen T, Ukonaho T, Lovgren T, Soukka T (2005) Homogeneous assay technol-
    ogy based on upconverting phosphors. Analytical Chemistry 77:7348–7355

Kuningas K, Ukonaho T, Pakkila H, Rantanen T, Rosenberg J, Lovgren T, Soukka T (2006) Upconversion fluorescence resonance energy transfer in a homogeneous immunoassay for estradiol. Analytical Chemistry 78:4690–4696

Lee KS, El-Sayed MA (2006) Gold and silver nanoparticles in sensing and imaging: sensitivity of plasmon response to size, shape, and metal composition. Journal of Physical Chemistry B 110:19220–19225

Lee TH, Gonzalez JI, Zheng J, Dickson RM (2005) Single-molecule optoelectronics. Accounts of Chemical Research 38:534–541

Li BX, He YZ (2007) Simultaneous determination of glucose, fructose and lactose in food samples using a continuous-flow chemiluminescence method with the aid of artificial neural networks. Luminescence 22:317–325

Liu X, Tan W (1999) A fiber-optic evanescent wave DNA biosensor based on novel molecular beacons. Analytical Chemistry 71:5054–5059

Miao WJ, Bard AJ (2003) Electrogenerated chemluminescence. 72. Determination of immobilized DNA and C-reactive protein on Au(111) electrodes using Tris(2,2'-bipyridyl)ruthenium(II) labels. Analytical Chemistry 75:5825–5834

Michelini E, Mirasoli M, Karp M, Virta M, Roda A (2004) Development of a bioluminescence resonance energy-transfer assay for estrogen-like compound in vivo monitoring. Analytical Chemistry 76:7069–7076

Neumann T, Johansson ML, Kambhampati D, Knoll W (2002) Surface-plasmon fluorescence spectroscopy. Advanced Functional Materials 12:575–586

Papkovsky DB, O'Riordan T, Soini A (2000) Phosphorescent porphyrin probes in biosensors and sensitive bioassays. Biochemical Society Transactions 28:74–77

Park HJ, Vak D, Noh YY, Lim B, Kim DY (2007) Surface plasmon enhanced photoluminescence of conjugated polymers. Applied Physics Letters 90:161107

Paulmurugan R, Gambhir SS (2007) Combinatorial library screening for developing an improved split-firefly luciferase fragment-assisted complementation system for studying protein-protein interactions. Analytical Chemistry 79:2346–2353

Peter LM (2007) Dye-sensitized nanocrystalline solar cells. Physical Chemistry Chemical Physics 9:2630–2642

Ray K, Badugu R, Lakowicz JR (2007) Sulforhodamine adsorbed Langmuir-Blodgett layers on silver island films: effect of probe distance on the metal-enhanced fluorescence. Journal of Physical Chemistry C 111:7091–7097

Richter MM (2004) Electrochemiluminescence (ECL). Chemical Reviews 104:3003–3036

Robelek R, Niu LF, Schmid EL, Knoll W (2004) Multiplexed hybridization detection of quantum dot-conjugated DNA sequences using surface plasmon enhanced fluorescence microscopy and spectrometry. Analytical Chemistry 76:6160–6165

Roda A, Pasini P, Mirasoli M, Michelini E, Guardigli M (2004) Biotechnological applications of bioluminescence and chemiluminescence. Trends in Biotechnology 22:295–303

Rose A, Zhu ZG, Madigan CF, Swager TM, Bulovic V (2005) Sensitivity gains in chemosensing by lasing action in organic polymers. Nature 434:876–879

Sabanayagam CR, Lakowicz JR (2007) Increasing the sensitivity of DNA microarrays by metal-enhanced fluorescence using surface-bound silver nanoparticles. Nucleic Acids Research 35:e13

So MK, Xu CJ, Loening AM, Gambhir SS, Rao JH (2006) Self-illuminating quantum dot conjugates for in vivo imaging. Nature Biotechnology 24:339–343

So PTC, Dong CY, Masters BR, Berland KM (2000) Two-photon excitation fluorescence microscopy. Annual Review of Biomedical Engineering 2:399–429

Somers RC, Bawendi MG, Nocera DG (2007) CdSe nanocrystal based chem-/bio-sensors. Chemical Society Reviews 36:579–591

Taitt CR, Anderson GP, Ligler FS (2005) Evanescent wave fluorescence biosensors. Biosensors & Bioelectronics 20:2470–2487

Tokuyama H, Nakamura M (2005) Acceleration of reaction by microwave irradiation. Journal of Synthetic Organic Chemistry Japan 63:523–538

Tovmachenko OG, Graf C, van den Heuvel DJ, van Blaaderen A, Gerritsen HC (2006) Fluorescence enhancement by metal-core/silica-shell nanoparticles. Advanced Materials 18:91–95

Varnavski OP, Ostrowski JC, Sukhomlinova L, Twieg RJ, Bazan GC, Goodson T (2002) Coherent effects in energy transport in model dendritic structures investigated by ultrafast fluorescence anisotropy spectroscopy. Journal of the American Chemical Society 124:1736–1743

Xie X, He X, Song Z (2007) A sensitive chemiluminescence procedure for the determination of carbon monoxide with myoglobin-luminol chemiluminescence system. Applied Spectroscopy 61:706–710

Xu SL, Cui H (2007) Luminol chemiluminescence catalysed by colloidal platinum nanoparticles. Luminescence 22:77–87

Xu Y, Piston DW, Johnson CH (1999) A bioluminescence resonance energy transfer (BRET) system: application to interacting circadian clock proteins. Proceedings of the National Academy of Sciences of the United States of America 96:151–156

Xu Y, Cai H, Liu QJ, Qin LF, Wang LJ, Wang P (2006) A novel structure of LAPS array for cell-based biosensor. Rare Metal Materials and Engineering 35:51–54

Yagai S, Kinoshita T, Higashi M, Kishikawa K, Nakanishi T, Karatsu T, Kitamura A (2007) Diversification of self-organized architectures in supramolecular dye assemblies. Journal of the American Chemical Society 129:13277–13287

Yakovleva J, Davidsson R, Bengtsson M, Laurell T, Emneus J (2003) Microfluidic enzyme immunosensors with immobilised protein A and G using chemiluminescence detection. Biosensors & Bioelectronics 19:21–34

Yashchuk VM, Gusak VV, Drnytruk IM, Prokopets VM, Kudrya VY, Losytskyy MY, Tokar VP, Gumenyuk YO, Yarmoluk SM, Kovalska VB, Balanda AO, Kryvorotenko DV (2007) Two-photon excited luminescent styryl dyes as probes for the DNA detection and imaging. Photostability and phototoxic influence on DNA. Molecular Crystals and Liquid Crystals 467:325–338

Yguerabide J, Yguerabide EE (1998) Light-scattering submicroscopic particles as highly fluorescent analogs and their use as tracer labels in clinical and biological applications - II. Experimental characterization. Analytical Biochemistry 262:157–176

Yoshinobu T, Iwasaki H, Ui Y, Furuichi K, Ermolenko Y, Mourzina Y, Wagner T, Nather N, Schoning MJ (2005) The light-addressable potentiometric sensor for multi-ion sensing and imaging. Methods 37:94–102

Zhang S, Zhang Z, Shi W, Eremin SA, Shen J (2006a) Development of a chemiluminescent ELISA for determining chloramphenicol in chicken muscle. Journal of Agricultural and Food Chemistry 54:5718–5722

Zhang YX, Aslan K, Previte MJR, Malyn SN, Geddes CD (2006b) Metal-enhanced phosphorescence: Interpretation in terms of triplet-coupled radiating plasmons. Journal of Physical Chemistry B 110:25108–25114

Zhang YX, Aslan K, Previte MJR, Geddes CD (2007) Metal-enhanced superoxide generation: A consequence of plasmon-enhanced triplet yields. Applied Physics Letters 91:023114

Zu YB, Ding ZF, Zhou JF, Lee YM, Bard AJ (2001) Scanning optical microscopy with an electrogenerated chemiluminescent light source at a nanometer tip. Analytical Chemistry 73:2153–2156

Chapter 9
The Sensing Devices

In this chapter we will concentrate on the problem of the *transformation of the fluorescence signal* obtained in the molecular sensing event into recordable information. The interplay between the two events, *signal transduction* occurring on a molecular level and the *detection* by man-made instrumentation on a macroscopic level is central to the construction of efficient sensor technologies. The sensor technique should optimally satisfy the requirements of selectivity, sensitivity, stability and, often preferably, convenience in performance. In addition, measurements in kinetic regime, scanning the image of a nano-structured material or living cell or simultaneous detection of hundreds and thousands of analytes put additional requirements on the instrumentation and on its versatility. Because of the innumerable amount of potential targets and inestimable variety of experimental conditions in which the measurements have to be made, no single solution could exist regarding assay format and proper instrumentation.

In Fig. 9.1 several formats, which are the most frequently used in sensor technologies, are depicted. *Cuvette*-type spectroscopy in solutions is applicable to any homogeneous assay, it allows recording of all possible fluorescence parameters and is applicable for polarization and time-resolved measurements. It allows the highest possible spectral resolution sensitivity but requires a large amount of sample and does not allow spatial resolution (recording of images) or simultaneous detection of a large number of samples.

The cuvette format does not satisfy the requirements of many applications in sensing. Therefore, many efforts have been made for the miniaturization of fluorescence sensing techniques, which has become the leading topic of technology development during the last decade. Various *lab-on-a-chip* techniques have been developed (Vinet et al. 2002). They allow solving in a consistent manner the problems of miniaturization, integration and systematization of various sensor types, with the possibility to apply modern micro-fabrication technologies. A variety of pocket-sized lab-on-a-chip instruments have appeared (Gardeniers and van den Berg 2004), they have started to occupy strong positions in practical use. Their development strongly follows the tendency of *multi-parameter* and *multi-analyte* monitoring in parallel. The use of nanoliter plates became possible, resulting in a reduced sample volume and reduced amount of necessary reagents, together with the improvement of the detection technique.

A.P. Demchenko, *Introduction to Fluorescence Sensing*,   371
© Springer Science + Business Media B.V. 2009

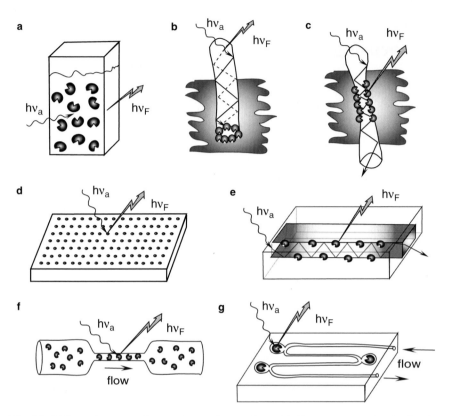

Fig. 9.1 Different instrument formats that are frequently used in fluorescence sensing (**a**) Sensing in a spectrofluorimeter cuvette. The most precise technique that needs a large amount of sample. The samples are detected one by one. Flow-through detection is possible. (**b**) Optical waveguides in tip configuration. Both the light source and fluorescence detection unit operate at a distal terminal of the waveguide. (**c**) Optical fiber sensor utilizing the evanescent wave excitation and collection of the fluorescence signal outside the fiber. (**d**) Flat supported microarray. It allows the attachment of a number of different sensor units. Scanning the excitation beam allows a read-out of fluorescence. (**e**) Planar optical waveguide. Miniaturized sample-saving technique that allows efficient collection of out-the-waveguide fluorescence. (**f**) Microsphere-based array. Sensor units are attached to micro-beads that are detected one by one, passing through a narrow capillary. (**g**) Microfluidic device. Fabricated microscopic channels and reservoirs allow in addition to sensing, preparation of the sample (separations, reagent additions)

9.1   Instrumentation for Fluorescence Spectroscopy

In this section we will provide a short description of optical elements that can be found in conventional instrumentation and are needed for the development of miniaturized devices. More detailed information on them can be found in the textbooks and reviews (Bacon et al. 2004; Lakowicz 2007; Valeur 2002). Their miniaturization and adaptation to the need of sensor applications will be the main focus.

The sources of error peculiar to fluorescence and suited fluorescence standards are well described in the literature (Resch-Genger et al. 2005).

9.1.1   Standard Spectrofluorimeter

Cuvette-type fluorescence detection is commonly provided by using the standard fluorimeter described in Fig. 9.2. It is usually equipped with excitation and emission monochromators, which allow easy setting of excitation and emission wavelengths and scanning them to obtain the spectra. Polarizers at excitation and emission beams allow recording the spectra at horizontal and vertical polarizations and the measurement of anisotropy.

These instruments use the *cuvettes* for liquid samples from 4 ml down to 0.04 ml. Though different samplers for automatic cuvette filling can be used with them, the

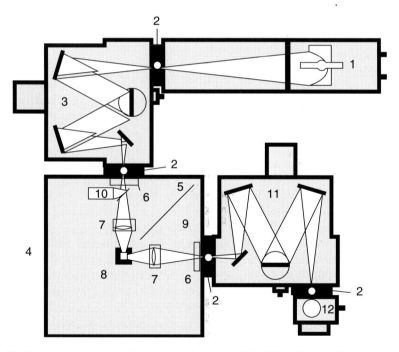

Fig. 9.2 Schematics of a conventional spectrofluorimeter. The light emitted by light source 1 with a continuous spectrum illuminates via slit 2 the monochromator 3 that enters the sample chamber 4 with inner space 5 and, passing through polarizer 6, is focused with lens 7 on cuvette 8. Emission light passing through similar optical elements enters monochromator 11, where after decomposition into the spectrum, is recorded by photomultiplier tube 12. Many instruments use the reference channel with photodetector 10 to compensate the fluctuations in intensity of the light source

samples are measured one by one or in a flow of one sample with a rather low productivity. Thus, these versatile and high-precision instruments are needed mostly for sensor development but not for sensing itself.

The *microplate readers* used for scanning spotted arrays are built based on the same schematics as spectrofluorimeters but with some modifications. The difference is mostly in the geometry of incidence and fluorescence beams. The microplate (e.g., array) is located horizontally. The incident beam illuminates the plate from the top, allowing its scanning in both **x** and **y** directions. The emission light is also collected from the top by using a special mirror that directs it to an emission monochromator or, through the filter, to the detector.

The miniaturization and simplification of such instruments may occur in many directions. Simplified instruments may contain instead of monochromators, wavelength-selective optical filters on excitation and emission beams and also miniaturized and economical light sources and light detectors. The size of the sample chamber can be dramatically reduced and, as we shall see below, substituted with an optical waveguide.

9.1.2   Light Sources

The fluorescence reporter is commonly excited by light, therefore the light source is a necessary part of spectroscopic instrumentation (luminescence excited by an electric field and in chemical reactions is discussed in Chapter 8). There may be different requirements for the light source dictated by the methods of observation of the response. Often, spectral resolution is needed, which can be provided by the light source itself (producing monochromatic light) or by a lamp with a broadband spectrum, with the aid of additional optical devices (monochromators and filters). Pulsed or frequency-modulated excitation is commonly used for time-resolved measurements and polarized light is needed for detection in anisotropy. Spectral, temporal or polarization resolution (or a combination of them) is used for reducing the background signal and reducing light-scattering effects.

The requirements for the light sources may be varied but all of them have to provide the basic function of delivering sufficient radiant energy at a particular wavelength or in the wavelength range of interest. They must maintain a constant or regulated light intensity over the whole time interval of measurement. This requires keeping a working balance between the intensity and degree of monochromaticity. Additional requirements for collimation, polarization, rejection of stray light, etc. should also be accounted for.

The present generation of scientists may not know that a very popular term '*laser*' first appeared as an acronym for 'light amplification by stimulated emission of radiation'. Pulsed solid-state lasers and wavelength-tunable dye lasers have made a strong contribution to spectroscopy. Now comes the era of semiconductor lasers. Their small size and simplicity of maintenance is just what is needed for sensor technologies.

With the rapid decrease of price for instrumentation, *time-resolved detection* methods have gained a lot of popularity. The power of this technique lies mainly in its important features allowing discriminating the scattered light and most of the background emission. Lasers for picosecond-nanosecond time-resolved spectroscopy should follow the following requirements. Pulse duration should be short compared to lifetime τ_F. This can allow high time resolution in the fluorescence signal. Meanwhile, the repetition rate should be at least several times longer than the lifetime; this is because the new pulse has to arrive only when the system returns to the ground state after excitation by a previous pulse. For common fluorescent dyes, 50–100 mHz is a reasonable compromise. If a long-living emission in the microsecond-millisecond time range has to be observed (such as lanthanide chelates, phosphorescent dyes and metalloporphyrin labels) the same principle should be maintained but on a much longer time scale, so that economical pulsed lamps can be used.

In addition to these requirements lasers for *two-photon microscopy* should possess high peak power for providing high instant photon density (since two-photon excitation depends as a square power of intensity) and low average power to prevent thermal damage. Since the image is scanned point-by-point, a sufficient number of laser pulses should be provided at every point to collect a sufficient number of emission quanta. Therefore, for most practical situations the femtosecond pulse laser with a pulse duration of 100–200 fs may be the best (but expensive) choice. Mode-locked Ti:sapphire lasers became very popular in this application. Neodymium-based systems excited by laser diodes have recently become strong competitors; the latter are more reliable and much simpler to adjust. For two-photon excitation of dyes excited by visible light, the laser light must be about twice (with some deviations) the wavelength of the one-photon peak absorption, roughly 700–1,200 nm.

Regarding miniaturization, gallium arsenide-based vertical-cavity surface-emitting lasers (VCSELs) can be considered as optimal excitation sources but they are presently developed into mature and reliable devices only for the near-IR region (850 nm and longer). The VCSEL chip is small ($0.3 \times 0.3 \times 0.1$ mm), requires only one-tenth of the operating current of edge-emitting lasers, can operate in a pulsed (<1 ns) regime, providing high peak intensity at low-power operation and does not require cooling.

A revolutionary change in fluorescence sensing instrumentation comes with the development of semiconductor *light emitting diodes* (LEDs), which can be miniaturized to occupy millimeter dimensions and feature reduced power of operation (Moe et al. 2005). They can efficiently provide electrical power conversion into monochromatic (bandwidth of tens of nanometers) visible or IR light emission. The reverse process, electrical power generation upon their illumination, is also possible, which allows using them not only as light sources but also as light detectors. The findings that materials with such properties can be made not only of inorganic semiconductors but also of organic polymers (Samuel and Turnbull 2007), opened the way for their easy and inexpensive fabrication. With their use, the whole spectrofluorimeter attains pocket-size dimensions. Because of a much wider emission bandwidth and inferior beam quality, the LEDs are not very competitive with the lasers in spectroscopic instrumentation but their apparent merits pose them as the light sources of choice in different sensor devices (Dasgupta et al. 2003), including those using time-resolved fluorescence detection.

A critical issue in the miniaturization of the light source is its power supply unit. Recently, it was suggested to avoid it totally by using the energy of radioactive decay. The light source can be the *plastic scintillator*, in which the light is produced by β-particles generated in radioactive decay of ^{90}Sr (Holthoff et al. 2005). With the lifetime of isotope ^{90}Sr of 28.5 years, these light sources can shine with no external supply of energy, being almost eternal within the scale of our lives.

9.1.3 Light Detectors

Photodetectors convert the energy of light to an electrical signal that can be recorded in analog or digital form. Their response should be a linear function of light intensity over at least several orders of magnitude. An adequate signal-to-noise ratio should also be maintained. Presently, in addition to vacuum photocathode devices, such as photomultiplier tubes (PMT), there has appeared a number of different types of semiconductor photodetectors. In general, vacuum technology being superior in absolute sensitivity and of low noise level shows its limitations in terms of size, operating voltage and robustness. Therefore, in an attempt to decrease the size and cost of instrumentation, the PMTs are currently more and more substituted by new semiconductor photodetectors, such as photodiodes (Yotter and Wilson 2003).

In *vacuum devices*, the photons are incident on an active region of photocathode and this triggers the flow of electrons, which is amplified in several cascade steps. These devices can be used both for steady-state detection of photocurrent and for counting single photons. When PMTs operate in a single-photon counting mode, they produce electron fluxes in response to the incidence of every single photon. Such devices exhibit very low noise levels, which can be further decreased by cooling. In microchannel plate photomultipliers, which are the PMT versions, the electrons from the photo-cathode pass through a narrow semi-conductive channel. Multiple secondary electrons are generated when they hit the channel and this can happen many times with an avalanche effect, producing a gain exceeding 10^8. These devices are expensive however and need a very high operating voltage of 2,400–3,000 V, which, together with their high cost, limits their application in miniaturized schemes.

Solid-state semiconductor devices use a different principle, a photon-induced electron-hole pair generation in a semiconductor material. When the energy of the absorbed photon is large enough to raise an electron from the valence band to a conduction band, they create positive charges, 'holes', in the valence band. The applied electric field, by separating the electrons and holes, produces an electric current that is proportional to the photon flux. The small size and low voltage requirement makes these detectors preferable in all cases of steady-state fluorescence measurement where the light intensity may be high enough and high precision of the measurement is not needed.

The low sensitivity of photodiodes is due to the fact that no amplification or any other gain mechanism is implied in their operation. A certain gain mechanism can be realized only in the so-called *'avalanche photodiodes'* (APD). These devices operate in a mode similar to a Geiger counter. When a bias above the diode breakdown voltage is applied, no current follows until a photon generates a photoelectron, which causes a so-called avalanche breakdown and an appearance of a large current. The bias across the diode is then reduced to terminate the current and is re-started again to count another photon. Therefore, these devices can operate in a single-photon counting mode. They allow low-voltage operation (on the level of 30–40 V), which allows miniaturization of the whole system, leading to such applications as even the implantation of an integrated wireless sensor into a human body.

Other silicon-based detectors, such as PIN (*p-i-n*) photodiodes can be made as small as thin films. But they are not sensitive enough for use with low-power light sources and faint fluorescence emission. For the improvement of PIN photodiodes, a signal processing algorithm called 'lock-in amplification' has been developed (Tung et al. 2004) and with this improvement they provide at least equally high signal-to-noise ratios as a PMT without signal processing.

A two-dimensional matrix of detectors is needed to produce an image. For *fluorescence imaging* the currently available instrumentation is based on one of two principles. One involves laser scanning and detection with PMT and the other – the combination of a lamp for excitation and a *charge-coupled device* (CCD) for detection (Spibey et al. 2001). The latter possibility is preferable when the multi-color applications require variations in excitation wavelength. Charge-coupled device (CCD) sensors, in the form of cameras and array detectors, are explored in many applications. They exhibit low noise, a broad spectral range and linearity of response over many orders of magnitude. They can be interfaced directly with the computer. Such CCD-based devices (intensified and electron-multiplying CCDs) are expensive, bulky and requiring cooling. Therefore, their application areas are mainly limited to basic research.

9.1.4 Passive Optical Elements

Light sources and light detectors are traditionally called active optics. In contrast, mirrors, lenses, optical filters and waveguides are passive optics. *Mirrors and lenses* improve detection by focusing the light in the desired illumination volume. So do *planar waveguides*, which can also allow beam splitting in the emission channel for multi-wavelength detection in combination with optical filters.

Precisely polished quartz lenses are commonly used in bench-top instrumentation. The fabrication of plastics allowed the reduction of their cost and also allowed their fitting to miniaturized sensor devices. Their fabrication directly into a glass chip allowed the substitution of the microscope (Roulet et al. 2002).

Optical fibers and waveguides are often much more convenient to use than traditional lens systems, since they allow extending the possibilities of realizing different instrument configurations (Monro et al. 2001). *Optical fibers* are preferable for measurements at a distance from analytical instrumentation, e.g., in aggressive media, for providing a connection between the sensor tip and an instrument and waveguides – in miniaturized systems. Both operate on the principle of total internal reflection (see Section 9.2).

Optical waveguides have become very important integrating elements that allow miniaturization of all instrumentation. Such integration may include miniature light sources and detectors to produce microfluidic devices (see Section 9.5), in which all optical elements can be aligned during fabrication. They may also serve as miniature support materials for the immobilization of sensor molecules on their surfaces. Fabricated micro-lenses can be located on top of optical waveguides to provide optimal light collection.

The application of *organic polymers* as materials for all passive optical elements used for various coupling, routing and filtering applications has become widespread. With their aid, many different instrument configurations have become possible. Inorganic materials such as oxidized porous silicon can also be used with great success (Mogensen et al. 2004). These new materials prove to be powerful basic building blocks for constructing miniature sensor modules. They can be fabricated in automated, low-cost processes such as replication and thin-film deposition.

In miniaturized devices it is especially important to optimize all elements of the optical train, including emission filters (Dandin et al. 2007). In the process of optimizing detection by removing the interfering light contributions, they should not remove a significant portion of a light signal.

9.1.5   Integrated Systems

The sensor devices for many practical applications should be mass produced. They should be different from analytical instrumentation used in research labs with respect to compactness, convenience of use and price. The miniaturization of all optical elements, their fabrication from inexpensive materials and in the form of integrated units are the present tendencies in instrument design. Therefore, the emerging generation of fluorescence sensors vaguely resembles the spectrofluorimeter depicted in Fig. 9.2. The miniaturization and application of direct contact between optical elements dramatically changes these instruments, though maintaining the basic function of exciting the sample with monochromatic light and detecting monochromatic emission.

An example is one of these developments in which the target binding to the sensor molecules attached to the immobilization layer results in detection provided by a thin film detector via thin films serving as the optical filters (Fig. 9.3). Such systems exclude the necessity of out-of-chip data acquisition (Fixe et al. 2004).

Such dramatic miniaturization occurs not without loss of absolute sensitivity. Maintaining high absolute sensitivity and high signal-to-noise ratio is a general requirement that should be observed. A productive idea for realizing this is to use on the level of detection a discrimination of non-random statistics given by the useful signal and the random signal produced by the noise (Mogensen et al. 2004).

9.1.6   Prospects

In contrast to classical spectroscopic instrumentation, the miniaturized integrated sensor devices evolve with a tremendous rate. The basic scheme of a spectrofluorimeter will probably remain for some time in the background of future devices but the most significant changes will occur with these elements themselves. These

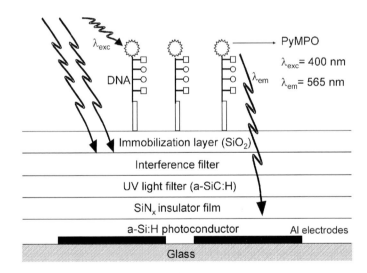

Fig. 9.3 Sensor unit based on an on-chip thin film amorphous silicon photodetector for the quantification of hybridized DNA. Schematic diagram showing assembly of layers. The unit integrates the target (DNA labeled with PyMPO dye) binding with the detection of its fluorescence using a-Si:H semiconductor thin film detector via the thin-film cut-off and interference optical filters transmitting only the emission light. (Reproduced with permission from Fixe et al. 2004)

devices can be fabricated by standard semiconductor and polymer technologies, be of a very small size and operate at low voltages. A variety of sensor configurations performing these functions is possible by selection from the diversity of optical and electronic elements and using them as construction blocks. By combining such functions as power supply, excitation, focusing, filtering, detection, signal processing and communication in the small integrated units, the price of these instruments will go down and the application areas will become extremely broad. It is expected that both the light sources and the detectors will be fabricated based on semiconductor technology and all other elements, such as filters, gratings and lenses, will be deposited during the fabrication. This will move optical instrumentation from bench-top through lap-top to finger-top size. The latter has already been achieved in pioneering works.

Still further miniaturization of devices will need the development of new tools and their association for manipulating and controlling the light onto very small areas (Monat et al. 2007). They may involve the application of photonic crystals and with their aid, the light paths can be confined to sub-wavelength scale. Novel optical nano-devices will provide the basic building blocks of completely new architectures and platforms that will have an impact on numerous applications.

9.2   Optical Waveguides, Optodes and Surface-Sensitive Detection

Fiber optics technology has found important application in sensing for the delivery of excitation light at a distance, for the acquisition of an optical signal and for its delivery to the detection system. These techniques develop rapidly, being stimulated by the progress in optical telecommunications. In both cases, the transmission of the optical signal along the fiber can occur without loss, if necessary, along hundreds or thousands of kilometers.

The light travels along the cores of optical waveguides almost without loss because of *total internal reflection*. Total reflection occurs at an interface of two optical phases when the index of the refraction of the outer phase is smaller than the index of refraction of the core (see Section 8.5). Fluorescence sensing technology benefits from using the fibers ranging in length from centimeters to hundreds of meters. The property of the light beams is such that they do not interact and do not interfere with each other if they propagate in the same or opposite directions. This is the great advantage of optical waveguides over electrical circuits, in which such independence is impossible to achieve.

The transmission properties of the fibers are known to be wavelength-dependent. Presently, most of the fibers are made of inexpensive materials such as plastic and silica; their transparency range occupies the whole visible and near-IR part of the spectrum. Since optical fibers transmit light of different colors, lifetimes or polarizations independently, this allows using the whole arsenal of fluorescence responses. In addition, fiber optical sensing allows using the fibers as the platforms for *evanescent-wave detection* (Section 8.5), thus providing essential enrichment to all sensing technology.

The *optical fiber sensing devices* are frequently applied in the continuous monitoring of different industrial processes and, on a more limited scale, in environmental monitoring of air and water quality (Bosch et al. 2007; Taitt et al. 2005). They are attractive for clinical diagnostics. Since they allow a high degree of miniaturization and considerable geometric versatility, this makes them appealing both for the field and for point-of-care analyses. In industrial applications, they are particularly useful for making measurements in potentially hazardous, radioactive or explosive environments, with the transfer of the optical signal via the fibers for measurement at a safe distance.

Sensing with waveguides and classical measurements with a spectrofluorimeter can employ the same principles. Moreover, it may use the same equipment, in which the delivery of excitation light 'by air' within the instrument is substituted by delivery in a waveguide from a distant 'probe'. The application of optical fibers and waveguides allows constructing the 'probes', in which the beam of excitation light is guided out from the instrument, provides distant target detection and then, as a fluorescence response, is guided in to reach the instrument for recording the signal.

This is only one possibility out of many others. Waveguides can also perform in a way, in which common spectrofluorimetry can never operate. Assembled together, the fibers can form optical arrays that are easily integrated with a multitude of different sensing schemes (Epstein and Walt 2003). These arrays can allow multiparametric analyte detection. Moreover, many such sensors put together can form an image. Imaging fiber bundles composed of thousands of fused optical fibers are the basis for an optically connected, individually addressable, parallel sensing platform (Yagai et al. 2007).

9.2.1   Optical Fiber Sensors with Optode Tips

There are two major possibilities to use optical waveguides in sensing. One is to use the fiber terminal for attaching the sensor molecules that have to be excited by light through the waveguide and then produce an emission signal that has to be returned by the same waveguide to the instrument detector. The other possibility is to use the sensor molecules or particles attached to the periphery of a waveguide for excitation by the evanescent wave.

The optical fiber sensing instrument based on the first principle may consist of three main parts: (a) the probe called an *optode*, which provides the fluorescence response by modulating the optical signal on target-sensor interaction; (b) an optical link (waveguide) that carries the optical signal from the instrument to the probe and *vice versa* and (c) an optoelectronic system that contains the hardware and software for interrogating the probe and processing the signal. The light modulated by the optode and carried by the fiber should bring the information on the presence of the target and allow its quantitative determination.

It is easy to fabricate an optode for sensing if the target analyte is fluorescent itself. In case the target molecules are optically silent, one has to develop the recognition process on the optode tip, which could involve fluorescence signal generation. In oxygen sensing, for instance, this could be the decrease of the intensity and lifetime on quenching by the oxygen of a long-lifetime luminescence reporter and in pH sensing – the change of spectra of an attached acid-base fluorescence indicator. In other cases, one has to use more complicated receptor-transducer-reporter combinations, like those described in previous chapters of this book.

Optical fibers have been used to develop sensors based on nucleic acids and cells (Brogan and Walt 2005). Sensors employing DNA probes have been developed for various genomics applications and microbial pathogen detection. Live cell-based sensors have enabled the monitoring of environmental toxins and have been used for fundamental studies on populations of individual cells. Both single-core optical fiber sensors and optical fiber sensor arrays have been used for these purposes. Many original papers and reviews describe various applications of this technique in biosensing, including the detection of enzymes, whole cells, antibodies, nucleic acids and biomimetic polymers (Marazuela and Moreno-Bondi 2002).

9.2.2   Evanescent-Field Waveguides

In addition to sensor devices based on optode tips, there are devices that use a part of the *optical fiber* (usually close to its distant end) as the *sensing surface*, with the operation based on evanescent-field transduction. Due to the evanescent field propagating out from the waveguide at a distance up to 100–200 nm, only the labeled molecules that are bound to the surface will be selectively excited. Tapered waveguide areas at its distant terminus can be used as the solid support for the proper sensors (Leung and Cavalieri 2003). It has been shown that this technique works well in a sandwich configuration and can compete with traditional immunoassays. The decrease of the fiber diameter (tapering) to the level of several microns increases dramatically the magnitude of the evanescent field and therefore the sensitivity of the assay. It can be stated that optical fiber-based sensors combine the use of a target recognition element with an optical fiber or optical fiber bundle. They can be used for a variety of analytes, ranging from metals and chemicals to physiologically active materials (Monk and Walt 2004).

Similar in performance to optical fibers are the *waveguides of planar geometry* (Bradshaw et al. 2005). They are also based on the evanescent wave excitation principle. Their advantage is that they can easily be fabricated in the form of thin films. Planar waveguides provide the possibility of immobilizing multiple capture biomolecules onto a single surface and therefore, offer the exciting prospect of multi-analyte detection. They have found important applications in microfluidic devices, as we will see below (Section 9.4).

The ultimate sensitivity of the optical waveguide sensing technique is achieved mainly due to the background signal suppression. It has been reported (Plowman et al. 1999) that the construction of a dual-channel fluoroimmunoassay, which was able to detect femtomolar (i.e., less than 1 part per trillion, w/w) target concentrations. It operated on an etched channel silicon oxynitride thin-film impregnated optical waveguide, which was tested both in direct (fluorescent-labeled protein binding to immobilized receptor) and sandwich-type assays.

Optical fiber-based techniques can be successfully used. They operate in a similar way to the planar waveguide sensors and in comparison with them they have certain advantages and disadvantages. Overall, they are flexible and light, easy to manipulate with ingoing and outgoing light, they can be operated from long distances and are very cheap. On the other hand, due to their core-cladding geometry, it is difficult to access from outside the evanescent field, which is the main fiber parameter that is needed for sensing. One way to overcome this problem is to taper the fiber (reduce its diameter to a few microns or less). The light guided at the outer boundary of a taper waist can interact with the external media via the evanescent field. Recently, tapered fibers or silica nanowires have attracted great attention for sensing applications (Domachuk and Eggleton 2004). Due to their unique waveguiding properties, they have the potential to become the most sensitive waveguide sensors.

9.3   Multi-Analyte Sensor Chips and Microarrays

Miniaturization is, of course, not the only aim in constructing new generations of sensors. Novel sensors can do much more – they can detect many analytes simultaneously, adapt to several analyte concentration ranges and provide multi-analyte comparison between the samples. To proceed in depth with these tasks and correspondent devices, we have to introduce the principle of multiplexing. A system is defined as *multiplexed* if the number of sensing elements is larger than their possible number in the case of their individual operation (i.e., supplied individually by light sources, detectors, etc.). The sensors can be multiplexed by location (sensing the same target in different areas) or by target (simultaneously sensing different targets).

Multiplexed operation by location can be performed by microarrays. The *spotted microarrays* are assemblies of a great number (hundreds or thousands) of sensor instruments in which every instrument is represented by a spot on a surface formed by the deposition of one type of sensor molecule or particle. The analyzed sample is exposed to the whole microarray, so the binding of different targets to correspondent sensor instruments occurs in essentially the same conditions. The read-out of the reporter signal is made by using specialized fluorescence microscope instruments – the *readers*.

In biosensing technologies, the term '*spotted microarrays*' is often substituted with a term of much broader meaning, '*biochips*'. Biochips are highly miniaturized micro-scale analytical devices for the detection of proteins and mainly, nucleic acids exhibiting massive parallelism of function. The microarray technologies have become very popular in biosensing applications and their development is stimulated by practical needs in genetic analysis (Dharmadi and Gonzalez 2004; Epstein and Butow 2000), detection of antigens and antibodies of different specificities (Luppa et al. 2001) and in drug discovery. The advantages of microarray technology are miniaturization, reduction of volume of reagents and of reaction volume. Unified automated procedures can be applied to all steps of manipulation with the arrays.

9.3.1   Fabrication

The design and fabrication of microarrays requires operating with materials with great precision on the micrometer and even on the nanometer scale. The rapidly developing semiconductor and integrated circuit research and industry has acquired a wealth of experience in nano-scale manipulations, which can be successfully applied in sensing devices. In a recently published review (Blattler et al. 2006), the leading techniques for *generating nanopatterns* with a biological recognition function are described. They include parallel techniques, such as extreme ultraviolet interference lithography, soft-lithographic techniques (e.g., replica molding and microcontact printing), nanoimprint lithography and nanosphere lithography (e.g., colloid lithography or colloidal block-copolymer micelle lithography). In addition, direct-writing techniques, including e-beam lithography, focused ion-beam lithography and dip-pen nanolithography can be used.

Details on how the *nanopatterns are generated*, how the sensing function is imparted to these patterns and examples of how these surfaces can and are being used for the immobilization of sensor molecules and particles, can be found in different publications (Bally et al. 2006; Venkatasubbarao 2004). The support materials could be either glass or plastic and of major importance is their surface chemistry. The methods of *sensor application* should allow the optimal exposure of sensors to capture targets but also minimize the non-specific adsorption of non-target components of the studied sample (Sobek et al. 2006). Following the progress in these activities, one may observe that every step brings substantial progress in increasing the spot density, by making them three-dimensional (using hydrogel support) and by increasing the sensitivity of the response.

Microarray fabrication allows different possibilities for the application of sensor molecules and nanoparticles. Deposition of oligonucleotides, PNA and peptides can be provided in two ways – by covalent attachment and by on-site synthesis (Venkatasubbarao 2004). In the latter case, the growth of the polymer chain occurs directly on the surface of the glass support. Direct spotting is used for the attachment of proteins and long-chain DNA. For DNA attachment, the slide surface can be modified with the attachment of amino groups and the DNA is kept bound by ionic interactions. Protein sensors are usually deposited on the array surface based on affinity interaction between biotin and streptavidin. Other partners for affinity coupling mentioned in Section 7.3 can also be of use.

9.3.2   Problems with Microarray Performance

The *spotted microarrays*, providing a powerful analytical tool for the simultaneous analyses of thousands of sensor responses in a single experiment, allow the application of all fluorescence detection methods described in Chapter 3. Meantime, many problems have appeared that are related to the sensitivity, accuracy, specificity and reproducibility of microarray results. These problems are of special concern when the technologies based on the preliminary labeling of all components of the analyzed sample are applied (Draghici et al. 2006). Since this is presently the dominant approach in the DNA hybridization assay and alternative methods (such as molecular beacons) are introduced slowly into practice, there remains considerable uncertainty and skepticism regarding the data obtained using this technology (Wilkes et al. 2007).

Comparison of the results of seemingly identical experiments from different laboratories or even from different days shows that they do not always match satisfactory. Even stronger inconsistencies can be observed when data from different array platforms are compared. For improving confidence in these data, optimization during every step of microarray performance and standardization of optimized procedures is needed.

In real systems, the concentrations of analyzed compounds vary within broad ranges. They may exceed the size of the analytical window determined by the binding constant in either direction, which is no more than two orders of magnitude (see Section 2.2). To covering an extended range of target concentrations, a number of sensor units with the same specificity but different affinities can be constructed to analyze every target. This has to increase the size of the array. In addition, attempts to extend the target detection to weaker affinities, with dissociation constant, K_d, lower than $\sim 10^{-7}$ M, leads the detection methods to direct sensing, which requires establishing the dynamic equilibrium between the free and bound analyte and limits the application of washing and other manipulations with the arrays.

A significant related problem arises from the fact that on target binding, on microarrays the true *association-dissociation equilibrium* is not reached (Carletti et al. 2006). The characteristic time for the DNA hybridization equilibrium, $t_{1/2}$, is estimated at 41 days for a target human genome. Similar hybridization times are expected for the whole-transcriptome chips hybridized with tissue DNA. These conditions are commonly not achieved. Non-equilibrium binding introduces a stochastic factor into hybridization dynamics and kinetic variables, such as a differential diffusion rated of sample components, which may influence the result. This issue was discussed in Section 2.3.

Since the spotted sensor concentration and optical alignment to the spot readout may vary from sample to sample, the problem of *internal calibration* is of special importance for microarrays. Similar problems also exist in the fluorescence imaging of cells, due to the commonly observed spatial variation of intracellular sensor concentration and in flow cytometry, in which a broad distribution of data is commonly due to variations of cell volume and of the extent of the fluorescence probe uptake. In all these cases, the sensor (and reporter dye) concentration is variable (or unknown) and it can change over time.

Often, the reporter dye responds not only to analyte binding but also to many other factors affecting its molecular environment due to the unavoidable sensitivity of its emission to various parameters of physical and chemical origin. Therefore, it is not possible to provide a response only to analyte binding and achieve suppression of all other influences. Because of that, the measured fluorescence intensities in microarrays may vary significantly and may not reflect the actual differences in target concentrations in analyzed samples.

The problem of the *calibration of the response signal* on arrays has been discussed (Demchenko 2005a, b) and it was shown (see Section 3.7) that different fluorescence sensing techniques are affected differently by these problems. Comparative analysis based on three important properties – insensitivity to instrumental factors, insensitivity to the amount/concentration of sensor molecules and insensitivity to intrinsic factors influencing the fluorescence, such as the fluorescence quenching, as given in Table 3.1, demonstrates convincing reasons for substituting intensity sensing by more advanced methods of detection. Internal calibration is possible with the application of two dyes (with the reference in intensity sensing and with the fluorescent acceptor in FRET) and with a single dye in anisotropy and lifetime sensing. Meantime, the most efficient could be the application of a single dye with a wavelength-ratiometric response.

9.3.3   Read-Out and Data Analysis

Each member of the array (spot) is easily identified by its position in the (x,y) coordinate space, which can be done with microscope-based two-coordinate readers. Meanwhile, the analysis of read-out information from microarrays is a non-trivial task because of the large data volume and the many levels of its variation introduced by different stages of experiments. Visually, it is very hard to analyze an array that consists of several tens of thousands of elements. The analysis can be further complicated by the heterogeneity in response, for instance, by variations of affinities of sensors binding the same target (Elbs et al. 2007). The comparison of different data sets obtained at a different time and with different instrumentation may appear even more difficult. Statistical analysis cannot improve the primary data; it can only suggest their most probable interpretation. Therefore, the primary data have to be of the best quality, they should be as little affected by systematic and experimental errors as possible.

During the step of analysis, *different algorithms* can be applied with the correspondent software. Their description can be found in the literature (Leung and Cavalieri 2003). Some of them are rather complicated but some are available to those users who do not have experience in programming. Such is visual programming (Curk et al. 2005), which offers an intuitive means of combining known analysis and visualization methods into powerful applications. For a detailed description of different algorithms for the analysis of microarray data (such as data filtering, hierarchical clustering and principal component analysis) the reader is advised to refer to a comprehensive review (Quackenbush 2001).

9.3.4   Applications of Microarrays

The applications of microarrays are well presented in other sections of this book (Chapter 10, in particular), so here we will provide only a brief listing. The primary application that this technology has found is in *DNA hybridization assays*, where a comparative method, based on the simultaneous application of studied and control samples labeled with different dyes, is commonly used (Dudley et al. 2002). Usually, whole-pool labeling is provided in these tests. In this respect, molecular beacons and tests using PNA and conjugated polymers have started to demonstrate their advantage mainly due to the absence of the necessity to label the whole analyte pool.

Protein-detecting microarrays (Tomizaki et al. 2005) must produce an enormous impact on the functional analysis of cellular activity and regulation, especially at the level of protein expression and protein-protein interaction and may become an invaluable tool in disease diagnostics. However, the biochemical diversity and the sheer number of proteins are such that an equivalent analysis is much more complex and thus difficult to accomplish. In addition, proteins do not demonstrate simple chemistry or uniformity of labeling, which requires the construction of microarrays based on different principles than analyte pool labeling before application.

Microarrays based on the *application of antibodies* (Kusnezow et al. 2003) have been applied for many purposes. The antibodies have strong competition from other molecules and particles used for molecular recognition, in particular from aptamers (Collett et al. 2005).

9.3.5   Prospects

In less than a decade, microarray technology has made tremendous progress in its development. It started from exposing analyte samples to droplets robotically produced on microscope glass. Presently, it actively uses methods derived from microelectronics in combination with advanced synthetic chemistry. Photolithography is used for the on-chip synthesis of highly dense arrays or pre-synthesized peptides and oligonucleotides. Alternatively, they are put down by electronically addressed deposition. Both approaches lead to densely located surface-exposed molecular sensor elements, which could detect hundreds or thousands of targets simultaneously.

The power of array technology is that obtaining such a huge amount of information is accomplished (almost) simultaneously and in the same conditions, which minimizes the systematic errors in their detection. Most of the recent applications of these devices are in DNA and RNA hybridization assays, in the detection of antigens and antibodies and in drug discovery. They extend rapidly into the studies of proteomes with the aim of constructing extended maps of protein-protein interactions (interactome). The technique is extending into the areas of diagnostics, forensic analysis and environmental monitoring. With this success, one has to take into account the weak points of this technology. Reading the microarrays requires bulky and expensive instrumentation and the sample application and labeling are quite laborious. The incubation time is long and the method is subject to many types of errors. These problems are addressed to future researchers.

Fluorescence reporting offers a very high sensitivity of detection within picoliter volumes. The further reduction of this volume leads to a natural limit of single molecules. On this pathway we observe the appearance of '*nanoarrays*', in which the size of every spot can be decreased by three orders of magnitude in comparison to common microarrays (Lynch et al. 2004). It is possible but not always reasonable to reach even smaller volumes, due to the necessity of applying much more expensive instrumentation for the reading. It is also important that determination of target concentrations is necessary with some statistical value. Meanwhile, in living cells the interactions of individual molecules occur. They may not always follow thermodynamic rules and it is interesting to observe what kind of new effects can then be detected.

9.4   Microsphere-Based Arrays

Plain spotted arrays on a solid support are not the only means for multiplex detection. Individual sensors that are simultaneously focused at the detection of an immense number of analytes need not be assembled into spots that are geometrically separated by attachment to the supporting plane. The sensor location may be unimportant and they may remain in solution (suspension), if there is a possibility to recognize any sensor in this system and distinguish it from the sensors of other specificities. *Microspheres* made of synthetic polymer or silica can serve as good carriers of such sensors. Samples containing a mixture of potential analytes together with non-analyte compounds can be exposed to the mixture of such sensing beads, so that specific binding of every analyte should occur to the appropriate bead.

Applying this technology seems to be easy but there remain two problems to be resolved. One is *to provide uniquely addressable coding* of microspheres carrying particular sensors, so that they can be recognized in the mixture. The other problem is to obtain the information on target binding to the sensors *on every bead individually* and to provide its quantitative analysis. Solutions to these problems will be discussed below.

9.4.1   Barcodes for Microsphere Suspension Arrays

The principle is that each target-specific element in the array has to be represented by a subpopulation of particles with distinct and well recognizable optical properties, their '*barcodes*'. Dyes (and nanoparticles) emitting at different wavelengths may serve as these barcodes. For instance, the simultaneous detection of multiple bacterial targets can be achieved by using antibodies marked by silica particles containing three co-located dyes in different proportions and coupled by FRET (Ma et al. 2007). The dyes allow excitation at a single wavelength (of the donor) and recognition of the particles by their different colors of fluorescence emission is provided by the acceptor. In this and similar cases the FRET mechanism is not used for the reporting function of a sensor but as a 'wavelength convertor', which allows obtaining a family of 'barcodes' that can be recognized by variable fluorescence spectra.

Coding with different dyes emitting at different wavelengths should not necessarily involve FRET. The code can be presented by a number of non-interacting dyes that differ in their wavelength position and intensity. In this case, the number of codes C can be estimated from the formula $C = N^{m-1}$, where N is the number of intensity levels and m = the number of colors (Wilson et al. 2006). In theory, six colors at six different intensities should provide nearly 40,000 unique codes. Such estimates, however, have to be considered as limits that are hard to reach in practice in view of the spectral overlaps between the dyes and the variability of fluorescence intensities introduced during both the production and reading steps. In this respect, Quantum Dots have much better prospects than organic dyes because they possess much narrower spectra, covering the whole visible range, they can be excited by a single wavelength and exhibit higher photostability. In addition, composites between Quantum Dots and organic dyes can easily be constructed. By employing a mixture of many microspheres, each containing a different target-binding site and distinctive fluorescent encoding, one can increase the array size enormously (Eastman et al. 2006).

Quantum Dots can be efficient FRET donors and, in combination with organic dyes, they can provide many possibilities for barcode constructions (Clapp et al. 2005). This allows the construction of an immense number of detection channels by variation of the QDs size and thus their emission spectra, together with variation of the QD/dye ratio.

In addition to the possibilities for spectral-fluorescence barcoding, other possibilities to identify the particles can be suggested. A two-dimensional graphical code similar to the barcodes found in supermarkets can be generated in a laminar flow in a microfluidic system (Pregibon et al. 2007). This method, based on continuous-flow lithography, combines particle synthesis and encoding with probe incorporation into a single process, initiated and controlled by photopolymerization, with the introduction of a photomask. Over a million of unique codes can be generated in this way.

9.4.2   Reading the Information from Microparticles

The marked sensor particles should be counted individually and not only their recognition but also their response to target binding should be recorded. There are two possibilities for this. One is to count and measure the microspheres *one by one*, as they pass through the detector unit. The other is to read them with a *microscope imaging system*. Both can be successfully realized with common equipment (flow cytometers and scanning microscopes, correspondingly) as well as with the aid of specially designed instrumentation (Wilson et al. 2006).

Flow cytometers are instruments developed for measuring cells and particles in the flow. In the heart of these instruments is a very thin glass capillary that can be illuminated with several lasers and allows detection of scattered light and fluorescence emission at several wavelengths from microscopic objects (particles or cells) passing through this capillary. The results in flow cytometry are usually displayed on a computer screen in two dimensions, with dots representing the counted objects as a function of light-scattering and fluorescence intensity or fluorescence intensities at two different wavelengths. Modern flow cytometers can measure and sort more than 20,000 cells or particles per second, with continuous detection of many (ten or more) light emission and laser-scattering signals. Since this method allows for large-scale screening applications, it can be very useful for multiplexed assays and is able to provide a combined output that involves barcode reading and a reporter signal. Using automated sample handling, these measurements can be provided in real time, with subsecond kinetic resolution.

Therefore, merging a microarray concept with flow cytometry looks very natural. Highly specialized instruments based on this principle can be constructed (Tung et al. 2004).

Reading barcodes with *imaging devices* is an alternative that can be realized with inverted epifluorescence microscopes equipped with CCD cameras. Decoding should be provided by image-recognition software.

9.4.3   Prospects

Microsphere suspension arrays, with the solution of the barcode problem, are becoming strong competitors to flat microarrays. They are potentially superior in speed of analysis, since the kinetics of target binding to them can easily be accelerated by an efficient mix in solution. There are additional benefits of this technology, such as a larger surface area for the functionalization of microparticles relative to flat supports, flexibility and simplicity of manipulation (e.g., microspheres can be made magnetic). Microspheres are much more easily reproducible in fabrication than the flat arrays. Whereas the same chemistry has to be used for sensor binding to flat arrays, microsphere arrays are free from this limitation. Moreover, one can easily make the necessary array composition by just mixing the stocked microspheres loaded with the necessary sensors. The availability of instruments for reading the microsphere arrays must stimulate further development of this technology. Regardless of the principle of fluorescence detection, no sample pre-treatment or after-treatment, such as washing, may be needed. This is common for flow cytometry and its ability to resolve free and bound probes (by realizing sandwich assays, target labeling, the change in light-scattering, etc.) allows different assay formats to be applied. Thus, suspensions may compete successfully with solid chips.

9.5 Microfluidic Devices

The reduction of the size and volume of sensor and chemical reactor devices by several orders of magnitude can be achieved with present technologies. This stimulates developing the technologies that allow sample manipulation together with sensing in a greatly reduced volume. *Microfluidic devices* that are often called *micrototal analysis systems* (μ-TAS) and *lab-on-a-chip devices* have already been developed in a number of laboratories and promise an even greater future (Dittrich et al. 2006) (Fig. 9.4). In addition to miniaturization, they offer many useful properties. It is the reduction of the diffusion lengths in the sample that increases the rates of target binding and of different chemical or biochemical transformations. Confined geometries prevent high dilutions and planar active surfaces allow a high surface-to-volume ratio and provide the high-density immobilization of the participants in the sensing process. The surface-related effects can be made dominant in sample handling and sensor-based applications.

Though many detection methods can operate together with microfluidic systems, fluorescence is of primary use. The recently developed technologies allow serial and inexpensive production of integrated microfluidic systems that incorporate the whole necessary range of optical detection elements, including light sources and optical detectors. A broad variety of sensors can be incorporated into these devices, from small molecules to nanoparticles and up to intact living cells.

9.5.1 Fabrication and Operation of a Lab-on-a-Chip

The '*lab-on-a-chip*' concept involves the fabrication of devices in which all elements will be miniaturized, including sample supply, excitation source and fluorescence readout unit. Both inorganic (silica) and organic polymers can be used as the bodies of these devices and different automatic micro-fabrication techniques, such as replica and injection molding, embossing and laser ablation, are used for producing fluidic channels and the introduction of reactive and optical elements (Dittrich et al. 2006). Their mass production can make them disposable and one-time use (Fiorini and Chiu 2005).

The inner surfaces of fluidic channels and volumes are modified for the capture of sorbent, catalyst and sensor molecules with the aid of laser-activated photo-

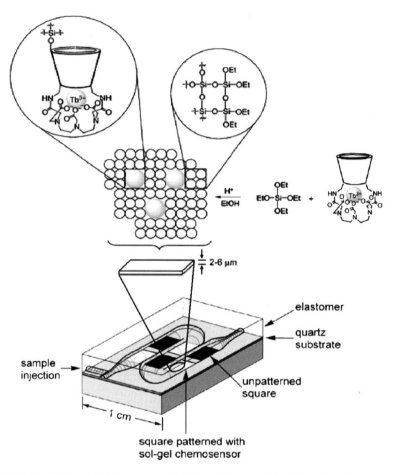

Fig. 9.4 Microfluidic optical chemosensor fabricated in a serpentine channel configuration. The white, sol-gel squares contain the supramolecular chemosensor, a cyclodextrin strapped by a DTPA macrocycle in which a Tb^{3+} ion resides. The conical bucket schematically represents the cyclodextrin receptor site. The binding of the ions is detected by the appearance of its characteristic luminescence bands (Reproduced with permission from Rudzinski et al. 2002)

chemical reactions, which allows high precision. They represent the solid phase, which can be combined with the moving phase, consisting of different types of molecules and nanoparticles of different sizes. Particularly Quantum Dots and other fluorescent nanoparticles can be used in this combination. Moreover, all the advantages of multiplex analysis using barcoded microspheres can be realized here (Lim and Zhang 2007).

The pumping and valving of fluid flow, the mixing of different reagents and the separation and detection of different chemical species have to be implemented in a microfluidic format and this is technically achievable. The filling and manipulation of these devices can be accomplished due to capillary forces and by centrifugation using a developed compact disc (CD) based on centrifuge platforms (Madou et al. 2006). Microscopic pumps and valves can be made based on different principles, such as pneumatic, electromagnetic (Wang et al. 2004) and photo-switchable with laser beams (Nagai et al. 2007). A strong effort is presently being made to make interfacing and interconnecting with macroscopic lab devices as simple and convenient as possible or even substituting them completely.

An important possibility for manipulation in microfluidic devices can be provided by optical forces that drive the particles of the higher refractive index, rather than the surrounding medium, to the region of maximum optical density. The application of this effect has led to the rapid development of '*optofluidics*' (Monat et al. 2008).

9.5.2 Microfluidic Devices as Microscale Reactors and Analytical Tools

The lab-on-a-chip devices allow *different manipulations* with the applied sample, so that extremely low sample and reagent quantities are consumed inside the chip's microspace. Sample preparation, including separation of components, can be provided directly inside the device and this process can be coupled with sensing. From the side of synthetic chemists, the interest in miniaturized analytical systems was stimulated by the fact that physical processes can be more easily controlled and harnessed when instrumental dimensions are reduced to the micrometre scale. Such systems define new operational paradigms that might revolutionize the fields of high-throughput synthesis and chemical production (deMello 2006).

An important characteristic feature of these devices is that *chemical reactions* and *sensing events* can be coupled or preceded by *sample preparation and separation steps*. Such a combination allows one not only to simplify the assay but also to minimize the non-specific adsorption and side reactions. The measurement is often coupled with preparation of the sample. The pre-treatment of the sample is commonly needed to remove sample components that do not contain the analyte but may complicate the analysis. This may be a separate procedure. Researchers often try to combine it with measurement, especially when dealing with such a heterogeneous species as human blood. As an example, steroid screening can be accomplished starting from the whole blood sample; a membrane filter can be incorporated directly into the micro-device (Thorslund et al. 2006).

Using the lab-on-a-chip concept, miniaturized devices for *DNA analysis* can be constructed that are able to provide dielectrophoresis for the purification of DNA, artificial gel structures for rapid DNA separation and nanofluidic channels for the direct visualization of single DNA molecules. The marriage of microfluidics with detection technologies that rely on highly selective nucleic acid hybridization provides improvements in bioanalytical methods for purposes such as the detection of pathogens or mutations and drug screening. Importantly, a complex and multi-step process such as the polymerase chain reaction (PCR), which requires not only a reagent change but also temperature cycling, can be conducted in continuous-flow microchips (Dettloff et al. 2008).

Microfluidic immunosensor systems can be realized in both homogeneous and heterogeneous assay formats (Bange et al. 2005), which is important for clinical applications. Sandwich immunoassays can also be adapted to a microarray format that allows the detection of many target compounds as antigens.

9.5.3 Fluorescence Detection in Microfluidic Devices

There are two tendencies in the realization of fluorescence detection in microfluidic devices. One is to make an integrated optical chip located in close contact with the microfluidic part but separated from it. This allows one to make the microfluidic part disposable but the optical part is re-used. As an illustration, one of the suggested devices (Chediak et al. 2004) contained a semiconductor LED, a CdS optical filter and a silicone photodiode integrated in the same unit and a microfluidic reactor located in close proximity.

The other tendency is to assemble all the elements that are necessary for efficient instrument performance in one chip, including sample separation, chemical modification and binding to immobilized sensor molecules with visualization of the result using the necessary assembly of the excitation and fluorescence detection technique. In this configuration, the microfluidic units have to become the true *micro-total-analytical-systems* ($\propto$TASs) that will be self-consistent devices independent of interfacing to external macro-scale equipment. A strong move in this direction has been observed. One of these lab-on-a-chip systems integrates five different components monolithically assembled on one substrate. They represent three main domains of microchip technology: optics, fluidics and electronics (Balslev et al. 2006). Interestingly, among these elements an optically pumped dye laser can be integrated into such a miniature device.

Multi-analyte sensing in these systems is possible in different configurations of the microfluidic chip. One, for instance, involves sensor molecules immobilized on the walls of *capillary channels*, so that a microliter droplet of sample fills the channels by capillary forces (Henares et al. 2007). An optical image is then obtained with a fluorescent microscope. The other possibility is to apply optical waveguides exposed to the sample filling the channels, with evanescent-field detection (Irawan et al. 2006). The planar waveguides can be monolithically integrated with the microfluidic channels. They also offer the possibility of creating multiple parallel assays on a single structure. Mobile-phase sensing can be performed with barcoded microspheres and this provides the link between micro(nano)arrays and microfluidics.

The marriage of microarray technologies with the emerging field of microfluidics is especially attractive. Combining the highly parallel analysis of many targets in the sample with the advantages of microfluidics, such as the reduction in reagent costs, reductions in hybridization assay times, high-throughput sample processing and also integration and automation capabilities of the front-end sample processing steps, is very promising (Situma et al. 2006). This potential marriage requires many additional steps, such as developing low-cost manufacturing methods for the fluidic chips and providing them with good interfaces to the macro-world, in addition to optimizing the interactions of the microspheres with the microchannel walls.

9.5.4 Prospects

Micrometer-scale analytical devices combining the separation of sample components, their chemical reactions and detection properties are more attractive than their macro-scale counterparts for various reasons. They use smaller volumes of reagents and are therefore cheaper, quicker, less hazardous to use and more environmentally appealing. The dramatically reduced volume and comparatively extended inner surface create unique possibilities for sample manipulation, separation of its components, chemical reactivity and sensing. Such devices are already capable of performing difficult chemical and biological manipulations combined with analytical detection. The possibility of their mass fabrication and making them disposable will be the key determinants in their future applications. No doubt, fluorescence will be the major detection technique in these devices. It will allow exploration of many possibilities that are especially attractive for miniaturized devices, such as evanescent-field and near-field excitation. Single-molecule and single-particle detection will provide the ultimate sensitivity.

Thus, a microfluidic device can serve the function of a whole analytical laboratory and, in addition to sensing, can provide sample preparation and separation of its components, which can be done with the active participation of the modified surfaces of the microchannels. The benefits of miniaturization are the low sample and reagent consumption and extremely fast analysis times. The operation of such devices requires very sensitive detection, which can be provided by the integrated fluorescence detection technique.

Portable microfluidic devices using the flow cytometry principle are expected to enjoy rapid development in the near future (Tatosian et al. 2005). This could allow the solving of a number of problems. The microspheres, nanoparticles or living cells will be studied in the flow, so there will be no need for their immobilization (adhesion) in the microchannel. Their maintenance and exposure to an analyzed sample can be performed in special reaction chambers and the recorded signals from the statistical analysis of their large numbers will allow average variables to be derived.

9.6   Devices Incorporating Whole Living Cells

The idea to use whole cells as highly integrated biosensors responding to different stimuli with their incorporation into microfluidic devices is especially attractive. This is because these devices offer an efficient and relatively easy way to incorporate the cells and to maintain them in native conditions by supplying the culture media. The inner volume of the microfluidic channel can be as small as to allow the location of one or several cells, so that their individual properties can be studied. The size of the miniaturized optical system becomes closer to that of the cell and therefore, the excitation and collection of emission light can be more efficient than with the microscope. However, such a device is not a substitute for a microscope because of the absence of inherent imaging possibilities. The cell is illuminated as a whole and its response is detected as an integrated signal. The selectivity of the response is thus determined by incorporated fluorescent dyes (or biosynthesis of reporters within the cells), their location in the cell and their response to applied stimuli.

In this technology the whole cell may be considered as a top-level biosensor. This is because it can respond to a wide variety of functionally important stimuli in a *highly integrated* but *specified* manner. It can enhance, integrate or suppress the effects of different chemical stimuli, which allows detecting them as targets. In many cases, the signal generated by target binding to cell receptors can be transformed into the *generalized and amplified signal of a cell response* and expressed in such integrated variables as *toxicity*. This opens many possibilities for determining and testing different hazardous compounds and drugs.

A lot of important information about cellular functions and their responses have been obtained with the use of two powerful methods – fluorescence microscopy and flow cytometry (see Chapter 11 for more details). Scanning confocal and two-photon microscopies can allow obtaining valuable information about the cells in two complementary aspects. Microscopy allows studying the cell ultrastructure and following the distribution of metabolites and ions, etc. But this can be done only for one cell at a time, so it is hard to derive the statistical distribution of the measured parameters within the cell population. Flow cytometry, in contrast, allows obtaining such statistical information but for a limited number of parameters. The introduction of the microfluidic concept into cellular research brought a new possibility: to study an extended number of parameters for one or a small number of living cells trapped in the lab-on-a-chip channels.

9.6.1   Cellular Microorganisms or Human Cultured Cell Lines?

Many different cells, microbial, yeast and mammalian cell lines have been suggested for whole-cell sensor applications (O'Shaughnessy and Pancrazio 2007; Yagi 2007). They may display different sensitivities to target compounds, so different cells may be needed for different ranges of variation of target concentration. Bacterial cells are of active use, particularly for environmental monitoring of pollutants. They are less demanding regarding cultivation conditions. Recent findings, however, showed that the cloned bacterial populations are always physiologically, phenotypically and genotypically heterogeneous. This emphasizes the need for sound statistical approaches for the interpretation of a reporter response in individual bacterial cells (Tecon and van der Meer 2006). In order to reduce this heterogeneous response and to determine a number of targets within the same format, sensor arrays with individual cells immobilized in fabricated, patterned, hydrogel microwells have been developed.

The advantages of *bacterial* and *yeast cells* follow from their specific properties:

(1) They can easily be *genetically engineered* with the incorporation of genes that are able to synthesize the required recognizing or reporting molecules.

(2) Some targets, such as pollutants, can activate their metabolic pathways by influencing the expression of single genes (Belkin et al. 1997). In a more general sense, they can offer insight into the *physiological effect* of an analyte (Stenger et al. 2001).

(3) They can be *bioluminescent* naturally or modified so that a bioluminescent reporting gene is fused to a promoter sequence of the gene of the relevant pathway. Such modified bacteria offer a solution to the problem of generating a reporter signal. Alternatively, fluorescent proteins of the GFP type can be fused with desired proteins to generate a fluorescence response.

Several approaches for the *transduction of cellular signals* are being explored. The versatility of fluorescent dyes together with the new possibilities offered by fluorescent nanoparticles allows tracking cellular response on the nuclear, mitochondrial, cytoplasmic and biomembrane levels. The methods, based on well-developed determinations of membrane potential (Farinas et al. 2001) or intracellular release of Ca^{2+} ions, can be successfully applied.

9.6.2   Living and Fixed Cells

In all analytical tools, *live cells* should be treated gently so that the incorporation of different molecules and particles as sensors should be compatible with cell viability. Whereas passive diffusion can be realized with small molecules, large molecules, such as polypeptide and protein sensors, need special cationic leader peptides (such as HIV-derived TAT peptide) to facilitate their entry into the cell. For larger particles, such as Quantum Dots, there is no general procedure for their incorporation and the strategy should be selected depending on the cell type. There could be uptake by endocytosis, direct microinjection of nanoliter volumes, electroporation (using the charge for physical delivery through the membrane) and mediated uptake (by incorporation into phospholipids vesicle or attachment of cationic peptide) (Derfus et al. 2004). All this can be realized in microfluidic devices. To maintain the cell's living conditions, a culture medium with all the necessary nutrients can be supplied through the microchannels.

In contrast, *fixed cells* that are often used in microscopic studies are in fact dead cells with chemically cross-linked components to maintain cellular architecture. Nanosensor particles of various sizes can enter the fixed cell easily through the pores, which can be created chemically. Different technologies in cellular research, such as flow cytometry, often involve the determination of intracellular antigens using corresponding antibodies. This procedure is possible only after cell fixation and permeabilisation treatment is needed to allow the antibodies to enter inside the cell. For the sake of biosensor applications, the researcher can make a selection between these two living or fixed conditions.

9.6.3   Single Cells in Microfluidic Devices

Microfluidic chips are ideal platforms for housing and handling small numbers of living cells. A large variety of microfluidic systems is available for cell analysis that allow not only a short-term manipulation with them but *maintaining their long-term culture* (for >2 weeks) and observing the whole range of their differentiation. This possibility has been shown for muscle cells, spanning the whole process of differentiation from myoblasts to myotubes (Tourovskaia et al. 2005). The microfluidic system provides accurate control of the perfusion rates and biochemical composition of the environment surrounding the cells, so that no difference is observed in differentiation between microfluidic and traditional cultures. Microfluidic systems allow performing a number of cell-based tests in parallel and monitoring them over the long term continuously (Hung et al. 2005).

We can observe that microfluidic devices used in research rapidly fill the gap between single-cell studies performed by microscopy and mechanistic studies of cellular processes, which are usually carried out with large populations of cells (Lidstrom and Meldrum 2003). This is important because the parameters that are measured as averages of large populations can be misleading. For instance, an apparently linear response to a signal could, in fact, reflect an increasing number of cells in the population that have switched from 'OFF' to 'ON', rather than a graded increase in response by all the cells. Studies in parallel of a *significant number of individual cells* are challenging but new technologies allow them to become a reality.

Optical waveguides are the basic and most important optical elements that must be integrated into lab-on-a-chip miscrosystems together with living cells. The common methods for waveguide fabrication are the conventional deposition techniques on glass substrates. However, polymers are more favorable substrates for microfluidic devices due to the fabrication flexibility that they offer and their low cost. Consequently, the fabrication of optical waveguides on polymeric substrates has the potential to solve a major integration problem (Leeds et al. 2004).

The deposition of living cells on top of waveguides in a microfluidic chip can potentially give valuable information about cell life and its perturbations. It has to be noted, however, that the evanescent field interaction depth is small (~200 nm) and decaying with distance to the waveguide, so in view of the larger cell size, the full integral response of the cell is difficult to achieve.

The *manufacture of microfluidic chips* for cell handling must meet several requirements. On the one hand, the design of the chip has to ensure the microfluidic functions (reproducible loading and unloading, washing filling and processing) and on the other, it has to ensure low costs and easy handling. Extremely important is to make a cell-friendly microenvironment that can be achieved by covering the device channels with peptides or biocompatible polymer hydrogels. There have been only a few attempts to integrate many functions together on a microfluidic platform and the challenge lies ahead. A solution can be found in building systems with a modular design. This design should still allow easy fabrication, for instance, by writing waveguides on polymers using UV treatment.

9.6.4 Bacterial Cells with Genetically Incorporated Sensors

Bacterial cells allow many possibilities for the synthesis of *genetically engineered guest proteins*. These proteins may serve as molecular sensors inside the cells. The microfluidic device in this case serves as a support and detection medium. A microsystem wherein *E. coli* cells were genetically engineered to express the desired capture proteins on the membrane surface and were spatially arrayed as sensing elements in a microfluidic device could serve as an example (Oh et al. 2006). Through the co-expression of peptide-based capture ligands on the cell surface and fluorescent protein in the cytoplasm, an effective means of directly linking the fluorescence intensity to the density of the capture ligands was demonstrated.

In line with inducing the ability of testing specific targets, strong attempts have been made to improve *general toxicity tests*. In this respect also, genetic engineering is becoming more and more important. Using the recombinant DNA technology, the toxicity-responsive gene promoter is fused to a gene coding a protein that can be easily quantified. Usually, this is either a GFP-family protein with detection in fluorescence or luciferase producing bioluminescent emission. Photosynthetic bacteria, with their built-in mechanisms of synthesis of fluorescent pigments, can also be actively used (Yagi 2007). Unfortunately, the promoter activation and generation of cellular response takes several hours, which is not very attractive for in-field analysis.

The examples of the application of the bacterial sensing system take advantage of the recognition of the regulatory protein, ArsR, for arsenite and antimonite to produce the reporter protein, which in this case is GFPuv (Rothert et al. 2005). The fluorescence emitted by the GFPuv in the cells can be directly related to the concentration of the analyte in the cell, making this sensing system useful in the detection of arsenite and/or antimonite in a variety of samples. Alternatively, the analyte may intervene in the metabolic processes of living cells and this provides the means for its highly sensitive detection.

9.6.5   The Cultured Human Cells

This technique, based on bacterial cells for the detection of toxic compounds, has some limits, since these cells do not respond to many compounds that are toxic to man. These are compounds that influence the nervous system, immunity and cell differentiation. In addition, some pollutants become toxic after processing in the liver. Therefore, the application of *cultured human cell lines* is very actual. Cardiac myocites and neurons are of primary concern and parameters such as membrane potential are attractive as the means of response (O'Shaughnessy and Pancrazio 2007).

9.6.6   Whole Cell Arrays

In many tests, in particular for pathogen detection, there were attempts to apply *microarray technology* with incorporated living cells (Ron 2007; Yoo et al. 2007). Genetically modified bacteria were used for this purpose. The modifications include the inducement or augmentation of recognition of the pathogen together with induction of the response in bioluminescence.

A significant problem is the viability of cells on arrays, since for long-term cellular monitoring, the cells need continuous perfusion. Special devices have been designed that allow performing many functions, including repeated cell growth/passage cycles, reagent introduction and real-time optical analysis (Hung et al. 2005).

9.6.7   Prospects

The living cell can be viewed as a constellation of potential molecular sensors with complex correlations between them. Cell-based biosensors can provide all the information that is available to molecular sensors regarding quantitative responses to analyte concentrations. In addition, their response may also contain integrated functionally relevant information, for instance, a response to a toxic compound by general stress. Being incorporated into microfluidic devices, the cells are able to respond dynamically and rapidly and allow long-term temporal observations. The detection of physiologically active compounds is the major field of application for whole-cell biosensors (Stenger et al. 2001). Importantly, in contrast to all other types of chemical sensors and biosensors, the cells can detect *unknown* threat agents by their functional response.

Sensing and Thinking 9: Optimizing Convenience, Sensitivity and Precision to Obtain the Proper Sensor Response

In practice, working with sensors is always dealing with devices that are required to connect the events occurring on a molecular and supramolecular or cellular level with the quantitative measure that we need to obtain and understand. These devices can be evaluated according to a number of parameters, such as sensitivity, speed of response, reversibility, repeatability, reproducibility, possibility of continuous monitoring, portability, ease of operation and price. Until now, many techniques in analytical chemistry have been adapted for the cuvette assay format (Lakowicz 2007; Valeur 2002), which is reasonable for detecting a single analyte and is not applicable for multi-analyte arrays. Still common is bench-top instrumentation that allows such precise but hardly productive measurements.

Presently, we shall witness dramatic changes. We observe that similar (or at least quite satisfactory) precision in analysis can be achieved with highly miniaturized instrumentation. Furthermore, its use can lead to a dramatic increase of efficiency of analysis regarding the possibility of sensing many targets together and thus to characterize samples of very complex composition. Furthermore, the new instrumentation demonstrates extreme flexibility in addressing different analytical tasks. Finally, it can be mass produced, and is cheap and available for personal use.

Questions and Problems

1. Explain the functioning of every element of the spectrofluorimeter. Find the miniaturized counterpart for each of these elements.
2. Explain the physical mechanism of light propagation in optical fibers and plain waveguides. Is there any difference? What will happen if $n_2 < n_1$? If $n_2 = n_1$? (see Fig. 8.9).
3. Which principles of incorporation of fluorescence reporters can and which cannot be realized with microarrays on a planar solid support (a) the same environmentally-sensitive reporters coupled to the sensors attached to all the spots, (b) competitors specific to different sensors labeled with the same dyes and (c) sandwich assays with detectors labeled with the same dyes? Can FRET be applied to these microarrays? In what configurations?
4. Compare the microsphere-based microarrays with spotted microarrays on a flat solid support with regard to their capacity and convenience of application.
5. Why is the equilibrium in sensor target interactions established much more quickly in microsphere-based microarrays than in microarrays on a solid support?
6. How do we detect the target binding to arrayed microspheres when the whole visible wavelength range is occupied with the emission needed for barcoding?
7. Estimate the optimal size of microspheres in view of the convenience of manipulation and detection.

8. What are the means to manipulate liquid flow in microfluidic channels?
9. Where should the sensors be located in lab-on-a-chip devices (a) attached to the inner volume of the channels, (b) in the moving phase, (c) on the incorporated optical waveguides, (d) in the outlet sample volume? Compare different possibilities.
10. How does one apply the optical units to lab-on-a-chip devices (a) unit providing monochromatic excitation, (b) unit providing the collection and detection of fluorescence?
11. What is the target-related difference between molecular and whole-cell sensors? Suggest the range of targets that are optimal for molecular sensing. Do the same for cell-based sensing.
12. How can the fluorescence response be introduced into/onto the cells captured in microfluidic devices?

References

Bacon CP, Mattley Y, DeFrece R (2004) Miniature spectroscopic instrumentation: applications to biology and chemistry. Review of Scientific Instruments 75:1–16

Bally M, Halter M, Voros J, Grandin HM (2006) Optical microarray biosensing techniques. Surface and Interface Analysis 38:1442–1458

Balslev S, Jorgensen AM, Bilenberg B, Mogensen KB, Snakenborg D, Geschke O, Kutter JP, Kristensen A (2006) Lab-on-a-chip with integrated optical transducers. Lab on a Chip 6:213–217

Bange A, Halsall HB, Heineman WR (2005) Microfluidic immunosensor systems. Biosensors & Bioelectronics 20:2488–2503

Basabe-Desmonts L, Reinhoudt DN, Crego-Calama M (2007) Design of fluorescent materials for chemical sensing. Chemical Society Reviews 36:993–1017

Belkin S, Smulski DR, Dadon S, Vollmer AC, Van Dyk TK, Larossa RA (1997) A panel of stress-responsive luminous bacteria for the detection of selected classes of toxicants. Water Research 31:3009–3016

Blattler T, Huwiler C, Ochsner M, Stadler B, Solak H, Voros J, Grandin HM (2006) Nanopatterns with biological functions. Journal of Nanoscience and Nanotechnology 6:2237–2264

Bosch ME, Sanchez AJR, Rojas FS, Ojeda CB (2007) Recent development in optical fiber biosensors. Sensors 7:797–859

Bradshaw JT, Mendes SB, Saavedra SS (2005) Planar integrated optical waveguide spectroscopy. Analytical Chemistry 77:28A–36A

Brogan KL, Walt DR (2005) Optical fiber-based sensors: application to chemical biology. Current Opinion in Chemical Biology 9:494–500

Carletti E, Guerra E, Alberti S (2006) The forgotten variables of DNA array hybridization. Trends in Biotechnology 24:443–448

Clapp AR, Medintz IL, Uyeda HT, Fisher BR, Goldman ER, Bawendi MG, Mattoussi H (2005) Quantum dot-based multiplexed fluorescence resonance energy transfer. Journal of the American Chemical Society 127:18212–18221

Collett JR, Cho EJ, Ellington AD (2005) Production and processing of aptamer microarrays. Methods 37:4–15

Curk T, Demsar J, Xu QK, Leban G, Petrovic U, Bratko I, Shaulsky G, Zupan B (2005) Microarray data mining with visual programming. Bioinformatics 21:396–398

Dandin M, Abshire P, Smela E (2007) Optical filtering technologies for integrated fluorescence sensors. Lab on a Chip 7:955–977

Dasgupta PK, Eom IY, Morris KJ, Li JZ (2003) Light emitting diode-based detectors absorbance, fluorescence and spectroelectrochemical measurements in a planar flow-through cell. Analytica Chimica Acta 500:337–364

Demchenko AP (2005a) The future of fluorescence sensor arrays. Trends in Biotechnology 23:456–460

Demchenko AP (2005b) The problem of self-calibration of fluorescence signal in microscale sensor systems. Lab on a Chip 5:1210–1223

deMello AJ (2006) Control and detection of chemical reactions in microfluidic systems. Nature 442:394–402

Derfus AM, Chan WCW, Bhatia SN (2004) Intracellular delivery of quantum dots for live cell labeling and organelle tracking. Advanced Materials 16:961–966

Dettloff R, Yang E, Rulison A, Chow A, Farinas J (2008) Nucleic acid amplification of individual molecules in a microfluidic device. Analytical Chemistry 80:4208–4213

Dharmadi Y, Gonzalez R (2004) DNA microarrays: experimental issues, data analysis and application to bacterial systems. Biotechnology Progress 20:1309–1324

Dittrich PS, Tachikawa K, Manz A (2006) Micro total analysis systems. Latest advancements and trends. Analytical Chemistry 78:3887–3907

Domachuk P, Eggleton BJ (2004) Photonics - shrinking optical fibres. Nature Materials 3:85–86

Draghici S, Khatri P, Eklund AC, Szallasi Z (2006) Reliability and reproducibility issues in DNA microarray measurements. Trends in Genetics 22:101–109

Dudley AM, Aach J, Steffen MA, Church GM (2002) Measuring absolute expression with microarrays with a calibrated reference sample and an extended signal intensity range. Proceedings of the National Academy of Sciences of the United States of America 99:7554–7559

Eastman PS, Ruan WM, Doctolero M, Nuttall R, De Feo G, Park JS, Chu JSF, Cooke P, Gray JW, Li S, Chen FQF (2006) Qdot nanobarcodes for multiplexed gene expression analysis. Nano Letters 6:1059–1064

Elbs M, Hulko M, Frauenfeld J, Fischer R, Brock R (2007) Multivalence and spot heterogeneity in microarray-based measurement of binding constants. Analytical and Bioanalytical Chemistry 387:2017–2025

Epstein CB, Butow RA (2000) Microarray technology - enhanced versatility, persistent challenge. Current Opinion in Biotechnology 11:36–41

Epstein JR, Walt DR (2003) Fluorescence-based fibre optic arrays: a universal platform for sensing. Chemical Society Reviews 32:203–214

Farinas J, Chow AW, Wada HG (2001) A microfluidic device for measuring cellular membrane potential. Analytical Biochemistry 295:138–142

Fiorini GS, Chiu DT (2005) Disposable microfluidic devices: fabrication, function, and application. Biotechniques 38:429–446

Fixe F, Chu V, Prazeres DMF, Conde JP (2004) An on-chip thin film photodetector for the quantification of DNA probes and targets in microarrays. Nucleic Acids Research 32:e70

Gardeniers JGE, van den Berg A (2004) Lab-on-a-chip systems for biomedical and environmental monitoring. Analytical and Bioanalytical Chemistry 378:1700–1703

Henares TG, Takaishi M, Yoshida N, Terabe S, Mizutani F, Sekizawa R, Hisamoto H (2007) Integration of multianalyte sensing functions on a capillary-assembled microchip: simultaneous determination of ion concentrations and enzymatic activities by a "drop-and-sip" technique. Analytical Chemistry 79:908–915

Holthoff WG, Tehan EC, Bukowski RM, Kent N, MacCraith BD, Bright FV (2005) Radioluminescent light source for the development of optical sensor arrays. Analytical Chemistry 77:718–723

Hung PJ, Lee PJ, Sabounchi P, Lin R, Lee LP (2005) Continuous perfusion microfluidic cell culture array for high-throughput cell-based assays. Biotechnology and Bioengineering 89:1–8

Irawan R, Tay CM, Tjin SC, Fu CY (2006) Compact fluorescence detection using in-fiber microchannels - its potential for lab-on-a-chip applications. Lab on a Chip 6:1095–1098

Kusnezow W, Jacob A, Walijew A, Diehl F, Hoheisel JD (2003) Antibody microarrays: an evaluation of production parameters. Proteomics 3:254–264

Lakowicz JR (2007) Principles of fluorescence spectroscopy. Springer, New York

Leeds AR, Van Keuren ER, Durst ME, Schneider TW, Currie JF, Paranjape M (2004) Integration of microfluidic and microoptical elements using a single-mask photolithographic step. Sensors and Actuators A-Physical 115:571–580

Leung YF, Cavalieri D (2003) Fundamentals of cDNA microarray data analysis. Trends in Genetics 19:649–659

Lidstrom ME, Meldrum DR (2003) Life-on-a-chip. Nature Reviews Microbiology 1:158–164

Lim CT, Zhang Y (2007) Bead-based microfluidic immunoassays: the next generation. Biosensors & Bioelectronics 22:1197–1204

Luppa PB, Sokoll LJ, Chan DW (2001) Immunosensors - principles and applications to clinical chemistry. Clinica Chimica Acta 314:1–26

Lynch M, Mosher C, Huff J, Nettikadan S, Xu J, Henderson E (2004) Functional nanoarrays for protein biomarker profiling. Nsti Nanotech 2004, Vol 1, Technical Proceedings:35–38

Ma Q, Wang XY, Li YB, Shi YH, Su XG (2007) Multicolor quantum dot-encoded microspheres for the detection of biomolecules. Talanta 72:1446–1452

Madou M, Zoval J, Jia GY, Kido H, Kim J, Kim N (2006) Lab on a CD. Annual Review of Biomedical Engineering 8:601–628

Marazuela MD, Moreno-Bondi MC (2002) Fiber-optic biosensors - an overview. Analytical and Bioanalytical Chemistry 372:664–682

Moe AE, Marx S, Banani N, Liu M, Marquardt B, Wilson DM (2005) Improvements in LED-based fluorescence analysis systems. Sensors and Actuators B-Chemical 111:230–241

Mogensen KB, Eriksson F, Gustafsson O, Nikolajsen RPH, Kutter JP (2004) Pure-silica optical waveguides, fiber couplers, and high-aspect ratio submicrometer channels for electrokinetic separation devices. Electrophoresis 25:3788–3795

Monat C, Grillet C, Domachuk R, Smith C, Magi E, Moss DJ, Nguyen HC, Tomljenovic-Hanic S, Cronin-Golomb M, Eggleton BJ, Freeman D, Madden S, Luther-Davies B, Mutzenich S, Rosengarten G, Mitchell A (2007) Frontiers in microphotonics: tunability and all-optical control. Laser Physics Letters 4:177–186

Monat C, Domachuk P, Grillet C, Collins M, Eggleton BJ, Cronin-Golomb M, Mutzenich S, Mahmud T, Rosengarten G, Mitchell A (2008) Optofluidics: a novel generation of reconfigurable and adaptive compact architectures. Microfluidics and Nanofluidics 4:81–95

Monk DJ, Walt DR (2004) Optical fiber-based biosensors. Analytical and Bioanalytical Chemistry 379:931–945

Monro TM, Belardi W, Furusawa K, Baggett JC, Broderick NGR, Richardson DJ (2001) Sensing with microstructured optical fibres. Measurement Science & Technology 12:854–858

Nagai H, Irie T, Takahashi J, Wakida S (2007) Flexible manipulation of microfluids using optically regulated adsorption/desorption of hydrophobic materials. Biosensors & Bioelectronics 22:1968–1973

Oh SH, Lee SH, Kenrick SA, Daugherty PS, Soh HT (2006) Microfluidic protein detection through genetically engineered bacterial cells. Journal of Proteome Research 5: 3433–3437

O'Shaughnessy TJ, Pancrazio JJ (2007) Broadband detection of environmental neurotoxicants. Analytical Chemistry 79:8838–8845

Plowman TE, Durstchi JD, Wang HK, Christensen DA, Herron JN, Reichert WM (1999) Multiple-analyte fluoroimmunoassay using an integrated optical waveguide sensor. Analytical Chemistry 71:4344–4352

Pregibon DC, Toner M, Doyle PS (2007) Multifunctional encoded particles for high-throughput biomolecule analysis. Science 315:1393–1396

Quackenbush J (2001) Computational analysis of microarray data. Nature Reviews Genetics 2:418–427

Resch-Genger U, Hoffmann K, Nietfeld W, Engel A, Neukammer J, Nitschke R, Ebert B, Macdonald R (2005) How to improve quality assurance in fluorometry: fluorescence-inherent sources of error and suited fluorescence standards. Journal of Fluorescence 15:337–362

Ron EZ (2007) Biosensing environmental pollution. Current Opinion in Biotechnology 18:252–256

Rothert A, Deo SK, Millner L, Puckett LG, Madou MJ, Daunert S (2005) Whole-cell-reporter-gene-based biosensing systems on a compact disk microfluidics platform. Analytical Biochemistry 342:11–19

Roulet JC, Volkel R, Herzig HP, Verpoorte E, de Rooij NF, Dandliker R (2002) Performance of an integrated microoptical system for fluorescence detection in microfluidic systems. Analytical Chemistry 74:3400–3407

Rudzinski CM, Young AM, Nocera DG (2002) A supramolecular microfluidic optical chemosensor. Journal of the American Chemical Society 124:1723–1727

Samuel IDW, Turnbull GA (2007) Organic semiconductor lasers. Chemical Reviews 107:1272–1295

Situma C, Hashimoto M, Soper SA (2006) Merging microfluidics with microarray-based bioassays. Biomolecular Engineering 23:213–231

Sobek J, Bartscherer K, Jacob A, Hoheisel JD, Angenendt P (2006) Microarray technology as a universal tool for high-throughput analysis of biological systems. Combinatorial Chemistry & High Throughput Screening 9:365–380

Spibey CA, Jackson P, Herick K (2001) A unique charge-coupled device/xenon arc lamp based imaging system for the accurate detection and quantitation of multicolour fluorescence. Electrophoresis 22:829–836

Stenger DA, Gross GW, Keefer EW, Shaffer KM, Andreadis JD, Ma W, Pancrazio JJ (2001) Detection of physiologically active compounds using cell-based biosensors. Trends in Biotechnology 19:304–309

Taitt CR, Anderson GP, Ligler FS (2005) Evanescent wave fluorescence biosensors. Biosensors & Bioelectronics 20:2470–2487

Tatosian DA, Shuler ML, Kim D (2005) Portable in situ fluorescence cytometry of microscale cell-based assays. Optics Letters 30:1689–1691

Tecon R, van der Meer JR (2006) Information from single-cell bacterial biosensors: what is it good for? Current Opinion in Biotechnology 17:4–10

Thorslund S, Klett O, Nikolajeff F, Markides K, Bergquist J (2006) A hybrid poly(dimethylsiloxane) microsystem for on-chip whole blood filtration optimized for steroid screening. Biomedical Microdevices 8:73–79

Tomizaki KY, Usui K, Mihara H (2005) Protein-detecting microarrays: current accomplishments and requirements. Chembiochem 6:783–799

Tourovskaia A, Figueroa-Masot X, Folch A (2005) Differentiation-on-a-chip: a microfluidic platform for long-term cell culture studies. Lab on a Chip 5:14–19

Tung YC, Zhang M, Lin CT, Kurabayashi K, Skerlos SJ (2004) PDMS-based opto-fluidic micro flow cytometer with two-color, multi-angle fluorescence detection capability using PIN photodiodes. Sensors and Actuators B-Chemical 98:356–367

Valeur B (2002) Molecular fluorescence. Wiley-VCH, Weinheim

Venkatasubbarao S (2004) Microarrays - status and prospects. Trends in Biotechnology 22:630–637

Vinet F, Chaton P, Fouillet Y (2002) Microarrays and microfluidic devices: miniaturized systems for biological analysis. Microelectronic Engineering 61–62:41–47

Wang Z, El-Ali J, Engelund M, Gotsaed T, Perch-Nielsen IR, Mogensen KB, Snakenborg D, Kutter JP, Wolff A (2004) Measurements of scattered light on a microchip flow cytometer with integrated polymer based optical elements. Lab on a Chip 4:372–377

Wilkes T, Laux H, Foy CA (2007) Microarray data quality - review of current developments. Omics-A Journal of Integrative Biology 11:1–13

Wilson R, Cossins AR, Spiller DG (2006) Encoded microcarriers for high-throughput multiplexed detection. Angewandte Chemie-International Edition 45:6104–6117

Yagai S, Kinoshita T, Higashi M, Kishikawa K, Nakanishi T, Karatsu T, Kitamura A (2007) Diversification of self-organized architectures in supramolecular dye assemblies. Journal of the American Chemical Society 129:13277–13287

Yagi K (2007) Applications of whole-cell bacterial sensors in biotechnology and environmental science. Applied Microbiology and Biotechnology 73:1251–1258

Yoo SK, Lee JH, Yun SS, Gu MB, Lee JH (2007) Fabrication of a bio-MEMS based cell-chip for toxicity monitoring. Biosensors & Bioelectronics 22:1586–1592

Yotter RA, Wilson DM (2003) A review of photodetectors for sensing light-emitting reporters in biological systems. IEEE Sensors Journal 3:288–303

Chapter 10
Focusing on Targets

In this Chapter the reader may not find exhaustive coverage of data on the sensing of all possible targets. This information is immense with the tendency of its exponential expansion day by day. The aim is different – to show the diversity of tasks and the diversity of possibilities for their solutions. Responding to intermolecular interactions, fluorescent dyes can recognize the medium and determine its physical parameters, such as temperature and pressure. Being incorporated into structures adsorbing gas molecules, they are able to identify the gas composition. Fluorescence sensors are also applicable in solutions, being efficient in the broadest possible range of target concentrations. Increasing the complexity of the target molecules requires adequate improvement of the binding and reporting properties of the sensors. Molecules of biological importance and especially proteins, nucleic acids and glycopolymers, are the targets, where the principles of biomolecular recognition are of great utility. Finally, we have to identify bacteria and viruses, achieving ultimate speed and ultimate sensitivity of their detection. This is also a resolvable task for fluorescence sensors.

10.1 Temperature, Pressure and Gas Sensing

There are many methods to measure the *thermodynamic properties of matter*, which seem to be simple, convenient and reliable. Fluorescence sensing occupies a strong niche in all cases when the response should be obtained on a molecular level: in microscopic volumes, at interfaces, in conditions of strong gradients of these parameters. They are of preference if we need to create an image in these parameters or if there is a necessity to make the measurements distantly or with high time resolution.

The design of fluorescence or luminescence sensors can be very simple, since their response can be formed by collective effects produced by collisional interaction of individual molecules due to their thermal motion. Temperature and pressure are the thermodynamic parameters that characterize these systems and have to be determined for many practical needs. Their gradients in solids and liquids as well as their surfaces and interfaces are also important in many research and industrial applications.

A.P. Demchenko, *Introduction to Fluorescence Sensing*,   407
© Springer Science+Business Media B.V. 2009

10.1.1 Molecular Thermometry

The measurement of temperature is ubiquitous in life, as well as in all fields of science, engineering and medicine. Nowadays most temperature measurements are based on temperature-dependent electrical signals and use thermocouples, thermistors or resistance thermometers. For remote measurements, the detection of IR radiation (pyrometry) is frequently used. These methods have essential limitations. For instance, the accurate measurement of temperature distribution, the 'thermal image', cannot be easily obtained with thermo-electrical devices and it is hard to measure IR radiation at low temperatures (due to a low signal), in water or in humid media (because of the strong absorbance of water molecules in the whole IR region). This range of applications offers a broad area of activity for the type of sensors described in this book. In all these conditions, methods based on luminescence demonstrate their advantage. They allow the sensing response (a) to be generated on a molecular level, (b) to be detected remotely and, if necessary, (c) to achieve very high spatial resolution.

Since the *thermal quenching* of all kinds of luminescence is a very common phenomenon, both fluorescent and phosphorescent dyes as well as luminescent metal-ligand complexes can be used for *sensing the temperature*. Thermal quenching is a dynamic process that reduces both the intensity and lifetime in a coupled manner, so both these parameters can be measured and calibrated on a temperature scale. The advantage of intensity measurement is the simplicity and lifetime detection – the independence of the response on instrumental and photo-bleaching effects.

Sensors can be constructed using the temperature-dependent structural properties of self-assembled or polymeric materials, into which the environment-sensitive dyes can be embedded. This idea was realized with a series of thermally responsive polymers (Ellison and Torkelson 2002).

The *calibration* of the probe response is complicated if only a single channel measurement of intensity is made. It could be natural to make the response signal ratiometric. For this, the intensity measurements need the application of a reference and such a reference signal can be introduced in different ways. As in other application areas, the best solution could be to use a single dye with two response channels, so that the two signals could exhibit different temperature dependence.

The dyes exhibiting *fluorescence and phosphorescence* simultaneously or the systems allowing the detection of both the fluorescence and long-lived luminescence of metal complexes deserve special attention. This is because for these two types of emissions the thermal quenching is different. Due to this, the ratio of their intensities is temperature-dependent and can be calibrated as an empirical ratiometric parameter in temperature units. Since the phenomenon of thermal quenching is very general, many possibilities exist in the optimal design of such temperature sensors. The complexes of rare earth ions incorporated into different glasses and crystals have been reported to be efficient thermometers, operating based on this principle. They can be efficient at temperatures up to 600°C, with a resolution of the order of 1°C (Wade et al. 2003).

The *monomer-exciplex equilibrium* between perylene and N-allyl-N-methyl-aniline embedded into the polymer matrix is temperature-dependent. The monomer emits a blue light and the exciplex a green light. On temperature increase, this equilibrium is shifted towards the monomer. The linear temperature dependence of the ratio of these emission intensities can be used to read the temperature (Chandrasekharan and Kelly 2001).

Other excited-state reactions, such as the *excited-state intramolecular proton transfer* (ESIPT), can also be used in sensing. Aggregates of 2-(2'-hydroxyphenyl) benzoxazole (HBO) show a fluorescent ratiometric change in a range of temperature from 15°C to 60°C. The reversibility and robustness as well as the stability of the aqueous dispersion of the aggregates show very good performances, which may be useful in applications of molecular thermometers (Huang et al. 2006a).

The reader may expect that thermometers using the molecular mechanism of response should, in principle, achieve a *molecular-scale spatial resolution*. Such molecular thermometers have the potential of measuring the temperature gradients inside living cells (Gota et al. 2008). In fact, there are two possibilities for such measurements. One is to use a protein or a nucleic acid with a defined temperature range of thermal transition and label it with environment-sensitive fluorescent dye. Such a construct can be a good molecular thermometer in the range of this transition on condition that no uncontrollable factors, such as ligand binding or local pH, can influence it. The other possibility is to incorporate temperature-responsive fluorescent nanoparticles calibrated in the wavelength shift, lifetime or two-wavelength ratiometric ratio.

Such molecular thermometers are particularly needed for sensing *on a tissue level*. They can provide control of the clinical treatment of tumors by hyperthermia. It is known that tumors are commonly more sensitive to high temperatures than normal cells. Because of this, they can be selectively destroyed by thermal treatment. In order to avoid destroying the normal cells, the temperature has to be kept locally within narrow ranges (at about 42.5°C). Dyes and nanoparticles with near-IR emission that fit the wavelength range of the transparency of human skin and tissues can be used for such monitoring. It has been shown that nanoscale gold thermometers can effectively determine the local temperatures inside or around the targeted cancer cell (El-Sayed et al. 2006; Huang et al. 2006b).

Luminescent temperature sensing has found many *industrial applications*, particularly in aerodynamics and space technologies and also in the polymer processing industry. This is mainly because of the ability to visualize the temperature gradients in the sample. This property may be useful in case of the necessity to test any industrial unit under gas or liquid flow conditions and to observe and quantify the thermal gradients that appear on testing. One can resolve such problems in a very simple and elegant way. The surface can be covered with luminescent dye using a brush or sprayer and after illumination, one can take pictures. The luminescence response of such a surface being calibrated and recorded provides information on the distribution of the temperature (Chandrasekharan and Kelly 2001).

10.1.2   Molecular Barometry

The luminescence quenching effect depends not only on the temperature but also, for long-living (microsecond-millisecond) emission, on the pressure-dependent concentration of molecular oxygen in the medium, which allows *measuring the pressure*. The mechanism of this quenching is collisional, which allows both spectroscopic and lifetime measurements (Gouterman 1997).

The oxygen diffusion and its permeability in a thin polymer film coating can be put into the background of developing pressure-responsive thin polymer films. Any object (even of the size of an airplane or bridge) can be covered with such a film and the pressure distribution on it following any applied strain can be detected by observing the luminescence response. Since the observed emission is also influenced by the temperature, a two-parameter temperature-pressure responsive layer can be suggested, based on the response of two dyes. One of them, for instance, can be a platinum-porphyrin type complex and the other a europium chelate complex. Then, one of these emitters responds to pressure (with a small temperature dependence) and the other responds to temperature (with no dependence on pressure).

10.1.3   Sensors for Gas Phase Composition

The sensor must provide efficient adsorption of gas molecules on its surface and produce the change in properties recorded as the response in fluorescence or other type of luminescence (Apostolidis et al. 2004). A simple and efficient *gas senso*r can be constructed based on a very thin polymer film or a film with a high level of porosity deposited on a solid support. Presently, the general tendency is to use the effect of quenching in a conjugated polymer (Kwan et al. 2004) or lifetime sensing, in which the reporter function is played by ruthenium complexes exhibiting long-lifetime luminescence. This allows developing a sol-gel complex for humidity sensing (McGaughey et al. 2006). The formation a FRET pair within such a complex with pH-active azo dye, allowed creating a sensor for carbon dioxide (von Bultzingslowen et al. 2003).

Gasses are composed of small neutral molecules and specially prepared surfaces are needed for their adsorption. In some cases, *hydrogen bonds* can be involved in their recognition. They are rather weak noncovalent interactions formed between two partners – the donor (e.g., hydroxyl) and the acceptor (e.g., carbonyl). The transduction of the signal on hydrogen-bonding interactions can be made with the direct involvement of a dye containing a proton acceptor group (Section 4.1) and this can be the method to generate the response. Such an effect can be facilitated by the incorporation of a strong hydrogen bond donor (e.g., hexafluoroisopropanol) into a conjugated polymer (Amara and Swager 2005). The target binding perturbs the electronic system of the polymer and activates

ICT quenching. Films made of this polymer show quenching upon exposure to pyridine vapor.

Monitoring the *oxygen* concentrations is particularly important in chemical or clinical analysis and environmental control. Recently, a variety of devices and sensors based on the quenching of the photoluminescent states of organic dyes have been developed to measure the partial pressure of oxygen on the solid surface. Many optical oxygen sensors are composed of organic dyes, such as polycyclic aromatic hydrocarbons (pyrene, pyrene derivative etc.), transition metal complexes (Ru^{2+}, Os^{2+}, Ir^{3+} etc.), metalloporphyrins (Pt^{2+}, Pd^{2+}, Zn^{2+} etc.) and fullerene (C-60 and C-70), immobilized in oxygen permeable polymer films (Amao 2003)

Many *explosives* are electron-deficient highly nitrated organic compounds, such as nitroaromatics, nitramines or nitrates. Their detection is a major security concern, therefore the tests need to be performed rapidly and 'non-invasively' in public places by detecting their trace amounts in vapors. Much effort has been made for this and the major trend is to exploit the electron-acceptor properties of these molecules into the systems, producing fluorescence quenching by a photoinduced electron transfer. Thin films of conjugated polymer have been suggested for this purpose. This approach allowed developing a device that is able to detect the vapors of *trinitrotoluene* (TNT) on the femtogram level (Thomas et al. 2005).

In conclusion, it should be added that fluorescence or luminescence-based gas sensors are often needed in a combination that would allow obtaining the recognition pattern for many ingredients and that the low-specificity binding can be useful for forming this pattern. Such an 'electronic nose' approach will be discussed in Section 12.5. It should be noted however, that in this respect the optical detection techniques are in strong competition with the technique based on semiconductor thin films, which allows direct detection of electrical signal. The future will show which of these technologies will demonstrate more potential for improvement and better ability for applications.

10.2 Probing the Properties of Condensed Matter

The choice of a proper medium is important for all the technologies that involve chemical synthesis and separation/purification of reaction products. Dye molecules exhibiting sensitivity of their emission to variations of many properties in their molecular environment, such as polarity, fluidity, molecular order and mobility, can be used to characterize these properties. The distinction between probes and sensors is not always clear in these cases. When these integrated parameters characterizing the studied medium can be obtained, we consider these dyes as *probes*. Meanwhile, the same parameters can be used for providing a response to particular targets in these media. In the latter cases, they can be employed as sensors.

10.2.1 Polarity Probing in Liquids and Liquid Mixtures

Fluorescence probes are frequently used for characterizing the *polarity* and *viscosity* of various condensed media. Since in this case the response is provided by molecular structures and on a molecular level, in some systems it may appear different from that obtained by the macroscopic technique because it may reflect variations of molecular structures of solvents and solutions. For instance, the commonly measured viscosity as the resistance to liquid flow may not perfectly match data obtained by the detection of mobility on a molecular level. Therefore, it is often called 'microviscosity' or '*nanoviscosity*'. In many applications a *continuous approximation* is applied, in which the fluorescence response of 'nanoviscosity' is calibrated in standard solvents with a known 'macroscopic' viscosity. This allows using the same poise units.

The same applies to the term '*polarity*'. The physical definition of polarity is quite different from that used by chemists and physical chemists. It is based on the analysis of intermolecular interactions (see Section 4.1) and can be studied with fluorescent dyes (Suppan and Ghoneim 1997). Physical description is very important for the development of new dyes as molecular reporters and probes, since their molecular parameters are included in the physical analysis of spectroscopic data. In contrast, chemists often use polarity scaling that is calibrated on the basis of the distribution of matter between water and hexane. Finally, physical chemists prefer empirical scaling, based on an empirical correlation of polarity with some spectroscopic property, usually the shifts of absorption or fluorescence spectra of the applied probes.

Specially designed or selected probe molecules possess the most remarkable properties to exhibit the solvent-dependent shifts of absorption or fluorescence spectra. On the basis of these effects, several *empirical solvent polarity scales* have been suggested (Catalan 1997; Reichardt 1990). They are not totally equivalent and the scatter of experimental data against these scales is significant, even within the families of structurally related solvents. It can be observed that the classical scale based on the absorption wavelength shifting for *non-fluorescent betaine*, $E_T(30)$ and its normalized version, E_T^N, (Reichardt 1990), remains the most reliable among many suggested scales. The newly suggested dyes as polarity calibrators based on absorption or fluorescence are neither universal enough to cover the broadest range of solvents nor specific enough to investigate the mechanisms of deviations from simple regularities provided by specific features of every solvent.

The use of *pyrene* as a probe allows exploring the polarity-dependent change in relative intensity of two vibronic bands in its fluorescence spectrum (Dong and Winnik 1984). Among the wavelength-shifting dyes, the strongest polarity-dependent effect is observed for Fluoroprobe, possessing a strong intramolecular charge transfer (Hermant et al. 1990). Next in efficiency are ketocyanines and dialkylaminonaphthalenes, such as Prodan. Unfortunately, Fluoroprobe is severely quenched in protic solvents and the two other types of dyes in these solvents exhibit shifts due to hydrogen bonding almost as strong as the whole polarity scale (Section 4.1).

Two-band emissive *3-hydroxychromone dyes* allow not only obtaining a highly sensitive ratiometric response to polarity, but also provide the means for discriminating polarity and hydrogen bonding effects. Their two-band ratiometric response is so sensitive that a series of different dyes had to be developed for the determination of polarity in its narrow ranges. Several dyes with a small variation in structure (the major variation in the ICT property of the N* state) can cover the whole polarity scale. An example of the application of one of these dyes responding in aprotic environments in the low-polarity range is presented in Fig. 10.1.

It must be acknowledged that empirical scaling is useful in studies of solvent mixtures, interfaces and micro-heterogeneous systems, such as micelles. In the broadest sense, *polarity* is an efficient variable describing the interaction forces of the probe with the solvent and its components. Their description must involve several molecular properties (electronic and nuclear polarizabilities, abilities to form specific bonding) and it is therefore difficult to provide a strict definition of polarity and provide its measurement in a straightforward manner. The empirical methods use the description of polarity with a single parameter and they are therefore simple in application. As they provide a satisfactory description of polarity-related properties in many systems, they are very popular.

The necessary condition for dyes to exhibit spectral shifts in response to variation of solvent polarity is a strong change of their dipole moments on electronic excitation. There occurs a response of the dye environment to this change, as a change of electronic and nuclear polarizations that determine the polarity. After the classical works of Gregorio Weber, most of the activities were focused on finding the dyes with a dipole moment *increase*. Dyes with a dramatic *decrease* of the dipole moment also exist. These are the cyanine dyes. These dyes, with red and near-IR absorption and emission, promise prospects for using two-photon absorption and as FRET acceptors in different sensor technologies, especially for *in-vivo* sensing. The long-chain cyanine dyes respond to changes of polarity by their two-band response in *excitation spectra* (Lepkowicz et al. 2004).

As shown in Fig. 10.2, an increase in the length of the polymethine chain leads to symmetry breaking and the appearance of two forms with symmetrical and asymmetrical distributions of the charge density in the ground state. This distribution depends on the substituents in the chain and for some dyes it is strongly polarity-dependent.

10.2.2   Viscosity and Molecular Mobility Sensing

For sensing *molecular dynamics* in condensed-phase systems such as liquids, polymers and glasses, a fluorescent dye should respond to the change of *translational or rotational diffusion* in this system by a change in its spectroscopic parameters. The oldest and simplest method to measure solvent viscosity is to apply dyes exhibiting molecular rotation with the detection of anisotropy and application of Eq. (3.9), connecting determined *rotational correlation time* with solvent viscosity.

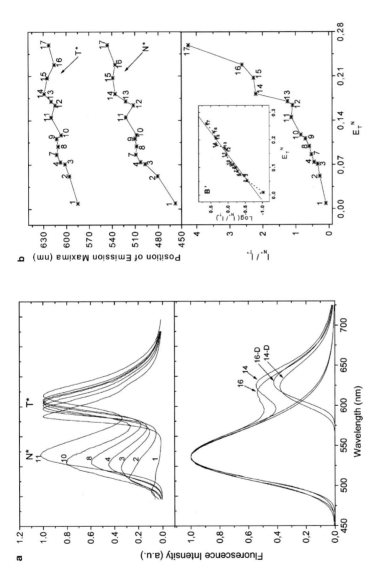

Fig. 10.1 The two-band ratiometric response of a 3-hydroxychromone dye FA to solvent polarity. (**a**) Fluorescence emission spectra in different solvents at excitation 430 nm. Above: the spectra were intensity-normalized at the T^* band maximum. Below: the spectra were normalized at the N^* band maximum and the contribution of the T^* band is shown by deconvolution. (**b**) The correlation of the positions of N^* and T^* emission maxima (above) and the ratios of the intensities at these maxima (I_N^*/I_T^*) with E_T^N polarity scale (below). Insert: the semi-logarithmic transform of this function. The solvents are (1) hexane; (2) CCl_4; (3) dibutyl ether; (4) 1,4-dimethylbenzene; (5) 1,3-dimethylbenzene; (6) 1,2-dimethylbenzene; (7) tetralin; (8) toluene; (9) benzene; (10) diethyl ether; (11) thiophene; (12) dioxan; (13) 1,2,4-trichlorobenzene; (14) bromobenzene; (15) tetrahydrofuran; (16) ethyl acetate; (17) trichloromethane (Reproduced with permission from Ercelen et al. 2002)

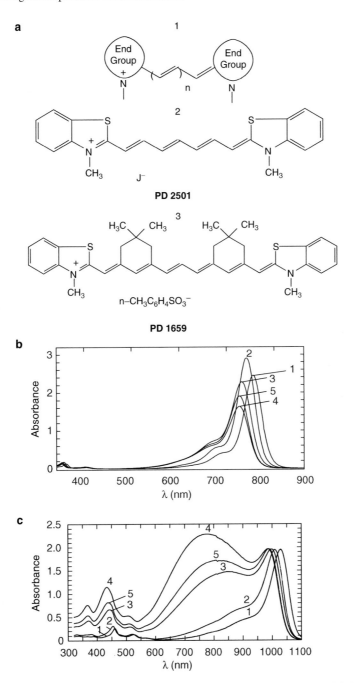

Fig. 10.2 Selected representatives of polarity-sensitive near-IR polymetine dyes. (**a**) The structures of PD 2501 and PD 1659 dyes. (**b** and **c**) Their absorption spectra for PD 2501 (**b**) and PD 1659 (**c**) in o-dichlorobenzene (1), methylene chloride (2), ethanol (3), acetonitrile (4) and methanol (5) (Reproduced with permission from Lepkowicz et al. 2004)

The other possibility is to use dyes possessing *mobility of their segments*. In low-viscosity medium these dyes are free to rotate and such rotation induces the quenching (Baptista and Indig 1998). In rigid media, such dynamics are frozen and a bright emission appears. Typical in this respect is the behavior of triphenylmethane dyes, such as Crystal Violet and Malachite Green. They possess three-blade propeller-like phenyl rings joined by the central carbon atom. The viscosity-dependent flexibility of these dyes comes from their peculiar structure, which allows mobility of the phenyl rings, resulting in fluorescence quenching. Therefore, the lifetimes τ_F depend strongly on the solvent viscosity and, due to the absence of groups forming specific noncovalent bonds, practically do not depend on any other solvent properties. Being about 40 ns in rigid environments, τ_F drops to several picoseconds in room-temperature methanol ($\eta \sim 0.6$ cP).

Isomerization dynamics can be coupled with the dynamics of formation-breaking of complexes with solvent molecules. Such effects have been observed in asymmetric polymethine dyes and applied to studies of microviscosity of membranes and living cells (Volovik et al. 1994). The two forms, one detected in solid, the other in liquid environments, differ strongly in fluorescence spectra. Thus, in fluorescence spectra excited at 600 nm the 'solid' band observed in solid media is located at 700–710 nm and the 'liquid' band is the only feature of the spectrum in liquid media, it is located at 760–770 nm. In viscous media (e.g., glycerol) both of these forms coexist, so that the ratio of their intensities can be calibrated as a function of viscosity. Amazingly, neither the position of these bands nor the intensity ratio show notable dependence on any solvent parameter, except viscosity (Volovik et al. 1994).

An attempt to provide a wavelength-ratiometric response to viscosity was recently made, also with the aid of a substituted styryl pyridinium probe (Wandelt et al. 2005). This dye is soluble and strongly fluorescent in an aqueous solution, displaying dual excitation peaks at 469 and 360 nm, detected at emission wavelengths 500–650 nm, indicating the formation of ground-state complexes with the solvent. The dye could be irreversibly loaded to a living cell, which allowed obtaining the distribution of viscosity in the cytoplasm using fluorescence imaging microscopy, with a spectrofluorimeter in dual-excitation single-emission mode.

An alternative way to provide viscosity sensing is to use highly dipolar 'polarity-sensitive' dyes and to study the *rates of dielectric relaxations* in their environment with the aid of time-resolved spectroscopy (Vincent et al. 1995) or by observation of the Red-edge effects (Demchenko 2002). The problem here is that high and wavelength-selective temporal resolution is needed in conditions where the dielectric relaxation rates match the time scale of the fluorescence lifetimes. For low-viscosity liquids it is on the scale of several picoseconds, which is very inconvenient for routine measurements. These methods can be used in highly viscous systems in which the relaxations are a nanosecond or longer.

A *viscosity*-sensitive response can be achieved with *molecular rotors*. They are fluorescent dyes exhibiting viscosity-dependent isomerization in the excited state that modulate a fluorescence response. An example of such molecules is 4-tricyanovinyl-[N-(2-hydroxyethyl)-N-ethyl]aniline (TC1). These molecules can be covalently bound to a fiber-optic tip without loss of viscosity sensitivity.

Covalently bound molecular rotors exhibit a viscosity-dependent intensity increase similar to molecular rotors in solution. An optical fiber-based fluorescent viscosity sensor may be used in real-time measurement applications, ranging from biomedical applications to the food industry (Haidekker and Theodorakis 2007).

The problem of introducing the reference for ratiometric measurements that is common to all intensity-based sensors, also applies to viscosity sensors. It can be solved, for instance, by fusing viscosity-sensitive and viscosity-insensitive dyes with resultant ratiometric detection (Fischer et al. 2007). Meanwhile, the best solution could be provided by viscosity-sensitive ratiometric dyes that require no reference, as shown in previous studies (Volovik et al. 1994).

In addition, the combination of molecular rotors with specific recognition groups allows binding them to target specific sites, for example the cell membrane or cytoplasm. Membrane viscosity characterizes the dynamic processes that determine membrane transport and cytoplasm viscosity is an important characteristic of the intracellular diffusion of enzymes and metabolites. Molecular rotors are therefore emerging as important functional biosensors for local microviscosity measurements (Nipper et al. 2008) but intensity sensing seems inappropriate, in view of the uncertain distribution of the probe dyes within the cells. For molecular rotors, the primary solution for obtaining a dye concentration independent response must be lifetime sensing (Kuimova et al. 2008).

Thus, fluorescence probes allow obtaining the polarity and viscosity scales of common liquids that can be used for the evaluation and selection of different liquid compositions and for the characterization of heterogeneous media and even condensed media with poorly understood molecular properties, such as ionic liquids and supercritical liquids.

10.2.3 Probing Ionic Liquids

Room-temperature ionic liquids have recently received a lot of interest in view of their attractive properties as media for chemical reactions involving synthesis, catalysis and extraction (Chiappe and Pieraccini 2005). They are usually composed of an organic cation (alkylammonium, alkylphosphonium, imidazolium, pyridinium, isoquinolinium) and an inorganic anion. They remain liquid over a wide temperature range, including room temperature and in normal conditions they behave as molten salts. Many ionic liquids are colorless and optically transparent over a wide spectral range, from UV to near-IR, which does not present any problems for their optical studies.

The high viscosity and the absence of a sharp freezing point suggest the 'supercooled' nature of ionic liquid media and their nano-structure. Therefore, the study of their molecular dynamics with fluorescence probes has attracted much attention. Time-resolved spectroscopy with excitation-wavelength selection (Funston et al. 2007) allowed the authors to show that in the origin of the high viscosity of these solvents is the very slow rate of dielectric relaxations, which are in the range of hundreds of nanoseconds. This is by a factor of 10^4–10^5 slower than in common liquids (Hu and Margulis 2006).

It has been shown that ionic liquids exhibit both static and dynamic disorder. *Static disorder* is the characteristic of the system when on a relatively long time scale each molecule reveals the peculiar local and stable characteristics of its environment, which may be different from that of another molecule situated within this system. *Dynamic disorder* appears when the probe molecules are located in different environments but these environments interchange dynamically on the time scale of emission. Static disorder is characteristic of solid systems (e.g., glasses) and for polymers in glassy states, whereas dynamic disorder is observed in molten polymer states. They can easily be distinguished and characterized with fluorescent dyes, using Red Edge wavelength-selective fluorescence spectroscopy (Demchenko 2002).

10.2.4 The Properties of Supercritical Fluids

Supercritical fluid is a unique state of condensed matter, which can be reached in liquids by a simultaneous increase of temperature (T) and pressure (P) above their critical points T_c and P_c. In these conditions, the liquid and gas phases coalesce into a single phase. Its properties, such as density, viscosity and diffusion rate can be tuned continuously in broad ranges by simply adjusting the T and P values, without changing the solvent's composition. The combination of liquid-like but tunable solvation properties with gas-like rates of molecular diffusion makes supercritical fluids a very attractive solvent media for chemical reactions. In this respect, supercritical CO_2, being both efficient and non-toxic, presents an attractive example of the medium for future 'green' chemical synthetic technologies.

A high level of *fluctuations of physical properties* is characteristic for critical phenomena. Dynamic molecular clusters are forming and breaking in these systems. Such clusters also have to surround solute molecules, which can be fluorescent dyes. As in liquid solvents, polarity may serve as an important characteristic for predicting the solubility and reactivity of different compounds.

Studies with solvent-sensitive fluorescent dyes have shown that close to the critical temperature and pressure, the local solvent density around a solute molecule is higher than the actual bulk value. Such enhanced solvent density has been explained as bearing witness to such 'clustering' or 'local density augmentation' (Abbott et al. 2007).

An interesting effect was observed in the pressure dependence of the fluorescence of wavelength-ratiometric 3-hydroxychromone dye exhibiting the ESIPT reaction in supercritical CO_2 (Barroso et al. 2006). Figure 10.3 shows that the fluorescence spectra, normalized at the normal emission maximum (422–428 nm) nm, demonstrate a dramatic increase in the tautomer band (535–538 nm) intensity with an increase in the pressure of the system. As the density exceeds 0.7 g/ml, the relative intensity of the two bands tends toward a constant value, indicating polarity similar to that of apolar organic solvents, such as toluene and di-*n*-butyl ether. At lower densities, the substantial decrease of the total fluorescence intensity (a 600-fold decrease as the pressure decreases from 100 to 80 bar) is accompanied by an even more accentuated decrease of the tautomer fluorescence.

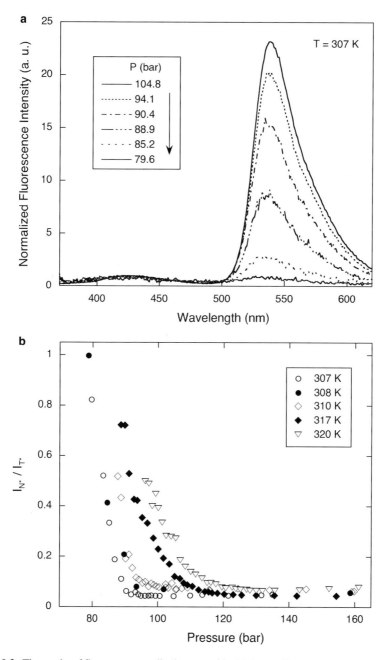

Fig. 10.3 The results of fluorescence studies in supercritical CO_2 at different pressures and T 307 K using a wavelength-ratiometric dimethylamino-3-hydroxyflavone dye. (**a**) The spectra normalized to the maximum of the normal band (428 nm) to emphasize the change in the relative intensities of the normal and tautomer bands. (**b**) The pressure dependence of the ratiometric response at different temperatures. The excitation wavelength was 350 nm (Reproduced with permission from Barroso et al. 2006)

If only the intensity ratio is considered, such behavior could mean a dramatic increase of polarity. The true explanation is different. Impressive dynamic fluorescence quenching at low pressures makes the fluorescence rates comparable or even shorter than the rates of the ESIPT reaction (see Section 3.7), so the excited species simply have no time for transition to a tautomer state exhibiting a long-wavelength band in emission. The origin of such dramatic quenching at low pressures is probably related to a more 'naked' environment of the fluorescent dye and its greater exposure to collisions with CO_2 molecules possessing high kinetic energies.

10.2.5   The Structure and Dynamics in Polymers

Polymers are well-known systems in which the estimation of molecular parameters characterizing segmental dynamics and interactions is important for understanding their macroscopic properties. This can be done by fluorescence probe techniques. Their application allows one to study the dynamic processes of interest, such as polymerization kinetics and mechanisms, thermal transitions, photodegradation, swelling in solvents and so forth (Bosch et al. 2005). In polymer studies, fluorescent probes that sense both polarity and viscosity (or molecular void volume in solid samples) are of special value. Usually these are the dyes that exhibit photoisomerization dynamics, such as 2-(4-dimethylaminostryl)benzoxazole and its butadienyl analogues (Fayed and Etaiw Sel 2006)

Fluorescence polarity-sensitive probes and molecular rotors as viscosity probes have found many applications for monitoring the reaction of polymerization (Quirin and Torkelson 2003) and for controlling the desorption of water from polymer films (Ellison et al. 2004). Their application for characterizing and monitoring the structural transitions in polymers is very important.

One of the most interesting applications of the fluorescence probe method is the study of the local dynamics of a polymer matrix at the onset of the *glass transition* (Deschenes and Vanden Bout 2001). In these experiments, molecules of the organic dye Rhodamine 6 G dispersed within a thin poly(methacrylate) film were observed at temperatures slightly above the melting temperature of the polymer. Fluorescence polarization was used to follow a slow rotational diffusion of individual molecules over several hours. This study demonstrated the presence of both a static and dynamic disorder in these conditions.

Some polymers can exist in a *highly elastic state*. Being macroscopically solid, they differ substantially in dynamic properties from those in a glassy state. The application of a polymethine dye as a probe and the investigation of static and time-resolved spectra allowed us to reveal and characterize the relaxation processes in one of them, polyurethane acrylate, on the nanosecond time scale (Przhonska et al. 1999). The dye developed and used in these studies has a unique property of ICT-dependent change in the conjugation length in the excited state, which leads to strong Stokes shifts, even in rigid polymer media. Its application allowed revealing the important features of molecular dynamics in elastic polymers, which could not be detected with common dyes.

A new field of applied research is the study of structural transformations (e.g., glass transitions) of polymers in the forms of thin films and nanoparticles (Masuhara et al. 2006). Among other areas, these data are needed for sensor technologies themselves.

An important application of fluorescent probes is in the optimization of polymer properties by rapid screening of many samples varying in composition, technology of preparation, etc., so that combinatorial discovery cycles can be developed (Potyrailo 2006). The selection of fluorescence dyes for this purpose depends upon the desired properties to be optimized.

10.2.6   *Fluorescence Probing the Interfaces*

In many cases, it is important to characterize the structure, dynamics and interactions at liquid-liquid, liquid-solid and solid-solid interfaces and at the liquid and solid surfaces exposed to air. Most straightforward in these cases is to locate the small environment-sensitive dye molecules (probes) at the interface region. This can provide information on static properties such as polarity, hydrogen bonding ability and electrostatic potentials and also on the dynamic properties, such as the rates of segmental motions and the rates of dielectric relaxations (rotations of dipoles surrounding the probe).

In *liquid-liquid interfaces*, the strongly *amphiphilic* dye molecules (containing both apolar and polar sites) can concentrate at the interface between two non-miscible liquids and report about the properties of their interface. Sulforhodamine dye located at the interface demonstrates intermediate polarity between two phases, as is witnessed by time-resolved spectroscopy with total internal reflection observation (Ishizaka et al. 2001).

In *solid-liquid interfaces* some interactions that are saturated in bulk liquid can remain unsaturated, with some new interactions being formed. The same applies to *solid-solid interfaces*. Commonly in these cases, the dye is attached to the surface of one component before depositing another component. Sometimes, a third component is used that serves to provide adhesion between the interacting surfaces and it can be used for the introduction of a fluorescence sensing dye. In studies of surfaces exposed to the gas phase, the dyes can be incorporated in a similar way.

Small-sized organic dye molecules are commonly used to study interfaces, including nanoparticles, the surfaces of protein molecules and biomembranes (Epand and Kraayenhof 1999). This is because the interfaces are the sites of the greatest gradients of all molecular properties and in view of the fact that the probe sizes should be commeasurable with the sizes of molecules and groups of atoms forming these gradients. The dyes are usually used as environment-sensitive probes. In addition to spectroscopic properties, dye solubility in both media and a strong preference to interface are particularly important. They can be enhanced by covalent modification with correspondent 'affinity' groups.

In *solid-solid compositions*, silane coupling agents are commonly applied to glass fibers to promote fiber/resin adhesion and to enhance durability in composite parts. In one of the studies (Lenhart et al. 2000) a multilayer coupling agent on glass was doped with trace levels of dimethylaminonitrostilbene (DMANS) dye. A 53-nm blue shift in the fluorescence spectrum from the position for immobilized DMSCA can be followed during the cure of an epoxy resin overlayer, which indicates a substantial decrease in polarity and an increase in the rigidity of the dye environment. Such effects were observed only in the case of very thin silane layers. Thicker layers showed smaller fluorescence shifts during cure, suggesting incomplete resin penetration into these layers (Lenhart et al. 2002). This result shows the potential of such an approach to monitor the properties of the fiber/resin interface during composite processing.

In another study (Olmos et al. 2003, 2005) the interfaces between glass fibers or silica particles coated with silanes and an epoxy polymer were studied using dansyl and pyrene labels. An increased heterogeneity of interactions and structural rigidity were indicated in comparison to epoxy bulk. Comparing the curing reaction in the polymer bulk and at the interface, it could be concluded that epoxy curing at the interface proceeded faster but only during the first stages of the reaction.

10.3   Detection of Small Molecules and Ions

The broadest diversity of approaches involving the design and operation of recognition units, reporters and transduction mechanisms are used in these applications. They involve both synthetic chemical sensors and biosensors.

10.3.1   pH Sensing

pH is the measure of the proton concentration in the studied medium. It is also an indicator of the ionization state of many compounds dissolved in water and a modulator of their reactions. It is also an essential parameter in cell physiology, determining the rates of many biochemical reactions that operate within a narrow (commonly, neutral) pH range. Therefore, many fluorescent pH indicators and sensors have been developed and are actively used.

Sensing pH is based on the fact that in some fluorescent dyes, the ionization of attached groups changes the spectroscopic properties dramatically, leading to the appearance of new bands in absorption and in emission. The titration ranges in absorption and in emission do not usually correspond. Because the acidity of these groups is much higher in the excited than in the ground state, the pH range of the sensitivity in emission becomes dramatically shifted to a lower pH. If the two forms are fluorescent, one can obtain with a single-band excitation the two emission bands that belong to neutral and proton-dissociated forms. During the gradual pH change with fluorescence detection, one of the forms disappears and the other one appears in a coupled manner. Equilibrium between these forms can be established on a very fast time scale (Davenport et al. 1986; Laws and Brand 1979). This allows wavelength-ratiometric recording.

Among the dyes suggested as fluorescent indicators of pH, benzo[c]xanthene derivatives such as C-SNAFL-1 (Mordon et al. 1995; Whitaker et al. 1991) are the most extensively used. Their excitation spectra may reflect the ground-state proton dissociation and in the emission spectra the pH-dependent appearance of a strongly red-shifted second band is due to the deprotonation of the hydroxy group in the excited state. A number of other dyes have been synthesized that have the ability to change their ionization state in the ground and excited states as a function of pH. Among them is the new water-soluble 3-hydroxychromone dye, 3,4′-dihydroxy-3′,5′-bis(dimethylaminomethyl)flavone, which exhibits a two-band ratiometric response in a broad pH range (Valuk et al. 2005).

The *range of detectable titration* of ionizable groups obeys the rule of equilibrium between two states (Section 2.3) and therefore the range of pH sensitivity in the titration of a single group is limited to two sequential digits of pH. To increase this, several ionisable groups ranging on a pH scale can be attached to a dye. Based on this principle, an optode-based sensor showing an almost linear pH response in the range from 2 to 10 was constructed (Li et al. 2006). It uses amino-functionalized corrole immobilized in a sol-gel glass matrix. Other researchers (Niu et al. 2005) constructed a ratiometric fluorescence sensor with a broad dynamic range based on two fluorescent dyes that were sensitive in different pH-ranges. The dyes were co-polymerized with acrylamide, hydroxyethyl methacrylate and triethylene glycol dimethacrylate on a silanized glass surface. The sensor covers a broad dynamic range of pH from 1.5–9.0.

Following the excitation in the range 370–405 nm, the emission spectrum of a cell-permeable macrocyclic Eu(III) complex incorporating an N-methylsulfonamide moiety changes with pH, allowing ratiometric pH measurements in the range from 6 to 8 (Pal and Parker 2007). A number of other pH sensors based on the emission of lanthanide chelates has been suggested (Bunzli and Piguet 2005). They use the change of Eu(III) fluorescence due to the pH-dependent change in the ionization state of chelating units.

Since many processes *in the living cell* depend on pH, its values (pH_i) are highly regulated and kept constant near the neutral values. This regulation fails in some pathological conditions, leading to acidosis or alkalosis. This is why an effort has been made to synthesize cell-permeable fluorescence sensitive dyes and to develop pH-sensitive GFPs. Such a pH-sensitive member of the GFP family has been described (McAnaney et al. 2002; McAnaney et al. 2005). It exhibits dual emission properties due to its pH-sensitive excited-state proton transfer (ESPT) rate.

10.3.2   Oxygen

Oxygen sensing is needed in many areas of research and industry. Oxygen is known as a very potent *collisional quencher* of long-living luminescence, so if the lifetime of a sensor emission is sufficiently long on the scale of oxygen diffusion, the quenching rate is proportional to its concentration in the studied medium. Such a decrease of the quantum yield and lifetime is used as the major principle applied in oxygen sensing (Nagl et al. 2007). The most typical are the lifetime-based methods, since they allow avoiding the problem of self-calibration of the output signal.

For the purpose of spectroscopic analysis of oxygen, probes exhibiting both *fluorescence and phosphorescence* on the same intensity scale are very attractive. The short-living fluorescence is not affected by the presence of oxygen and may serve as the reference, whereas the long-living phosphorescence can provide the necessary response in quenching. Since the phosphorescence band is strongly shifted to longer wavelengths, a very precise, convenient and self-calibrating detection can be achieved in a steady-state recording at two emission wavelengths. This provides an alternative to lifetime measurements and may offer similar precision, but needs simpler instrumentation (Fig. 10.4).

Of course, the two dyes, phosphorescent and fluorescent, incorporated into one nanoparticle, can also be used for wavelength-ratiometric sensing of oxygen. A disadvantage in this case could be their different photostabilities. There is a possibility of photobleaching of one of the dyes, which would change the ratio of the two signals in the emission unrelated to oxygen sensing.

Different detection formats need to be realized in oxygen sensing. Thus, for application in cell culture bioreactors an external *fiber-optic probe* has to be incorporated (O'Neal et al. 2004). The suggested probe takes advantage of the oxygen-stimulated fluorescence quenching of dichloro(tris-1,10-phenanthroline) ruthenium (II) hydrate. This fluorophore was immobilized in a photo-polymerized hydrogel made of poly(ethylene glycol) diacrylate. The sensor showed a high degree of reproducibility across a range of oxygen concentrations typical for cell culture experiments.

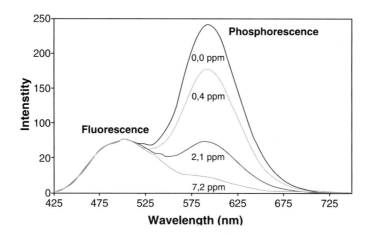

Fig. 10.4 An example of a wavelength-ratiometric phosphorescent oxygen indicator, which displays an oxygen-quenchable phosphorescence and a fluorescence emission that does not depend on the presence of oxygen (Hochreiner et al. 2005)

10.3.3 Heavy Metals

Industrial pollution has raised the necessity of determining the concentrations of *heavy metal ions* on a very low level of concentration (down to nanomoles), since even in such concentrations these ions can accumulate and display a negative effect. The most needed ions to be determined are Hg^{2+}, Cu^{2+}, Cd^{2+}, Co^{2+} and Pb^{2+}.

The heavy metal ions are known as potent *fluorescence quenchers* and therefore the first-generation sensors were based on this effect. Further developments have suggested 'turn-on' sensors with an increase of intensity on ion binding. Such sensors were developed for copper (Konishi and Hiratani 2006) and mercury (Huang and Chang 2006) in aqueous solutions. Metal nanoparticles and Quantum Dots are becoming involved in the construction of nanosensors and they offer an increase in sensitivity of detection. Sensors that could fully satisfy the practical needs have still to appear.

One of the most important applications of fluorescent detection methods is in the control of heavy metal contamination of food and beverages. A series of organic dyes, combining ion recognition with fluorescence reporting, have been synthesized for this purpose. Some of them possess nanomolar sensitivity of detection or sensitivity quite sufficient for distinguishing non-toxic and toxic levels of contaminants (Domaille et al. 2008).

Fluorescence-based sensing systems for heavy metal ions have started to be developed based on *genetically engineered* living yeast cells (see Section 9.6).

10.3.4 Glucose

Glucose sensing is important both in the food processing industry and in clinical chemistry. Close monitoring of *blood glucose* on the level of 10 mM is needed for diabetes patients and an increase requires immediate treatment. The existing devices, based on enzyme sensors and electrochemical detectors, have shown their limitations and this fact encourages alternative approaches, one of which is fluorescence sensing (Pickup et al. 2005).

The chemical receptor approach is represented by *boronic acid derivatives* (which bind the diols of sugars) coupled to organic fluorescence dyes, so that the direct mechanisms of their response based on PET or ICT can be used with the detection in intensity, spectral shift or lifetime. *Bioanalytical recognition units* include the lectin concanavalin A (Con A), enzymes such as glucose oxidase, glucose dehydrogenase and hexokinase/glucokinase and the bacterial glucose-binding protein. In these cases, the most popular mechanism of response is FRET between a fluorescent donor and an acceptor within a protein that undergoes glucose-induced changes in conformation or re-arrangement of subunits.

Competitive displacement assays have also been developed. They use fluorescently labeled dextran as a competitor in glucose binding to concanavalin

A (Ballerstadt et al. 004) or apo-glucose oxidase (Chinnayelka and McShane 2004). Plasmonic amplification of such assays based on gold colloid particles has been reported (Aslan et al. 2004).

Brightly fluorescent nanoparticles such as *Quantum Dots* have started to be used in the development of glucose sensors. One of the recent examples (Fig. 10.5) is the use of a two-component glucose-sensing system comprised by a boronic acid substituted viologen quencher and fluorescent, CdSe/ZnS QDs. The modulation of the fluorescence signal occurs by competitive binding of glucose with the boronic acid receptor moiety. Unfortunately, only the fluorescence intensity increase can be detected with this type of sensor.

Though simple when performed in a test tube, many of these methods are still far from use on patients. They require taking a blood sample or invasion into the human tissue. Attempts to provide *noninvasive glucose monitoring* by the measurement of cell autofluorescence due to NAD(P)H and fluorescent markers of mitochondrial metabolism, are of interest. Two alternative approaches have strong potential for development. One is to use biological fluids, the collection of which does not need invasion, such as saliva, sweat and tears. The introduction of contact lenses that light-up the increase of the glucose level in tears is a remarkable step in this direction (Badugu et al. 2003).

The other approach is an attempt to provide *in vivo* monitoring based on the response of the safe near-IR dyes (Ballerstadt et al. 2004). The near-IR light can transmit through the skin and may provide an optical excitation and readout.

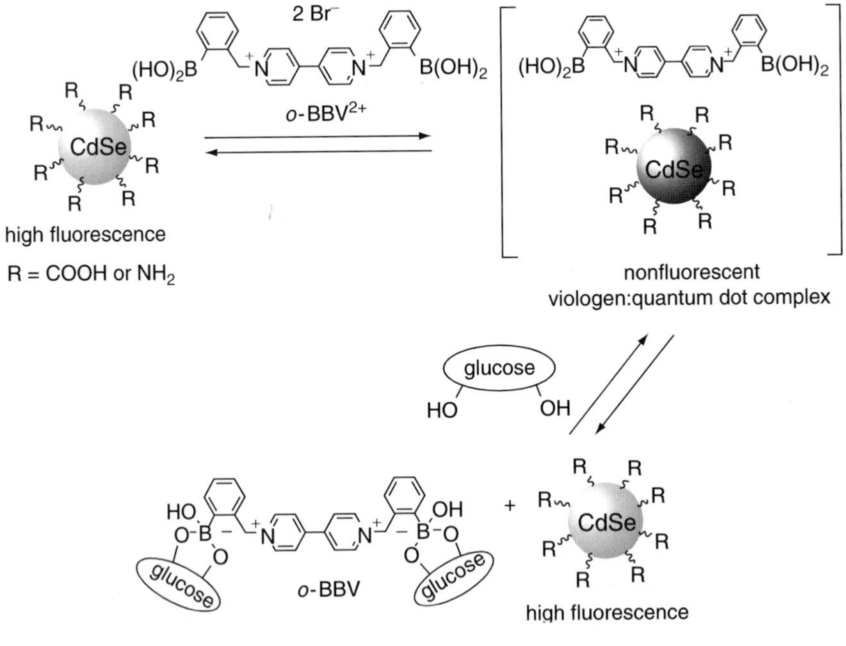

Fig. 10.5 An example of using fluorescent Quantum Dots for glucose sensing (Reproduced with permission from Cordes et al. 2005)

10.3.5   Cholesterol

Cholesterol is an important parameter in clinical diagnosis, due to a strong correlation between the total cholesterol in human blood serum and heart and blood vessel diseases. A large number of clinical disorders are related to arteriosclerosis, in which the role of cholesterol is very important. Presently, the detection of cholesterol is based on the action of the enzymes cholesterol oxydase and cholesterol esterase, so that hydrogen peroxide liberated in this reaction is the target for amperometric detection (Vidal et al. 2003).

Attempts to establish *direct sensing* methods that avoid manipulation with enzymes include the application of cyclodextrin dimers labeled with a fluorescent reporter dye (Liu et al. 2004) and europium-tetracycline complex (Silva et al. 2008).

The present author has made an attempt to use the well-known property of cholesterol to incorporate it into lipid bilayers and to form very strong complexes (called 'rafts') when this lipid is sphingomyelin. We produced sphingomyelin vesicles with embedded wavelength-ratiometric 3-hydroxychromone dyes and observed dramatic changes in florescence color upon incorporation of cholesterol molecules into these bilayer structures (Turkmen et al. 2005). These experiments were reproduced with the application of a number of different dyes and the results of one of them are depicted in Fig. 7.11.

10.4   Nucleic Acid Detection and Sequence Identification

The detection and characterization of DNA and RNA is needed in many applications, which include all branches of biology, medicine, environmental control and forensic detection. Such an extremely broad range of applications requires the search for diverse methodologies that should be optimal for every particular task. The last decade saw the active development of DNA high-throughput technologies. Their principles were discussed in Chapter 9. The microarrays were successfully applied for genotyping and gene expression studies. The limitations in the performance of this technology and the insufficient reproducibility of its data required finding new solutions. The high performance techniques offered by modern biotechnology and nanotechnology will be discussed here.

10.4.1   Detection of Total Double-Stranded DNA

The demand for the detection of nucleic acids in natural samples is often limited to the detection of *double-strand DNA* (dsDNA). In a living cell, DNA commonly exists in a double-helical form, which allows its easy detection by fluorescence dyes. A number of cell-permeable and cell-impermeable dyes have been developed for this purpose. The most attractive among them are the dyes that exhibit almost no fluorescence in water but a very high emission when they are bound with high affinity to DNA (with enhancement by a factor of 10^3). Particularly acridine dyes and ethidium bromide possess this property (Haugland 2005).

The affinity of these dyes towards DNA can be dramatically (by 10^3–10^4 times) increased upon the use of the *dimers of these dyes* instead of the previously applied monomers. These dimers, such as EthD, YOYO and TOTO (Thiasole Orange dimer), carry four positive charges, allowing a strong electrostatic interaction with the negatively charged DNA. Their binding is so strong that the electrophoresis of DNA labeled with nanomolar quantities of these dyes can be performed without dye loss (Glazer et al. 1990). The affinity towards the DNA and the brightness of the fluorescence response are sufficient for single-molecular detection.

There are two major sites of application of these dyes on DNA molecules. One is the site of the intercalation between the nucleic acid bases. In the double helix, the latter form planar couples stabilized by hydrogen bonds. *Intercalation* is the inserting and stacking of planar small-sized molecules between the base pairs of the DNA double helix, due to the hydrophobic and van der Waals interactions. Intercalation is a noncovalent binding, in which the molecule is held perpendicular to the helix axis. This type of binding is characteristic for acridine, ethidium and thiasole dyes and their dimers. The planar dye molecule of the same size as the base pair accommodates between the neighboring base pairs. The major effect of the fluorescence enhancement of these dyes is their screening from the quenching effect of water. In the cases of the dye dimers, the two heterocyclic parts intercalate between the base pairs and the linker accommodates to the minor groove.

The other type of DNA-staining dyes are the dyes that bind directly to the *minor groove* of the double helix. In this case, the binding of the dye molecules is mostly electrostatic and great enhancement of their fluorescence upon the binding can be due to both displacement of water from their contact areas with DNA and to the suppression of their internal segmental mobility. The dyes, known as Hoechst dyes (e.g., Hoechst 33258 or Hoechst 33342) possess three positive charges per molecule and represent examples of this type of binding. They are cell membrane permeable and their positive feature is the strong Stokes shifts that allow to excite fluorescence at about 350 nm and observe it in the 500 nm region.

The highly popular in nucleic acid research *cyanine dyes* demonstrate remarkable diversity of binding patterns. The short-chain monomethine dyes act as intercalators, while the increase in the polymethine chain length leads to increasing the potency for groove binding (Kovalska et al. 2006; Yashchuk et al. 2007).

Different cyanine dyes (Section 4.1) are frequently used for DNA detection. Their advantage is the presence of low-polar groups together with the positive charge that provides specific affinity towards nucleic acids, together with the high brightness necessary for ultrasensitive detection. Fluorescence enhancement on their binding is due to dye fixation in trans-conformation. Their sensitivity can be even further increased by the formation of cyanine dye aggregates on this interaction (Ogul'chansky et al. 2000; Yarmoluk et al. 2002).

10.4.2 Detection of Single-Stranded DNA and RNA

Many fluorescent dyes bind not only to double helices but also to *single-stranded DNA* (ssDNA) and RNA. Acridine orange is one of the earliest examples for which such binding has been described. The mechanisms of binding to double and single-stranded nucleic acids are so different that there is a difference in emission color. When this dye binds to DNA, it exhibits a green fluorescence, while when bound to RNA it yields a red fluorescence. Thus, the DNA in the nucleus stains green, while the RNA in cytoplasm – are red. These observations were recently extended by the application of a number of acridinium derivatives and these were found to bind to single strands of DNA and RNA, with negligible binding to double-stranded DNA (Kuruvilla and Ramaiah 2007).

Due to the presence of four positive charges, the ethidium dimer and also YOYO and TOTO dyes can also bind with ssDNA and RNA.

10.4.3 Sequence-Specific DNA Recognition

When dyes bind to nucleic acids by intercalation, they do not exhibit sequence specificity. This is conceivable, since no specific interactions of the DNA bases participate in the dye binding. The required sequence specificity can be observed on *groove binding*. Here, in addition to the hydrophobic and van der Waals interaction, the electrostatic interaction between a dye molecule and charged phosphate groups can take place, as well as hydrogen bond formation with the base pairs and hydroxyl groups of sugar residues. Groove binders can also be *sequence-specific* because of the possibility of aggregation, which extends the interactions within the groove (Vercoutere and Akeson 2002).

Efficient recognition of specific DNA sequences can be achieved with the proteins functioning as specific DNA binders and with their fragments retaining these properties. Their function requires structural motifs known as '*zinc fingers*'. The HIV-derived NCp7 protein is one of these examples. In an experiment discussed below, the truncated NCp7 peptide fragment (11–55) was prepared using solid-phase peptide synthesis. At the last step of the synthesis, a carboxylic acid analog of wavelength-ratiometric 3-HC dye was covalently attached at the N-terminus of this peptide. This allowed obtaining a direct reporting signal on the oligonucleotide-peptide interaction (Color Plate 10).

It was shown that the addition of oligonucleotide dTAR to the solution of the peptide conjugate, which is known to interact with NCp7, leads to significant changes of the fluorescence spectra: the N^*/T^* intensity ratio decreases and the T^* band (located at longer wavelengths) shifts to the red (Color Plate 10a). According to the data on the parent fluorophore in model solvents, the observed decrease in the N^*/T^* ratio suggests a decrease in the environment polarity. In addition, the binding of the peptide to dTAR is accompanied by a small hypochromic effect and some red shift in the absorption maximum, which could be explained by partial stacking of the label with nucleic acid bases. Obtained results show that the 3-HC dye senses NCp7-oligonucleotide complex formation by reporting on the decrease in the polarity of the interaction site.

The magnitude of the spectral effect depends strongly on the oligonucleotide nature (Color Plate 10). For example, the 1:1 complex formation of this peptide with oligonucleotide PBS responds by only a 25% decrease of the N^*/T^* ratio, which is much less then that observed upon binding to another oligonucleotide, SL2. These differences suggest that in the latter complex the label is more exposed to solvent water and interacts less with oligonucleotide than in a complex with SL2. This is in line with the known 3D NMR structure of these complexes, showing that in a complex with SL2, the N-terminus of NCp7 is in close proximity to the nucleic acid bases, while with PBS, the N-terminus locates close to the phosphate groups of the oligonucleotide backbone. Thus, using the environment-sensitive 3-HC dye, one can observe not only the act of specific binding to a nucleic acid sequence and determine the binding affinity (in titration experiments) but also obtain some information on the mechanism of molecular recognition.

10.4.4 'DNA Chip' Hybridization Techniques

Nothing is more specific than the recognition between specific DNA or RNA sequences that leads to the formation of double-helical structures. Such recognition is always in the background of *hybridization assays*. The two strands recognize each other by the formation of very selective hydrogen bonds between the complementary nucleic acid bases. Three such bonds are formed between guanine (G) and cytosine (C) base pairs and two bonds are formed in complementary pairs of adenine (A) and thymine (T), for DNA, or uracyl (U) for RNA. Due to the collective effect produced on the affinity of interaction by multipoint binding, the formation of the first three–four base pairs is sufficient to trigger the process of hybridization. The formed hybrid double helices are highly specific and stable.

In order to possess a detection method that could satisfy many demands (such as in sensing in microbial detection and/or clinical gene expression profiling), we have to resolve three technological problems. First, we need an *adequate transduction method* to generate a physically measurable signal from the hybridization event. Second, this method should be *extremely sensitive* to allow detection on the picomolar level of concentrations of formed hybrids and below. Third, this method should allow the *application in microarrays* to allow essentially the same way of detection for all the spotted (or otherwise marked) sensor sequences.

With this perspective, we can critically analyze the 'DNA chip' technology that presently dominates in biomedical research and clinical diagnostics. It relies on a combination of amplification by the *polymerase chain reaction* (PCR) of the whole pool of tested sequences and on the application of organic dyes as the labels for the amplified products in this pool.

The common DNA chip technology consists in three essential steps (Fig. 10.6):

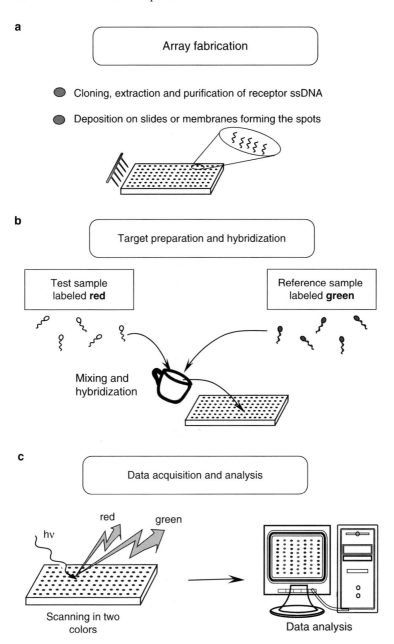

Fig. 10.6 Schematic showing the steps involved in fabrication (**a**), hybridization (**b**) and analysis (**c**) of a spotted DNA microarray with target sequences from a test sample and a reference sample that has been fluorescently labelled with different fluorophores, mixed and hybridized to the same array

(a) *The array fabrication.* The sensing ssDNA sequences possessing no label are immobilized on the surface (slices or membranes) in multiple copies, forming the microarray. Each type of ssDNA sequence occupies a separate spot.

(b) The *preparation of the target and the reference.* A comparative study is commonly made (*competitive hybridization*). The pool of DNA or RNA fragments that contain potential targets are amplified by using the polymerase chain reaction (PCR) and are covalently labeled with the dye. (Usually, total RNA is extracted from the cell and used as the starting material.) The same is done with the reference sample. In this case, the labeling is done with the dye emitting a different color. Both labeled samples are then mixed and hybridized to an array. The target sequences compete for the binding sites and those that are present in greater quantities in the solution are predominantly bound.

(c) *Data acquisition and analysis.* This is provided by a special reader, recording two fluorescence emissions. The formed image is analyzed with the aid of special computer software.

Thus, the elementary units of DNA flat microarrays (biochips) are the spots containing numerous ssDNA chains of identical sequence terminally anchored to the support surface. The pattern of the spots' location allows finding each DNA sequence at a unique site. These chains hybridize selectively with free ssDNA chains with a complementary sequence (Fig. 10.7a). The microarray is exposed to a solution containing the labeled ssDNA chains. The presence of specific sequences (targets) is signaled by hybridization on the corresponding spot, as monitored by correlating the strength of the label signal with the position of the spot (Dharmadi and Gonzalez 2004).

Based on this principle, the functional and convenient DNA biochips have been applied to analyze genomes, to detect inherited and acquired diseases, such as cancer and to detect cellular responses to different stimuli, such as the effects of drugs on a genomic level. They have proved to be very useful for the rapid detection of mutations, single nucleotide polymorphism (SNP) and also for gene discovery and expression monitoring.

Modern technology allows depositing as much as 10^5 of different oligonucleotides on a square centimeter chip, which shows the capability of extension of the whole *transcriptome profiling* (see Section 12.1). Arrays containing many thousands of unique probe sequences have been constructed and sophisticated algorithms have been developed for fluorescence readout.

This technique is still far from being ideal and there exist at least two important critical steps that have to be improved. One is the use of PCR and the other – the target labeling. Both refer to manipulations with the studied sample that need to be avoided in all cases where the time of analysis and the costs of reagents and labor are the critical points.

The PCR technique allows producing many nucleic acid sequences using the target nucleic acid fragment as a template. With its aid, the detection limit reaches the picomolar range. This reaction is time-consuming, it requires temperature cycling (for producing dissociation and re-formation of double helices) and expensive manipulation with the enzymes. Furthermore, the selective and nonlinear target

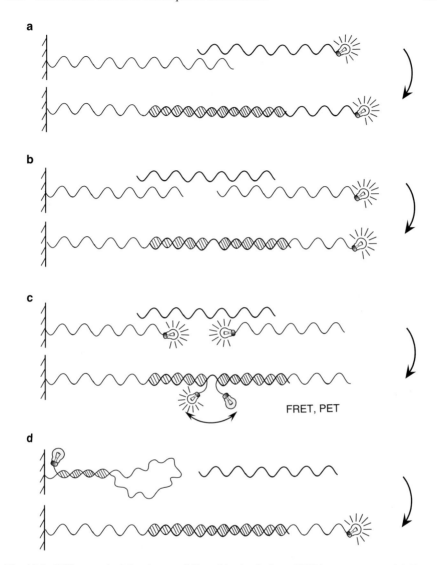

Fig. 10.7 Different principles that are followed in the design of DNA sensor arrays. (**a**) Target pool labeling. The immobilized sensor chain is unlabeled. (**b**) The sandwich assay. Neither the target pool nor the sensor chain is labeled. The indicator chain brings the label. (**c**) Both sensor and indicator sequences are labeled. Binding the target brings two labels together and allows realizing PET quenching or FRET emission wavelength switching effects. The sensor and the indicator can be connected by a flexible linker. (**d**) Molecular beacon. The target binding to the complementary loop structure disrupts the stem and removes the label from quenching by the neighboring groups or by the surface

amplification in PCR may distort the gene expression. The portability and miniaturization of this reaction remains a problem. Avoiding the PCR reaction can be
possible only when the necessary level of sensitivity is achieved with alternative
procedures.

The incorporation of sensor DNA into hydrogels makes them three-dimensional,
greatly increasing their capacity. This development has found many applications in
research and in clinical diagnostics (Rubina et al. 2004).

The simple 'DNA chip' technology has opened exciting possibilities for applied
genetic analysis, for the diagnosis of infections, identification of genetic mutations
and forensic inquiries. Meantime, we have to admit that the target pool labeling
used here is the most primitive concept in sensing (Section 1.2). Its success in DNA
hybridization assays is due to a lucky fact that the same chemistry can be used for
the labeling of all nucleic acids present in the pool. Attempts to substitute it with
the techniques based on more advanced concepts (as described in Chapter 1) will
be discussed below.

10.4.5   Sandwich Assays in DNA Hybridization

A *sandwich assay* avoids the necessity of pool labeling but requires recognition of
the target at two distinct sites (Section 1.4). In nucleic acid detection, one of the
DNA terminal sequences can be recognized by the capture sequence (that is usually
attached to the support) and the remaining single-chain segment exposed to the
solvent can be hybridized with an indicator sequence that contains a fluorescence
label (Fig. 10.7b). An illustration of this principle is presented in Fig. 1.12. As in
the case of immunoassays, this label can not only be the fluorescent dye but also a
much brighter nanoparticle. Dye-silica doped nanoparticles were successfully used
for DNA detection in this assay (Zhao et al. 2003). An impressive detection limit
of 0.8 fM (femtomoles) and a selectivity ratio of 14:1 against one-base mismatches
have been achieved.

In general, the sandwich technique has found a more limited application than the
DNA chips. Its advantage is the possibility of using relatively large fluorescent
nanoparticles coupled with the indicator, which increases dramatically the sensitivity of response and makes PCR amplification unnecessary.

This approach can allow one to make one more step forward and develop the
wavelength-ratiometric fluorescence nanosensor, based on double labeling (Fig.
10.7c). The capture sequences, instead of being attached to a plain support, can be
bound to fluorescent nanoparticles that serve as FRET donors. The indicator DNA
is modified with the attachment of a fluorescent dye that serves as a FRET acceptor.
When both of them are present in the test medium without target DNA, they do not
interact and we observe the emission of the donor. In the presence of the target
DNA sequence, it hybridizes with both capture and indicator sequences, the FRET
donor and acceptor approach and we observe the emission of the acceptor (Zhang

et al. 2005). Many constructions like this can be devised but their extension to high-throughput screening needs the precise identification of nanoparticles carrying specific sensors, 'barcoding'.

Can the intercalating dye substitute the labeled indicator DNA sequence? Traditionally, they have been very useful to detect the formation of double-stranded DNA in hybridization assays signaling by a dramatic increase of fluorescence intensity. Meantime, the background emission of the unbound intercalator, though small, remains an obstacle. An improvement of their response can pave their way to microarray technologies.

10.4.6   Molecular Beacon Technique

A quite different methodology is behind the molecular beacon techniques. They operate according to the principle of direct sensing, in which the *recognizing sequence also possesses the reporting function*. Reporting is coupled to conformational change in this sequence. In Section 6.3 the reader can find a detailed description of this mechanism and below, we concentrate on different possibilities for its realization and on some of the obtained results.

Molecular beacons are nucleic acids that contain two structural components, a loop and a stem, forming a 'molecular hairpin' (Fig. 6.12). The loop is the recognition element for complementary nucleic acid sequences of the target. The stem is formed of two complementary sequences that flank the loop and that are end-labeled with a reporter dye and a quencher (Tyagi and Kramer 1996). The hybridization of the target nucleotide sequence to the loop opens the stem-loop, inducing a separation between them, removing the quenching effect (Fig. 10.7d). Different signal transduction mechanisms, in addition to fluorescence de-quenching have been suggested to provide the response to analyte ssDNA or RNA binding (Section 6.3). Moreover, the hairpin is the simplest but not the only possible structure of a molecular DNA sensor that combines target binding and reporting. Hybrid structures have been proposed that contain DNA segments connected by a flexible polymer linker (Yang et al. 2006). Upon hybridization, the labeled DNA segments appear in close proximity with the response in the generation of FRET.

The process of recognition in molecular beacons is associated with the *disruption of intramolecular contacts* between DNA bases in the stem in favor of intermolecular contacts with the target, formed by the loop. Because of this competition, this method is much more sensitive to single nucleotide substitutions than the simpler hybridization assays (Demidov and Frank-Kamenetskii 2004).

A great advantage of molecular beacons is the possibility of its use for intracellular studies. The labeled hairpins are molecular instruments that do not need any support and contain everything for both recognition and reporting. There is a problem, however, that any nonspecific disruption of the stem-loop structure would give a false positive result. This can occur due to nuclease degradation, protein binding or other conditions making the stem structure unstable.

10.4.7   *DNA Sensing Based on Conjugated Polymers*

A cationic conjugated polymer (such as polythiophene derivatives) exhibits strong electrostatic binding to nucleic acids with the change of its fluorescence properties. It is fluorescent in its native random coil configuration but the fluorescence is quenched when it combines with a single-stranded DNA chain to adopt a highly conjugated, planar conformation. It becomes fluorescent (λ_{exc} = 420 nm, λ_{em} = 525 nm) when a complementary oligonucleotide target is added to this molecular system. This interaction is different with the double-stranded DNA. In this case, the polymer binds in a helical and non-planar configuration (a 'triplex') with the negatively charged phosphate backbone of the double-stranded DNA (see Fig. 10.8a).

Thus, the *conjugated polymer* can serve as a very sensitive indicator of hybridization. Moreover, since it exhibits remarkable properties of the antenna effect in FRET (Section 6.2) and superquenching (Section 4.6), the sensitivity of

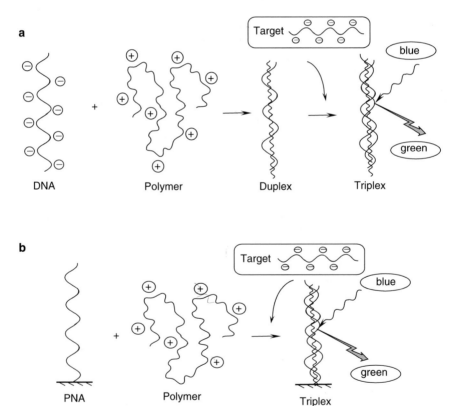

Fig. 10.8 Schematic representation of the transduction mechanism produced by a conjugated polymer in cases where the hybridization of the target DNA is produced with DNA in solution (**a**) and with PNA on solid support (**b**) (After Dore et al. 2006)

the response can be dramatically increased with the inclusion into this system of an additional organic dye. It has been shown that tagging the sensor chain with *fluorescent dye serving as a FRET acceptor* that emits with the wavelength shift to 572 nm, allows detecting as much as only five copies of dsDNA in 3 ml in the presence of the entire human genome, in only 5 min (Ho et al. 2005). This approach is suitable for the rapid assessment of the identity of single nucleotide polymorphisms (SNPs) and of different genes and pathogens, without the need for nucleic acid amplification (Dore et al. 2006).

When the neutral *peptide nucleic acid* (Section 5.6) is immobilized as the recognition unit on the surface of a planar array, it does not interact with the cationic polymer (Fig. 10.8b). The triplex formation occurs upon hybridization with the target nucleic acid. The exploration of the antenna effect and superquenching are also efficient in this case, with the introduction of the fluorescent dye serving as the quencher or as the FRET acceptor (Baker et al. 2006; Liu and Bazan 2005).

10.4.8 Concluding Remarks and Prospects

DNA analytic techniques have led to revolutionary changes in modern biology and related areas. They allow both single target detection and the whole-cell expression of genes. They have brought unique possibilities for the identification of personality involving the characteristics of single nucleotide polymorphism (SNP), recognition of cancer cells and tissues and detection of pathogenic microorganisms. They allow the comparison of closely related species in plant and animal worlds. This huge amount of tasks requires the development of different competitive technologies, which is actually being observed.

The highly needed multiplexed microarray platforms provide parallel detection capabilities enabling the measurements of many thousands of simultaneous responses. DNA microarrays are especially attractive due to the achieved combination of high throughput, parallelism, miniaturization, speed and automation. The inclusion of the expression profile of the whole genome eliminates the bias related to any preliminary selection. Moreover, the global nature of the DNA microarray technique holds tremendous promise for the discovery of complex genetic and metabolic networks. In this rapidly developing technology, measurement is ahead of analysis and sensitivity is ahead of accuracy. Overcoming these difficulties is expected soon.

Methods that offer avoiding PCR amplification and labeling of the target pool are expected to complement and in many important areas, substitute the standard 'DNA chip' approach. These methods are expected to make DNA tests considerably simpler, cheaper and quicker alternatives and therefore more applicable to large-scale applications. This goal can be achieved on one condition: they have to demonstrate superior sensitivity. This condition can be realized when ultra-bright nanoparticles will substitute or complement organic dyes and when enzymatic amplification mechanisms are substituted by photophysical ones. Dye-doped nanoparticles (Zhao et al. 2003), Quantum Dots as FRET donors (Zhang et al. 2005), conjugated polymers as self-amplified systems (Dore et al. 2006; Liu and Bazan 2005), metal-enhancement support (Aslan et al. 2006), as well as enhancing the support of porous semiconductor materials (Dorfman et al. 2006), are the trends pointed forward by pioneering research.

10.5   Recognition of Protein Targets

Proteins are biological macromolecules that are the most essential constituents of cells, the building blocks of all their organelles and the catalysts of all biochemical reactions. In different conditions and for different purposes, there is a need to determine (a) the total protein content, (b) the specific proteins, such as a particular enzyme or antibody, (c) all protein composition in a given system and (d) all human proteome. We will start with the simplest tasks and proceed with the more complicated ones.

10.5.1   Total Protein Content

The classical protein assays known from textbooks on biochemistry, such as the Biuret and Lowry methods, have a strong tendency to be substituted by rapid, sensitive and precise methods employing biosensors. The problem with broad-scale protein determination is in its great diversity and even the uniqueness of every protein, which complicates its recognition 'as a protein'. Nevertheless, such attempts continue, using different recognition units such as fluorescent dyes themselves (Harvey et al. 2001) or modified cyclodextrins combined with attached reporting dyes (Zhu et al. 2007).

10.5.2   Specific Protein Recognition

Presently, in sensing the most requested recognition elements for the detection of proteins are antibodies (Kusnezow et al. 2003) that recognize linear segments or conformationally unified clusters of 3–5 amino acids on the protein surface. In addition, the designed binding proteins, peptides and aptamers demonstrate their increased utility (Chapter 5). Because of the dramatic variations of protein structures, their surface topologies and their surface modifications, there is great difficulty in unifying the strategies for protein-receptor interactions. A new, interesting approach to protein recognition is based on the concept of 'hot spots' on the protein surface responsive to protein interactions (Peczuh and Hamilton 2000). This concept is illustrated in Fig. 10.9.

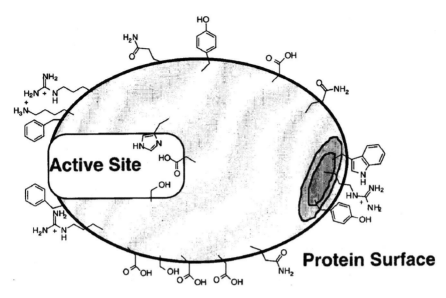

Fig. 10.9 Schematic representation of a protein (enzyme) interior, exterior and hot spot (Reproduced with permission from Peczuh and Hamilton 2000)

A *hot spot* is a defined locale of ca. $600\,\text{Å}^2$ on the surface of a protein at or near the geometric center of the protein-protein interface. The residues that comprise the hot spot contribute significantly to the stability of the protein-protein complex. Moreover, mutations of single residues in these areas modulate the protein interactions dramatically. A survey of amino acids appearing on hot spots showed a predominance of Trp, Tyr and Arg residues. The surface area of the hot spot may be of a size that is critical to make a water-excluding seal around the energetically favorable interactions. The usefulness and limitations of this concept will be seen in subsequent studies.

Complementary to the highly specific recognition described above, is the strategy of cross-responsive arrays that are inspired by the mammalian olfactory system (Section 12.5). The high level of cross-reactivity still allows precise recognition of a particular target. This is because each target generates a unique composite pattern of interactions with receptors. In analogy, an array of nonspecific or weakly interacting agents would give rise to distinctive fingerprints in response to a target. This strategy can be the basis for the identification of "hot spots" on protein surfaces.

10.5.3 Protein Arrays

A strong demand for high-throughput analysis on a proteomic level stimulates rapid development of *protein arrays* for the simultaneous detection of many proteins. Due to their high throughput, they must have great potential to reduce the cycle time of drug discovery and to improve the efficiency of medical diagnostics. For real world use, a protein array has to become simple and inexpensive to manufacture, its fabrication amenable to automation, the size and shape of their spots must be controlled, they must be durable and reproducible in application. Their sensitivity should be on a competing level with that of ELISA, which means a picomolar detection limit.

Making these arrays by using the traditional methods for DNA chips is much more difficult than DNA or RNA microarrays because of three major problems. One is related to the conformational lability of protein molecules and the necessity to immobilize them on support in native conformation. The other is the fluorescent labeling of a protein pool. This requires from the researchers a broader view and revision of the basic concept of array operation. The third problem is related to the variable and often low output signal.

(a) The concept that the *best recognition partner of a protein is a protein* has to be revised. Individual proteins differ in stability, surface charge and reactivity of surface groups and other properties. This creates difficulties in their immobilization in the native form and orientation that could allow target binding. Protein stability in storage and application remains an important concern. Novel strategies to fabricate sensor molecules include single chain antibodies displayed by phage, protein- oligonucleotide conjugates, synthetic peptides and aptamers.

Protein sensors should be used on account of new possibilities that exist for their application as recognition units (Section 5.3). It is also possible to construct arrays of full-length, functional proteins from a library of expression clones (Bertone and Snyder 2005). They may be of great use when global observations of biochemical activities are needed and hundreds or thousands of proteins can be simultaneously screened for protein-protein, protein-nucleic acid and small molecule interactions.

(b) The *concept of pool labeling* should be abandoned since it cannot reproduce the true protein concentration. Usually, such labeling is provided by amino-reactive dyes in the form of their active ester or isothiocyanate derivatives (see Section 7.3). Proteins differ so much in the amount and chemical reactivity of their surface amino groups that no uniform fluorescent labeling is possible. Moreover, the site-dependent variation of the fluorescence properties is observed for many fluorescent dyes.

(c) Direct reading of the intensity of labeled dyes does not provide the necessary sensitivity. Therefore, nanoparticles with a brighter emission (Sections 4.3 and 4.4) or some signal amplification mechanism should be implemented into the process of fluorescence response.

These facts explain why, despite the offer of different ligand-binding proteins, peptides and aptamers, *antibodies* are still very much in vogue, as well as recombinant antibody fragments (Pavlickova et al. 2004). Many problems with the design and production of microarrays based on them have been resolved, including their immobilization in 3D hydrogels (Rubina et al. 2003). The prospects of their application on a proteomic level will be critically discussed in Section 12.1.

The *aptamers*-based protein microarray uses immobilized DNA and RNA aptamers, selected against many different protein targets, allowing detecting simultaneously and quantifying the levels of individual proteins in complex biological mixtures. Aptamers can each be fluorescently labeled and immobilized on a glass substrate. Different methods, including fluorescence polarization (anisotropy), can be used for detecting the target protein binding. The latter detection method can operate both in solution and on solid support.

Water-soluble *conjugated polymers* with pendant-charged residues provide an excellent scaffold for the design of sensor arrays, since they combine the binding and responsive properties (Thomas et al. 2007). The introduced side groups can provide specific protein binding through multivalent interactions and their optical properties are sensitive to minor changes in their conformation, with the possibility of fluorescence superquenching (Section 4.6).

Sensor arrays based on non-covalent interactions with *metal* and *semiconductor nanoparticles* and their conjugates with organic dyes and fluorescent polymer conjugates have started to be developed.

10.6 Polysaccharides, Glycolipids and Glycoproteins

Known under the common name of *glycans*, polysaccharides and carbohydrate portions of glycolipids and glycoproteins constitute a large family of compounds that involve sugar elements (Jelinek and Kolusheva 2004). These molecules are present both inside and on the surface of cells and are integral constituents of a vast number of proteins, which are involved in a myriad of cellular events. Carbohydrates and glycoconjugates of proteins and lipids participate in such processes as normal and pathogenic cell adhesion, inflammation, signal transduction, etc. Therefore, a new field of *glycomics* has to follow in the development of the more advanced fields of genomics and proteomics, with a strong extension into clinical diagnostics. In addition, control on carbohydrate levels is needed in industrial biotechnologies and in many other industrial processes, such as food processing and storage.

Before these highly needed arrays come into practice, many technical problems have to be resolved (Shin et al. 2005). These problems are much more difficult than with DNA and protein arrays. They can be outlined as follows:

(a) Glycans are rather *diverse chemical structures* and there are no unified methods for their synthesis, modification and immobilization on a surface. Moreover, the designed sensors must be able to probe protein-carbohydrate, carbohydrate-carbohydrate, nucleic acid-carbohydrate and intact cell-carbohydrate interactions. This makes it difficult to develop a unified strategy of sensor design. The questions to be resolved involve proper surface preparation, proper linking chemistry and immobilization with proper orientation.

(b) The glycan interactions display *diverse affinities* and are often rather weak ($K_d \sim$ µM-mM). Definitely, such strength of sensor-target interactions does not allow using in full the sandwich techniques, which require washing steps. In addition, the researcher should often address the problem of nonspecific adsorption effects and false-positive signals. The application of competition assays can also be problematic.

(c) There is *no unified procedure* for fluorescent labeling and the existing methods of introduction of reactive groups are not specific enough. Also, close fluorescent analogs of carbohydrates do not exist and the conjugation of fluorescent dyes may intervene strongly into target recognition.

These difficulties do not detract researchers and, with the implementation of new ideas and new technologies, they have to be overcome. Several examples are presented below.

It has been shown that with the aid of combinatorial methods of sensor design, a library can be constructed for the specific binding of *lectins*, which are the specific carbohydrate binding proteins (Kolomiets et al. 2007). A 15,625-membered short peptide-dendrimer combinatorial library was produced and acylated with an alpha-C-fucosyl residue at its four N-termini and screened for binding to fucose-specific lectins. In this way, the targets can be identified on a sub-micromolar level.

Similarly to antibodies recognizing surface antigens, *glycan sensors* can recognize a carbohydrate pattern on the surfaces of pathogenic microbes. It is known that a large number of bacterial toxins, viruses and bacteria target carbohydrate derivatives on the cell surface to attach and gain entry into the cell and glycan sensors can detect them (Ngundi et al. 2006b). Because of this, they find important an application for the determination of microbial contaminants in food.

Being made of *lectins*, glycan sensors provide the profiling of carbohydrate expression on the surface of human cells (Zheng et al. 2005). Sensing the glycan markers of diseases is an important approach in clinical diagnostics (Dai et al. 2006).

Therefore, it is expected that glycan microarrays will rapidly join the DNA and protein microarrays in biomedical research and diagnostics and in industrial practice.

10.7   Detection of Harmful Microbes

Rapid, sensitive, and selective detection of pathogenic bacteria and viruses is extremely important for the proper diagnosis and treatment of diseases. Global biosecurity threats such as the spread of emerging infectious diseases (i.e., avian influenza) and bioterrorism have dramatically increased the importance of this problem. The control for microbial contamination is not only a medical problem. It plays an important role in pharmaceutical clean room production and in food processing technology. What does the future hold and which biosensor technology platform is suitable for the real-time detection of infectious microbes?

10.7.1   Detection and Identification of Bacteria

The role of sensors in the analysis of microbes is primarily in the simplification and speeding up of the assays. Conventional microbiological classification tests involving colony growth in culture media require tedious cultivation for several days and highly professional analysis of only one bacterial pathogen at a time. The sensors should substitute these tests and allow detecting multiple microbes, obtaining a response in minutes and in the simplest possible way. Therefore, many assays, mostly immunoassays have been suggested for practical use.

Traditionally, these assays are based on the recognition properties of *antibodies* against surface cell receptors. Critical analysis of these assays can be found elsewhere (Sapsford et al. 2006). During the present phase of technology development, their essential improvement can be provided. In the detection of individual bacterial cells they are developing into potent biosensor technologies that allow high-affinity binding to the bacterial surface together with a very efficient fluorescence response. *Silica nanoparticles* responding in different colors have proved to be promising for the simultaneous and sensitive detection of multiple bacterial targets (Wang et al. 2007).

Antimicrobial peptides are different types of recognition units. These peptides represent an innate immune system as a first line of defense against microbial invasion that evolved in bacteria, plants and higher and lower animals. They exert their antimicrobial activity by binding to components of the microbe's surface and disrupting the membrane. Their recognition power to detect *E. coli* and *Salmonella* has been demonstrated (Ngundi et al. 2006a).

A novel approach for the detection of bacteria is the use of *carbohydrate-functionalized fluorescent polymers* (Disney et al. 2004). It is based on the fact that many pathogens that infect humans use the cell surface carbohydrates as receptors to facilitate the cell-cell adhesion. The hallmark of these interactions is their multivalency, or the simultaneous occurrence of multiple interactions. Fluorescent polymers that occupy an increasingly strong position in fluorescence sensing (Section 4.6) can be modified with different functional substituents. A carbohydrate-functionalized fluorescent polymer, which displays many carbohydrate ligands on a single polymer chain, was used to allow for multivalent detection of pathogens. With their application, the brightly fluorescent aggregates of bacteria can easily be observed.

The sensitive detection of pathogens via their *nucleic acid sequences* is another possibility for research and development. A microfluidic biosensor with disposable microchannels possessing defined areas for the capture and detection of a target pathogen RNA sequence has been demonstrated (Kwakye and Baeumner 2003). Two different DNA probes complementary to unique sequences on the target pathogen RNA serve as the biorecognition elements. For signal generation and amplification, one probe is coupled to dye-encapsulated liposomes, while the second probe is coupled to superparamagnetic beads for target immobilization. The probes hybridize to target RNA and the liposome-target-bead complex is subsequently captured on a magnet. The amount of liposomes captured correlates directly to the concentration of the target sequence and is quantified using a fluorescence microscope.

The special developments of instrumentation for environmental, clinical and food monitoring or for the detection of bioterrorist agents, include an automated apparatus based on the total internal reflection reading of fluorescence arrays weighing only 5.5 kg (Sapsford et al. 2004).

10.7.2   Bacterial Spores

Many harmful microbes form protective spores (endospores) that allow them to survive through hard times and to transform into active microbes in conditions favorable for their development. Such dormant forms can be very dangerous since they can survive for extremely long periods of time. The spores are composed of DNA, DNA-stabilizing agents (such as calcium dipicolinate) and a protective shell that can be recognized by fluorescence sensors.

A novel fluorescence assay was developed for the detection of Bacillus thuringiensis (BT) spores based on *aptamer-Quantum Dot* binding (Ikanovic et al. 2007). The assay is based on fluorescence observed after binding an aptamer-quantum dot conjugate to BT spores. The SELEX technique was used in order to select and identify the DNA aptamer sequence specific for BT. The 60-base aptamer was then coupled to fluorescent zinc sulfide-capped, cadmium selenide Quantum Dots. The assay is semi-quantitative, specific and can detect BT at concentrations of about 1,000 colony forming units/ml.

10.7.3   Detection of Toxins

Harmful bacteria can be detected by the products they synthesize. These products, the *toxins*, provide poisoning effects. This refers not only to the most dangerous infections. For instance, staphylococcal enterotoxins are a major cause of food poisoning. Here, the role of sensor technologies is not only to provide rapid and sensitive tests, but also to avoid extensive pretreatment or concentration of the sample prior to analysis. Though not fully conforming to these needs, immunosensor arrays based on immobilized antibodies have become a common approach for the detection of toxins (Rucker et al. 2005) and the most popular for detection is the competition format (Ngundi et al. 2006a).

Glycan arrays discussed in the previous section are emerging as very powerful tools for the detection of toxins that contain a carbohydrate component. Arrays of N-acetyl galactosamine (GaINAc) and N-acetylneuraminic acid (Neu5Ac) derivatives were immobilized on the surface of a planar waveguide and used as receptors for protein toxins with a sensitivity on the level of 100ng/ml (Ngundi et al. 2006b).

10.7.4   Sensors for Viruses

Viruses are sub-microscopic infectious agents that are able to grow or reproduce only inside a host cell. Each viral particle consists of genetic material in the form of DNA or RNA and a protective coat made of proteins, called a capsid. Both these nucleic acids and capsid proteins can be the targets of viral infection. Of diagnostic importance are also the antibodies raised in the organism in response to virus infection.

The early detection of the *human immunodeficiency virus* (HIV) is an extremely important problem in all efforts to stop epidemic propagation of the AIDS disease. Currently, the standard diagnostic tests for HIV infection are based on the ELISA technique, which is an expensive and inconvenient multi-step procedure. For this and other viral diseases, the important method of detection is a serial determination of anti-viral antibodies. A great variety of synthetic and phage-displayed peptides have been collected as test antigens (Palacios-Rodriguez et al. 2007), so what remains is to develop simpler methods that could be equal or superior in precision.

The recent suggestion for the detection of anti-HIV antibodies based on *peptide beacons* (Oh et al. 2007) is interesting. These peptides contain a sequence recognized by the antibodies and adopt different conformations in the free and antibody-bound forms, which influences the effect of the bound quencher on the fluorescence of the dye bound in a different position in the peptide sequence. The application of techniques providing signal amplification up to the sensitivity of ELISA is expected to follow.

Methods relying on *viral antigen* detection are expected to experience revolutionary changes with the application of novel fluorescence sensing technologies. The application of the time-resolved detection of Eu^{3+}-doped luminescence nanoparticles has been shown to improve the sensitivity of the adenovirus detection dramatically (Valanne et al. 2005).

10.7.5 Conclusions and Prospects

In view of a very strong theoretical background and novel sensor technologies, it is surprising that so little has been done to develop novel sensors with improved reporter properties for the detection of harmful bacteria and viruses. This situation of infancy should rapidly change. The new methods enable fast or real-time detection, portability and multi-pathogen detection for both field and laboratory analysis and soon they will find ways for broad-scale practical use. Rapid and sensitive detection can provide almost immediate interactive information about the sample tested, enabling users to take a timely clinical decision, corrective measures before the consumption of a product or take the necessary measures in the case of pathogens spreading in the environment.

Sensing and Thinking 10: Adaptation of Sensor Units for a Multi-scale and Hierarchical Range of Targets

By analyzing different applications of fluorescence sensors, we witness the interesting crossing points of microscopic and macroscopic concepts. Molecules that respond to thermal collisions with their neighboring molecules allow detecting the temperature and pressure gradients on the surface of airplanes and ion sensors are used for detecting the pollution of large tracts of land. Regarding the size of instrumentation, it will go down irrespective of complexity and application. Simultaneously, with the increase of complexity in tasks and efficiency in response, the size of elementary sensing units will grow from rather small and simple molecules to self-assembled supramolecular ensembles, involving biomacromolecules and the structures formed by them.

Aiming at the development of a sensor for a particular compound or series of compounds the researcher is able to select the optimal solution out of a broad variety of possibilities offered by synthetic organic and polymer chemistry and also by the chemistry and biochemistry of proteins, peptides and nucleic acids. In addition, all the recently developed techniques of manipulation with genes and gene products can be applied in this field. Though fragmentary, the analysis presented above involves the major classes of inorganic, organic and biological compounds, for which a strong demand exists for their detection. By providing analogies, the reader can find the optimal approach in the search for a rationally designed sensor that needs to be made.

It is expected that the reader will share with the author a strong belief that the described systems and their newly created analogs have the full potential to complement or replace the classical analytical methods. This can be done by simplifying or eliminating the sample preparation protocols and making field-testing easier and faster with a significant decrease in costs per analysis together with an increase in fidelity and precision.

Questions and Problems

1. What are the advantages and disadvantages in the application of luminescence to temperature and pressure sensing compared to other popular methods? Which method would you select for sensing temperature inside an oven? Inside a cryogenic device? Inside a water pipe? Inside targeted cells in a living body?
2. Temperature and pressure are parameters, of which the background is thermodynamic and connected with averaging over great molecular ensemble. Explain how to understand their meaning on a molecular level.
3. Why is oxygen so easily determined in gas mixtures? What are the means to use fluorescence detection for quantitative analysis of other gasses?
4. Explain the principle that can be used in the background of determining explosive vapors.
5. Explain the difference between the different definitions of the term 'polarity'. What is the mechanism behind fluorescence sensing of a solvent polarity? Should the polarity of the probe molecule itself correspond to the polarity of the studied medium?
6. How can the sensor response to polarity and viscosity be made ratiometric? Analyze the presented examples.
7. What is static and what is dynamic disorder in fluid systems? Does dynamic disorder depend on observation time window?
8. What is the major effect discovered with the wavelength-ratiometric probe in supercritical CO_2 (a) change of polarity? (b) change of solvent reactivity? (c) change of size and dynamics of solvent clusters?

9. What probes and what measurements can you suggest for studying the dynamic properties of polymers?
10. How can one define the polarity at an interface?
11. What determines the width of the pH range detected by fluorescence pH sensors? What are the means to extend this range?
12. Explain how to achieve wavelength-ratiometric oxygen sensing in solutions. Can this approach be applied to other gasses?
13. On what major principles do the sensors operate for heavy metal ions?
14. Why are optimal *in vitro* glucose sensors not the best when applied *in vivo*?
15. What are the differences in affinity, selectivity and mechanisms of binding to ssDNA and dsDNA between monomers and dimers of acrydine dyes?
16. Explain the advantages and disadvantages in the application of planar arrays with target pool labeling (DNA chips). Why it is hard to apply the same methodology to make protein arrays? What are the major problems in making glycan arrays?
17. For the detection of bacteria, what in your view are the best recognition units (choose between antibodies, antimicrobial peptides, aptamers, etc.) and reporting units (choose between dye-doped nanoparticles, conjugated polymers, systems exhibiting FRET, etc.)? Suggest the best composition of receptors and reporters.

References

Abbott AP, Hope EG, Palmer DJ (2007) Probing solute clustering in supercritical solutions using solvatochromic parameters. Journal of Physical Chemistry B 111:8119–8125

Amao Y (2003) Probes and polymers for optical sensing of oxygen. Microchimica Acta 143:1–12

Amara JP, Swager TM (2005) Synthesis and properties of poly(phenylene ethynylene)s with pendant hexafluoro-2-propanol groups. Macromolecules 38:9091–9094

Apostolidis A, Klimant I, Andrzejewski D, Wolfbeis OS (2004) A combinatorial approach for development of materials for optical sensing of gases. Journal of Combinatorial Chemistry 6:325–331

Aslan K, Lakowicz JR, Geddes CD (2004) Tunable plasmonic glucose sensing based on the dissociation of Con A-aggregated dextran-coated gold colloids. Analytica Chimica Acta 517:139–144

Aslan K, Huang J, Wilson GM, Geddes CD (2006) Metal-enhanced fluorescence-based RNA sensing. Journal of the American Chemical Society 128:4206–4207

Badugu R, Lakowicz JR, Geddes CD (2003) A glucose sensing contact lens: a non-invasive technique for continuous physiological glucose monitoring. Journal of Fluorescence 13:371–374

Baker ES, Hong JW, Gaylord BS, Bazan GC, Bowers MT (2006) PNA/dsDNA complexes: site specific binding and dsDNA biosensor applications. Journal of the American Chemical Society 128:8484–8492

Ballerstadt R, Polak A, Beuhler A, Frye J (2004) In vitro long-term performance study of a near-infrared fluorescence affinity sensor for glucose monitoring. Biosensors & Bioelectronics 19:905–914

Baptista MS, Indig GL (1998) Effect of BSA binding on photophysical and photochemical properties of triarylmethane dyes. Journal of Physical Chemistry B 102:4678–4688

Barroso M, Chattopadhyay N, Klymchenko AS, Demchenko AP, Arnaut LG, Formosinho SJ (2006) Dramatic pressure-dependent quenching effects in supercritical CO2 assessed by the fluorescence of 4′-dimethylamino-3-hydroxyflavone. Thermodynamic versus kinetics control of excited-state intramolecular proton transfer. Journal of Physical Chemistry A 110:13419–13424

Bertone P, Snyder M (2005) Advances in functional protein microarray technology. FEBS Journal 272:5400–5411

Bosch P, Catalina F, Corrales T, Peinado C (2005) Fluorescent probes for sensing processes in polymers. Chemistry-A European Journal 11:4314–4325

Bunzli JCG, Piguet C (2005) Taking advantage of luminescent lanthanide ions. Chemical Society Reviews 34:1048–1077

Catalan J (1997) On the E-T (30), pi*, P-y, S′, and SPP empirical scales as descriptors of nonspecific solvent effects. Journal of Organic Chemistry 62:8231–8234

Chandrasekharan N, Kelly LA (2001) A dual fluorescence temperature sensor based on perylene/exciplex interconversion. Journal of the American Chemical Society 123:9898–9899

Chiappe C, Pieraccini D (2005) Ionic liquids: solvent properties and organic reactivity. Journal of Physical Organic Chemistry 18:275–297

Chinnayelka S, McShane MJ (2004) Resonance energy transfer nanobiosensors based on affinity binding between apo-enzyme and its substrate. Biomacromolecules 5:1657–1661

Cordes DB, Miller A, Gamsey S, Sharrett Z, Thoniyot P, Wessling R, Singaram B (2005) Optical glucose detection across the visible spectrum using anionic fluorescent dyes and a viologen quencher in a two-component saccharide sensing system. Organic and Biomolecular Chemistry 3:1708–1713

Dai Z, Kawde AN, Xiang Y, La Belle JT, Gerlach J, Bhavanandan VP, Joshi L, Wang J (2006) Nanoparticle-based sensing of glycan-lectin interactions. Journal of the American Chemical Society 128:10018–10019

Davenport LD, Knutson JR, Brand L (1986) Excited-state proton transfer of equilenin and dihydro equilenin: interactions with bilayer vesicles. Biochemistry 25:1186–1195

Demchenko AP (2002) The red-edge effects: 30 years of exploration. Luminescence 17:19–42

Demidov VV, Frank-Kamenetskii MD (2004) Two sides of the coin: affinity and specificity of nucleic acid interactions. Trends in Biochemical Science 29:62–71

Deschenes LA, Vanden Bout DA (2001) Single-molecule studies of heterogeneous dynamics in polymer melts near the glass transition. Science 292:255–258

Dharmadi Y, Gonzalez R (2004) DNA microarrays: experimental issues, data analysis, and application to bacterial systems. Biotechnology Progress 20:1309–1324

Disney MD, Zheng J, Swager TM, Seeberger PH (2004) Detection of bacteria with carbohydrate-functionalized fluorescent polymers. Journal of the American Chemical Society 126:13343–13346

Domaille DW, Que EL, Chang CJ (2008) Synthetic fluorescent sensors for studying the cell biology of metals. Nature Chemical Biology 4:168–175

Dong DC, Winnik MA (1984) The Py scale of solvent polarities. Canadian Journal of Chemistry-Revue Canadienne De Chimie 62:2560–2565

Dore K, Leclerc M, Boudreau D (2006) Investigation of a fluorescence signal amplification mechanism used for the direct molecular detection of nucleic acids. Journal of Fluorescence 16:259–265

Dorfman A, Kumar N, Hahm JI (2006) Highly sensitive biomolecular fluorescence detection using nanoscale ZnO platforms. Langmuir 22:4890–4895

Ellison CJ, Torkelson JM (2002) Sensing the glass transition in thin and ultrathin polymer films via fluorescence probes and labels. Journal of Polymer Science Part B-Polymer Physics 40:2745–2758

Ellison CJ, Miller KE, Torkelson JM (2004) In situ monitoring of sorption and drying of polymer films and coatings: self-referencing, nearly temperature-independent fluorescence sensors. Polymer 45:2623–2632

El-Sayed IH, Huang X, El-Sayed MA (2006) Selective laser photo-thermal therapy of epithelial carcinoma using anti-EGFR antibody conjugated gold nanoparticles. Cancer Letters 239:129–135

Epand RM, Kraayenhof R (1999) Fluorescent probes used to monitor membrane interfacial polarity. Chemistry and Physics of Lipids 101:57–64

Ercelen S, Klymchenko AS, Demchenko AP (2002) An ultrasensitive fluorescent probe for hydrophobic range of solvent polarities. Analytica Chimica Acta 464:273–287

Fayed TA, Etaiw Sel D (2006) Fluorescence characteristics and photostability of benzoxazole derived donor-acceptor dyes in constrained media. Spectrochim Acta Part A-Molecular and Biomolecular Spectroscopy 65:366–371

Fischer D, Theodorakis EA, Haidekker MA (2007) Synthesis and use of an in-solution ratiometric fluorescent viscosity sensor. Nature Protocols 2:227–236

Funston AM, Fadeeva TA, Wishart JF, Castner EW, Jr. (2007) Fluorescence probing of temperature-dependent dynamics and friction in ionic liquid local environments. Journal of Physical Chemistry B 111:4963–4977

Glazer AN, Peck K, Mathies RA (1990) A stable double-stranded DNA-ethidium homodimer complex: application to picogram fluorescence detection of DNA in agarose gels. Proceedings of the National Academy of Sciences in United States of America 87:3851–3855

Gota C, Uchiyama S, Yoshihara T, Tobita S, Ohwada T (2008) Temperature-dependent fluorescence lifetime of a fluorescent polymeric thermometer, poly(N-isopropylacrylamide), labeled by polarity and hydrogen bonding sensitive 4-sulfamoyl-7-aminobenzofurazan. Journal of Physical Chemistry B 112:2829–2836

Gouterman M (1997) Oxygen quenching of luminescence of pressure sensitive paint. Journal of Chemical Education 74:1–7

Haidekker MA, Theodorakis EA (2007) Molecular rotors-fluorescent biosensors for viscosity and flow. Organic and Biomolecular Chemistry 5:1669–1678

Harvey MD, Bablekis V, Banks PR, Skinner CD (2001) Utilization of the non-covalent fluorescent dye, NanoOrange, as a potential clinical diagnostic tool. Nanomolar human serum albumin quantitation. Journal of Chromatography B-Biomedical Sciences Applications 754:345–356

Haugland RP (2005) The handbook. A guide to fluorescent probes and labeling technologies. Tenth edition. Invitrogen Corp., Eugene, OR

Hermant RM, Bakker NAC, Scherer T, Krijnen B, Verhoeven JW (1990) Systematic studies of a series of highly shaped donor-acceptor systems. Journal of the American Chemical Society 112:1214–1221

Ho HA, Dore K, Boissinot M, Bergeron MG, Tanguay RM, Boudreau D, Leclerc M (2005) Direct molecular detection of nucleic acids by fluorescence signal amplification. Journal of the American Chemical Society 127:12673–12676

Hochreiner H, Sanchez-Barragan I, Costa-Fernandez JM, Sanz-Medel A (2005) Dual emission probe for luminescence oxygen sensing: a critical comparison between intensity, lifetime and ratiometric measurements. Talanta 66:611–618

Hu ZH, Margulis CJ (2006) A study of the time-resolved fluorescence spectrum and red edge effect of ANF in a room-temperature ionic liquid. Journal of Physical Chemistry B 110:11025–11028

Huang CC, Chang HT (2006) Selective gold-nanoparticle-based "turn-on" fluorescent sensors for detection of mercury(II) in aqueous solution. Analytical Chemistry 78:8332–8338

Huang J, Peng AD, Fu HB, Ma Y, Zhai TY, Yao JN (2006a) Temperature-dependent ratiometric fluorescence from an organic aggregates system. Journal of Physical Chemistry A 110:9079–9083

Huang X, Jain PK, El-Sayed IH, El-Sayed MA (2006b) Determination of the minimum temperature required for selective photothermal destruction of cancer cells with the use of immuno-targeted gold nanoparticles. Photochemistry and Photobiology 82:412–417

Ikanovic M, Rudzinski WE, Bruno JG, Allman A, Carrillo MP, Dwarakanath S, Bhahdigadi S, Rao P, Kiel JL, Andrews CJ (2007) Fluorescence assay based on aptamer-quantum dot binding to Bacillus thuringiensis spores. Journal of Fluorescence 17:193–199

Ishizaka S, Kim HB, Kitamura N (2001) Time-resolved total internal reflection fluorometry study on polarity at a liquid/liquid interface. Analytical Chemistry 73:2421–2428

Jelinek R, Kolusheva S (2004) Carbohydrate biosensors. Chemical Reviews 104:5987–6015

Kolomiets E, Johansson EMV, Renaudet O, Darbre T, Reymond JL (2007) Neoglycopeptide dendrimer libraries as a source of lectin binding ligands. Organic Letters 9:1465–1468

Konishi K, Hiratani T (2006) Turn-on and selective luminescence sensing of copper ions by a water-soluble Cd10S16 molecular cluster. Angewandte Chemie-International Edition in English 45:5191–5194

Kovalska VB, Volkova KD, Losytskyy MY, Tolmachev OI, Balanda AO, Yarmoluk SM (2006) 6,6′-Disubstituted benzothiazole trimethine cyanines-new fluorescent dyes for DNA detection. Spectrochim Acta Part A-Molecular and Biomolecular Spectroscopy 65:271–277

Kuimova MK, Yahioglu G, Levitt JA, Suhling K (2008) Molecular rotor measures viscosity of live cells via fluorescence lifetime imaging. Journal of the American Chemical Society 130:6672–6673

Kuruvilla E, Ramaiah D (2007) Selective interactions of a few acridinium derivatives with single strand DNA: study of photophysical and DNA binding interactions. Journal of Physical Chemistry B 111:6549–6556

Kusnezow W, Jacob A, Walijew A, Diehl F, Hoheisel JD (2003) Antibody microarrays: an evaluation of production parameters. Proteomics 3:254–264

Kwakye S, Baeumner A (2003) A microfluidic biosensor based on nucleic acid sequence recognition. Analytical and Bioanalytical Chemistry 376:1062–1068

Kwan PH, MacLachlan MJ, Swager TM (2004) Rotaxanated conjugated sensory polymers. Journal of the American Chemical Society 126:8638–8639

Laws WR, Brand L (1979) Analysis of two-state excited-state reactions. The fluorescence decay of 2-naphthol. Journal of Physical Chemistry 83:795–802

Lenhart JL, van Zanten JH, Dunkers JP, Zimba CG, James CA, Pollack SK, Parnas RS (2000) Immobilizing a fluorescent dye offers potential to investigate the glass/resin interface. Journal of Colloid and Interface Science 221:75–86

Lenhart JL, van Zanten JH, Dunkers JP, Parnas RS (2002) Using a localized fluorescent dye to probe the glass/resin interphase. Polymer Composites 23:555–563

Lepkowicz RS, Przhonska OV, Hales JM, Fu J, Hagan DJ, Van Stryland EW, Bondar MV, Slominsky YL, Kachkovski AD (2004) Nature of the electronic transitions in thiacarbocyanines with a long polymethine chain. Chemical Physics 305:259–270

Li CY, Zhang XB, Han ZX, Akermark B, Sun L, Shen GL, Yu RQ (2006) A wide pH range optical sensing system based on a sol-gel encapsulated amino-functionalized corrole. Analyst 131:388–393

Liu B, Bazan GC (2005) Methods for strand-specific DNA detection with cationic conjugated polymers suitable for incorporation into DNA chips and microarrays. Proceedings of the National Academy Sciences of the United States of America 102:589–593

Liu Y, Song Y, Chen Y, Li XQ, Ding F, Zhong RQ (2004) Biquinolino-modified beta-cyclodextrin dimers and their metal complexes as efficient fluorescent sensors for the molecular recognition of steroids. Chemistry 10:3685–3696

Masuhara H, Asahi T, Hosokawa Y (2006) Laser nanochemistry. Pure and Applied Chemistry 78:2205–2226

McAnaney TB, Park ES, Hanson GT, Remington SJ, Boxer SG (2002) Green fluorescent protein variants as ratiometric dual emission pH sensors. 2. Excited-state dynamics. Biochemistry 41:15489–15494

McAnaney TB, Shi X, Abbyad P, Jung H, Remington SJ, Boxer SG (2005) Green fluorescent protein variants as ratiometric dual emission pH sensors. 3. Temperature dependence of proton transfer. Biochemistry 44:8701–8711

McGaughey O, Ros-Lis JV, Guckian A, McEvoy AK, McDonagh C, MacCraith BD (2006) Development of a fluorescence lifetime-based sol-gel humidity sensor. Analytica Chimica Acta 570:15–20

Mordon S, Devoisselle JM, Soulie S (1995) Fluorescence spectroscopy of pH in vivo using a dual-emission fluorophore (C-SNAFL-1). Journal of Photochemistry and Photobiology B-Biology 28:19–23

Nagl S, Baleizao C, Borisov SM, Schaferling M, Berberan-Santos MN, Wolfbeis OS (2007) Optical sensing and imaging of trace oxygen with record response. Angewandte Chemie-International Edition in English 46:2317–2319

Ngundi MM, Kulagina NV, Anderson GP, Taitt CR (2006a) Nonantibody-based recognition: alternative molecules for detection of pathogens. Expert Review of Proteomics 3:511–524

Ngundi MM, Taitt CR, McMurry SA, Kahne D, Ligler FS (2006b) Detection of bacterial toxins with monosaccharide arrays. Biosensors & Bioelectronics 21:1195–1201

Nipper ME, Majd S, Mayer M, Lee JC, Theodorakis EA, Haidekker MA (2008) Characterization of changes in the viscosity of lipid membranes with the molecular rotor FCVJ. Biochimica Et Biophysica Acta 1778:1148–1153

Niu CG, Gui XQ, Zeng GM, Yuan XZ (2005) A ratiometric fluorescence sensor with broad dynamic range based on two pH-sensitive fluorophores. Analyst 130:1551–1556

O'Neal DP, Meledeo MA, Davis JR, Ibey BL, Gant VA, Pishko MV, Cote GL (2004) Oxygen sensor based on the fluorescence quenching of a ruthenium complex immobilized in a biocompatible poly(ethylene glycol) hydrogel. IEEE Sensors Journal 4:728–734

Ogul'chansky T, Yashchuk VM, Losytskyy M, Kocheshev IO, Yarmoluk SM (2000) Interaction of cyanine dyes with nucleic acids. XVII. Towards an aggregation of cyanine dyes in solutions as a factor facilitating nucleic acid detection. Spectrochim Acta Part A-Molecular and Biomolecular Spectroscopy 56:805–814

Oh KJ, Cash KJ, Hugenberg V, Plaxco KW (2007) Peptide beacons: a new design for polypeptide-based optical biosensors. Bioconjugate Chemistry 18:607–609

Olmos D, Aznar AJ, Baselga J, Gonzalez-Benito J (2003) Kinetic study of epoxy curing in the glass fiber/epoxy interface using dansyl fluorescence. Journal of Colloid and Interface Science 267:117–126

Olmos D, Aznar AJ, Gonzalez-Benito J (2005) Kinetic study of the epoxy curing at the silica particles/epoxy interface using the fluorescence of pyrene label. Polymer Testing 24:275–283

Pal R, Parker D (2007) A single component ratiometric pH probe with long wavelength excitation of europium emission. Chemical Communications:474–476

Palacios-Rodriguez Y, Gazarian T, Rowley M, Majluf-Cruz A, Gazarian K (2007) Collection of phage-peptide probes for HIV-1 immunodominant loop-epitope. Journal of Microbiological Methods 68:225–235

Pavlickova P, Schneider EM, Hug H (2004) Advances in recombinant antibody microarrays. Clinica Chimica Acta 343:17–35

Peczuh MW, Hamilton AD (2000) Peptide and protein recognition by designed molecules. Chemical Reviews 100:2479–2493

Pickup JC, Hussain F, Evans ND, Rolinski OJ, Birch DJS (2005) Fluorescence-based glucose sensors. Biosensors & Bioelectronics 20:2555–2565

Potyrailo RA (2006) Polymeric sensor materials: toward an alliance of combinatiorial and rational design tools? Angewandte Chemie-International Edition 45:702–723

Przhonska O, Bondar M, Gallay J, Vincent M, Slominsky Y, Kachkovski A, Demchenko AP (1999) Photophysics of dimethylamino-substituted polymethine dye in polymeric media. Journal of Photochemistry and Photobiology B-Biology 52:19–29

Quirin JC, Torkelson JM (2003) Self-referencing fluorescence sensor for monitoring conversion of nonisothermal polymerization and nanoscale mixing of resin components. Polymer 44:423–432

Reichardt C (1990) Solvents and solvent effects in organic chemistry. Second edition. VC Publishers, Germany

Rubina AY, Dementieva EI, Stomakhin AA, Darii EL, Pan'kov SV, Barsky VE, Ivanov SM, Konovalova EV, Mirzabekov AD (2003) Hydrogel-based protein microchips: manufacturing, properties, and applications. Biotechniques 34:1008–1014, 1016–1020, 1022

Rubina AY, Pan'kov SV, Dementieva EI, Pen'kov DN, Butygin AV, Vasiliskov VA, Chudinov AV, Mikheikin AL, Mikhailovich VM, Mirzabekov AD (2004) Hydrogel drop microchips with immobilized DNA: properties and methods for large-scale production. Analytical Biochemistry 325:92–106

Rucker VC, Havenstrite KL, Herr AE (2005) Antibody microarrays for native toxin detection. Analytical Biochemistry 339:262–270

Sapsford KE, Medintz IL, Golden JP, Deschamps JR, Uyeda HT, Mattoussi H (2004) Surface-immobilized self-assembled protein-based quantum dot nanoassemblies. Langmuir 20:7720–7728

Sapsford KE, Ngundi MM, Moore MH, Lassman ME, Shriver-Lake LC, Taitt CR, Ligler FS (2006) Rapid detection of foodborne contaminants using an Array Biosensor. Sensors and Actuators B-Chemical 113:599–607

Shin I, Park S, Lee MR (2005) Carbohydrate microarrays: an advanced technology for functional studies of glycans. Chemistry 11:2894–2901

Silva FR, Samad RE, Gomes L, Courrol LC (2008) Enhancement of europium emission band of europium tetracycline complex in the presence of cholesterol. Journal of Fluorescence 18:169–174

Suppan P, Ghoneim N (1997) Solvatochromism. Royal Society of Chemistry, Cambridge, UK

Thomas SW, Amara JP, Bjork RE, Swager TM (2005) Amplifying fluorescent polymer sensors for the explosives taggant 2,3-dimethyl-2,3-dinitrobutane (DMNB). Chemical Communications : 4572–4574

Thomas SW, 3rd, Joly GD, Swager TM (2007) Chemical sensors based on amplifying fluorescent conjugated polymers. Chemical Reviews 107:1339–1386

Turkmen Z, Klymchenko AS, Oncul S, Duportail G, Topcu G, Demchenko AP (2005) A triterpene oleanolic acid conjugate with 3-hydroxyflavone derivative as a new membrane probe with two-color ratiometric response. Journal of Biochemical and Biophysical Methods 64:1–18

Tyagi S, Kramer FR (1996) Molecular beacons: probes that fluoresce upon hybridization. Nature Biotechnology 14:303–308

Valanne A, Huopalahti S, Soukka T, Vainionpaa R, Lovgren T, Harma H (2005) A sensitive adeno-virus immunoassay as a model for using nanoparticle label technology in virus diagnostics. Journal of Clinical Virology 33:217–223

Valuk VR, Duportail G, Pivovarenko VG (2005) A wide-range fluorescent pH-indicator based on 3-hydroxyflavone structure. Journal of Photochemistry and Photobiology A-Chemistry 175:226–231

Vercoutere W, Akeson M (2002) Biosensors for DNA sequence detection. Current Opinion in Chemical Biology 6:816–822

Vidal JC, Garcia-Ruiz E, Espuelas J, Aramendia T, Castillo JR (2003) Comparison of biosensors based on entrapment of cholesterol oxidase and cholesterol esterase in electropolymerized films of polypyrrole and diaminonaphthalene derivatives for amperometric determination of cholesterol. Analytical and Bioanalytical Chemistry 377:273–280

Vincent M, Gallay J, Demchenko AP (1995) Solvent relaxation around the excited-state of Indole - analysis of fluorescence lifetime distributions and time-dependence spectral shifts. Journal of Physical Chemistry 99:14931–14941

Volovik Z, Demchenko A, Skursky S (1994) Solvent-dependent photophysics of non-symmetric polymethine dyes as fluorescence probes: dual emission and inhomogeneous broadening. Proceedings of SPIE - the International Society of Optical Engineering 2137:600–607

von Bultzingslowen C, McEvoy AK, McDonagh C, MacCraith BD (2003) Lifetime-based optical sensor for high-level pCO(2) detection employing fluorescence resonance energy transfer. Analytica Chimica Acta 480:275–283

Wade SA, Collins SF, Baxter GW (2003) Fluorescence intensity ratio technique for optical fiber point temperature sensing. Journal of Applied Physics 94:4743–4756

Wandelt B, Cywinski P, Darling GD, Stranix BR (2005) Single cell measurement of micro-viscos-ity by ratio imaging of fluorescence of styrylpyridinium probe. Biosensors & Bioelectronics 20:1728–1736

Wang L, Zhao WJ, O'Donoghue MB, Tan WH (2007) Fluorescent nanoparticles for multiplexed bacteria monitoring. Bioconjugate Chemistry 18:297–301

Whitaker JE, Haugland RP, Prendergast FG (1991) Spectral and photophysical studies of benzo[c]xanthene dyes: dual emission pH sensors. Analytical Biochemistry 194:330–344

Yang CJ, Martinez K, Lin H, Tan W (2006) Hybrid molecular probe for nucleic acid analysis in biological samples. Journal of the American Chemical Society 128:9986–9987

Yarmoluk SM, Losytskyy MY, Yashchuk VM (2002) Nonradiative deactivation of the electronic excitation energy in cyanine dyes: influence of binding to DNA. Journal of Photochemistry and Photobiology B-Biology 67:57–63

Yashchuk VM, Gusak VV, Drnytruk IM, Prokopets VM, Kudrya VY, Losytskyy MY, Tokar VP, Gumenyuk YO, Yarmoluk SM, Kovalska VB, Balanda AO, Kryvorotenko DV (2007) Two-photon excited luminescent styryl dyes as probes for the DNA detection and imaging. Photostability and phototoxic influence on DNA. Molecular Crystals and Liquid Crystals 467:325–338

Zhang CY, Yeh HC, Kuroki MT, Wang TH (2005) Single-quantum-dot-based DNA nanosensor. Nature Materials 4:826–831

Zhao X, Tapec-Dytioco R, Tan W (2003) Ultrasensitive DNA detection using highly fluorescent bioconjugated nanoparticles. Journal of the American Chemical Society 125:11474–11475

Zheng T, Peelen D, Smith LM (2005) Lectin arrays for profiling cell surface carbohydrate expression. Journal of the American Chemical Society 127:9982–9983

Zhu X, Sun J, Hu Y (2007) Determination of protein by hydroxypropyl-beta-cyclodextrin sensitized fluorescence quenching method with erythrosine sodium as a fluorescence probe. Analytica Chimica Acta 596:298–302

Chapter 11
Sensing Inside Living Cells and Tissues

The full power of the fluorescence technique can be demonstrated in sensing with and within the living cells. In this application, its most important advantage, which will probably be never be beaten by any other technique, is the high spatial resolution obtained in a noninvasive or low invasive manner. The methods for achieving such a resolution, to be primarily discussed, are based on the concept of forming an image within the microscope's *focal plane* with simultaneous suppression of out-of-plane emission. The other constructive idea in cellular studies is scanning with *focal volume* when this volume is dramatically reduced. This pathway leads to the detection of single molecules. Meantime, these essential advancements are still insufficient for the localization of macromolecules and the formation of an image with molecular resolution because the size of the molecules is smaller than the wavelength of light. This fundamental restriction known as the *diffraction limit* can be overcome by novel techniques.

All these achievements become very useful when they allow addressing fundamental problems of cellular biology (which is outside the scope of this book) and also when they allow characterizing the species inside the cell that are of diagnostic and prognostic value. Extremely important are the interactions between macromolecules and their biocatalytic transformations. These issues will be discussed here. Manipulation with fluorescent dyes and nanoparticles allows many possibilities for their use as tags, probes and sensors. These species can be intrinsic or extrinsic to the cell; they can become not only silent observers but also participants, modulators or disruptors of specific activities that form biological functions. They can be successfully studied and quantitatively characterized with fluorescence techniques.

11.1 Modern Fluorescence Microscopy

The basic function of a *fluorescence microscope* is to provide the image of a studied object in fluorescent light with the full rejection of incident light. Advancements in such instrumentation have stimulated great progress in imaging. For forming an image that is informative for sensing in fluorescent light, a variety of different dyes have been suggested as labels, tags, probes and sensors.

A.P. Demchenko, *Introduction to Fluorescence Sensing*,
© Springer Science+Business Media B.V. 2009

There is a choice to compose an image based on a selected parameter characterizing their emission. Commonly it is the light intensity but it can also be the anisotropy, lifetime and ratio of intensities at selected wavelengths. Microscopes allow the selection of excitation wavelengths (often, in a limited range) and detection at different emission wavelengths that are usually selected in a broader range by the filters. Thus, the studied object is illuminated by the light of a (almost) desired wavelength and then the much weaker emitted fluorescence is separated from the excitation light and an image is formed. This image can be recorded with a high contrast against the background.

There are two sources of this *background* and instrument makers and researchers combine their efforts to eliminate (or, at least, to reduce) them. One is the intrinsic fluorescence of cellular components (*autofluorescence*). It is caused by pigments such as reduced nicotinamide dinucleotide (NADH) and flavine adenine dinucleotide (FAD), which are always present in a living cell. Such a background signal can be reduced by a proper selection of excitation and emission wavelengths (usually, by shifting from near-UV and violet to green and longer wavelengths). Of course, the applied fluorescence reporter should have the highest brightness in these conditions.

The other source of background fluorescence is fundamentally unavoidable but can be dramatically reduced by optical means. It is the excitation of the emitting species that are illuminated by an excitation beam but are *out of focus*. These emitters do not participate in the formation of the image but spoil it. Actually, here the aims in microscopy are the opposite of those in common photography, where a photographer tries to keep both close and distant objects in focus. Here various optical principles and instrument constructions are employed to restrict the excitation and detection of fluorophores to a thin region of the sample (focal plane) and to make the fluorescence from this region much stronger than that coming from the regions outside it. Below, we will overview three technologies that provide a dramatically increased response from the dyes located within the focal plane. These are *total internal reflection microscopy*, *confocal microscopy* and *two-photon microscopy*. The elimination of background fluorescence from outside the focal plane can dramatically improve the signal-to-noise ratio and consequently, the contrast of the obtained images.

With all these advancements, spatial resolution remains the most important factor to be considered. The fundamental limit is given by the length of the light wave and is known as the *diffraction limit*. It is associated with the name of Ernst Abbe, who in 1873 formulated this limit as the ability of a lens-based optical microscope to discern only those details that are larger than a half of the wavelength of light. The diagram presented below (Fig. 11.1) may help the reader in orienting the dimensions of the cellular components and whole cells in comparison with the wavelength of the light. The objects with dimensions smaller than 200–400 nm (presented to the left of the grey vertical band) are not resolvable by conventional image-making microscopy.

Thus, two or more orders of magnitude separate the best resolution that can be obtained by conventional diffraction-limited optical techniques and the molecular level. However, there are two possibilities to overcome this limit in producing the fluorescence images. One is to make a waveguide with a pointed edge of nanometer dimension and to provide scanning with a 'nano-beam'; this is *near-field fluorescence microscopy*. Recently, another possibility to break the diffraction

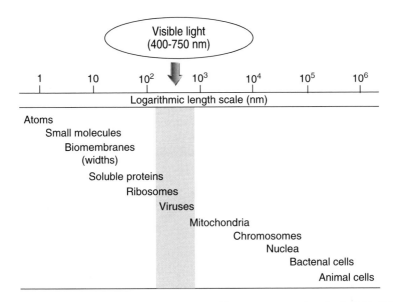

Fig. 11.1 The sizes of cells and their components with respect to wavelength of visible light

limit was suggested and received the name *far-field optical nanoscopy*. Both these techniques will also be discussed below.

11.1.1   Epi-fluorescence Microscopy

The wide-field *epi-fluorescence microscopy* is the most popular fluorescence method in cellular research. In this configuration the excitation and observation of the fluorescence are from above (epi) the specimen, in contrast to common microscopes that collect transmitted light. With different fluorescent stains, this technique allows identifying cells and cellular components with a high degree of specificity. This microscopy has become an important tool in the field of biology, opening the doors for more advanced technical designs. For example, with labeled antibodies, different disease conditions can be identified.

The construction of the epi-fluorescence microscope (Fig. 11.2) allows a very efficient collection of fluorescent light that forms the image. The epi-configuration allows the microscope to let the excitation light illuminate the specimen and then sort out the much weaker emitted light from the scattered excitation light to make up the image. For this, the microscope has a filter that allows only the radiation with the desired wavelength that matches the fluorescing material to pass. The emitted light is separated from the much brighter excitation light with a second filter. Here, the fact that the emitted light has a spectrum at longer wavelengths (being Stokes-shifted) is used. The bright fluorescing areas can be observed in the microscope against a dark background with a high contrast.

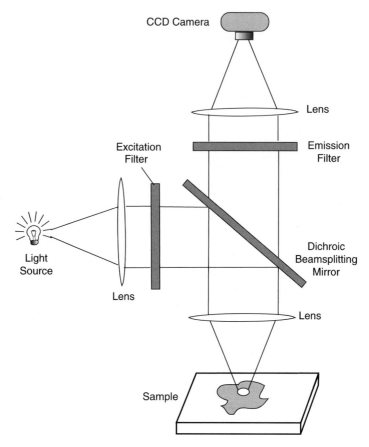

Fig. 11.2 The schematic of the simplest epi-fluorescence microscope. Incident excitation light, after passing the optical filter, is focused on the specimen from above. Fluorescent light is directed to the detector after passing the filter, which rejects the reflected and scattered excitation light

The major disadvantage of this microscope is that it allows collecting the image of a specimen with a three-dimensional distribution of fluorescent species as a two-dimensional picture formed by *both in-focus* and *out-of-focus* located fluorescence emitters. It produces 'photographic' imaging of the object in which sometimes it is hard to judge what is in front and what is behind. A significant part of the spatial information can thus be lost.

11.1.2 Total Internal Reflection Microscopy

The *total internal reflection microscopy* takes advantage of the evanescent wave that exists when the light is totally internally reflected at the interface between two media having dissimilar refractive indices (see Section 8.5 and Fig. 8.9). In this technique, a prism of sapphire or special glass with a high refractive index is used to illuminate the sample located on top of it. If the light is directed into the prism at an angle higher than the critical angle, the beam will not enter the studied specimen in a low-refractive aqueous environment and will be totally internally reflected at the interface. However, some of the light energy can propagate at a short distance and excite fluorescent dyes close to the interface. The reflection phenomenon develops a so-called *evanescent wave* at the interface that permeates about 100–200 nm outside the interface (Fig. 11.3). The light intensity of the evanescent wave is sufficiently high to excite the dyes but remains located within this short distance. Because of such a shallow penetration depth of excitation energy, the *x*-*y* plane close to the interface becomes in fact the focal plane.

Depicted in Fig. 11.3 is the simplest configuration of a microscope based on this principle that allows many modifications. They are well described in the literature and in information materials provided by the instrument manufacturers. Because the excitation of the fluorophores in the bulk of the object is avoided and the fluorescence emission is confined to a very thin region, a much higher signal-to-noise ratio is achieved compared to conventional epi-fluorescence imaging. Some instruments allow varying the illumination incidence angle and consequently, the penetration depth of the evanescent wave. This allows increasing the resolution along the **z** axis and distinguishing the depths of the dye location on a nanometer scale. These variations are limited to the pre-surface area and do not allow seeing the whole cells.

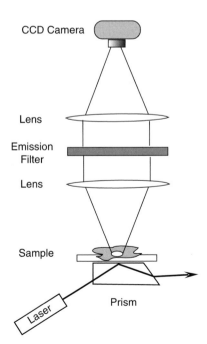

Fig. 11.3   The schematic of a total internal reflection microscope with illumination 'from below' and detection 'from above'. The incident beam does not penetrate into the medium where the specimen is located and excites its fluorescence by producing an evanescent field in a narrow pre-surface area

11.1.3   *Confocal Microscopy*

Confocal microscopy offers a different principle for forming sharp images. The conventional focusing system is applied here, in which the excitation of out-of-focus located dyes by the incident beam is not avoided but an optical configuration of the instrument allows rejecting their emission. This is achieved with the application of scanning with the use of a *confocal pinhole* that provides geometric restriction to the passing of the out-of-focus emission (Fig. 11.4).

The application of this principle allows rendering very sharp fluorescence images due to a dramatically improved signal-to-noise ratio. In confocal microscopy, the resolution along the z axis is on the level of $1\,\mu m$, which is one order of magnitude less than in total internal reflection microscopy. However, the great advantage here is the possibility to *move the focal plane* enabling scanning the object at different z levels, so that cross-sections at various depths can be obtained. If necessary, by computational

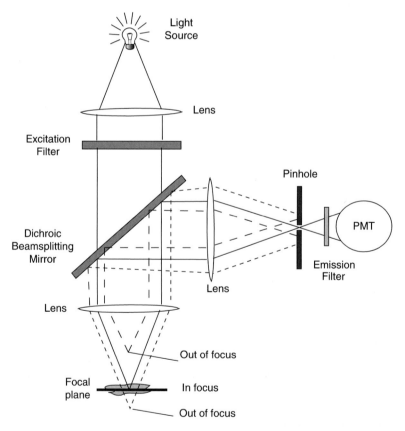

Fig. 11.4   Schematic of a typical confocal microscope showing the beam geometry that allows rejection of photons emitted from outside the focal plane

means these cross-sections can be transformed into a three-dimensional image of the object. A remarkable explosion in the popularity of confocal microscopy in recent years is due in part to the relative ease with which the high-quality cross-sectional images can be obtained from samples prepared for conventional optical microscopy.

The weak point of confocal microscopy is the need to use high-intensity lasers for excitation and to illuminate with them a large volume, not only within the focal plane. Since the illumination is not limited to a focal plane, it generates not only fluorescence but also produces photobleaching and phototoxicity in the whole illuminated volume throughout the object. The involvement of these factors strongly depends on the properties of the used dyes.

Laser scanning can be substituted by the formation of an image by the simultaneous illumination of many pinholes. Confocality is here provided by the synchronous rotation of two discs (called Nipkow disks). One of these disks contains about 20,000 microlenses and the other is placed with the same number of pinholes (Straub et al. 2000). Passing through the array of microlenses and focusing upon correspondent pinholes, the excitation laser beam can create about 1,000 independent focal volume elements. By rotating the disk pair, a full high resolution confocal image can be acquired within 50 ms. The disadvantage of rotating the disks is that the pattern of illumination and detection cannot be adjusted at will for a particular purpose.

Remarkable versatility can be achieved with a new version of the confocal microscope, which is called the *programmable array microscope* (PAM). To scan a specimen, it uses the elements of a spatial light modulator. It operates with two cameras or a single dual-view camera to detect in-focus and out-of-focus images separately. This allows the PAM to provide point and line scans and to adapt the spacing of the pattern and the aperture size to a particular experimental need (Hagen et al. 2007).

11.1.4 Two-Photon and Three-Photon Microscopy

In Section 8.3 of this book, we described the physical background of a phenomenon that brings the electronic system to an excited state by the *absorption of two photons* simultaneously. Two-photon excitation can occur only when the photon density is very high, which can be realized by confining the light flux on a spatial and temporal scale. Very short pulses of a well-focused laser can do this. In this case, the *laser focal point* is the only location along the optical path where the photon density is high enough to generate a significant occurrence of a two-photon excited species.

The generation of two-photon excitation in a fluorophore-containing specimen at the microscope focal point is illustrated in Fig. .11.5. Above and below the focal point, the photon density is not sufficiently high for two photons to pass at the same instant within the absorption cross section of a single fluorophore. Only at the focal point can this density be so high that the two photons will be absorbed simultaneously with sufficient probability.

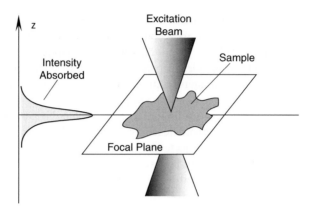

Fig. 11.5 Illustration of the principle of two-photon and three-photon microscopy. The focused light provides the high photon density that can be absorbed by only in-plane dyes, thus forming a sharp image. The near-IR laser can excite 'blue-green' and 'near-UV' emissions by two-photon and three-photon absorption, respectively

   This phenomenon has found a very important application in microscopy because it brings many benefits:

(a) In contrast to confocal microscopy, two-photon microscopy allows concentrating all the excitation in a narrow focal plane. Outside this plane, the dyes are not excited simply because the photonic flux is insufficient. This brings not only a sharp image but also the absence of photochemical reactions outside the focal plane, which would lead to photobleaching and phototoxicity.

   Thus, if we want to express in a few words the key difference between confocal and two-photon microscopy, we must say: *in confocal microscopy the out-of-focus background is rejected but in two-photon microscopy it does not appear!*

(b) In two-photon microscopy the wavelengths of excitation and emission differ dramatically (for instance, dyes with a normal absorption band maximum at 520 nm can be excited by two photons of a 1,050 nm laser and an emission detected at 560 nm). Therefore, *the scattered excitation light is easily rejected.*

(c) The cells and tissues are relatively transparent in the near-IR, so an intensive light for exciting two photons *can penetrate without essential loss* to substantial depths into the living tissue or even inside intact animal specimens. Additionally, because there is no absorption in out-of-focus specimen areas, more of the excitation light penetrates through the specimen to the plane of focus. Together with the formation of focal planes at substantial depths and with the possibility to move these planes, this provides unique prospects in imaging.

Three-photon excitation has interesting prospects in cellular research not only by producing a sharper image. Such three-photon excitation occurs in much the same way as the two-photon process, except that three photons must interact with the same dye simultaneously to produce emission. One of the benefits of this is the possibility to provide the excitation of two dyes absorbing light at two quite different wavelengths with the same laser. With the 1,050 nm laser, the blue-green dye can be excited by two photons at about 520 nm, whereas a near-UV absorbing (at about 350 nm) dye can be excited simultaneously by three photons. The fluorescence of these two dyes can form a two-color image.

New possibilities appear with the intracellular applications of reporting dyes operating by variation of lifetime. Two-photon excitation microscopy utilizes pulsed lasers, therefore it can be readily extended into a combination with fluorescence lifetime imaging. The detection of various analytes and establishing proximity relations between molecules often use the FRET methodology (Section 3.5) and lifetime imaging is best applicable to cellular research FRET detection.

11.1.5   Time-Gated and Time-Resolved Imaging

Time-resolved imaging has become one of the most powerful techniques in cellular research. Lifetime is a fluorescence parameter that is extremely sensitive to intermolecular interactions. In sensing it can be used in different aspects, mostly in configurations that allow contact fluorescence quenching and FRET. Lifetime is a dye concentration-independent parameter but a parameter that requires clever operation, since totally quenched species cannot be detected (Section 3.3). The purpose of fluorescence lifetime imaging (FLIM) is to detect the variations of intermolecular interactions by measuring the lifetime at each point of the image (Marriott et al. 1991). These variations can be correlated with the presence of target species and their interactions.

This technique is especially applicable to working with visible fluorescent proteins, such as GFP, since these proteins easily exhibit the quenching effects but do not respond so easily by spectral changes.

Time-resolved microscopy is implemented in one of two ways. Both of them come from time-resolved spectroscopy. One uses the *time-domain* measurement and detection with the *single-photon counting* technique. Its simplest version is the *time-gated* technique, which allows collecting the quanta within a selected time window. Close in methodology and simpler in performance is the *stroboscopic* technique, which is based on the formation of the time window by opening and closing the detector voltage with synchronized pulses (strobs). Frequency domain measurement uses a different principle. It is based on excitation by *frequency-modulated light*, so that the lifetime information can be obtained from the *phase shift* and *demodulation* of excitation light (Lakowicz 2007).

The application of time-gated microscopy is reasonable when there is a need to distinguish between emissions possessing well separated lifetimes and this necessitates the acquisition and analysis of two sets of images (at short and long lifetimes). This technique is best suited for the images formed by lanthanide chelates and metal-ligand complexes, which possess long lifetimes. Time-resolved imaging allows more flexibility and a combination of intensity and lifetime imaging allows improving the application of FRET (Hanley et al. 2002), which is a key approach to elucidating the proximity relations and dynamics inside the cell.

11.1.6 Breaking the Diffraction Limit: Near-Field Microscopy

Improving the spatial resolution is a strong demand from the side of researchers and practical users of intracellular sensing. Diffraction limit was for many years both a technological and a psychological barrier and only recently the attempts to lift this barrier resulted in new technologies. One of them is *near-field microscopy*. Here, the underlying idea is to reduce the size of a light source to the dimensions of molecules, which is far below the diffraction limit. With such a nano-scale source, the scanning of a microscopic object can be done to obtain an improved image.

What is the *near field*? According to a commonly used definition, the optical near field is the region of space with dimensions of less than a wavelength of the light source, within which the optical diffraction does not occur. In these conditions the light interacts with the matter before diffracting. Because of this interaction, the resolution does not depend upon the wavelength but is a function of only the size of the light source and the distance from the sample.

Scanning near-field optical microscopy (SNOM) is currently seen as a powerful tool for nano-scale imaging (De Serio et al. 2003; Rasmussen and Deckert 2005). It promises high spatial resolution capability down to a few tens of nanometers. Of course, everything depends on the skills of the researcher to provide the optical waveguide with a nanometer-sized tip and manipulate it. The diameters of fiber optics tips can be made as small as 20–40 nm (Voronin et al. 2006), which is by one order of magnitude smaller that the wavelength of light used for fluorescence excitation. Such thin fibers attain very interesting properties that become useful in cellular studies. The photon cannot escape normally from the tip; it travels some distance along the fiber and is further substituted by an exciton or evanescent wave. This allows the excitation of fluorescent species that are present in a small sensing volume within about 100 nm from the tip surface. As a result, only the species within this small volume can be excited (Fig. 11.6).

SNOM is a proximity method that requires scanning the sample relative to an optical probe of sub-wavelength size at a distance of a few nanometers. It is very difficult to combine high resolution with a single fluorophore sensitivity, however. The gap between the probe and sample is controlled by measuring the force interactions, so SNOM can provide both optical and topographical images of the sample surface simultaneously. The latter function is similar to that of the *atomic force microscope* (AFM). A number of examples demonstrate the efficiency of SNOM. In one of the recent studies (Ma et al. 2006) sub-10 nm resolution was achieved in the detection of a single DNA molecule.

Due to these technical restrictions, all the presently developed SNOM types are focused on scanning the surface of a sample and do not allow the three dimensional possibilities of a confocal set-up. This technique is conceptually close to atomic

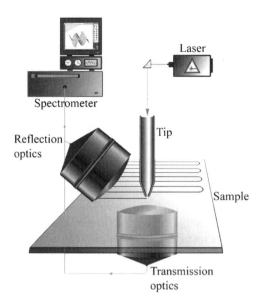

Fig. 11.6 The SNOM set-up operated in illumination mode: laser light is coupled into the sub-wavelength aperture, which illuminates the sample placed in the near field of the tip; scattered or fluorescent light from the sample is collected in the far field through an objective placed in either the reflection or transmission (Reproduced with permission from De Serio et al. 2003)

force microscopy. It requires scanning the surface of the object by using some shear-force-feedback device allowing a three-dimensional representation of its surface, this time with a fluorescence response. Since it is not diffraction-limited, such *near-field excitation* can provide spatial selectivity and resolution that is much higher than that achieved with common fluorescence excitation.

The application of this technique for sensing inside the cell, however, meets many problems associated with the multitude of reactions occurring within it and the mobility of its constituents.

11.1.7 Stimulated Emission Depletion Microscopy in Breaking the Diffraction Limit

Can the resolution be substantially improved in the *far-field*, in which the image is formed on an optically created focal plane? In other words, can improvements of the resolution to overcome the diffraction limit be made based on already popular platforms of confocal and two-photon microscopies? The answer is yes. Single emitters can be localized within a focal plane with a precision higher than the wavelength.

Some improvements in resolution can be achieved by computational techniques. The emitters can be approximated as ideal point sources of fluorescent light and their intensity distribution can be fitted to two-dimensional parameters of a point-spread function (Sharonov and Hochstrasser 2006). The application of such an approach improves the image but requires definite assumptions and therefore cannot provide a general solution of the problem. The recently developed *stimulated emission depletion microscopy* (STED microscopy), being an experimental technique, shows such a possibility (Donnert et al. 2006).

The STED microscopy (Color Plate 11) reduces the size of the excited region by using a very short excitation pulse, which is immediately followed by a 'depletion' pulse, tuned to a red-shifted emission line of the fluorescent dye that forms an image. This depletion pulse causes *stimulated emission* (see Section 8.3), moving electrons from the excited state (from which fluorescence occurs) to the ground energy state. The profile of the depletion beam is altered in such a way that it is focused onto a ring instead of a spot, featuring a dark spot of zero laser intensity in the beam center. The only region of the sample that is allowed to emit light is the much smaller central portion that is not irradiated by the depleting pulse. Scanning this sub-diffraction spot across the specimen delivers images with a sharpened sub-diffraction resolution. A one-order improvement in the resolution over a confocal microscope (down to 15–20 nm) has been reported.

Recent studies have shown that an equally high resolution of images can be achieved even without the synchronization of lasers providing the excitation and depletion and allows one to achieve a three-dimensional sub-diffraction structural resolution (Willig et al. 2007). The high temporal resolution of the image allows observing the fast motions of nanoparticles (Westphal et al. 2007).

STED microscopy is considered as the first implementation of a more general concept, which under ideal conditions offers *unlimited improvement of spatial resolution*. By making the STED doughnut very intense, it is in principle possible to shrink the fluorescent spot to molecular size, thus attaining molecular resolution – an exciting goal for the near future. However, the range of present applications of STED is limited by the high cost of instrumentation and, most importantly, by the necessity to use high light intensities to induce stimulated emission, which imposes more stringent demands than usual on the photostability of the applied fluorescence dyes.

Thus, STED microscopy radically distinguishes itself from conventional far-field light microscopy in the fact that its resolution is no longer fundamentally limited by the wavelength of light. It opens a new chapter in the story of light microscopy, one in which the fundamental questions of biological processes at the nanoscale can potentially be answered with focused light.

11.1.8   Considerations on the Problem of Photobleaching

From discussions in this section we can derive that *photobleaching* and *phototoxicity* are two of the most severe limitations in fluorescence microscopy of living cells and tissues. Despite the fact that individual dyes and nanoparticles exhibit dramatic variations in these properties, they all contribute to this general problem.

The irreversible loss of emitters in time occurs because many chemical reactions proceed much more easily in the excited than in the ground state. They ruin the fluorophore, making it non-fluorescent. Moreover, the products of these reactions may be toxic to the cells.

The rate of fluorophore destruction as a result of *photobleaching* determines the observed duration of the fluorescence emission. Usually, the textbooks present the example of fluorescein. In an oxygenated saline solution, each fluorescein molecule can emit on average only about 36,000 photons before being destroyed. When dissolved oxygen is removed from this solution, the rate of photobleaching diminishes about tenfold, so 360,000 photons can be emitted per fluorescein molecule. Thus, the *quantum yield of photodestruction* (the ratio of the number of photobleaching events to the whole number of absorbed quanta) will be 2.8×10^{-4} to 2.8×10^{-5}. The photon fluxes used in confocal microscopy are of the order of 10,000–100,000 photons per second per molecule. The reader may calculate that in these conditions the dye will burn out in seconds. The dyes that are used for cellular imaging and sensing can be more photostable or less photostable than fluorescein. The reader must be aware of these problems. A lot of promise appears from the side of Quantum Dots and other nanoparticles, since they are expected to display much higher photostability.

The other major problem with fluorescence microscopy is *phototoxicity* (Hoebe et al. 2007). When a fluorophore (endogenous or exogenous) is excited, there is a probability that instead of decaying to a ground state with the emission of light, it will exhibit transition to a triplet state. These long-lived states are very reactive and produce species that can damage the cells. One of the most significant damage mechanisms is the generation of highly reactive *singlet oxygen*. The decrease of the concentration of dissolved oxygen in the medium and the reduction of the light flux (Nishigaki et al. 2006) are known to reduce these effects.

The photobleaching of reporter dyes used in sensing can create significant problems in obtaining good-quality images, especially with the application of FRET. However, there are at least two possibilities to benefit from photobleaching effects. One is the rather old technique known as *fluorescence recovery after photobleaching* (FRAP), which allows determining the rates of translational diffusion in biomembranes of integral and membrane-bound compounds, including lipids and proteins (Meyvis et al. 1999; Hagen et al. 2005). When these labeled species are distributed in the membrane and an intensive laser light focused on its small spot bleaches all these species, this laser, which remains focused on the same spot, is switched to a low-intensity fluorescence excitation. This allows monitoring the recovery of the fluorescence emission due to the diffusion of the labeled species from outside the spot, thus determining their diffusion rate.

The other interesting application is the *modulation of FRET*. The idea is the following. The image is obtained in the presence of a donor and acceptor and then the acceptor can be bleached by direct application of the laser light, exciting only the acceptor. In the same image, this allows one to remove FRET and observe the distribution of the fluorescence intensity of only the donor (Van Munster et al. 2005).

11.1.9   Critical Comparison of the Techniques

The era when optical microscopy was purely a descriptive instrument or an intellectual toy is past. It plays a very important role in the study of life on a cellular level. The microscope accomplishes the first step toward data analysis in conjunction with electronic detectors, image processors and display devices that can be viewed as extensions of the imaging system. The computerized control of focus, stage position, optical components, shutters, filters and detectors is in widespread use and enables experimental manipulations that are absolutely impossible with mechanical microscopes that allow simple observation by the human eye. Microscopy now depends heavily on electronic imaging to rapidly acquire information at undetectable low light levels or at visually undetectable wavelengths. The information content of this image becomes the key factor. It is not enough to localize the cell organelles and determine their number and sizes. The amount and localization of at least some of the many thousand types of molecules are of major interest and this constitutes intracellular sensing.

Several trends of optical imaging have demonstrated a strong potential for development. Whereas a simple epi-fluorescence microscope cannot achieve a high structural resolution, this can be done with more advanced techniques. In a confocal microscope, the fluorescence is excited throughout the illuminated volume of the specimen but only the signal originating in the focal plane passes through the confocal pinhole, allowing one to collect background-free data but greatly reducing the number of quanta reaching the detector. By contrast, two-photon excitation generates fluorescence only at the focal plane and since no background fluorescence is produced, a pinhole in front of the detector is not required. In both methods the spatial resolution along the z axis is similar, it is about 500–600 nm. In this respect they are behind total internal reflection microscopy, in which this resolution is about 100 nm but without the possibility to move the focal plane. All three methods are adaptable for time-resolved measurements. Time-resolved imaging, with or without using FRET, allows making important steps towards real intracellular sensing of many analytes.

A further increase of resolution by overcoming the diffraction limit can be achieved with the near-field technique by using mechanical scanning with an ultra-thin probe serving as a nano-sized light source. In the formation of an optical image (the far-field) such ultra-high resolution can be achieved with a special, recently suggested technique called stimulated emission depletion (STED) microscopy. This technique demonstrates that in principle, the resolution in confocal images can approach a molecular level. Such new 'molecular nanoscopy' offers bright prospects for further development.

11.2   Sensing on a Single Molecule Level

Every time we discuss intermolecular interactions we build in our imagination individual molecules. We construct models of their structures and the modes of their interacting interfaces. The real world is different. Commonly, we deal with an immense number of one type of molecule that similarly interacts with a great number of molecules of another type in the presence of molecules of many additional types. Therefore, we obtain the *picture averaged in both the molecular ensemble and time*. Operating with excitation and emission light we can sometimes select subpopulations in this ensemble (Demchenko 2002) but its behavior may still be remote from the behavior of individual molecules. So, can we detect fluorescence from individual molecules?

This possibility has recently become a reality. Due to advances in spectroscopic techniques we can observe that each molecule 'lives its own life', it diffuses, collides with other molecules, exhibits intramolecular transformations, etc. Many of these processes occur as *thermal fluctuations* and their averaging may allow better understanding of molecular reactivity.

Fluorescence probing of *individual sensor molecules* and events occurring on their target binding and/or their image formation needs an ultimate degree of sensitivity. It is not achievable in a common sperctroscopic or microscopic experiment, which operates with micromolar to nanomolar concentrations. Under ideal conditions, it is possible to detect fluorescence emission from a single molecule. As discussed above, a single fluorescein molecule could emit as many as 300,000 photons before it is destroyed by photobleaching. Assuming a 20% collection and detection efficiency, about 60,000 photons can be detected. Therefore, researchers are able to monitor the behavior of single molecules for several seconds and even minutes.

The observation of single molecules became real because scientists managed to fulfill three important conditions.

(a) The illuminated volume has to be dramatically reduced, to the level of femtoliters $(1\,fl = 10^{-15}\,l)$.
(b) The density of incident light has to be dramatically increased to collect a sufficient number of quanta from a single emitter.
(c) Dyes with high photostability should be used in this technique; they should allow a great number of excitation-emission cycles to be achieved without decomposition (photobleaching).

The detection of individual molecules is not a matter of mere curiosity. The possibility appeared to not only detect and identify the freely diffusing or immobilized molecules but also to detect the dynamic changes occurring within them (Tinnefeld and Sauer 2005). We will retain an interest in intermolecular interactions, which are important for sensing.

11.2.1 Single Molecules in Sensing

Single molecule detection is a technique that can be used to record rare events occurring with individual molecules. On this level, such details of intermolecular interactions can be revealed that cannot be seen on the level of molecular ensembles. They can be important for sensing. For instance, is the sensor-target binding a single-step process? If not, what are these individual steps? Success in the studies on a single-molecular level of interactions of such complex molecules as the molecular rotors kinesins (Xie et al. 2006), promise strong progress in the studies of other systems.

During the initial steps of single-molecular studies, the fluorescent molecules were fixed by adsorption to the surface or in a volume of glass in cryogenic conditions and the microscopic technique best fitting for these studies was total internal reflection microscopy. With the development of confocal and two-photon microscopy they began to be more actively used and more researchers started to study the diffusing and interacting single molecules in solutions. This opened the road for the extensive development of single-molecular sensing.

A typical setup for these studies is depicted in Fig. 11.7. A collimated laser beam is focused on a sample using a high numerical aperture objective. The fluorescence is collected through the same objective and the emission light is separated from the excitation light by an appropriate dichroic mirror. A 3D-gaussian observation volume (usually <1 fl) is achieved by using a pinhole (10–100 μm) on the detection path. After filtering, the emitted photons are detected using avalanche photodiodes (APDs), which are point detectors with fast time-resolution (down to the sub-nanosecond timescale). With such a detection geometry, when the dye concentration is sufficiency low (<200 pM), it is possible to detect fluorescence bursts arising from single diffusing molecules.

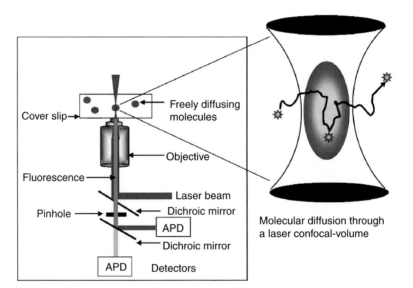

Fig. 11.7 The principle and experimental setup for single-molecule fluorescence measurements of diffusing molecules in solutions, using a confocal microscopic setup. For FRET measurements, photon bursts are separated from laser excitation light by a dichroic mirror and detected by using two avalanche photodiodes (APDs). Molecular diffusion through a laser illuminated confocal volume gives rise to fluorescence bursts (Mukhopadhyay and Deniz 2007)

To record the fluorescence from an individual molecule, one has to achieve the presence of this molecule as the *only emitter in the detection volume*. In liquid solvents, this can be done by multi-fold dilution of the sample. In studies inside the living cells, such a possibility does not exist and one has to manipulate only the decrease of the detection volume.

The major problem is in providing an adequate suppression of the *optical background noise* and the detector noise, which should be kept at a very low level. Fluorescence microscopy techniques, particularly those utilizing total internal reflection, provide the desired combination of a low background and high excitation light flux. Therefore, it is the method of choice in many techniques detecting single molecules. The rapid motion of molecules can be retarded by their absorption on the surface or attachment to supramolecular structures (Gai et al. 2007).

The *fluorescent dyes* and the mechanisms of their response in sensing have to be most carefully selected. The perylene 3,4,9,10-tetracarboxyl bisimide was reported as a very prospective dye that can be excited by a 488 nm line of argon ion laser and exhibits almost 100% quantum yield. With this dye, both PET and FRET can be realized as the sensor transduction mechanisms (Sauer 2003).

There are many examples of the application of the single-molecule technique. The results on probing the local properties of DNA are very interesting. They allow identifying the presence of knots, loops or folds in its structure. Similarly, DNA may be used as a local sensor. The range of environment conditions in which DNA exhibits structural transformations is determined by such factors as ionic strength, the presence of binding proteins, etc., therefore, by employing single-molecule fluorescence methods it is possible to learn about these conditions (Metzler and Ambjornsson 2005).

The elementary steps in different enzyme mechanisms can be detected on a single molecular level (Smiley and Hammes 2006) and this information is important for sensor development. Remarkable are examples of the application of single molecular detection for better understanding the mechanisms of photophysical processes occurring in molecules such as conjugated polymers (Barbara et al. 2005) and dendrimers (De Schryver et al. 2005), which are of special interest for different sensing technologies.

Single molecular detection allows studying the *interactions* between individual molecules directly by fluorescence microscopic techniques. They are based on the detection of FRET between two labeled partners if they are in close contact. The idea in these studies is simple: to detect the fluorescence color change from that of the donor to that of the acceptor, together with microscopic detection of the location of this pair in space (e.g., in a cell) at a high level of resolution. The use of two lasers for differential excitation of the donor and acceptor with the latter operating by producing time-gated flashes, makes this approach more informative (Doose et al. 2007).

Single-molecule detection is an important achievement not only because it allows an ultimate level of sensitivity in any sample analysis. It allows providing a conceptually important move from *ensemble analysis* to *individual event analysis*. Ensemble analysis is what is commonly measured and analyzed in chemistry and molecular science. It deals with ensemble-averaged properties. In contrast, the single-molecule detection allows catching the elementary events, including those that are concealed from ensemble analysis. Fluctuations and statistical distributions that are sometimes seen on the level of molecular ensembles, are seen as distinct events on the level of single molecules (Deniz et al. 2008).

The transition from ensemble-averaged detection to single-molecule detection can be illustrated with the aid of the scheme presented in Fig. 11.8. With a high concentration of emitters in a large detection volume we always detect an ensemble-averaged response. Lowering concentration and squeezing down the observation volume (confocal volume, for example) gives rise to fluorescence fluctuations for the in-and-out diffusion of molecules. Further lowering the concentration to ensure primarily one molecule diffusing through at a time, gives rise to fluorescence bursts. These bursts contain information about the structure and dynamics of individual molecules (Mukhopadhyay and Deniz 2007).

11.2.2 Detection of Single Molecules Inside the Living Cells

The living cell is a very special chemical reactor. Any biochemical reaction in a cell may have different thermodynamic and kinetic properties from the same reaction when studied in a test tube. This is because on a genomic level, a particular gene represented by a DNA fragment exists in only one or a small number of copies. Many mRNA and enzyme molecules also exist in a small number of copies.

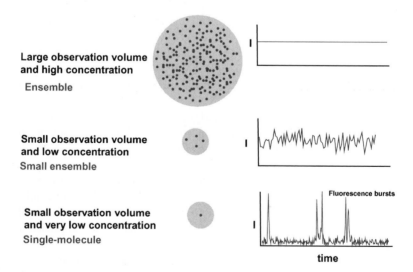

Fig. 11.8 Schematic illustrating the transition from ensemble fluorescence to single-molecule quantum bursts by lowering the concentration and decreasing the detection volume (Mukhopadhyay and Deniz 2007)

Their reactions do not follow common thermodynamic and kinetic rules that are based on the statistics of many molecules. These reactions often occur in *non-equilibrium* conditions being part of complex multi-step processes with a constant supply of free energy and reactants and with the consumption of reaction products. They proceed in *localized compartments* inside the cell, so the diffusion of the reaction partners may be very limited in space between the different binding-release and catalytic events. Therefore, sensing inside the cell at a single molecular level is a very challenging and very important task (Moerner 2007).

Recent developments in cellular research have demonstrated that in such complex systems a single molecular level can be reached. This can be done in different experiments and they can provide the means for sensing different targets. Here are several examples.

1. The *single-channel recording of transmembrane ionic current*. The ionic current is observed when the ion channel formed by transmembrane protein is open. The opening and closing of ionic channels can be detected by single-channel recording of the electrical current and also by the application of a fluorescence reporter (Claydon and Fedida 2007). Since the operation of ionic channels is one of the strongest modulators of cellular activity and the site of the application of different drugs, the possibility of their study on a single molecular level is extremely important.

2. The detection and tracking of *single mRNA molecules* in a living cell using multiple copies of a fluorescent mRNA-binding protein (Rafalska-Metcalf and Janicki 2007). With this approach it becomes possible to characterize the movement of single mRNA-protein complexes in the cell nucleus in real time (Shav-Tal et al. 2004). It is surprising but it was found that these motions are ATP-independent and are consistent with a diffusional model. Such achievements open new possibilities to study gene expression during the very early steps.

3. The accurate quantitation of *microRNA (miRNA) gene expression* in human tissues. Such analysis shows the capability of the quantitative detection of miRNA expression from as little as 50 ng of total RNA. Furthermore, by incorporating the locked nucleic acid (LNA)-DNA oligonucleotide probes, this method has been shown to be highly specific and capable of discrimination between miRNA targets that differed by as little as a single nucleotide. Future extensions of single-molecular miRNA assays and their integration with the current microfluidic devices and nanotechnologies may prove important for understanding the spatiotemporal regulation of miRNA expression across various tissue and cell types. This information is highly necessary for clinical diagnostics (Puleo et al. 2006).

4. *Co-synthetic tracking of single proteins*. Using fast-maturing and membrane-targeted visible fluorescent proteins as reporters, the expression of single protein molecules can be provided on a translational level. The immobilization by the membrane attachment of a protein reporter allows overcoming the problem of fast protein diffusion during fluorescence signal acquisition time, which would spoil the image.

5. *Detection of single cytoplasmic proteins*. Particularly, this can be done if they are labeled with fluorescent reporters. Microfluidic devices allow achieving

such levels of confinement that single molecules from a single cell can be separated and individually counted, even those that are present in a small number of copies (Huang et al. 2007).

The single molecules inside the cells can be 'recognized' based on both the wavelength and polarization of their emission. This information is especially useful when the co-localization of proteins (indicating their interaction) is studied. This can be shown by the presence of FRET between single pairs of fluorescent molecules bound to signaling receptors in the plasma membrane of live cells (Webb et al. 2006).

11.2.3   Fluorescence Correlation Spectroscopy and Microscopy

Fluorescence correlation spectroscopy (FCS) is a technique in which the temporal fluctuations of fluorescence parameters are analyzed to obtain information about the processes that give rise to these fluctuations. This information is provided by the *temporal autocorrelation function* of fluorescence fluctuations (with one-channel recording) and the *cross-correlation function* (with recording using two or more channels).

The *autocorrelation function*, G(t), is a measure of the self-similarity of the signal after a lag time (t). It displays the conditional probability of finding a molecule in the focal volume at a later time, t, given it was there at t = 0. Decaying with time, this function measures the average duration of a *fluorescence fluctuation*. This knowledge allows deriving a number of parameters characterizing the fluorescent molecules, their motions and interactions. The *cross-correlation function* can be used to study intermolecular interactions between two types of molecules labeled in different colors and providing two-channel detection. The interaction is indicated by the coincidence of signal fluctuations.

FCS is a rapidly developing field and it is already recognized as an essential tool for the *in vivo* characterization of absolute concentrations, molecular interactions and kinetic processes, such as diffusion and chemical reactions (Haustein and Schwille 2007). In FCS the light emitted from a few (~1–10) molecules is recorded with microsecond–second time resolution, detecting the processes much longer than the fluorescence lifetime (~1 ns). On this time scale, which is typical of *molecular diffusion*, the fluorescence emission can be thought of as an instantaneous relaxation process (Krichevsky and Bonnet 2002).

When used as an *in vitro* technique, the FCS measurements are easy to conduct and they can be made on simplified instrumentation. The applicability of FCS for sensing is demonstrated in the Evotec technology. Single molecules are measured here as they diffuse through an extremely small volume of about 1 fl. Because of increased molecular mass, the complexes of the fluorescent sensor with the target diffuse more slowly than the sensor molecules alone. This can easily be detected using confocal fluorescence read-out techniques.

The *microfluidic single-molecule detection* setup allows decreasing the sample volume below the illuminated volume and thus reduces the out-of-focus background emission substantially (Foquet et al. 2004). Such an experimental setup resembles a highly miniaturized flow cytometer. The cross-section of the microchannels can be made so narrow that the examined molecules are allowed to diffuse in only one dimension and thus to be detected. The fact that in such systems a very high dilution can be avoided and the single molecules can be detected even when their total concentration is in the micromolar range is important.

Fluorescence correlation microscopy (FCM) is an image analog of FCS. It is based on a spatial autocorrelation analysis of fluctuations in fluorescence intensity produced within a very small volume in a microscopic sample illuminated by a laser beam. In these conditions, FCS records information about the absolute concentrations of the fluorescent molecules, their state of aggregation and the dynamic properties of these molecules in a cellular environment. Based on this, information on molecular interactions including the binding constants can be obtained. Again, in a cell the diffusion of many molecules does not follow the law for free diffusion, it can be a process with different and possibly interesting, regularities. Researchers possess the means to study them, though limitations exist. The minimal illumination volume (~1 fl) is still comparable to the volumes of small bacterial cells.

11.2.4 Additional Comments

In common experiments in molecular physics or biology the researcher in reality operates with large molecular ensembles, so that only the ensemble-averaged properties are detected. They are used for system description and also for evaluating its behavior based on the rules of thermodynamics. These rules are also based on the statistics in large systems; they do not necessarily operate in small systems. The living cell represents a combination of a large and small system, in which some compounds are in high sub-molar quantities and others are represented by only a few copies. Inside the cell, the biocatalytic transformations occur in cascades so that the concentrations of many compounds are not at equilibrium. Therefore, it may be doubtful that all regularities derived in Chapter 2 for the conditions of equilibrium will be valid here. New physics of small systems should be developed for the description of all these effects when applied to living cells.

With this perspective, we have to evaluate the single-molecule experiments. They can be adequate to ensemble-averaged properties of target molecules when their ensemble is dramatically reduced. They can also represent the properties of their sub-ensemble in a particular location or at a particular step after their synthesis or before transformation. Finally, they can detect molecules with properties uniquely deviating from the ensemble. The combination of ensemble-averaged and single-molecular experiments is an important and probably the only way to shed light on these regularities.

11.3 Site-Specific Intracellular Labeling and Genetic Encoding

The field of cellular fluorescence is developing so rapidly that many techniques suggested a decade or so ago and still actively used by histologists and cytologists may be considered as 'pre-historic'. In some of these applications, fluorescent dyes are used as staining agents to identify the cells and to distinguish small organelles. In doing so they succeed in achieving a very high contrast. Their staining specificities, however, often rely on poorly understood interactions. It is the understanding of these interactions, the active manipulation of them and even more, inducing the cell to synthesize fluorescent interacting partners, that characterize the frontier of modern research.

Success in the detection and characterization of specific interactions in the sensing and imaging domains depends totally on the researcher's ability to provide the specific fluorescence reporter to a desired site. Many possibilities exist for this. There can be a covalent attachment that requires a chemical reaction with the correspondent groups of atoms and also a high-affinity noncovalent binding. In the latter case, the formed complex should be highly stable in a cellular milieu, with the dissociation constant lower than nanomolar. The interacting partners have to find and recognize each other among tens of thousands of different types of molecules constituting the cells.

11.3.1 Attachment of a Fluorescent Reporter to Any Cellular Protein

The general solution of the problem of intracellular labeling has not been found and probably does not exist at all. The researcher has to select among different possibilities:

(a) *Labeling in vitro* with subsequent incorporation into the cell. This is the simplest procedure, which has been frequently used in the past. The labeling occurs in the controlled conditions of a chemical lab. The labeled protein, nucleic acid or nanoparticle can penetrate into the cell through a variety of *cell-loading techniques*, such as endocytosis, permeabilization or microinjection. Different methods have been developed to facilitate such incorporation. Thus, the attachment of oligoarginine peptides facilitates spontaneous entry into the cell of proteins and nanoparticles (Bullok et al. 2006). These membrane-permeant peptides are commonly based on the HIV-1 Tat basic domain sequence, GRKKRRQRRR.

DNA molecules can be modified with the shell of cationic lipids or polymers and many of these techniques have been developed for the demand for targeted gene delivery.

The *in-vitro* labeled species cannot necessarily be biomolecules but also large particles, such as viruses. In this case, the fluorescence imaging allows detecting all the steps of virus transformation inside the cell, as shown for the adeno-associated virus (AAV) during its infection pathway into the living HeLa cell (Seisenberger et al. 2001) (Color Plate 12).

(b) *Labeling 'in situ'* inside the cell. Such labeling should be based on a strong and specific complex formation. In view of tens of thousands of different molecules existing inside the cell in a high total concentration, only reagents of exceptional affinity can avoid nonspecific binding. The general protein labeling strategy in this case is to introduce a specific recognition pattern into the intracellular synthesized protein sequence and to provide the labeling reagent with high affinity to this sequence.

Several technologies have been developed for this. One is the introduction into the target protein of the oligohistidine segment of six and more His residues, the *histidine tag*, which can be recognized by an interaction partner (Guignet et al. 2004). This partner has to be a metal-ion-chelating nitrilotriacetate (NTA) moiety, which can be attached to a fluorescence dye. In addition to the transition-metal complexes used earlier, the application of lanthanide complexes can also lead to a successful result (Tsukube et al. 2007).

Another technology requires genetic incorporation into the protein of the unique hexapeptide sequence -Cys-Cys-X-X-Cys-Cys-, where X is any other amino acid (Adams et al. 2002). Inside the cell it can be specifically recognized by a membrane-permeant fluorescein derivative with two arsenical As(III) substituents. This method received the acronym 'FlAsH' (*fluorescein arsenical helix binder*). A derivative of the fluorescein was selected that is fluorescent only after the arsenics bind to the cysteine thiols. A very low dissociation constant (~10^{-11} M) allows highly-specific binding. Meanwhile, the selection of fluorescein is not optimal in view of its relatively low photostability and pH dependence for emission. The application of the fluorinated fluorescein derivative significantly improves these properties (Spagnuolo et al. 2006). Many bi-arsenical dye analogues have been synthesized (such as ReAsH), including a resorufin derivative excitable at 590 nm and fluorescing in the red. Different biarsenicals enable the localization of tetracysteine-tagged proteins, together with their co-localization based on FRET (Martin et al. 2005).

Highly-specific intramolecular labeling with a fluorescent dye is needed in determining the mechanisms of pathological protein aggregation associated with Parkinson's and Alzheimer's disease and other related pathologies. In the case of Parkinson's disease the key protein that exhibits pathogenic folding is α-synuclein, a single-chain 15 kDa protein. Labeling it with a 27 kDa fluorescent protein is not reasonable, since the large-sized label can interfere with the aggregation process. Meantime, the introduction of a short tetracystein-containing insertion with intracellular binding of a biarsenical dye derivative (FlAsH or ReAsH) allows intracellular labeling without observably affecting the aggregation process (Roberti et al. 2007). Fluorescence microscopy established the intracellular distribution of a labeled species in an aggregate. It was shown that the topology of the aggregate can be studied by FRET between two attached biarsenical dyes, so that one of them can be selectively photobleached (Color Plate 13).

(c) *Enzyme-based transfer* of a fluorescent dye to a specific protein. Several technologies for such labeling are known. One is based on the enzyme alkylguanine transferase (AGT), which reacts with benzylguanine derivatives to form a covalent linkage, allowing rapid and irreversible labeling of AGT-fusion proteins. This technique is known as SNAP tag (Gautier et al. 2008). The presence of an additional polypeptide, a so-called tag, allows the binding of a fluorescent dye in a highly-specific manner. This peptide can be endowed with sensor properties and, in combination with a responsive dye, an efficient sensor can be designed.

The small (~8 kDa) acyl carrier protein (ACP) can also be used as the tag. In this system, the modified coenzyme A-based probes are covalently linked to ACP-tagged targets by an engineered synthase enzyme (George et al. 2004). Such an approach is restricted to cell-surface proteins but by using different synthase enzymes, allows multiple-color labeling.

More detailed information on the developments of these techniques can be found in recent publications (Keppler et al. 2003; Marks and Nolan 2006).

11.3.2 Genetically Engineered Protein Labels

The problem of the generation of fluorescence emitters directly inside the cells by using their biosynthetic mechanisms can be resolved with *visible fluorescent proteins* (Section 4.7). The genetically engineered green fluorescent protein (GFP) and other proteins of this family, which provide emissions in different colors, can be *synthesized inside the cells*. Moreover, they can be fused to other proteins on a genetic level. This allows one to establish the proximity relations between the structure-forming proteins, their interactions with soluble proteins and also the interactions between these proteins in cytoplasm and organelles (Giepmans et al. 2006). The ultimate goal in these studies is to establish the whole magnitude of intermolecular interactions in the cell (often called '*interactome*', Section 12.2). The Förster resonance energy transfer (FRET) (see Section 3.5) can be a straightforward approach in these studies. Thus, a significant effort has been made to evaluate the fluorescent proteins as FRET donors and acceptors.

FRET has proved to be the major tool for establishing the target protein location and interactions. The labeling of the FRET partners (donor and acceptor) is usually done with organic dyes but pairs of fluorescent proteins of different colors or the combinations of dye and fluorescent protein have started to be actively used for this purpose. Presently, most of the suggested biosensor applications in cell imaging of the GFP-like proteins are based on the FRET mechanism and a continuous effort is observed in optimizing their donor and acceptor properties (Shaner et al. 2004). The range of these applications is still narrow but continues to grow, with the discovery of many clever possibilities offered by protein fusions.

Remarkable in this respect has been the construction of 'chameleons' that are able to sense Ca^{++} ions. Such fused protein sensors consist of linear fusions of two

fluorescent proteins serving as the FRET donor and acceptor, which flank the calmodulin and calmodulin-binding peptide (Miyawaki et al. 1997, 1999). Calmodulin
is known as a protein that changes its conformation dramatically upon binding calcium ions. Upon Ca^{++} binding, the calmodulin wraps around the peptide so that the
distance between the flanking fluorescent proteins is reduced, producing the proper
spectroscopic response (Fig. 11.9).

   There have been other attempts to develop protein sensors that could be fully
synthesized and that can attain the binding and reporting functions inside the living
cell. Interesting is the possibility to modify the maltose binding protein, which is the
classical protein sensor scaffold, by fusion with two fluorescent proteins. One of
them can be fused to its N-terminal and serve as the FRET donor and the other one
is fused to the C-terminal and serves as the acceptor (Fehr et al. 2002). Maltose binding protein is known to undergo a dramatic conformation change (the hinge-bending
motion) upon ligand binding (Section 6.3). This change in the sensing event brings
the two fluorescent protein moieties together with the appearance of FRET. Since
the acceptor is fluorescent, the ratio of the two emission intensities can be recorded.
Moreover, an image can be obtained displaying this ratiometric signal as the maltose
distribution inside the cell. The uptake of maltose into living yeast cells can be
directly visualized with this sensor. Using this elegant methodology, glucose and

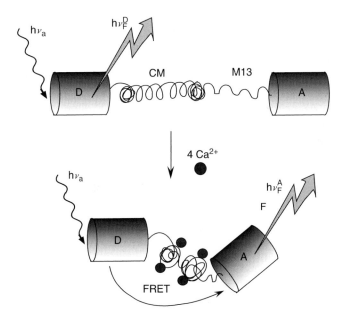

Fig. 11.9 Fluorescence sensor for calcium ions based on FRET between visible fluorescent
proteins that are fused to the N and C terminals of calcium-binding protein calmodulin (CM)
via calmodulin-binding peptide M13 (Miyawaki et al. 1997, 1999). The binding of Ca^{2+} ions
induces conformational change in calmodulin that brings together the donor (D) and acceptor
(A) domains

ribose fluorescent sensors have also been constructed with glucose and ribose binding proteins, respectively. Though the dynamic range of the fluorescence response of the proposed sensors is not great, it allows obtaining images of distribution of these metabolites and following their dynamics inside the cells.

Because the extent of change in fluorescence parameters for a particular sensor construction is unpredictable and the knowledge required for designing new binding sites in sensor molecules is not always available, the development of efficient sensors generally requires the construction and evaluation of a significant number of conjugates. The difficulties associated with the labeling can be overcome by screening for transduction properties (Doi and Yanagawa 1999). After fusing a protein domain containing a ligand binding site into a surface loop of GFP (Fig. 11.10), the fusion protein can mutate randomly to search for a variant that changes the fluorescence upon target binding. This change in fluorescence should occur as a result of the conformational change in GFP at a distance from the target binding site. Such an approach led to the development of a GFP biosensor for the inhibitory protein of TEM1 beta-lactamase.

Genetically encoded probes for the optical imaging of excitable cell activity have been constructed by fusing the fluorescent proteins to functional proteins. These proteins can be involved in physiological signaling systems, such as those that control membrane potential, free calcium, cyclic nucleotide concentrations and pH. Using specific promoters and the means for targeting, such probes can be introduced into an intact organism. They can be directed to specific tissue regions, cell types and subcellular compartments, thereby extracting specific signals more efficiently and in a more relevant physiological context because of the ability to follow the sensor location (Miyawaki 2003).

The green fluorescent protein (GFP) and its colored relatives (blue, cyan, yellow among others) revolutionized the imaging in living cells (Campbell et al. 2002; Tsien 1998). However, these tools are far from being ideal. Their large sizes (e.g., 238 amino acids in length and 28 kDa in mass of GFP) limit the possibilities for their integration into cellular structures. They should be used with caution under

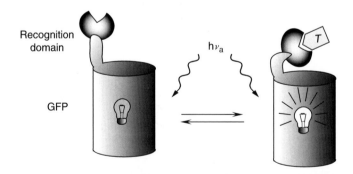

Fig. 11.10 The suggested principle of a generic GFP-based sensor (Doi and Yanagawa 1999). A molecular recognition domain is inserted into a loop of GFP. In the absence of a target, the GFP fluorescence is quenched due to conformational fluctuations in the inserted domain. Target binding stabilizes the inserted domain, which results in fluorescence enhancement

strong illumination light, since they may undergo photochemical transformations with the change of color that are independent of target binding or of response in FRET (Creemers et al. 2000). Their biosynthesis needs hours of incubation.

11.3.3   The Co-synthetic Incorporation of Fluorescence Dyes

The methods of labeling the cellular proteins discussed above can be applied only to known proteins, the genetic manipulation of which is possible. A more global analysis of protein synthesis and transport requires a different approach that excludes genetic manipulation but allows *protein labeling coupled with synthesis.*

Recently, visualization of newly synthesized proteins in bacterial and mammalian cells was demonstrated through a co-translational introduction of an alkynyl amino acid followed by selective Cu(I)-catalyzed ligation of the alkynyl side chain to the fluorescent dye 3-azido-7-hydroxycoumarin (Beatty et al. 2006). Protein tagging with homopropargylglycine (Hpg) in the absence of Met synthesis in mammalian cells and the promiscuity of the methionyl-tRNA synthetase, make it straightforward to incorporate Hpg into mammalian proteins in competition with Met. After incorporation, Hpg is susceptible to labeling with a membrane-permeant dye for *in situ* imaging (Fig. 11.11).

Noncanonical amino acid tagging offers a facile means of labeling newly synthesized proteins in mammalian cells and this approach is extendable to visualizing proteins of unknown sequence, structure, or function. A new technique for the selective and efficient biosynthetic incorporation of a low-molecular-weight fluorophore into proteins at defined sites has been reported (Summerer et al. 2006). The synthetic fluorescent amino acid dansylalanine was genetically encoded in *Saccharomyces cerevisiae* by using an amber non-sense codon and corresponding orthogonal tRNA – aminoacyl-tRNA synthetase pair. This dye exhibits wavelength-shifting environmental sensitivity, though its short-wavelength (335 nm) excitation is not very attractive. However, it is believed that such a strategy should be applicable to a number of different fluorophores in both prokaryotic and eukaryotic organisms and this should facilitate both biochemical and cellular studies and sensing applications.

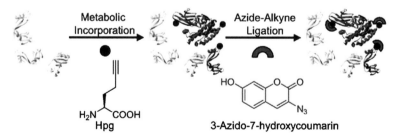

Fig. 11.11 Bioorthogonal labeling of newly synthesized proteins for fluorescence visualization in mammalian cells (Reproduced with permission from Beatty et al. 2006)

11.3.4 Concluding Remarks

'Affinity labeling' inside the cells is a powerful strategy that requires dramatic improvement with a substantial contribution from synthetic chemistry. It is essential to devise chemistries that render the reagents fluorescent or dramatically change their emission (color, polarization or lifetime) only after binding to the target site. This advancement should go hand in hand with improvements in the specificity and affinity of the tags.

For the successful application of visible fluorescent proteins, a different type of improvement is needed. Besides the complexity of the genetic fusion technique, the GFP-like proteins add considerable bulk. In addition, their fusion is often limited by prolonged maturation and increased risk of misfolding or aggregation. Therefore, steps must be made towards a dramatic miniaturization of these proteins, probably with the change in chemistry of spontaneously formed fluorophores. Alternative strategies for live-cell labeling deserve rapid development. Prospective in this respect are genetically encoded fusion tags that are not themselves fluorescent but act as targets for labeling with chemical probes that incorporate fluorophores or other functional groups.

One of the most expected trends in cellular imaging is the incorporation of unnatural amino acids that can serve as fluorescent reporters themselves or that allow highly specific modifications.

11.4 Advanced Nanosensors Inside the Cells

Sensing with and inside the cells requires new tools. The properties that should be primarily improved for proceeding further into cellular studies are the brightness and photostability. The *dye-doped, metallic* or *semiconductor nanoparticles* (Chapter 4) conform to these needs better than organic dyes and therefore, they are becoming increasingly popular in cellular research (Aylott 2003; Grecco et al. 2004).

Much like fluorescent organic dyes, there is no general solution of the problem of the delivery of nanoparticles into the cells. Currently used methods include picoinjections, conjugation with cationic peptides, liposomal delivery, sequestration into macrophages and gene gun (Webster et al. 2007). Every one of the described methods can be tried but not one of them is perfect or generally applicable.

11.4.1 Fluorescent Dye-Doped Nanoparticles in Cell Imaging

Cell imaging and intracellular sensing are important fields in the application of fluorescence emitters in the form of nanosensors. *Nanosensors* are considered here as nano-scale hybrid systems that integrate recognition and reporting functionalities and their support matrix for the detection and quantitative analysis of target binding. Their advantages over molecular sensors can be summarized as follows:

(a) An *increased number of analytes* can be measured with one type of sensor, because the nanosensors are not limited to using a single recognition or reporter unit and can utilize cooperative interactions between ionophores, enzymes, reporter dyes, etc.

(b) Nanosensors of composite design can be *prepared in vitro* by highly reproducible fabrication technologies. The calibration of nanosensors that can be easily made *in vitro* is commonly valid for *in vivo* measurements.

(c) The dyes immobilized with the matrix of nanoparticles can be *protected* from potential interferences in the cellular environment, e.g., from non-specific binding proteins and organelles.

(d) The particles can exhibit *size-exclusion selectivity*. The porosity of such composite nanoparticles can be sufficient to allow easy penetration into contact with the recognition and detection units of ions and small molecules, while large proteins are blocked from entering the nanoparticle matrix and from interacting with the dyes.

(e) Due to a relatively large size, the nanosensor particles exhibit a *decreased selective sequestration* into cellular compartments or leaking from the cells. The large size of nanosensors limits some applications but also offers important advantages.

(f) *Ratiometric measurements* are made easy by incorporating a reference dye into a single nanosensor together with a reporting dye. The principle of optical encoding (Section 12.2) becomes applicable to cellular research with an expansion of the possibilities for simultaneous sensing numerous targets.

Nanoparticles with such important properties can be composed by imbedding fluorescent dyes into the body of polymeric or sol-gel nanospheres. The particles, in which the dyes are only attached to their surfaces, can also be used but in this case not all of the advantages listed above can be realized. If the functional dyes are designed for sensing small molecules and ions and retain this property upon incorporation into nanoparticles, their bright spots in the fluorescence cell image may signal many cellular processes. For instance, we can learn more about the storage and release in cells of zinc and calcium ions (Buck et al. 2004).

11.4.2 Quantum Dots Applications in Imaging

Semiconductor QDs have rapidly found a broad range of applications as optical imaging agents (Grecco et al. 2004; Lidke et al. 2007). They are characterized by high brightness, improved photostability and multicolor size-dependent light emission (Section 4.4). A special feature of QDs that is very attractive for imaging is their fluorescence lifetime of 10–40 ns, which is at least one order of magnitude longer than that of typical organic dyes and also of intracellular pigments that cause autofluorescence. Therefore, a combination of pulsed excitation with time-gated detection can produce images with greatly reduced levels of background emission.

An additional and very significant advantage of Quantum Dots is their significant two-photon absorption that is two to three orders of magnitude greater than that of organic dyes. Therefore, it is natural to observe their increased level of application in two-photon microscopy, with all the benefits that are offered by this method.

These remarkable properties suggest many opportunities for imaging. In the studies of living cells and in operation with *in vivo* animal models, they bring unprecedented sensitivity and spatial resolution (Smith et al. 2006). Combined with biomolecular engineering strategies for tailoring the particle surfaces at the molecular level, the bio-conjugated Quantum Dot probes are well suited for imaging the single-molecule dynamics in living cells, for monitoring protein-protein interactions within specific intracellular locations and for detecting diseased cells and tissues. The explosive increase in the use of QDs in biological imaging has been triggered by an improved synthesis of water-stable QDs, the development of approaches to introduce them efficiently into the cells and the improvements in conjugating QDs to specific biomolecules.

Although the success in using QDs for applications in cellular research is apparent, the routine use of these powerful tools meets with some unresolved problems, related mainly to their conjugation (Michalet et al. 2005). To provide the binding specificity, the QDs must be conjugated to biomolecules that can label specific cellular proteins. Here, their surface properties are important and are determined by their shells. Conjugation with functional molecules can be provided in one step by covalent attachment to shell-forming molecules or in two steps by attachment of the self-assembly partners (such as avidin, oligohistidine or leucine zipper peptide) and following the self-assembly of a functional complex. The use of avidin permits a stable conjugation of the QDs to ligands, antibodies or other molecules that can be biotinylated, whereas the use of proteins fused to a positively charged peptide or oligohistidine peptide obviates the need for biotinylating the target molecule (Rajh 2006).

The solutions to the problems of QD delivery into the cells and their functionalization should be found simultaneously. One of the possibilities could be the incorporation onto their surface of a bifunctional oligoarginine cell penetrating peptide (based on the HIV-1 Tat protein motif), bearing a terminal polyhistidine tract (Delehanty et al. 2006). The polyhistidine sequence allows the peptide to self-assemble onto the QD surface via metal-affinity interactions while the oligoarginine sequence allows specific QD delivery across the cellular membrane and intracellular labeling. By incorporating the differently colored QDs into a variety of live cells, there is a possibility to generate a unique and spectrally resolvable code for each cell type (Mattheakis et al. 2004).

The coating of CdSe/ZnS QDs with neurotransmitter molecule dopamine (Clarke et al. 2006) allows obtaining a redox-sensitive pattern in living cells. Dopamine may serve as a photoinduced electron transfer quencher. Under reducing conditions, the fluorescence is only seen in the cell periphery and lysosomes. As the cell becomes more oxidizing, the QD signal appears in the perinuclear region, including mitochondria. With the most-oxidizing cellular conditions, QD labeling throughout the cell is seen. This principle is extendable to any QD conjugated to an electron donor.

Research of signaling pathways between and within the cells relies heavily on bright and sensitive fluorophores. QDs have begun to play a major role in this field. Thus, using serotonin-linked QDs it is possible to target the neurotransmitter receptor on the cell surface. The QD probes not only recognize and label the serotonin-specific neurotransmitters on cell membranes but also inhibit the serotonin transportation in a dose-dependent manner (Tomlinson et al. 2007). Although being one to two orders of magnitude less potent at inhibiting the receptor than free serotonin, the behavior of QD conjugates was similar to that of free serotonin, making QDs valuable probes for exploring the serotonin transportation mechanism.

Though the concern of the toxicity of QDs remains, most experiments conducted to date have shown that QDs do not interfere with normal cell physiology and cell differentiation.

11.4.3   Self-Illuminating Quantum Dots

QDs commonly require excitation from an external light source. They can also be excited by the energy transfer from other molecules or particles that serve as the FRET donors (Section 6.2). This mechanism of transfer allows using the excited species obtained in the bioluminescence reaction as the donors (Section 8.2). The luciferase needed for this reaction can be covalently attached to QDs and the substrate can be incorporated spontaneously into the cells and tissues. This construction allows the QDs to be 'self-illuminated' without any light source, by converting the chemical energy into the light emission of QDs as the FRET acceptors. The QDs can emit light in the spectral region 600–800 nm, which are wavelengths convenient for observation. Luciferase emission at 480 nm fits the QDs absorption band, which is very broad and increases dramatically in intensity on transition to a shorter wavelength.

This idea has been realized recently (So et al. 2006). It was shown that the self-illuminating QD conjugates possess a greatly enhanced sensitivity in imaging of not only the cells but also of small living animals. This allows avoiding many problems existing in the common illumination of the cells by light, such as strong autofluorescence and light-scattering.

11.4.4   Extending the Range of Detection Methods

Summarizing the results discussed in this section, we must indicate a growing interest from the side of researchers to novel functional brightly emitting nanoparticles. Despite their greater dimensions, they are superior to organic dyes in many respects and allow overcoming some limitations in their use for cellular imaging. Fluorescent Quantum Dots have proved to be a useful alternative for studies that require long-term and multicolor imaging of cellular and molecular interactions. Very positive results in this respect can also be achieved with nanoparticles composed of organic dyes and metal chelating complexes hosted by silica or organic polymeric matrices. These nanocomposites can explore the versatility of already developed molecular sensors, which is highly needed in the study of biosystems.

11.5   Sensing the Cell Membrane

In this section, without attempting to go deeply into membrane biology, we will outline some remarkable applications of fluorescence sensing and probing techniques. The membrane surrounding the living cell has the important functions of maintaining the specific composition and properties of the cell interior and also of communication with other cells and with the surrounding media. The cell membrane contains a specific recognition pattern related to its function. Fluorescence techniques offer many possibilities for studying these properties.

11.5.1   Lipid Asymmetry and Apoptosis

Normal cells exhibit remarkable asymmetry of lipid distribution between inner and outer leaflets of cell membranes, which is lost during the early steps of *apoptosis* (programmable cell death), when the cell integrity is not yet disrupted. The most characteristic in this change is the exposure to the cell surface of anionic phospholipids, such as phosphatidylethanolamine (PE) and phosphatidylserine (PS). This exposure is functionally important; it provides the signal for recognition and elimination of apoptotic cells by macrophages.

The methods based on molecular recognition of *surface-exposed* PS and PE allow one to identify and characterize apoptotic cells. The most popular of them are based on the property of the calcium-binding protein annexin V to interact with exposed PS in a Ca^{2+}-dependent manner. Because of this selectivity, the binding of the labeled protein occurs specifically only to apoptotic cells. Different variants of this method have been developed. For instance, annexin V labeled with fluorescein is used for flow cytometry, while its labeling with red and near-IR dyes is efficient for tissue imaging (Ntziachristos et al. 2004; Petrovsky et al. 2003). The recently reported conjugation of annexin V with Quantum Dots allowed an increased sensitivity of detection (Koeppel et al. 2007).

There can be different approaches to the same problem. They can be based on detecting the *change of membrane order, surface potential* and *hydration* that must be the result of the loss of transmembrane lipid asymmetry that is associated with the appearance of anionic lipids in the outer membrane leaflet. A very high sensitivity of 3-hydroxychromone probes to the variation of these parameters allowed developing a molecular sensor for apoptosis that could incorporate spontaneously into the outer leaflet of the membrane and report on apoptotic cell transformation in a wavelength-ratiometric manner (Color Plate 14).

This approach can be applied in three formats – in fluorescence spectroscopy of cell suspensions, in confocal microscopy of individual cells and in flow cytometry (Shynkar et al. 2007). In the latter case, we observe a much smaller dispersion of the measured parameter within the sub-populations of living and apoptotic/dead cells and a combination with cell-permeable dye (e.g., propidium iodide) allows obtaining a quantitative measure of living, apoptotic and dead cells.

The proposed methodology is not limited to apoptosis. Since the appearance of anionic lipids was detected on the surface of different cancer cells, it will probably find application for studying the development of cancer tissues and of the efficiency of applied anti-cancer drugs. The appearance on the cell surface of anionic lipids is also a remarkable event during the activation of thrombocytes and may be useful for monitoring blood-clotting events.

11.5.2 Sensing the Membrane Potential

Transmembrane potential (V_m) is the most dynamic electrostatic component of cell plasma membranes. It plays an important role in a variety of cellular functions related to bioenergetics, ion transport, motility and cell communication. It is highly important to monitor this potential, notably in neuronal cell assemblies, because this can provide clues to understand the mechanisms of brain activity. Therefore, V_m has to be measured with high time resolution, in milliseconds. Being reactive on the much shorter time scale of ESIPT, the 3HC dyes can provide a fast response. The high two-band ratiometric sensitivity of these dyes to the electric field (Klymchenko and Demchenko 2002; Klymchenko et al. 2003) is the background for these applications. Recently, for this purpose, a new 3HC dye possessing deep vertical incorporation into the lipid bilayer was developed (Klymchenko et al. 2006).

In order to achieve the highest sensitivity to the electric field associated with V_m, the fluorophore must be located in the phospholipid environment of the membrane as deeply as possible in its hydrophobic area, where the dielectric constant is minimal, which means a smaller dielectric screening. This requirement cannot easily be satisfied with any hemicyanine or styryl pyridinium dyes that are also used in sensing the membrane potential (Gross et al. 1994) because in these dyes the positive charge is part of the fluorophore. In contrast, 3-hydroxychromones are uncharged molecules of relatively low polarity, which can be introduced at any position within the bilayer. The rigid skeleton of the probe and the presence of two segments bearing negative charges allow fixing the position and orientation of the fluorophore part FA (Fig. 11.12).

The new dye exhibits a fast response to the transmembrane potential by *variation of the relative intensities of the two emission bands*. Its sensitivity is ca. 15% in the change of the ratiometric signal per 100 mV, which is one of the best results obtained so far in a fast-response fluorescence probing. This dye is a prospective for a variety of applications for monitoring the membrane potential in cell suspensions and its imaging in single cells, especially in neurons having fast electrical activity.

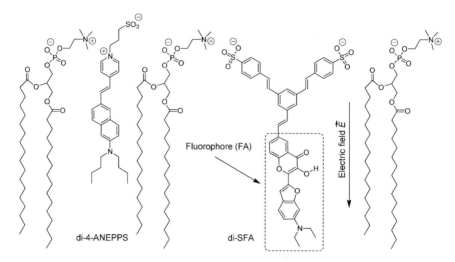

Fig. 11.12 Assumed location of di-4-ANEPPS (a styrylpyridinium dye) and di-SFA (a 3-hydroxychromone derivative) in lipid bilayers. The fluorophore part of the probe di-SFA is boxed. Only half of the bilayer is shown, represented by phosphatidylcholine structures (Reproduced with permission from Klymchenko et al. 2006)

The broad field of studying the spatio-temporal pattern of neuronal activity *in vivo* is open for the application of fast-responding voltage-sensitive dyes. It needs the involvement of two-photon microscopy and the first steps for the evaluation of aminostyryl voltage-sensitive dyes at two-photon excitation have already been taken (Fisher et al. 2008)

It is known that natural membranes generate a very strong electric field gradient known as the *dipole potential* (V_d) (Gross et al. 1994). This potential arises due to oriented hydrated carbonyl groups of phospholipids and is well reproduced in phospholipid biomembrane models. Neutral dipolar molecules, when incorporated into the bilayer, can provide a strong modulation of (V_d). It has been demonstrated that the compounds increasing or decreasing the dipole potential on their incorporation into the bilayers (Klymchenko et al. 2003) and cell membranes (Shynkar et al. 2005) induced opposite effects on the response of two-color ratiometric fluorescence probes. The effect on this incorporation of V_d modulator 6-ketocholestanol can be clearly seen in the confocal microscope (Color Plate 15). It has also been observed that the incorporation of the dye in the opposite orientation produces a reversal of the spectroscopic effects, which allows distinguishing vectorial electric field effects from orientation-independent effects, such as polarity and hydration.

Here, we have to stress that with two-color wavelength-ratiometric probes the absolute intensity forming the image of the cell membrane in the confocal microscope is not very informative, since it depends on local probe concentration and its location with respect to fluorescence quenching groups. In contrast, the ratio of the intensities of the two fluorescence bands is independent of these factors and reflects the true magnitude of the electrostatic effects.

11.5.3 Membrane Receptors

The examination of signaling pathways that involve the interaction of specific cellular receptors with their targets requires the labeling of both the receptor and the target. A productive strategy could be a co-synthetic labeling of the receptor with a visible fluorescent protein of the GFP family and the target with brightly fluorescent nanoparticles. Such a possibility was demonstrated recently (Lidke et al. 2005) for the epidermal growth factor receptor (EGFR) and its corresponding ligand, epidermal growth factor (EGF), see Color Plate 16.

It was shown that EGF conjugated to Quantum Dots is still capable of binding to and activating its receptor, which triggers the internalization of both the EGF-QD and its receptor via endocytosis (Hagen et al. 2008). The high photostability of QD allows the study of EGF-QD binding and internalization kinetics in great detail. The family of these receptors is implicated in the development and progression of cancer and is expressed in many human epithelial malignancies. Such quantitative understanding of the transduction mechanism is essential for receptor-targeted therapeutics. Quantum Dots become a valuable tool for this kind of investigation because of their much higher brightness and photostability than organic dyes.

The results of the studies on other types of receptors, e.g., glycine receptors in neuronal membranes (Dahan et al. 2003) are good illustrations of this. In these studies, tracking individual dots and analyzing their trajectories is possible, with the evaluation of QD diffusion coefficients.

11.5.4 Future Directions

The cellular membrane is the means for a cell to communicate with the whole world, so it has enough room for communicating with our fluorescent labels and probes. The cell membrane is a very complex structure, demonstrating both site-specific and integrated properties. The *integrated properties* allow mediating the functional response via cooperative structural transitions of the lipid component forming the bilayer. The change of lipid mobility and lipid segregation (e.g., formation of 'rafts') are a part of this mechanism. The response of this system to external stimuli is great and their study is relatively easy, since the amphiphilic dye derivatives (such as those depicted in Figs. 7.10 and 11.11) can incorporate spontaneously into a membrane structure from an external medium. In these studies, dyes exhibiting a multiparametric approach and that are able to characterize the membrane in terms of effects on its hydration, polarity and membrane potential, are of great importance. We expect their rapid exploration not only in basic research but also in applications using 'whole-cell biosensors' for the detection and testing of differently biologically active compounds.

The *specific properties* of membranes are provided by receptor proteins that provide either a direct response (such as the opening-closing of the ion channels) or transduce the signal into the cell interior, which can be mediated by second messengers and their receptors. For membrane receptor proteins, the most important methodology based on fluorescence involves their covalent dye labeling, for establishing proximity relations in the receptor system and the labeling of the ligands, for finding the stimulus-response correlations.

11.6   Molecular Recognitions in the Cell's Interior

The living cell demonstrates a high level of organization among the tens of thousands of types of molecules that constitute it. Here, we will demonstrate the applicability of the microscopic imaging technique for sensing some of its important constituents. We will also discuss the possibilities for studying the functional events in the cell, such as differentiation and apoptosis. The more general problems of characterizing genome, proteome and interactome will be discussed in Section 12.1.

11.6.1   Ion Sensing

Calcium is the most common target for intracellular imaging and sensing because this divalent cation is probably the most ubiquitous and important second messenger in cells. It is involved in key cellular events controlling muscle contraction, nerve cell communication, hormone secretion and the activation of immune cells.

The selective measurement of the *free calcium concentration* in living cells requires sensor molecules with selectivity for calcium by many orders of magnitude higher than for other cations. This is because in living cells, the Ca^{+2} ion concentration is maintained on a level of 10^{-7} M in the resting state and, upon transmission of cellular messages, it rises to 10^{-6}–10^{-5} M. This is in strong contrast to its high extracellular concentration, which can be on the level of 10^{-3} M. The calcium sensing should be discriminative from other ions that may interfere with the assay. Therefore, the intracellular concentrations of Ca^{+2} should be determined on the level of 10^{-7}–10^{-5} M in the presence of 10^{-3} M Mg^{+2}, 10^{-2} M Na^+ and 10^{-1} M K^+ ions. This is possible with calcium indicators that combine specific ion binding with a strong light-emission response. Presently, the researcher is able to select between calcium indicators built on the basis of several different principles.

1. *Fluorescent dyes* with attached ion chelator groups. In current use are the dyes Fura-2 (ratiometric in excitation) and Indo-1 (ratiometric in emission). Their major disadvantage is their near-UV excitation and emission, which overlaps the emission of some cellular pigments. A detailed investigation of Indo-1 shows that its ratiometric signal can be influenced by factors unrelated to the analyte concentration. These are photobleaching with the production of fluorescent but Ca^{++}-insensitive forms (Scheenen et al. 1996) and temperature (Oliver et al. 2000), which is probably due to the difference in thermal quenching

of ligand-bound and ligand-free forms of the sensor. Many probes respond to calcium ions by the change in intensity of a single fluorescence band. It is not reasonable to apply them in view of the possible calcium-dependent change of their location sites and thus, their local concentration, which is also proportional to the intensity.

2. Molecular sensors based on the conformation change of a *calcium binding protein calmodulin*. Calmodulin mutations can alter the binding affinities for calcium in a concentration-dependent manner, such that with a combination of sensors, Ca^{++} can be detected in the range of 10^{-8}–10^{-2} M (Miyawaki et al. 1999). The sensor, together with two fluorescent FRET donor and acceptors can be fully synthesized inside the cell (see Fig. 11.9).

3. *Luminescent calcium binding protein aequorin* that can be genetically expressed in the cells and synthesized in the form fused to different proteins (Rogers et al. 2005). To emit light, aequorin needs an imidazopyrazine cofactor (or its synthetic analog), which is membrane permeable and can be added to intact cells containing an engineered aequorin.

4. Sensors based on the known *calcium-binding protein troponin C*, which can be fused to a variant of green fluorescent protein to provide a response based on FRET (Mank et al. 2006). Such a chimeric structure can be synthesized inside the cell. It shows a maximum fractional fluorescence change of 400% in its emission ratio and linear response properties over an expanded calcium regime.

The determination of calcium ions on the level of single cells allows demonstrating an inhomogeneous distribution of these ions inside the cells, rhythmic oscillations of their levels, the so-called spikes and waves and other interesting phenomena that are in the background of the mechanisms of cell regulation. Therefore, all the applied detection methods should allow submillisecond temporal resolution.

Zinc, after calcium, is the second most important cation in biological systems. It is an essential structural element of a number of DNA binding proteins and enzymes (over 300, including carbonic anhydrase and alcohol dehydrogenase). It plays an important role in neurotransmission, signal transduction and gene expression. Perturbation in its homeostasis is related to a number of disorders, including Alzheimer's disease, diabetes and cancer. The lack of understanding of the molecular mechanisms of zinc physiology and pathology is partially due to the lack of tools for measuring changes in its intracellular concentrations with high spatial and temporal fidelity.

The total concentration of zinc ions in human blood serum is ~12 µM. But the major part of zinc ions is tightly bound to proteins and cell organelles, so that the amount of free or rapidly exchangeable zinc is very low, on a *picomolar level*. Measuring these concentrations with picomolar sensitivity in the presence of much higher and variable Mg^{+2} and Ca^{+2} levels requires a highly specific and sensitive fluorescent sensor.

Recently, a single-excitation, dual-emission ratiometric zinc probe was suggested (Chang et al. 2004). It uses the mechanism of controllable Zn^{2+}-induced switching between the fluorescein- and naphthofluorescein-like tautomer forms of the dye. The probe can be excited at a single wavelength of 499 nm and exhibits two fluorescence maxima at about 530 and 600 nm. In the presence of nanomolar concentrations of zinc ions, the long-wavelength band shifts to 624 nm, with a dramatic increase of relative intensity. The ratio of naphthofluorescein- to fluorescein-like emission intensities (F_{624}/F_{528}) upon 499-nm excitation varies from 0.4 to 7.1 in the absence of Zn^{2+}, when the zinc ions are bound. Such a substantial 18-fold dynamic range is not accessible to common wavelength-shifting probes. For ratiometric zinc measurements, iminocoumarin as a fluorophore and (ethylamino)dipicolylamine as a Zn^{2+} chelator have also been suggested (Komatsu et al. 2007).

In addition to small molecules, a series of *peptide-based zinc sensors* have been developed. Their structures are borrowed from the so-called 'zinc-finger' sections of the DNA-binding proteins. The sensing is based on the conformational change in peptide upon Zn^{2+} binding (Walkup and Imperiali 1997). The peptide wraps around the ion, which results in the change of environment of the attached dye or brings two dyes attached to the peptide terminals together, with the observation of FRET (see Fig. 6.11).

Although a variety of synthetic fluorescent sensors for zinc have been developed and applied in recent years, their affinity does not fully match the required concentration range. Therefore, attempts have been made to build a sensor based on the scaffold of one of the zinc binding proteins. The ratiometric fluorescent sensor based on enzyme carbonic anhydrase (Thompson et al. 2002) explores the binding of zinc in its active site and the generation of the wavelength-ratiometric response due to a zinc-dependent FRET between the two dyes. Exhibiting a picomolar sensitivity, this sensor is well adapted to zinc imaging in a physiologically important concentration range.

Sodium and *potassium* ions are also important for understanding cellular life. The living cells actively pump-out sodium ions by substituting them into potassium ions. In the reverse process their exchange occurs in membrane ionic channels. Therefore, determining the sodium and potassium ion concentrations is important for cell physiology. Dyes incorporating crown ethers and responding to ion binding by PET or ICT mechanisms are actively used at present and for the foreseeable future. A typical example is a commercial sodium indicator Sodium Green (Fig. 11.13), which can be used to signal the presence of sodium ions *in vivo* and quantify them on a millimolar scale of concentration (Haugland 2005). This and similar dyes typically respond to ion binding by the change (increase) of fluorescence intensity without observable spectral shifts, so they share all the disadvantages of intensity measurements, which have been discussed many times throughout this book.

11.6.2 Tracking Cellular Signaling

The determination of factors that provide *intercellular signaling* (first messengers) and *intracellular signaling* (second messengers) is very important for cell biology and pharmacology. The *first messengers* are neurotransmitters, cytokines

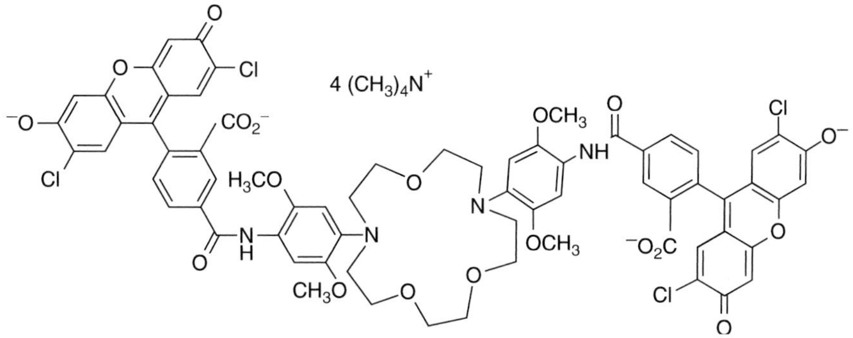

Fig. 11.13 The cellular indicator of sodium ions Sodium Green. It comprises two 2 ,7 -dichloro-fluorescein dyes that are linked to two nitrogen atoms of a crown ether that serves as an ion recognition unit with K_d of 6.0 mM

and hormones functioning respectively in nerve, immune and endocrine systems. They bind either to ion-channel, kinase or G protein-coupled membrane receptor proteins of different cells, transmitting the signals to them and trigger the respective downstream intracellular signaling.

Small molecules that are known as *second messengers* play a central role in signal transduction, modulating many physiological and pathological processes in living cells. Therefore, many attempts have been made to detect and track them inside cells. Calcium ions and cyclic nucleotides are second messengers. One of them is also *nitric oxide* (NO).

The direct recognition of such small molecules on a sub-nanomolar level of detection is problematic, since it is hard to develop a sensor of sufficiently high affinity. Therefore, one of the most interesting suggestions has been, instead of detecting NO, to determine the product of the NO-activated reaction producing cyclic guanosine monophosphate (cGMP). Here, the principle of catalytic amplification is used. Because a single NO molecule leads to the production of many cGMP molecules, this test system possesses the required very high sensitivity. The cGMP sensor is based on a change in the relative intensity of two visible fluorescent proteins that are coupled via the FRET mechanism, which allows a wavelength-ratiometric response in microscopic images (Sato et al. 2005).

Attempts to accomplish direct NO sensing using functional fluorescent dyes have also been reported. Thus, a highly sensitive and pH-independent fluorescence probe for NO was suggested, based on 4-methoxy-substituted BODIPY dye. This sensor explores the PET mechanism of an intensity-based response (Gabe et al. 2006).

Since the use of a FRET in sensor reporting usually requires a change in distance between two labels, this approach can also be applied to a sensor protein composed of two or several subunits, if the sensing event involves changing the distance between them. This possibility was demonstrated for the *detection of cAMP* in smooth muscle cells by labeling the catalytic and regulatory subunits of c-AMP-dependent protein kinase with two different dyes (Adams et al. 1991).

11.6.3   Location of Metabolites and Tracking Metabolic Events

The distribution of metabolites and their transport within and between the cells can be understood only if we know the subcellular distribution of metabolites obtained in living (not fixed) cells obtained in a nondestructive way. In addition, we need to observe their dynamics, especially under the influence of different stimuli. Novel techniques addressing this problem based on fluorescence imaging are emerging.

Attempts to suggest a flexible strategy for designing the intracellular sensors for a wide spectrum of solutes and their concentrations have been made by different researchers. A platform based on bacterial periplasmic binding proteins (PBPs) uses the ability of these proteins to bind different ligands, the number of which can be expanded by genetic engineering. The response to binding such targets is based on the ability of these proteins to transform their hinge-bend movement (see Color Plate 8) into increased fluorescence resonance energy transfer (FRET) between two coupled visible fluorescent proteins (Fehr et al. 2002). This allows obtaining a wavelength-ratiometric response. By using the maltose-binding protein as a prototype, nanosensors were constructed allowing *in vitro* determination of FRET changes in a concentration-dependent fashion. For physiological applications, mutants with different binding affinities were generated, allowing dynamic *in vivo* imaging of the increase in the *cytosolic maltose concentration* in single yeast cells. This approach is promising for the successful development of a myriad of PBPs, recognizing a wide spectrum of different substrates.

11.6.4   In Situ Hybridization

Fluorescent in situ hybridization (FISH) is a method for gene mapping and the identification of chromosomal abnormalities that utilizes fluorescently labeled DNA probes. FISH provides the means to visualize and map the genetic material in cells, including specific genes or portions of genes. In malignant cells, it allows quantifying gene copy numbers that have abnormal gene amplification.

The *in situ* hybridization procedure requires the application of probing DNA or its fragments that have to be conjugated with fluorescent labels. Different labels can be used on condition that the labeling does not interfere with the recognition between complementary oligonucleotide sequences. Moreover, dyes have been developed for staining each of the human chromosomes in different colors (Geigl et al. 2006). Fluorescent nanobeads and Quantum Dots have shown their applicability in these assays and, due to their higher brightness and photostability, they offer important advantages over organic fluorescent dyes.

11.6.5   Looking Forward

It is hard to imagine modern cell biology without fluorescence methods and the examples selected and presented above are good illustrations of this. We observe that more and more sophisticated fluorescence sensors are applied in cellular research. They have become brighter, possess increased stability and have become amendable for application in novel high-resolution technologies.

In addition to molecules and particles that are applied to the cell by diffusion or injection, there is an increasing number of visible fluorescent proteins that can be made inside the cell by using their biosynthetic machinery. The most attractive and possibly prospective for future development are the hybrid systems composed of incorporated segments of protein structure that contain recognition units for functional dyes and of these dyes, that are able to self-assemble with these segments inside the cells.

It is expected that the structural aspect in cellular studies will soon give priority to dynamic studies that will involve tracking and modulating the regulatory mechanisms of the cell. The key instruments in these studies will be luminophores forming nano-particles with extremely long-lifetime emission (milliseconds), which could allow exploring the time-resolved capability coupled with elementary cell signaling events. Dealing with the dynamic nature of living systems and new, important steps for understanding cell signaling and regulation will be essential for further progress.

11.7 Sensing the Whole Body

In this section we have to find a response to a number of important questions. In the era of widely applied magnetic resonance imaging and even positron imaging tomography, what are the prospects for the fluorescence sensing of tissues and whole living bodies? Will this sensing have any prospects in application to humans, addressing the needs of clinical diagnostics and treatment? What are the techno-logical challenges here and the possibilities for their solution?

11.7.1 Optimal Emitters for the Human Body

It should be stated primarily that the human body is highly protected against the penetration of UV and visible light. This is because of the presence in our tissues of many pigments. The most important of them is the skin pigment melanin, which absorbs light in the whole visible range down to about 800 nm and the porphyrins possessing strong absorption bands in the visible. In the near-IR, starting from 1,100–1,200 nm, there appears a strong light absorption due to collective vibrations of water molecules, so this range also becomes opaque in living systems. Therefore, there only remains a *window of relative transparency* between 800 and 1,100 nm, in which the energy of light quanta is not sufficient to excite the electronic transi-tions but it is already too high to excite vibrations.

Why is the transparency 'relative'? Because the cells and tissues possess many structures, the sizes and refractive index gradients of which lead to efficient *light-scattering*. These light-scattering effects are not as large as in the UV or visible region but they still have to be accounted for.

These restrictions still allow two possibilities for fluorescence sensing. One is to use *near-IR dyes* for imaging and reporting. Such dyes exist; their molecules are composed of many aromatic heterocycles or a long polymethine chain (Zaheer et al. 2002). These dyes can attain important functional properties. They can be used for targeted fluorescence imaging of bacterial infection in a living whole animal (Leevy et al. 2006). A near-IR probe has been found that has a strong and selective affinity for the surfaces of bacteria and after intravenous injection, it selectively accumulates at the sites of localized bacterial infections.

The other possibility is to use *two-photon microscopy* and labeling with fluorescent dyes with near-IR excitation. The near-IR light penetrates deeply into the tissues but the fluorescence response is in the visible, being subject to stronger reabsorption and scattering.

Recent studies demonstrate efficient ways to realize these possibilities with semiconductor *Quantum Dots*. Those that absorb light in the near-IR can be selected and applied in conditions of either single-photon or two-photon excitation. A variety of such semiconductor nanoparticles exists, for instance, those composed of a cadmium-tellurium core with a cadmium-selenium shell. In conditions of both single-photon or two-photon excitation, Quantum Dots have a brightness 100–1,000 times higher than that of typical organic dyes. The other positive features of QDs, which we marked when discussing cellular studies, will also be required here and the long-wavelength excitation conditions still further decrease the possibility of photobleaching and photodamage. Upon excitation with two photons in the near-IR, they can serve not only as reporters but also as sensitizers, by providing photodynamic properties to energy acceptors. Some of the published results illustrate their application (Dayal and Burda 2008).

11.7.2 Contrasting the Blood Vessels

Safe and non-invasive fluorescence tracers and reporters find different applications in research and clinical diagnostics *in vivo*. The examples presented below illustrate different possibilities for their use.

The dye *indocyanine green* has been serving in clinics, for visualizing blood vessels, for more than 50 years already (Frangioni 2003). It is one of the least toxic agents ever administered to humans. Its fluorescence emission at 830 nm allows minimizing the re-absorption and light-scattering effects in tissues. Many researchers have reported on the improvement of dye performance, on its modification and incorporation into nanoparticles (Rajagopalan et al. 2000). Dyes of this type are presently used to image heart vasculature and myocardial perfusion. Near-IR fluorescence imaging (*fluorescent angiography*) is used in the common practice of ophthalmology clinics. With the aid of a confocal microscope, it allows one not only to study blood microcirculation in detail but also to predict the appearance of small melanomas (Mueller et al. 2002).

Quantum Dots have only started to find their way into clinical practice and they have still to pass a careful toxicity test. Their use in *surgical procedures* on animals has been demonstrated in the procedure of mapping the sentinel lymph nodes. The surgeon quickly identified their positions in a precise and rapid surgical procedure (Kim et al. 2004).

The blood-brain barrier represents a significant impediment to a large variety of central nervous system-active agents. In order to study its permeability following cerebral ischemia and reperfusion, fluorescent polystyrene nanoparticles were used (Yang and Zhou 2006). An injection of these markers into the blood stream and their collection by microdialysis in the cerebral cortex, allowed observing transient accumulations of extracellular nanospheres in the brain throughout the course of cerebral ischemia/reperfusion.

11.7.3   *Imaging Cancer Tissues*

The most important field of application of fluorescence tracers is in the detection, imaging and control of treatment of cancer tissues (Gao and Dave 2007; Sharrna et al. 2006). The ability of QDs to target tumor vasculature has been impressively demonstrated in experiments on mice (Akerman et al. 2002). Being conjugated to several peptides that differentially recognized blood vessels located in the lung, tumor blood vessels and tumor lymphatic vessels, these nanoparticles were injected into mice. Guided by these peptides, QDs were delivered to the appropriate sites *in vivo*, allowing one to visualize and to characterize the tumor.

Targeting and imaging the cancer tissue *in vivo* can also be provided using QDs in conjugation with antibodies (Gao et al. 2004). The QDs were conjugated to the antibody specific for the prostate cancer cell marker. After injection into mice that contained transplanted prostate cancer, the QD-tagged antibodies recognized the tumor site, which allowed clearly observing its image *in vivo*. The advantage of QDs is obviously evidenced in comparison with GFP, the insufficient brightness of which does not allow obtaining any image recognizable from the background.

The progression of cancer is strongly connected with the development of metastasis. It has been shown (Voura et al. 2004) that metastatic cells can be identified *in vivo* by using the two-photon response of functional QDs. No toxicity of QDs was detected and spectral imaging allowed the simultaneous identification of five different populations of cells.

Killing cancer cells with a noninvasive treatment such as *photodynamic therapy* (a treatment that involves the photodamage of cells produced by the bound dyes) can be combined with the identification and imaging of these cells. Despite the complexity of this problem, serious progress has been achieved in its solution (Stefflova et al. 2007). The so-called 'killer beacons' have been developed to combine the functions of producing photodamage and fluorescence reporting in hybrid nanoparticles. The modular configuration of these beacons allows one more function – observing the results of anti-cancer treatment by detecting their apoptotic transformations. The latter can be detected by the response of the attached fluorescence substrate of caspase 3, the apoptosis marker.

11.7.4   Surgical Operations Under the Control of Fluorescence Image: Fantasy or Close Reality?

Such control is primarily needed in oncology. The surgeon has to remove malignant tissue without touching the healthy one. To serve as a guide, the image should appear in real time (seconds) with a sufficiently high spatial resolution. Efficient illumination and the collection of the fluorescent image with on-line computer analysis are technically achievable. What remains?

(a) We need to have nanoparticles with sufficiently high brightness and photostability. Presently, there is a clear understanding that the increased brightness requires instead the application of dye-doped particles or semiconductor Quantum Dots instead of dye molecules. Experiments with small animals show much promise for their use.
(b) These particles should possess high affinity towards cancer cells and strong discriminative power against healthy cells. The rapid progress in the identification of cancer cell surface markers (Jain 2007) and the development of sensors recognizing them show much promise. Based on these findings, providing the recognition function to nanoparticles is a relatively easy task.
(c) The particles injected into the body should be absolutely non-toxic and should allow rapid clearance after the operation is over. The present experience in this respect is not sufficient and some semiconductor QDs are considered to be potentially toxic. More research in this area needs to be completed before QDs can be used as probes in the human body. Whether the QDs can ultimately be cleared from the body is often not known. There are reports that injected QDs can accumulate in the kidney, liver and spleen. Meantime, it was reported that the clearance of nanoparticles depends on their size and with a size of less than 5.5 nm, they can easily pass through the kidney membranes (Choi et al. 2007).

Thus, the above-cited studies have shown the great potential of using QDs as *in vivo* probes for noninvasive whole-body imaging and, particularly, for cancer studies with good prospects for clinical diagnostics and treatment. If the problems with QDs toxicity and retention in the body remain unresolved, there remains the possibility of using organic dyes incorporated into nanocolloid particles (Ohnishi et al. 2005). A family of such particles, containing near-IR organic heptamethine indocyanine dyes, can optimally fulfill these demands.

Concluding, we respond to the questions posed at the beginning of this section. The established NMR and positron tomography-based platforms for tissue and whole-body imaging, even with their prospective development, cannot make fluorescence imaging unnecessary. Site-specific labeling and target-specific reporting allows not only an exquisite resolution of anatomical features but also provides their non-invasive diagnosis. This is because they are able to recognize the target molecules that are pathology markers, especially those existing in small quantities, such as cell surface antigens. This is the strength of fluorescence-based sensing.

Sensing and Thinking 11: Intellectual and Technical Means to Address Systems of Great Complexity

The improved resolution, sensitivity and versatility of fluorescence microscopy and the development of fluorescent targeting and labeling of proteins and nucleic acids in live cells and probing the membranes and organelles, have yielded a clearer understanding of the dynamics of intracellular networks, signal transduction and cell-to-cell interactions. New demands for portraying whole proteomes, metabolomes and interactomes require new tools, which could be applicable to studies on a cellular level. Analysis of the presently observed tendencies allows suggesting that future progress will be connected with the following steps forward in methodology.

1. A dramatic increase of spatial resolution in microscopy, down to tens of nanometers. Recent studies demonstrate that this can be achieved by several independent approaches, both in near-field and far-field microscopy.
2. The extensive use of single-molecule methods adapted to cellular research. The complex milieu encountered inside live cells requires substantial adaptations of common *in vitro* techniques. This adaptation will proceed step-by-step, together with progress in high-resolution microscopy.
3. The dramatic increase of the brightness and photostability of fluorescence reporters to satisfy the demands of high-resolution microscopy and single-molecule detection. In this sense, organic fluorophores alone have probably reached the limit of their perfection but nanocomposites, promise a great future with their participation.
4. Solving the problem of incorporation into the cell of any desirable (nano)probe and the attachment of a fluorescent label (or even, sensor) to any cellular protein.

What remains is to explore these fascinating possibilities. Living cells are systems of immense complexity. The methodology developed in cell studies must find a broader scale of application, also in the analysis of man-made systems of variable complexity.

Questions and Problems

1. Explain the basic principles that are applied for eliminating out-of-focus emissions and the constructions of devices that operate based on these principles.
2. Can we obtain a sharp image of the nucleus of a typical human blood cell of diameter 10–15 µM with an internal reflection microscope?
3. Can we observe a sharp image of a cellular membrane in a confocal microscope?
4. Will a microscope that combines the principles of a two-photon and confocal microscope be efficient?
5. Explain the advantages of time-resolved imaging.

6. Explain how to make the light source smaller than the wavelength of light.

7. In near-field scanning microscopy, is there a possibility to substitute the optical probe with a nano-sized electrode, to generate localized emission by using electrogenerated chemiluminescence (Section 8.1)?

8. What is the diffraction limit and how do we overcome it in wide-field microscopy?

9. Why does dye photobleaching depend on the presence of oxygen? Can photobleaching be guided by the researcher? Is there a useful application of this phenomenon?

10. What is a reasonably achieved minimal volume for allowing the single-molecular detection? How does it compare with the volumes of typical cells?

11. What are auto-correlation and cross-correlation? How can they be used to study intermolecular interactions?

12. What requirements are needed to observe specific labeling of molecules inside living cells by an externally applied dye? What controls have to be applied to prove correct labeling?

13. Compare the sensors based on organic dyes and on nanoparticles and their advantages and disadvantages for cellular applications.

14. Can luminescent nanoparticles be used inside the cell without optical excitation?

15. What is the mechanism of the response of the sensor based on a 3-hydroxychromone dye to the changes of cellular membranes on apoptosis?

16. Can the molecular sensors developed for use in the detection of Ca^{+2} ions inside cells be used to detect them in blood? In drinking water?

17. Explain how to obtain an image of maltose distribution inside a cell.

18. Why is the optical tomography of human tissue so difficult? What is the wavelength range and physical mechanism allowing their transparency?

19. The injection of organic dye into the human blood vessels, what does this allow one to detect?

20. What requirements have to be observed for the optical (fluorescent) imaging of cancer tissues?

References

Adams SR, Harootunian AT, Buechler YJ, Taylor SS, Tsien RY (1991) Fluorescence ratio imaging of cyclic AMP in single cells. Nature 349:694–697

Adams SR, Campbell RE, Gross LA, Martin BR, Walkup GK, Yao Y, Llopis J, Tsien RY (2002) New biarsenical ligands and tetracysteine motifs for protein labeling in vitro and in vivo: synthesis and biological applications. Journal of the American Chemical Society 124:6063–6076

Akerman ME, Chan WC, Laakkonen P, Bhatia SN, Ruoslahti E (2002) Nanocrystal targeting in vivo. Proceedings of the National Academy of Sciences of the United States of America 99:12617–12621

Aylott JW (2003) Optical nanosensors - an enabling technology for intracellular measurements. Analyst 128:309–312

Barbara PF, Gesquiere AJ, Park SJ, Lee YJ (2005) Single-molecule spectroscopy of conjugated polymers. Accounts of Chemical Research 38:602–610

Beatty KE, Liu JC, Xie F, Dieterich DC, Schuman EM, Wang Q, Tirrell DA (2006) Fluorescence visualization of newly synthesized proteins in mammalian cells. Angewandte Chemie-International Edition in English 45:7364–7367

Buck SM, Koo YEL, Park E, Xu H, Philbert MA, Brasuel MA, Kopelman R (2004) Optochemical nanosensor PEBBLEs: photonic explorers for bioanalysis with biologically localized embedding. Current Opinion in Chemical Biology 8:540–546

Bullok KE, Gammon ST, Violini S, Prantner AM, Villalobos VM, Sharma V, Piwnica-Worms D (2006) Permeation peptide conjugates for in vivo molecular imaging applications. Molecular Imaging 5:1–15

Campbell RE, Tour O, Palmer AE, Steinbach PA, Baird GS, Zacharias DA, Tsien RY (2002) A monomeric red fluorescent protein. Proceedings of the National Academy of Sciences of the United States of America 99:7877–7882

Chang CJ, Jaworski J, Nolan EM, Sheng M, Lippard SJ (2004) A tautomeric zinc sensor for ratiometric fluorescence imaging: application to nitric oxide-induced release of intracellular zinc. Proceedings of the National Academy of Sciences of the United States of America 101:1129–1134

Choi HS, Liu W, Misra P, Tanaka E, Zimmer JP, Itty Ipe B, Bawendi MG, Frangioni JV (2007) Renal clearance of quantum dots. Nature Biotechnology 25:1165–1170

Clarke SJ, Hollmann CA, Zhang ZJ, Suffern D, Bradforth SE, Dimitrijevic NM, Minarik WG, Nadeau JL (2006) Photophysics of dopamine-modified quantum dots and effects on biological systems. Nature Materials 5:409–417

Claydon TW, Fedida D (2007) Voltage clamp fluorimetry studies of mammalian voltage-gated K(+) channel gating. Biochemical Society Transactions 35:1080–1082

Creemers TM, Lock AJ, Subramaniam V, Jovin TM, Volker S (2000) Photophysics and optical switching in green fluorescent protein mutants. Proceedings of the National Academy of Sciences of the United States of America 97:2974–2978

Dahan M, Lévi S, Luccardini C, Rostaing P, Riveau B, Triller A (2003) Diffusion dynamics of glycine receptors revealed by single-quantum dot tracking. Science 302:442–445

Dayal S, Burda C (2008) Semiconductor quantum dots as two-photon sensitizers. Journal of the American Chemical Society 130:2890–2891

Delehanty JB, Medintz IL, Pons T, Brunel FM, Dawson PE, Mattoussi H (2006) Self-assembled quantum dot-peptide bioconjugates for selective intracellular delivery. Bioconjugate Chemistry 17:920–927

Demchenko AP (2002) The red-edge effects: 30 years of exploration. Luminescence 17:19–42

Deniz AA, Mukhopadhyay S, Lemke EA (2008) Single-molecule biophysics: at the interface of biology, physics and chemistry. Journal of the Royal Society Interface 5:15–45

De Schryver FC, Vosch T, Cotlet M, Van der Auweraer M, Mullen K, Hofkens J (2005) Energy dissipation in multichromophoric single dendrimers. Accounts of Chemical Research 38:514–522

De Serio M, Zenobi R, Deckert V (2003) Looking at the nanoscale: scanning near-field optical microscopy. Trac-Trends in Analytical Chemistry 22:70–77

Doi N, Yanagawa H (1999) Design of generic biosensors based on green fluorescent proteins with allosteric sites by directed evolution. FEBS Letters 453:305–307

Donnert G, Keller J, Medda R, Andrei MA, Rizzoli SO, Lurmann R, Jahn R, Eggeling C, Hell SW (2006) Macromolecular-scale resolution in biological fluorescence microscopy. Proceedings of the National Academy of Sciences of the United States of America 103:11440–11445

Doose S, Heilemann M, Michalet X, Weiss S, Kapanidis AN (2007) Periodic acceptor excitation spectroscopy of single molecules. European Biophysics Journal with Biophysics Letters 36:669–674

Fehr M, Frommer WB, Lalonde S (2002) Visualization of maltose uptake in living yeast cells by fluorescent nanosensors. Proceedings of the National Academy of Sciences of the United States of America 99:9846–9851

Fisher JA, Barchi JR, Welle CG, Kim GH, Kosterin P, Obaid AL, Yodh AG, Contreras D, Salzberg BM (2008) Two-photon excitation of potentiometric probes enables optical recording of action potentials from Mammalian nerve terminals in situ. Journal of Neurophysiology 99:1545–1553

Foquet M, Korlach J, Zipfel WR, Webb WW, Craighead HG (2004) Focal volume confinement by submicrometer-sized fluidic channels. Analytical Chemistry 76:1618–1626

Frangioni JV (2003) In vivo near-infrared fluorescence imaging. Current Opinion in Chemical Biology 7:626–634

Fulwyler M, Hanley QS, Schnetter C, Young IT, Jares-Erijman EA, Arndt-Jovin DJ, Jovin TM (2005) Selective photoreactions in a programmable array microscope (PAM): photoinitiated polymerization, photodecaging, and photochromic conversion. Cytometry A 67:68–75

Gabe Y, Ueno T, Urano Y, Kojima H, Nagano T (2006) Tunable design strategy for fluorescence probes based on 4-substituted BODIPY chromophore: improvement of highly sensitive fluorescence probe for nitric oxide. Analytical and Bioanalytical Chemistry 386:621–626

Gai HW, Griess GA, Demeler B, Weintraub ST, Serwer P (2007) Routine fluorescence microscopy of single untethered protein molecules confined to a planar zone. Journal of Microscopy-Oxford 226:256–262

Gao X, Dave SR (2007) Quantum dots for cancer molecular imaging. Advances in Experimental Medicine and Biology 620:57–73

Gao X, Cui Y, Levenson RM, Chung LW, Nie S (2004) In vivo cancer targeting and imaging with semiconductor quantum dots. Nature Biotechnology 22:969–976

Gautier A, Juillerat A, Heinis C, Correa IR Jr., Kindermann M, Beaufils F, Johnsson K (2008) An engineered protein tag for multiprotein labeling in living cells. Chemistry & Biology 15:128–136

Geigl JB, Uhrig S, Speicher MR (2006) Multiplex-fluorescence in situ hybridization for chromosome karyotyping. Nature Protocols 1:1172–1184

George N, Pick H, Vogel H, Johnsson N, Johnsson K (2004) Specific labeling of cell surface proteins with chemically diverse compounds. Journal of the American Chemical Society 126:8896–8897

Giepmans BNG, Adams SR, Ellisman MH, Tsien RY (2006) Review - the fluorescent toolbox for assessing protein location and function. Science 312:217–224

Grecco HE, Lidke KA, Heintzmann R, Lidke DS, Spagnuolo C, Martinez OE, Jares-Erijman EA, Jovin TM (2004) Ensemble and single particle photophysical properties (two-photon excitation, anisotropy, FRET, lifetime, spectral conversion) of commercial quantum dots in solution and in live cells. Microscopy Research and Technique 65:169–179

Gross E, Bedlack RS, Loew LM (1994) Dual-wavelength ratiometric fluorescence measurement of the membrane dipole potential. Biophysical Journal 67:208–216

Guignet EG, Hovius R, Vogel H (2004) Reversible site-selective labeling of membrane proteins in live cells. Nature Biotechnology 22:440–444

Hagen GM, Roess DA, de Leon GC, Barisas BG (2005) High probe intensity photobleaching measurement of lateral diffusion in cell membranes. Journal of Fluorescence 15:873–882

Hagen G, Caarls W, Thomas M, Hill A, Lidke K, Rieger B, Fritsch C, van Geest B, Jovin T, Arndt-Jovin D (2007) Biological applications of an LCoS-based Programmable Array Microscope (PAM). Proceedings of SPIE – International Society of Optical Engineering 6441:1–12

Hagen G, Lidke K, Rieger B, Caarls W, Arndt-Jovin D, Jovin T (2008) Dynamics of membrane receptors: single molecule tracking of quantm dot liganded epidermal growth factor. In: Ishii Y, Yanagida T (Eds.). Single Molecule Dynamics. Orlando: Wiley. in press

Hanley QS, Arndt-Jovin DJ, Jovin TM (2002) Spectrally resolved fluorescence lifetime imaging spectroscopy. Applied Spectroscopy 56:155–156

Haugland RP (2005) The handbook. A guide to fluorescent probes and labeling technologies. Tenth edition. Invitrogen corp, Eugene, OR

Haustein E, Schwille P (2007) Fluorescence correlation spectroscopy: novel variations of an established technique. Annual Review of Biophysics and Biomolecular Structure 36:151–169

Hoebe RA, Van Oven CH, Gadella TWJ, Dhonukshe PB, Van Noorden CJF, Manders EMM (2007) Controlled light-exposure microscopy reduces photobleaching and phototoxicity in fluorescence live-cell imaging. Nature Biotechnology 25:249–253

Huang B, Wu HK, Bhaya D, Grossman A, Granier S, Kobilka BK, Zare RN (2007) Counting low-copy number proteins in a single cell. Science 315:81–84

Jain KK (2007) Cancer biomarkers: current issues and future directions. Current Opinion in Molecular Therapeutics 9:563–571

Jaiswal JK, Goldman ER, Mattoussi H, Simon SM (2004) Use of quantum dots for live cell imaging. Nature Methods 1:73–78

Keppler A, Gendreizig S, Gronemeyer T, Pick H, Vogel H, Johnsson K (2003) A general method for the covalent labeling of fusion proteins with small molecules in vivo. Nature Biotechnology 21:86–89

Kim S, Lim YT, Soltesz EG, De Grand AM, Lee J, Nakayama A, Parker JA, Mihaljevic T, Laurence RG, Dor DM, Cohn LH, Bawendi MG, Frangioni JV (2004) Near-infrared fluorescent type II quantum dots for sentinel lymph node mapping. Nature Biotechnology 22:93–97

Klymchenko AS, Demchenko AP (2002) Electrochromic modulation of excited-state intramolecular proton transfer: the new principle in design of fluorescence sensors. Journal of the American Chemical Society 124:12372–12379

Klymchenko AS, Duportail G, Mely Y, Demchenko AP (2003) Ultrasensitive two-color fluorescence probes for dipole potential in phospholipid membranes. Proceedings of the National Academy of Sciences of the United States of America 100:11219–11224

Klymchenko AS, Stoeckel H, Takeda K, Mely Y (2006) Fluorescent probe based on intramolecular proton transfer for fast ratiometric measurement of cellular transmembrane potential. Journal of Physical Chemistry B 110:13624–13632

Koeppel F, Jaiswal JK, Simon SM (2007) Quantum dot-based sensor for improved detection of apoptotic cells. Nanomedicine 2:71–78

Komatsu K, Urano Y, Kojima H, Nagano T (2007) Development of an iminocoumarin-based zinc sensor suitable for ratiometric fluorescence imaging of neuronal zinc. Journal of the American Chemical Society 129:13447–13454

Krichevsky O, Bonnet G (2002) Fluorescence correlation spectroscopy: the technique and its applications. Reports on Progress in Physics 65:251–297

Lakowicz JR (2007) Principles of fluorescence spectroscopy. Springer, New York

Leevy WM, Gammon ST, Jiang H, Johnson JR, Maxwell DJ, Jackson EN, Marquez M, Piwnica-Worms D, Smith BD (2006) Optical imaging of bacterial infection in living mice using a fluorescent near-infrared molecular probe. Journal of the American Chemical Society 128:16476–16477

Lidke DS, Lidke KA, Rieger B, Jovin TM, Arndt-Jovin DJ (2005) Reaching out for signals: filopodia sense EGF and respond by directed retrograde transport of activated receptors. Journal of Cell Biology 170:619-626

Ma ZY, Gerton JM, Wade LA, Quake SR (2006) Fluorescence near-field microscopy of DNA at sub-10nm resolution. Physical Review Letters 97:260801

Mank M, Reiff DF, Heim N, Friedrich MW, Borst A, Griesbeck O (2006) A FRET-based calcium biosensor with fast signal kinetics and high fluorescence change. Biophysical Journal 90:1790–1796

Marks KM, Nolan GP (2006) Chemical labeling strategies for cell biology. Nature Methods 3:591–596

Marriott G, Clegg RM, Arndt-Jovin DJ, Jovin TM (1991) Time resolved imaging microscopy. Phosphorescence and delayed fluorescence imaging. Biophysical Journal 60:1374–1387

Martin BR, Giepmans BN, Adams SR, Tsien RY (2005) Mammalian cell-based optimization of the biarsenical-binding tetracysteine motif for improved fluorescence and affinity. Nature Biotechnology 23:1308–1314

Mattheakis LC, Dias JM, Choi YJ, Gong J, Bruchez MP, Liu J, Wang E (2004) Optical coding of mammalian cells using semiconductor quantum dots. Analytical Biochemistry 327:200–208

Meyvis TK, De Smedt SC, Van Oostveldt P, Demeester J (1999) Fluorescence recovery after photo-bleaching: a versatile tool for mobility and interaction measurements in pharmaceutical research. Pharmaceutical Research 16:1153–1162

Michalet X, Pinaud FF, Bentolila LA, Tsay JM, Doose S, Li JJ, Sundaresan G, Wu AM, Gambhir SS, Weiss S (2005) Quantum dots for live cells, in vivo imaging, and diagnostics. Science 307:538–544

Miyawaki A (2003) Fluorescence imaging of physiological activity in complex systems using GFP-based probes. Current Opinion in Neurobiology 13:591–596

Miyawaki A, Llopis J, Heim R, McCaffery JM, Adams JA, Ikura M, Tsien RY (1997) Fluorescent indicators for Ca2+ based on green fluorescent proteins and calmodulin. Nature 388:882–887

Miyawaki A, Griesbeck O, Heim R, Tsien RY (1999) Dynamic and quantitative Ca2+ measurements using improved cameleons. Proceedings of the National Academy of Sciences of the United States of America 96:2135–2140

Moerner WE (2007) New directions in single-molecule imaging and analysis. Proceedings of the National Academy of Sciences of the United States of America 104:12596–12602

Mueller AJ, Freeman WR, Schaller UC, Kampik A, Folberg R (2002) Complex microcirculation patterns detected by confocal indocyanine green angiography predict time to growth of small choroidal melanocytic tumors: MuSIC Report II. Ophthalmology 109:2207–2214

Mukhopadhyay S, Deniz AA (2007) Fluorescence from diffusing single molecules illuminates biomolecular structure and dynamics. Journal of Fluorescence 17:775–783

Nishigaki T, Wood CD, Shiba K, Baba SA, Darszon A (2006) Stroboscopic illumination using light-emitting diodes reduces phototoxicity in fluorescence cell imaging. Biotechniques 41:191–197

Ntziachristos V, Schellenberger EA, Ripoll J, Yessayan D, Graves E, Bogdanov A Jr., Josephson L, Weissleder R (2004) Visualization of antitumor treatment by means of fluorescence molecular tomography with an annexin V-Cy5.5 conjugate. Proceedings of the National Academy of Sciences of the United States of America 101:12294–12299

Ohnishi S, Lomnes SJ, Laurence RG, Gogbashian A, Mariani G, Frangioni JV (2005) Organic alternatives to quantum dots for intraoperative near-infrared fluorescent sentinel lymph node mapping. Molecular Imaging 4:172–181

Oliver AE, Baker GA, Fugate RD, Tablin F, Crowe JH (2000) Effects of temperature on calcium-sensitive fluorescent probes. Biophysical Journal 78:2116–2126

Petrovsky A, Schellenberger E, Josephson L, Weissleder R, Bogdanov A, Jr. (2003) Near-infrared fluorescent imaging of tumor apoptosis. Cancer Research 63:1936–1942

Puleo CM, Liu K, Wang TH (2006) Pushing miRNA quantification to the limits: high-throughput miRNA gene expression analysis using single-molecule detection. Nanomedicine 1:123–127

Rafalska-Metcalf IU, Janicki SM (2007) Show and tell: visualizing gene expression in living cells. Journal of Cell Science 120:2301–2307

Rajagopalan R, Uetrecht P, Bugaj JE, Achilefu SA, Dorshow RB (2000) Stabilization of the optical tracer agent indocyanine green using noncovalent interactions. Photochemistry and Photobiology 71:347–350

Rajh T (2006) Bio-functionalized quantum dots: tinkering with cell machinery. Nature Materials 5:347–348

Rasmussen A, Deckert V (2005) New dimension in nano-imaging: breaking through the diffraction limit with scanning near-field optical microscopy. Analytical and Bioanalytical Chemistry 381:165–172

Roberti MJ, Bertoncini CW, Klement R, Jares-Erijman EA, Jovin TM (2007) Fluorescence imaging of amyloid formation in living cells by a functional, tetracysteine-tagged alpha-synuclein. Nature Methods 4:345–351

Rogers KL, Stinnakre J, Agulhon C, Jublot D, Shorte SL, Kremer EJ, Brulet P (2005) Visualization of local Ca2+ dynamics with genetically encoded bioluminescent reporters. European Journal of Neuroscience 21:597–610

Sato M, Hida N, Umezawa Y (2005) Imaging the nanomolar range of nitric oxide with an amplifier-coupled fluorescent indicator in living cells. Proceedings of the National Academy of Sciences of the United States of America 102:14515–14520

Sauer M (2003) Single-molecule-sensitive fluorescent sensors based on photoinduced intramolecular charge transfer. Angewandte Chemie-International Edition in English 42:1790–1793

Scheenen WJ, Makings LR, Gross LR, Pozzan T, Tsien RY (1996) Photodegradation of indo-1 and its effect on apparent Ca2+ concentrations. Chemistry & Biology 3:765–774

Seisenberger G, Ried MU, Endress T, Buning H, Hallek M, Brauchle C (2001) Real-time single-molecule imaging of the infection pathway of an adeno-associated virus. Science 294:1929–1932

Shaner NC, Campbell RE, Steinbach PA, Giepmans BNG, Palmer AE, Tsien RY (2004) Improved monomeric red, orange and yellow fluorescent proteins derived from Discosoma sp red fluorescent protein. Nature Biotechnology 22:1567–1572

Sharonov A, Hochstrasser RM (2006) Wide-field subdiffraction imaging by accumulated binding of diffusing probes. Proceedings of the National Academy of Sciences of the United States of America 103:18911–18916

Sharrna P, Brown S, Walter G, Santra S, Moudgil B (2006) Nanoparticles for bioimaging. Advances in Colloid and Interface Science 123:471–485

Shav-Tal Y, Darzacq X, Shenoy SM, Fusco D, Janicki SM, Spector DL, Singer RH (2004) Dynamics of single mRNPs in nuclei of living cells. Science 304:1797–1800

Shynkar VV, Klymchenko AS, Duportail G, Demchenko AP, Mely Y (2005) Two-color fluorescent probes for imaging the dipole potential of cell plasma membranes. Biochimica et Biophysica Acta 1712:128–136

Shynkar VV, Klymchenko AS, Kunzelmann C, Duportail G, Muller CD, Demchenko AP, Freyssinet JM, Mely Y (2007) Fluorescent biomembrane probe for ratiometric detection of apoptosis. Journal of the American Chemical Society 129:2187–2193

Smiley RD, Hammes GG (2006) Single molecule studies of enzyme mechanisms. Chemical Reviews 106:3080–3094

Smith AM, Ruan G, Rhyner MN, Nie SM (2006) Engineering luminescent quantum dots for In vivo molecular and cellular imaging. Annals of Biomedical Engineering 34:3–14

So MK, Xu CJ, Loening AM, Gambhir SS, Rao JH (2006) Self-illuminating quantum dot conjugates for in vivo imaging. Nature Biotechnology 24:339–343

Spagnuolo CC, Vermeij RJ, Jares-Erijman EA (2006) Improved photostable FRET-competent biarsenical-tetracysteine probes based on fluorinated fluoresceins. Journal of the American Chemical Society 128:12040–12041

Stefflova K, Chen J, Zheng G (2007) Killer beacons for combined cancer imaging and therapy. Current Medicinal Chemistry 14:2110–2125

Straub M, Lodemann P, Holroyd P, Jahn R, Hell SW (2000) Live cell imaging by multifocal multiphoton microscopy. European Journal of Cell Biology 79:726–734

Summerer D, Chen S, Wu N, Deiters A, Chin JW, Schultz PG (2006) A genetically encoded fluorescent amino acid. Proceedings of the National Academy of Sciences of the United States of America 103:9785–9789

Thompson RB, Cramer ML, Bozym R (2002) Excitation ratiometric fluorescent biosensor for zinc ion at picomolar levels. Journal of Biomedical Optics 7:555–560

Tinnefeld P, Sauer M (2005) Branching out of single-molecule fluorescence spectroscopy: challenges for chemistry and influence on biology. Angewandte Chemie-International Edition 44:2642–2671

Tomlinson ID, Warnerment MR, Mason JN, Vergne MJ, Hercules DM, Blakely RD, Rosenthal SJ (2007) Synthesis and characterization of a pegylated derivative of 3-(1,2,3,6-tetrahydro-pyridin-4yl)-1H-indole (IDT199): a high affinity SERT ligand for conjugation to quantum dots. Bioorganic & Medicinal Chemistry Letters 17:5656–5660

Tsien RY (1998) The green fluorescent protein. Annual Review of Biochemistry 67:509–544

Tsukube H, Yano K, Ishida A, Shinoda S (2007) Lanthanide complex strategy for detection and separation of histidine-tagged proteins. Chemistry Letters 36:554–555

Van Munster EB, Kremers GJ, Adjobo-Hermans MJ, Gadella TW Jr (2005) Fluorescence resonance energy transfer (FRET) measurement by gradual acceptor photobleaching. Journal of Microscopy 218:253–262

Voronin YM, Didenko IA, Chentsov YV (2006) Methods of fabricating and testing optical nanoprobes for near-field scanning optical microscopes. Journal of Optical Technology 73:101–110

Voura EB, Jaiswal JK, Mattoussi H, Simon SM (2004) Tracking metastatic tumor cell extravasation with quantum dot nanocrystals and fluorescence emission-scanning microscopy. Nature Medicine 10:993–998

Walkup GK, Imperiali B (1997) Fluorescent chemosensors for divalent zinc based on zinc finger domains. Enhanced oxidative stability, metal binding affinity, and structural and functional characterization. Journal of the American Chemical Society 119:3443–3450

Webb SED, Needham SR, Roberts SK, Martin-Fernandez ML (2006) Multidimensional single-molecule imaging in live cells using total-internal-reflection fluorescence microscopy. Optics Letters 31:2157–2159

Webster A, Coupland P, Houghton FD, Leese HJ, Aylott JW (2007) The delivery of PEBBLE nanosensors to measure the intracellular environment. Biochemical Society Transactions 35:538–543

Westphal V, Lauterbach MA, Di Nicola A, Hell SW (2007) Dynamic far-field fluorescence nanoscopy. New Journal of Physics 9: doi:10.1088/1367-2630/9/12/435

Willig KI, Harke B, Medda R, Hell SW (2007) STED microscopy with continuous wave beams. Nature Methods 4:915–918

Xie XS, Yu J, Yang WY (2006) Perspective - living cells as test tubes. Science 312:228–230

Yang Z, Zhou DM (2006) Cardiac markers and their point-of-care testing for diagnosis of acute myocardial infarction. Clinical Biochemistry 39:771–780

Zaheer A, Wheat TE, Frangioni JV (2002) IRDye78 conjugates for near-infrared fluorescence imaging. Molecular Imaging 1:354–364

Chapter 12
Opening New Horizons

It is expected that in the XXI century sensor science and technology will address two major problems that present a great challenge. One is related to the biological world and, particularly, to the analysis of genomic and proteomic information. Possessing this information is an important step for starting active operations with it. This will open new, presently even unpredictable, possibilities in many areas: victory over human diseases, safety of the environment, superior agricultural products. The other problem refers to the chemical world and is related to new synthetic products and new materials. Millions of new compounds are synthesized by the chemical industry and many of them find industrial application and finally improve our lives. However, the reactivity of many of them is largely unknown, neither is their safety in production and use and their degradation as waste products. Therefore, evaluation of their properties and their quantitative assay should go together to provide optimal use and optimal protection.

In this chapter we will concentrate on the most rapidly developing application areas and try to make prognoses for future developments. The problem of lifting the limitations on the detection of any required target will be reviewed. This will allow putting forward an ambitious task of simultaneous detection of not only the whole genome (which is tens of thousands of genes presented by DNA sequences) but also of proteomes comprising all the results of their expression as synthesized proteins. Their spatial location and interactions are coming into the mainstream of present research. The sensors of an emerging generation, responding to strong demands from medicine and pharmacology, will be reviewed. With the appearance of this new level of personal diagnostics, it is not only an *improvement in health care* that is expected but also an *influence on everyday life*, in a profound way. Monitoring all kinds of industrial processes and all aspects of the human environment – all this will have a deep influence on human society, which will become a 'sensorized' society.

12.1 Genomics, Proteomics and Other 'Omics'

The recent completion of the human *genome* mapping, which contains about 30,000 genes, was achieved with the aid of fluorescence–based techniques. Now this research continues in the assessment of the genomes of individuals, finding

A.P. Demchenko, *Introduction to Fluorescence Sensing*,
© Springer Science+Business Media B.V. 2009

mutations and markers of diseases. Fully or in part, the genomes of many microbes, plants and animals have become available. Genomics has become a broad field for studying the organism's entire genome and the identification of its variations, including those that may lead to inherited diseases.

The first step of the expression of this information is the *transcriptome*, which is the set of all messenger RNA (mRNA) molecules, or "transcripts", produced in one or a population of cells. By using high-throughput techniques based on DNA microarray technology, the study of the expression level of mRNAs (transcriptomics) can be provided. The transcriptome can vary with the functional state of the cell and under the action of external stimuli.

Genomic information is not sufficient to understand the functioning of living cells. We have to understand the roles of all proteins and their mutual interactions in a given organism. Therefore, completing its *proteome* is a new, formidable task that will require new approaches and tools. Proteomics is the large-scale study of proteins, particularly of their structures and functions. The number of synthesized proteins in a human body is much larger than the genes; it is estimated to be on the level of one million. Such protein diversity is due to their alternative splicing and post-translational modifications. The amount of particular protein does not correlate directly with the amount of coding mRNA, so the expression level of every protein should be determined in the analysis of proteomes. Proteomics is much more complicated than genomics because also, in contrast to genome constancy, a proteome differs from cell to cell and exhibits substantial changes upon perturbations on a cellular level.

The terms metabolome and interactome have become increasingly popular. *Metabolome* represents metabolic profiling, the collection of all metabolites, which are the end products of its gene expression, into a biological organism. It also includes the consumed substances and those produced by symbiotic microorganisms. *Interactome* is the whole set of molecular interactions in cells. It includes the protein-protein interaction network and interaction networks with DNA, RNA and different metabolites. The study of metabolomes and interactomes produces a further dramatic increase in the massive amount of information to be collected.

Understanding complex functional mechanisms requires the global and parallel analysis of different cellular processes on several structural levels, schematically presented in Fig. 12.1.

A new epoch is approaching in which this information will be the most actively used in medicine, agriculture, environmental protection and other areas. Fluorescence sensing has become one of the major tools for exploring these new possibilities in full.

12.1.1   Gene Expression Analysis

After the sequencing of the human genome was achieved, DNA microarrays became the major tools in *gene-expression studies*. The level of synthesized

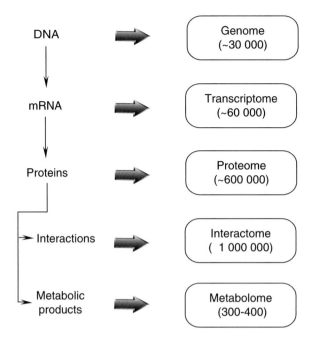

Fig. 12.1 Schematics of the flow of information from the genes to functional constituents of the living cell and the relations between genome, proteome, interactome and metabolome

specific mRNA sequences on DNA templates is not only a reflection of genome but also a reflection of the functional state of the cell and the whole organism. This information is called the *gene expression profile* – the pattern of gene expression produced by the studied sample. Genes exhibit different expression levels at different stages of growth, cell cycle, differentiation, response to stimulation, etc. Therefore, the quantitative determination of specific RNA molecules from biological samples is important for the study of regulated gene expression, which may indicate changes in many metabolic processes and in the synthesis of many proteins using mRNAs as synthetic templates. Expression profiles may help to understand how these processes occur and how, on the background of a normal process, the disease may start. In addition, RNA detection and quantification present an efficient strategy for the rapid identification of unknown biological agents of bacterial and viral origin.

The standard microarray technology, also known as DNA chips, allows obtaining the pattern of mRNA molecules synthesized in the cell, which can be referred to as *transcriptome profiling*. It involves the production of an amplified number of DNA copies of mRNAs, their labeling with a fluorescent dye and their deposition on the DNA array formed by single-chain oligonucleotides (Section 10.4). Commonly, it allows only the identification of the difference in composition between two samples that can be labeled with two dyes of different colors and hybridized together. Such a *differential expression pattern* is obtained based on the two-color response from every array spot representing competitive binding of the DNA of two samples. Polychromatic microarrays based on a larger number of colors offered by Quantum Dots have also been suggested (Shepard 2006). Meantime, the reported simultaneous hybridization of eight samples is probably the limit that can be reached with the basic pool-labeling technology.

The sandwich hybridization platform has a larger potential for development, since it is not limited to comparative studies. This potential is harder to achieve, reaching a comparative level of scalability with double the size of recognizing sequences. Meantime, there is a strong demand for a dramatic increase of the sensor scale in microarray technology. This requires the assembly of a great number of sensor sites responsive to different analytes. If we take as a reference the human genome content of 30,000 genes, some applications may require much smaller capacity but for some – the increased capacity also needs to include the genes of harmful microbes, viruses, etc. This justifies the search for new analytical methods.

Thus, for many applications the problem of determining the expression levels of all the genes coded by the genome can be reduced to determining a smaller number of characteristic genes. When these special genes are screened-out from the human genome, the analytical procedure with them becomes more efficient. They become particular targets in gene expression analysis and corresponding sensor sequences have to be spotted on a microarray. For reliable identification of these genes, micro-arrays monitoring the expression of 10^3–10^4 genes simultaneously in one hybridization experiment can be sufficient. By comparing a series of such experiments, one can construct the profiles of differentially expressed genes.

Regarding total transcriptome analysis, new solutions have to be found. It is expected that they will involve a further dramatic reduction of spot sizes, down to 100 nm and less, to produce *nanoarrays*. Their operation could be based on different technologies. Attractive in this respect is a fiber optic platform with the highest packing density of any array format (Lafratta and Walt 2008).

12.1.2 The Analysis of Proteomes

Proteome profiling is the identification of all proteins expressed in a sample. The proteome cannot be represented in full by the transcriptome, since the correlation between the extent of determined mRNA and that of synthesized (and active) proteins in a cell is not linear and the protein lifetimes in a cell vary over a broad range of values. Moreover, the proteins need to pass many post-translational modifications, such as phosphorylation, glycosylation or oxidation, which finally determine their functional activity. Therefore, in the protein world, not only the expressed proteins but also their post-translationally modified forms may need to be determined.

Presently, the leading positions in the study of proteomes are occupied by the methods of 2D electrophoresis, chromatography and mass spectrometry. Meantime, the protein diversity, estimated as ten billion species on Earth (Frishman 2007), needs new methods. They have to allow an extremely high throughput and have a comparative analysis capability and it is expected that fluorescent protein arrays will be developed to an extent satisfying the needs of researchers. It is desirable that the new methods should allow obtaining, in a parallel way, a very high capacity of information, which requires a high density of location of the sensor units. Such methods still do not exist and current attempts to transfer the technology developed in the hybridization of DNA are not very successful. The major difficulties are the following:

(a) Each protein has a unique chemistry and biochemistry. Unlike nucleic acids, there is no simple possibility for molecular recognition based on the complementarity of target and sensor receptor sequences.
(b) There are no possibilities to achieve a homogeneous labeling of proteins with a standard protocol. A broad variation in the extent of labeling is observed when the labeling reagent is introduced into the protein pool.
(c) In a cell, blood or any other biosystem, the divergence of concentrations between individual proteins is tremendous, covering six orders of magnitude and more.
(d) An even larger divergence exists in protein affinities towards their natural substrates and ligands. These affinities do not correlate with the protein size or the number of groups available for intermolecular interactions.

Still, there are many attempts to develop 'protein chips'. One of operational concepts for this trend is borrowed from DNA hybridization assays. It is the comparison of proteomic maps of healthy and diseased cells (Borrebaeck 2000). It uses broad-scale detection based on antibodies as the recognition units (Fig. 12.2). Such antibody arrays are applied for specific tasks, for instance, for the identification of disease biomarkers

Such assays should not be blind; they have to become expandable to the whole proteome or to its essential parts. Because of this, they should be preceded by an analysis of protein content in the studied system. This can be done by high-resolution protein separation (*2D electrophoresis*), identification of the obtained protein fractions (*mass spectrometry*), screening for functional activity and bioinformatics data analysis, supported by statistics (Freire and Wheeler 2006).

The combination of the separation and analytical techniques is a general approach for the elucidation of the whole proteome. However, the practical use of this information does not necessarily need the determination of hundreds of thousands of components. The detection of specific proteins or some number of them is the most frequently requested. A specific binder to a specific protein can then be developed and incorporated into an array. This long and difficult method takes a lot of effort. In the long run, it will allow scientists to understand cell signaling and metabolic networks, which will further stimulate the rapid development of future therapeutics.

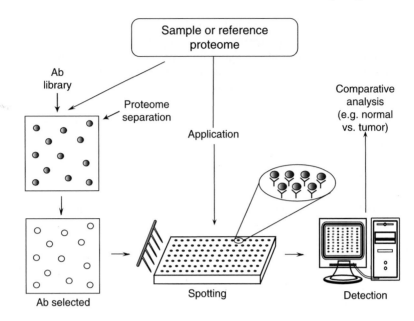

Fig. 12.2 Scheme illustrating one of the strategies for comparative global analysis of the proteome based on the antibody technique (Borrebaeck 2006). The proteomes derived from normal and tested (e.g., malignant) tissues are separated. These proteomes are used to select the antibodies, subsequently spotted on the chips. The selection is based on the availability of a highly functional phage display antibody library. The samples to be analyzed are then allowed to react with the functionalized microarray, with subsequent read-out and analysis

On this pathway, the selection of specific binders (receptors) for the recognition of every protein on a proteomic scale is extremely important. This issue is actively discussed in the literature (Taussig et al. 2007). In evaluating the technologies for the production of the binders that could satisfy the demand, it is argued that it will not be the antibodies but a number of different binders that will allow *in-vitro* evolution (Fig. 12.3).

Systems based on the principle of *recombinant selection in vitro* link the genotype and phenotype of the binding molecule. In contrast to antibodies, they possess important features, such as:

(a) The absence of any limit on the scope of targets, which for antibodies is limited by the response of the immune system.
(b) The ability to achieve, by multiple mutagenesis-selection rounds, binders that can be matured to the often necessary picomolar affinity.
(c) The ability to make the binders polyvalent and polyfunctional. Functional domains and tags can be fused to allow combining the binding abilities to different targets or for binding targets together with assembly to a supramolecular structure or immobilization on a support.

It is believed that this will allow establishing a comprehensive, characterized and standardized collection of specific binders directed against all individual human

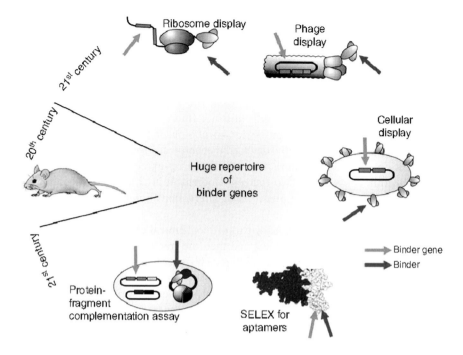

Fig. 12.3 The methods for generating proteomic-scale binders prospective for technologies of the XXI century (Reprinted with permission from Taussig et al. 2007)

proteins, including different variant forms and modifications. Establishing a binder collection will allow the development of a new generation of high-throughput assay systems and protein-detection technologies.

To be able to address the protein function, expression and localization on a proteome-wide scale efficiently, a lot of effort has been and will be directed toward the development of miniaturized and parallel assays (Uttamchandani et al. 2006). It is probably not the antibodies but the synthetic peptides and oligo-nucleotide aptamers selected from very large libraries (see Chapter 5) that will offer the solution to the problem of sensing on a scale of hundreds of thousands of species.

12.1.3 Addressing Interactome

Intermolecular interactions between receptors and their ligands, nucleic acids and proteins, as well as between various proteins, are tightly regulated within living cells. They compose innumerable molecular-level mechanisms that perform all biochemical processes and provide their regulation. Therefore, the identification of particular molecular interactions is always a crucial step in cellular research. When improperly established, these interactions lead to pathology; they are the points of application of various drugs. In this respect, classical biochemical *in vitro* studies are not sufficient. The cellular environment cannot adequately be modeled *in vitro* and the localization of different functions in the cell is extremely important.

The straightforward technique to track these interactions is to label two molecules of interest with spectrally distinct fluorescent markers and to study the cell images using either confocal, two-photon or total internal reflectance fluorescence microscopy, with spectrally separated image acquisition in two detection channels. The interactions are then revealed by observing signal co-localization between the two images (Comeau et al. 2006). This information can be considered as a strong predictor of interactions but still not proof of their existence.

There were times when the major part of biochemical experiments started from the preparation of cell homogenates and then the fractions containing individual molecules were extracted from this mixture. The information about intermolecular interactions (except those forming cell organelles) has been lost. Presently, all efforts are being made to localize the molecules inside the cells, to estimate their intracellular amount and to find their interaction partners.

A general approach for addressing this problem is absent. What is presently done is the *double labeling* of potential interaction partners. Their interaction can then be manifested from their co-localization in microscopy or from the observation of FRET. The labeling of both partners can be accomplished in different ways. Presently, two strategies are popular. (a) Fusion of the partner proteins with genetically incorporated emitters (visual fluorescent proteins and luciferase). (b) Affinity labeling with cell-permeating dyes targeted to genetically engineered sequences incorporated into partner proteins.

The combination of these two approaches is often employed. They are described in some depth in Section 11.3. Meantime, there are alternative possibilities that are less frequently used. They are based on genetic encoding into potentially interacting partners of complementary parts of a fluorescent or enzyme protein (*split proteins*). In this case, the fluorescent moiety or the catalytic activity generating a fluorescent product, can be formed on their self-assembly (Piehler 2005) and their detection could witness molecular interactions inside the cell.

The most popular mechanism that is presently employed for detecting the interaction between the fluorescent partners is FRET. This version, based on the bioluminescent enzyme luciferase as the FRET donor (Section 8.2), is becoming increasingly popular (Pfleger and Eidne 2006).

The networks of protein interactions mediate cellular responses to environmental stimuli and direct the execution of many different cellular functional pathways. Small molecules synthesized within the cells or incoming from the external environment mediate many protein interactions. The study of small molecule-mediated interactions of proteins is important to understand abnormal signal transduction pathways in cancer; they are important for drug development and validation. Combining the FRET response with flow cytometric analysis is one of the pathways for achieving an efficient and versatile methodology (Chan and Holmes 2004).

Despite great progress, structural information on the majority of proteins in the proteome is still unknown, which limits the application of assays based on genetic engineering. Therefore, the study of interactions *in vitro* by presently established methods remains actual. Meantime, it is currently impossible to design a huge amount of molecules that could constitute arrays for their detection and for the characterization of their interactions. The high throughput screening (HTS) approach that will be discussed in Section 12.4 in relation to drug discovery does not require prior information on the protein structure and function and can be extensively used in interactome studies. The whole yeast proteome was deposited on a microarray and many unexpected interactions were found upon screening for their ability to interact with proteins and phospholipids (Zhu and Snyder 2003). These studies are expected to be extended to larger proteomes. Microarrays can also be used to screen protein-drug interactions and to detect posttranslational modifications.

12.1.4 Outlook

It is hard to imagine how challenging the goal of complete characterization of transcriptome, proteome and interactome is, in view of the massive amount of information that has to be obtained and analyzed in parallel. But the process has started, the information is highly needed and the tools, which may be still on a basic level, already exist. The next generations of these tools may speed up progress tremendously. A broader scaling and higher level of miniaturization will definitely be required, up to the transition from microarrays to nanoarrays and further to manipulations with single molecules and their complexes.

Fortunately, the tools already exist for both collecting and processing large amounts of high-resolution data rapidly (Lafratta and Walt 2008). Electronic components such as CCD chips, CMOS devices and high-density integrated circuits provide the ability to collect enormous amounts of data on short time scales. Data storage capacity has increased to enable these data to be collected and stored.

Immense data sets, with hundreds of thousands to millions of data points, necessitate a new level of computer processing. Drawing meaningful correlations from data of this size is beyond human capability, therefore, new, powerful and user-friendly algorithms have to be developed.

Comparative analysis will continue to be an efficient approach to these studies, at least in the foreseeable future. Therefore, the introduction of 'standard' transcriptome, proteome and interactome will be needed for every type of human cell and on their background a new step in research will receive a new impulse.

12.2   Sensors to Any Target and to an Immense Number of Targets

In order to be ready to address new challenges and new goals, we need highly specific targeting of particular analytes together with the collection of a great mass of information about related and unrelated analytes. This information should be collected in a systematic way and in a reasonably short time, to provide timely decisions and actions.

12.2.1   The Combinatorial Approach on a New Level

The *trial-and-error* approach is still widely used for chemical sensor design, though our ability to predict the structural requirements for a perfect sensor for each analyte is limited and, in view of a great number of potential analytes, impossible to realize. Therefore, the implementation of the *combinatorial concept* remains the only possibility for the discovery of both the binding and fluorescent building blocks. This concept is based on the production of large libraries of compounds or devices and providing their multi-step selections. Only during the initial phase of establishing the library could the concepts of 'classical' *rational design* be applied but they do not lead directly to individual design and production (Lavigne and Anslyn 2001).

Many requirements have to be satisfied for the successful application of the combinatorial approach. First is the building of a *proper library*, the size of which should be large enough to incorporate successful hits and building blocks that are easy to construct and reproduce. Then, there should be an efficient *search algorithm* and its practical realization in screening techniques, an optimal methodology for evaluating successful hits, a formation of sub-libraries of a second generation, etc. The testing of successful candidates should be multiparametric, with the use of optimization algorithms. Despite all these difficulties, it should be realized that this is the only methodology foreseen to satisfy the task in the title of this section.

Combinatorial methodology has started to be successfully applied in various fields, such as drug discovery, catalysis, bimolecular interaction studies or discovery of sensor receptors (Gauglitz 2000). *Combinatorial chemistry* has revolutionized approaches to drug candidate synthesis and screening by the possibility of simultaneous testing of a great number of related compounds. These activities are stimulated by the pharmaceutical industry, which is oriented to the development of new drugs that are commonly the inhibitors of different metabolic pathways. Interestingly, the platforms that were originally exploited in the field of biosensors (such as microtiter plates, microarray technologies or fiber optic tips), have been applied to facilitate high-throughput drug screening (Section 12.3).

The other actual field of research is the development of new fluorescent dyes with improved reportive properties. This field was overviewed in Section 4.1.

The only remark that has to be made here is that the combinatorial concept is applicable in searches within quite different types of reporters, such as various fluorescent dyes with an ion-chelating function (Szurdoki et al. 2000). It is equally well applicable to the selection of support materials (Apostolidis et al. 2004).

Solid-phase organic synthesis is a well-established tool for the production of combinatorial libraries. Being mainly used for drug discovery, it now finds application for the design of resin-bound chemosensors. Thus, for the discovery of ATP binding sensors, the split and pool method has been used to generate a combinatorial library of more than 4,000 different resin-bound tripeptides (Schneider et al. 2000). After selecting the ATP binding receptor, the fluorophores were appended to the ends of the peptide chain in order to produce a reporter signal (Fig. 12.4).

Antibodies can be considered as the products of directed evolution 'in a body' (Section 5.2). They develop guided by a particular target, antigen, in several amplification-selection steps. Presently, they are the most available molecules that can allow the detection of a huge variety of targets. The SELEX procedure used in the

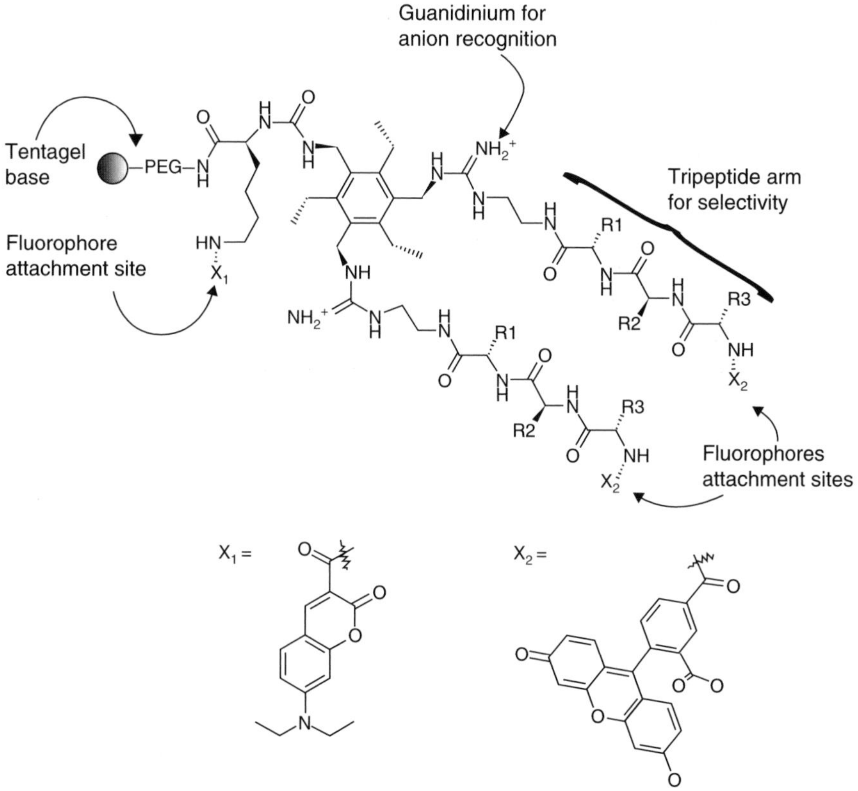

Fig. 12.4 Example of some probes for ATP sensing based on resin-bound tripeptides generated by combinatorial methods (Reprinted with permission from Schneider et al. 2000)

generation of aptamers is based on the same selection-amplification principle but is performed by human hand (Section 5.5). This principle can be further explored. New tools are needed in conditions in which old tools such as antibodies have reached the limit of their perfection.

The use of so-called *protein scaffolds* (Section 5.3) for the generation of novel binding proteins via combinatorial engineering has recently emerged as a powerful alternative to natural or recombinant antibodies. This concept requires an extraordinarily stable protein architecture, tolerating multiple substitutions or insertions at the primary structural level. With respect to broader applicability, it should involve a type of polypeptide fold, which is observed in differing natural contexts and with distinct biochemical functions, so that it is likely to be adaptable to novel molecular recognition purposes. It is expected that such artificial receptor proteins should be based on monomeric scaffolds and small polypeptides that are robust, easily engineered and efficiently produced in inexpensive prokaryotic expression systems.

Today, the progress in protein library technology allows for the parallel development of both antibody-clone and scaffold-based affinity units. Both biomolecular tools have the potential to complement each other, thus expanding the possibility to find an affinity reagent suitable for a given application. The repertoire of protein scaffolds hitherto recruited for combinatorial protein engineering purposes will probably be further expanded in the future, including both additional natural proteins and *de novo* designed proteins, contributing to the collection of libraries available at present (Taussig et al. 2007).

We also note that combinatorial approaches have started to be used to increase the performance of diagnostic devices for both clinical and field uses with optimization of the level of performance (Moats and Sullivan 2004). This offers substantial improvements in the detection of many analytes. The research efforts in this direction are motivated by a general need to detect as many different pathogens or hazardous materials as there may exist in the studied medium, using the smallest, most inexpensive and fastest system possible. These devices would be powerful if the effective library schemes were invented. It is important that the mathematical frameworks be more and more actively used in the development of novel combinatorial biosensor systems, for optimization and performance prediction. In addition to providing a linear response to particular targets, the novel sensors will be able to perform logical operations, as described in Section 6.5.

12.2.2 Toxic Agents and Pollutants Inconvenient for Detection

It could be primarily mentioned that the present researchers in sensor development favor addressing one of two types of targets, 'convenient' and 'needed'. Convenient are those that demonstrate in the simplest way the advantages of newly developed technology. Needed are those for which there is a range of potential applications and market. We will not discuss the cases 'convenient but not needed' but concentrate on those that are highly needed but very inconvenient. One of them is the sensor for *dioxin*, which is a potent industrial pollutant, producing a strong cumulative effect on human health. Until recently, this toxic agent was determined only by the combination of mass spectrometry and gas chromatography, which is time-consuming and very costly. The problem is the extremely low solubility of this compound in common solvents and the absence of the ability to form strong and specific non-covalent bonds that could be efficiently used in recognition. The solution of this problem was found recently by synthesizing a pentapeptide for its highly sensitive on-bead detection assay (Nakamura et al. 2005).

To find this peptide, a full peptide library consisting of 2.5 million possible amino acid combinations was constructed by a solid-phase split synthesis approach, using 19 natural amino acids. The beads with immobilized peptides were then subjected to a competitive binding assay using 2,3,7-trichlorodibenzo-p-dioxin and N-NBD-3-(3′,4′-dichlorophenoxy)-1-propylamine (NBD-DCPPA) as a competitor. Two almost identical pentapeptides, FLDQI and FLDQV, which could bind dioxin were identified from the combinatorial library. The peptides synthesized on resin beads in combination with NBD-DCPPA constitute an attractive assay to determine dioxin concentrations. The fluorescence intensity of the beads measured using fluorescence microscopy reports on the dioxin-dependent binding of the fluorescent competitor to the beads, after making a calibration curve for the dioxin concentrations. During the next step, to optimize the peptide sequence, a one-amino acid-substituted library was prepared with the inclusion of nonnatural amino acids. It was found that the amino acids internal in sequence, LDQ, could not be substituted without loss of affinity, which indicates their essential role in the recognition of dioxins. From this result one may derive an unexpectedly strong potential for short peptides as practical sensor materials, targeting low molecular weight compounds such as the 'inconvenient' dioxin.

This detection system was tested on real environmental samples (Inuyama et al. 2007). About 0.5 nM (150 pg ml⁻¹) of the 2,3,7,8-tetrachlorodibenzo-p-dioxin (2,3,7,8-TeCDD) could be detected under optimized conditions.

On a general scale, *endocrine disruptors* represent a large family of industrial pollutants that in extremely low concentrations may bring irreversible changes in human health and also affect domestic animals and wildlife. They have to be identified and determined at extremely low concentrations in the environment and in the human body (Rodriguez-Mozaz et al. 2004). Presently, tens of thousands of chemicals are under consideration for screening as potential endocrine disruptors, so it is essential that rapid, sensitive and reproducible high-throughput screening systems targeting them be developed. This is a difficult but feasible task.

12.2.3   The Problem of Coding and Two Strategies for Its Solution

The rapid, multiplexed, sensitive and specific simultaneous detection of immense numbers of analytes is in great demand for gene profiling, drug screening, clinical diagnostics and environmental analysis. In all systems that serve for this purpose, the individual result on a particular analyte binding-reporting should be recognizable from other results. Therefore, it is necessary to supply each sensor element with a different 'barcode'. There are two possibilities for doing this (Fig. 12.5).

One is *positional encoding*. It requires providing spatial resolution in the location of the sensor elements, so the coding is based on the *spatial location of sensors* deposited on a solid support. This possibility is used in spotted microarray technologies (Section 9.3). In this case, the sensor elements should be necessarily immobilized on the surface and spatial resolution is needed in reading the response signal. This approach is good with all 'responsive' sensing methodologies. If 'irresponsive' labeling is applied, there are two possibilities. First, an *additional washing* operation is needed to remove the unbound reporter dye. Presently, this is commonly applied. The other, more progressive method, offers excitation of only those reporters that are *located in a pre-surface layer* by exploring electrochemiluminescence (Section 8.1), evanescent wave excitation

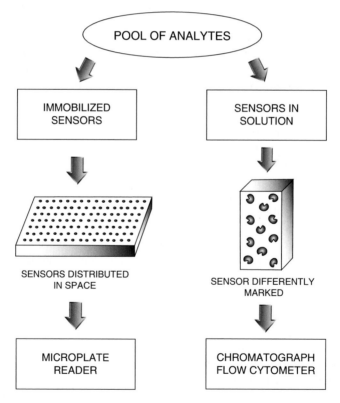

Fig. 12.5 The two strategies of simultaneous detection in multi-sensor systems. (A) Positional encoding. The sensor unit is recognized by its spatial location (B) Optical encoding. The sensor unit is recognized by the recognition element, 'barcode'

(Section 8.5) or plasmonic enhancement upon interaction with a metal surface (Section 8.6). Meantime, this strategy is not applicable or convenient if the target size is larger than the pre-electrode area or layer excited by the evanescent wave (such as the living cells) and if the response is limited to slow target diffusion to the surface. It is also not beneficial if the response units are large particles themselves. This can cause steric difficulties and does not allow exploring the particle surface in full.

The other possibility is to provide *sensing in solutions*, so that many sensor molecules or particles are dissolved together in an analyzed system. The reaction of target binding in solutions has its own advantages (such as faster diffusion rates) but without additional coding, the resolution of the fluorescent signal from different sensors in the system is commonly insufficient. In this case, every type of sensor unit should contain a specific fluorescent 'barcode'. This is the strategy of *optical encoding*. Following this strategy, we have to be able to read a huge number of different barcodes. The barcode can be connected or not connected with a specific response to target binding. The best, for the purpose of selective detection, is flow cytometry and its microscopic versions (possibly realized in microflow devices), in which the sensor particles pass through a detector in line, one by one and during this passage, different types of reading signals can be obtained. With flow cytometry, barcodes based on fluorescence intensity can be used efficiently. If particles with two or three emissions are used, this expands the possibility of barcode recognition to many millions (Wang 2000). In some simple cases, other methods based on chromatographic separation can also be used.

In parallel with barcoding the sensor units, the problem with the reporter signal should be solved. For every sensor, the two signals identifying the sensor and identifying target binding to this sensor should not interfere.

The two strategies depicted in Fig. 12.5 have developed into two different types of technologies of multiplex analysis. One is *spotted arrays* on a solid support and the other *suspension arrays* with barcoding. They develop in parallel, satisfying increasing demands for practical use.

12.2.4 Prospects

In this section we outlined the areas in which rapid progress in the nearest future is expected in response to a strong demand for more efficient sensors. The reader must be convinced that during the present phase of research and technology development, there is already the technical possibility to satisfy any demand for detecting any target and any (reasonable) number of targets. In almost all areas where sensor technologies are applied, our present ability to obtain the results is ahead of our ability to properly analyze them. The proper balance in this respect is expected to be achieved soon.

12.3 New Level of Clinical Diagnostics

Revolutionary changes are emerging in clinical diagnostics, in which the methods based on fluorescence detection have already become the most frequently used (Jain 2007a).

Invasive techniques have to be substituted for *noninvasive* or *low invasive* ones. The traditional samples of blood and urine taken for analysis can be strongly reduced in size. Together with the possibility of dramatic miniaturization, the biological fluids available in small quantities, such as microscopic skin cuts, saliva, sweat and tears have become available for diagnostics. Multiplexing techniques allow sensing many targets together and thus determine many clinically useful parameters. Their combined evaluation leads to a more reliable diagnosis and clinical decisions.

The ideal diagnostic instrument would be a device that is simple to use, hand-held and inexpensive, combining high specificity with sensitivity in the presence of a dense organic material such as blood and capable of detecting simultaneously more than one agent. The technological progress described in this book allows predicting the appearance of such instruments in a broad-scale use already in the near future. In this respect, of special interest are the problems that, at present, seriously limit these applications. They are the speed of the analysis and sensing in whole blood. We will discuss their solutions and after that, concentrate in particular on the interesting problem of disease biomarkers.

12.3.1 The Need for Speed

An important factor is the duration of the analytic procedure from sample application to obtaining the result. Direct sensing can allow its dramatic reduction by eliminating the steps additional to that of target binding. The miniaturization of the system can further increase the speed by reducing the time for target diffusion. Meantime, even in these cases the analysis may take many minutes and hours. This is because most biological recognition events, such as the antigen–antibody interactions and DNA hybridization, are kinetically slow. Time is needed not only for target diffusion in space but mostly for the structural adaptation of two interacting partners (molecular recognition).

In this regard, a new platform technology based on the *microwave acceleration* of metal-enhanced fluorescence (described briefly in Section 8.6) deserves close attention. It has recently been introduced (Aslan and Geddes 2005). It combines the benefits of metal-enhanced fluorescence with the use of low-power microwaves to accelerate the sensor-target molecular recognition. The metal-enhancement significantly increases the fluorescence signals of fluorophores in close proximity (~10 nm) to metallic nanostructures. Its combination with low-power microwave heating leads to the development of assays that can be completed within a few seconds. This new model platform system can be potentially applied to many other important assays, such as to the clinical assessment of myoglobin (see below), where both speed and sensitivity is paramount for the assessment and treatment of acute myocardial infarction.

12.3.2 Whole-Blood Sensing

The fluorescence signal of interest is often corrupted by high background luminescence and often it is the major factor affecting the sensitivity of fluorescence-based medical diagnostic tests. Several approaches have been suggested to minimize the background signal caused by the sample matrix (e.g., serum or whole blood in clinical diagnostics). The most efficient of them are based on the application of the time-resolved or time-gated emission of luminophores with long lifetimes (Rodriguez-Diaz et al. 2003) or two-photon excitation (Baker et al. 2000). Because of the high optical density, light-scattering and fluorescence background of the blood, clinical tests commonly start from serum or plasma, in which the blood cells have been removed. Alternatively, washing steps have to be applied after the application of the sample (Von Lode et al. 2004). Therefore, there is a continuous and unequivocal need for the development of better reporter molecules and/or detection methods for the direct analysis of whole blood samples.

Several pathways can be outlined for this. Primarily, it is the improvement of methods based on *two-photon detection* and using *upconverting luminophores*. This allows decreasing the light-scattering background and the absorption of pigments. The second important possibility is to raise the reporter signal over the background by using emitters with a brighter response and longer lifetimes.

12.3.3 Testing Non-invasive Biological Fluids

Non-invasive sample collection is very important for the multiparametric monitoring of health status. *Human saliva* is an attractive body fluid for disease diagnosis because its collection is simple, safe and noninvasive. Whereas many point-of-care diagnostic methods already exist for blood-borne analytes, the development of similar saliva-based techniques is still in its infancy. However, the use of saliva has been hindered by the inadequate sensitivity of previously applied methods to detect the lower salivary concentrations of many constituents, compared to serum. Meantime, the information content of saliva-based testing is significant and is not limited to oral infectious diseases. It is known that the clinically relevant analyte concentrations in saliva mirror the tissue fluid levels. Furthermore, recent studies have recognized the relevance of its protein composition to the development and progression of oral (Samaranayake 2007), as well as systemic diseases, including HIV infection (Yapijakis et al. 2006).

It is expected that the comprehensive analysis and identification of the proteomic content in human whole saliva will not only contribute to the understanding of oral health and disease pathogenesis but also form a foundation for the discovery of saliva protein biomarkers for human disease detection (Hu et al. 2007). With the rapid development of proteomic technologies, saliva protein biomarkers will be developed for the clinical diagnosis and prognosis of human diseases in the future.

One such biomarker, whose measurement in saliva could potentially contribute in a significant manner to the understanding and diagnosis of disease, is the *C-reactive protein*. This is an important inflammation marker and cardiovascular disease predictor, which is derived from the liver and whose production is regulated by cytokines. It was found to be a stronger predictor of coronary heart disease than is a high-density lipoprotein cholesterol level. Microsystem-based diagnostics of this protein with the use of saliva is at the stage of active development (Christodoulides et al. 2005).

Diagnostically important concentrations of *glucose* can be measured not only in blood but in different body fluids, such as tears, saliva and sweat (Mitsubayashi et al. 1995). This opens many possibilities for designing simple and convenient sensors for the continuous monitoring of this and other metabolites.

12.3.4   Gene-Based Diagnostics

Gene-based molecular diagnostics is changing the practice of medicine and will continue to do so for the foreseeable future (Finan and Zhao 2007). Two major fields are actively explored now: *detection of pathogens* and finding *disease-related gene mutations*. The major underlying principle of these diagnostic tests is the comparison of two nucleic acid sequences for their presence in a tested sample. The sequence used for comparison is labeled and introduced into the system as described in Section 10.4. Microarray recordings of mRNA expression profiles allow not only detection but also the classification of tumors, based on a distinct set of signature genes that comprise differently expressed transcripts between the cancer types. The gene expression patterns and the factors that control them represent potential and potent targets for therapeutic intervention. The application of a similar approach to the diagnosis of other diseases is expected to have great prospects.

Automation appears to be the current trend for high-volume molecular testing of *infectious diseases*. Molecular profiling of various diseases using genomic or proteomic approaches opens up a molecule wonderland with the promise and emergence of new molecular-based testing. It will likely impact the practice of medicine and allow the introduction of highly needed personalized testing.

Once a complex pattern of gene expression valuable for diagnostics can be reduced to a few genes, biosensor-based detection becomes practical and advantageous for cancer clinical testing, since it is faster, more user-friendly, less expensive and less technically demanding than microarray or proteomic analyses. However, a significant technical development is still needed, particularly for protein-based biosensors.

12.3.5   Protein Disease Biomarkers

With the progress in proteome research, the finding of specific *protein biomarkers* (molecular signatures of disease) to diagnose the health status of individuals and disease treatments will rise to a new level of perfection. The results of the present studies show much promise for this. For some diseases, the characteristic biomarkers have been established and physicians can use them for diagnosis. These biomarkers can be produced either by the malignant tissues and cells themselves or by the body in response to the presence of disease. They include DNA modifications and changed levels of specific mRNAs. However, proteins are often the most characteristic and the most easily detected.

Currently, commonly used cancer diagnostic and prognostic indicators are the morphological and histological characteristics of tumors. Single protein biomarkers, such as *prostate specific antigen* (PSA) have already started to be used in practice. Meantime, since tumor development involves many biological changes, the signatures of this disease can be very complex and characteristic for a particular type of tumor (Jain 2007b; McShane et al. 2005). Proteomics approaches based on comparative detection of increased levels of a few among many thousands of proteins have been used to generate differential protein expression maps of the normal cells. They can be compared to that of cancer cells with the detection of proteins, whose levels change significantly (Kuramitsu and Nakamura 2006) with subsequent development of microarrays for their detection.

Heart disease, for correct diagnostics, needs to use several key protein-based biomarkers. The standard protein biomarkers include interleukin-6, interleukin-8, serum amyloid A protein, fibrinogen fragments and troponins. Several cardiac markers detected in blood, such as myoglobin, troponin I, troponin T and creatine kinase-MB, are widely recognized as diagnostic tools for myocardial infarction (Yang and Zhou 2006). The problem is not only to determine these proteins quantitatively above their background levels but to do this rapidly to provide a correct, timely clinical decision (Storrow and Gibler 1999).

While not cardiac specific, myoglobin is one of the very early markers that increases in concentration directly after an acute myocardial infarction and therefore, it can be used as a suitable marker for rapid diagnosis using the new microwave-accelerated metal-enhanced fluorescence approach (Aslan and Geddes 2006). Presently, it is developed for immunoassay and offers an ultra-rapid and sensitive platform. With this approach it has been shown that myoglobin can be determined rapidly (within 20 s) and with high sensitivity (on the level of 100 ng/ml).

In *kidney-related diseases*, urine is a potential source for such biomarkers. The proteomic analysis of kidney cells and cancerous kidney cells is producing promising leads for biomarkers for renal cell carcinoma and developing assays to test for this disease. Recently, it has been shown that the identification of urinary polypeptides as biomarkers of kidney-related diseases allows one to diagnose the severity of the disease several months before the manifestation of pathology (Devarajan 2007).

Urinary excretion of *human serum albumin* (HSA) can be used to diagnose incipient renal disease and also type II diabetes and its progression is one of the most studied clinical indices using biosensor devices. HSA specifically binds a number of dyes and simple and reliable methods for its identification and quantitative assay are based on this principle (Harvey et al. 2001; Kessler et al. 1997).

12.3.6   Prospects

Biological fluids are a source of rich information about the health condition of a human organism. They contain proteins that can serve as specific biomarkers of the disease. A number of them have already been identified but a much larger number awaits discovery. The traditional and most frequently used diagnostic fluids are blood and urine. The miniaturization of sensor devices causes not only a dramatic reduction in the volume of collected test samples but also increases the speed and precision of their analysis, which can be still further improved by the application of low-energy microwaves. Saliva, sweat and tears, which can be collected in a noninvasive way, were previously considered as biological fluids of minor diagnostic value. The situation changes dramatically with the miniaturization and increase in the precision of diagnostic tools.

12.4   Advanced Sensors in Drug Discovery

Nowadays, the field of drug discovery has passed the period of empirical search and relies strongly on the achievements of molecular and cellular biology. The *in-vitro* search for particular activities and in-cell testing has become common in seeking for optimal drug candidates. Combinatorial libraries generated by chemical synthesis together with an immense number of natural compounds that are mostly extracted from plants, form a multi-million number of drug candidates. Therefore, *high-throughput screening* (HTS), which is the process of testing a large number of diverse chemical structures against disease targets to identify 'hits', has become a major tool. This approach is based on a concept that assumes that when a sufficiently large library of compounds is tested, the chances of discovering a new active compound are increased. Traditional drug screening methods cannot compete with HTS in simplicity, speed, cost and efficiency.

12.4.1   High-Throughput Screening

Fluorescence-based techniques are among the most important detection approaches used for HTS due to their many features, such as high sensitivity and amenability to automation, simplification, miniaturization and speeding up assays (Bosch et al. 2007). As a general technology, HTS involves a highly sensitive testing system based on several constituents. They are an automated operation platform, a specific screening model (*in vitro*) and an abundant components library together with a data acquisition and processing system. This combination makes possible the screening of more than 100,000 samples per day. Therefore, highly multiplexed optical-biosensor arrays have become the most important tools in this technology. The capacity of microarrays is growing and presently, the application of plates with 1,536 wells is not uncommon.

High-throughput screening is not the only problem to be resolved in efficient drug discovery. The screening should be *multiparametric* and involve many variables, such as:

(a) *Specific activity*, which is commonly the binding to a particular target in the organism leading to the inhibition of some enzyme system or antagonism to some receptor. The array composed of targets or target analogs is an appropriate strategy, at least during the first step of testing. In order to estimate the drug dissociation constant K_d, a series of targets with different affinities should be applied. In view of the substantial dilution of drugs upon administration into the patient's body, of pharmacological interest are commonly only those of them that have K_d in the concentration range of 10^{-7}–10^{-9} M or lower.

(b) *Toxicity* (or selectivity in toxicity between normal and malignant cells). The arrays of whole-cell biosensors may be a great help for this (Durick and Negulescu 2001). For instance, a disruption of cell integrity can be most easily revealed with fluorescence dyes such as propidium iodide.

(c) *Pharmacokinetics*, which is the determination of the fate of a drug in the body. It allows and requires many preliminary *in vitro* tests. Fluorescence and long-lifetime luminescence-based techniques provide a proper response for acquiring this information (Gomez-Hens and Aguilar-Caballos 2007).

Pharmacokinetics itself includes many parameters, such as chemical (biochemical) stability in biological media, ability to cross biological membranes, affinity towards serum albumin and other carriers in blood, ability to be destroyed by microsome oxidation system, etc. To some extent this information can be obtained in simple fluorescence tests. For instance, the binding to a serum albumin can be provided based on competitive substitution with fluorescent dyes (Ercelen et al. 2005). The liposome-based biosensors are useful by their ability to model the drug's affinity towards a biological membrane and its permeability through it (Przybylo et al. 2007).

Regarding the fluorescence detection method, it is the ultimate means in operation with high-capacity microarrays. Here, the preference is currently given to polarization assays and to lifetime-based or lifetime-gated detection, with long-lifetime luminophores. However, other detection methods, such as bioluminescence, offer good potential, primarily because of their ability to provide low background emission. Fluorescence microscopy-based methods that allow decreasing the sample volume to femtoliters have also found many applications in HTS (Eggeling et al. 2003) and the scope of these applications is expected to increase.

12.4.2   Screening for Anti-cancer Drugs

Screening for anti-cancer substances is one of the most important activities in drug discovery. Primarily, the tests that are usually done are on the *cell toxicity* and on the difference in toxicity between the lines of normal and cancer cells using viability assays. An inherent problem with this approach is that all the compounds that are toxic and growth inhibitory, irrespective of the mechanism of action, will score positively. Therefore, the search for anti-cancer drugs should involve tests for specific binding to cancer cells and for their ability to induce *apoptosis* (programmed cell death).

There are two major approaches to detect apoptosis during its early stages. One is targeted at the detection of a protease, which is activated on apoptosis, called Caspase 3. An interesting assay method available for HTS has been recently developed (Tian et al. 2007). It employs a stable HeLa cell line expressing a FRET-based biosensor protein. The activation of *Caspase 3* cleaves this sensor protein with the disruption of a FRET donor-acceptor pair, which causes its fluorescence emission to shift from a wavelength of 535 nm (green) to 486 nm (blue). A decrease in the green/blue emission ratio thus gives a direct indication of apoptosis.

The other approach for detecting the apoptosis is focused on the cell membrane and is based on its property to expose anionic lipids to the outer leaflet upon apoptotic transformation. This was discussed in Section 11.5. Fluorescence techniques are extensively used for the detection of these changes.

12.4.3   Future Directions

High-throughput screening, (HTS), which includes automated preparation of a large number of samples and then screening their properties in multi-well plates or multi-point arrays, allows both molecular and cellular testing. It improves the efficiency of research not only in drug discovery but in many other areas, from chemical synthesis to food processing and toxicology. There are two current limits that will gradually recede with the advancement of technology. One is the size (scale) of arrays, which has every chances of being increased to 10^4 spots and higher. The other limit is imposed by the huge amount of information in view of the huge number of low-specific interactions and 'cross-talks'. Some solutions for these problems were prompted by sensory systems used by Nature. They will be discussed in the next section.

12.5   Towards a Sensor that Reproduces Human Senses

For many years, scientists have discussed the possibility of creating artificial intellect. Whether this goal is real or not, it is sure that such intellect cannot operate without connections with the outer world. These connections have to be provided by artificial senses. Since the dawn of time, ideal sensing systems were developed by Nature and imposed on living organisms. We observe them being developed into sensory nerve cells of our eyes, ears and nose to detect scents and those of the tongue to taste food and drink. The sensual perception is formed from many simultaneous responses to the presence of different compounds forming a recognizable bunch of tastes and odors (Deisingh et al. 2004). Thus, the artificial sensors reproducing human senses should be multiparametric in collecting primary information and we need to use sophisticated algorithms for its analysis. Seeing the smells and tastes in fluorescent colors is very insufficient for the recognition of the real odor of perfumes or the real taste of wine.

12.5.1   Electronic Nose

The *olfactory system* recognizes the smell when the gas-phase molecules interact with the surface of specialized nerve cells that generate an electrical signal. The natural receptors sense different odors as a complex signal composed of responses to many substances, often existing simultaneously in very small concentrations. The primary responses are probably not very selective but they form a pattern that allows a high level of integration on the pathway to odor recognition. Understanding the operation of this system may provide a clue for the development of artificial olfactory sensing.

Hitherto developed optical *gas sensors* also exhibit low specificity in their ability to recognize different odors. These facts prompt us to the strategy to simulate odor perception. The sensors for different gases should be developed and combined as arrays. Employing chemical sensors in an array format with a proper *pattern recognition* should provide a highly reliable identification of odors, even if individual sensors are low specific. The systems based on this approach are termed '*electronic noses*'. In addition to simulating human perceptions, their design will lead to an extensive range of applications. These applications will range from the food industry and medicine to environmental monitoring and process control (James et al. 2005).

The field of *artificial olfaction* and gas identification is one of the fastest growing areas in sensing. This active field of development uses many of the sensor materials and technologies described in this book. These include conjugated polymers, metal oxide semiconductors, porous silica and other materials and are based on the response in fluorescence. The sensing material is usually presented as a thin film or a film of high porosity, to facilitate analyte diffusion for its detection. Polymeric materials with incorporated organic dyes and composites employing different metalloporphyrins and phthalocyanins lead this application

(Maggioni et al. 2007). The sensor designers may benefit from the fact that the luminescence of different metalloporphyrins is quenched by different vapors in a different manner (Rakow and Suslick 2004). Many of the suggested sensors are not sensitive to humidity and allow sensing in the air (Rakow and Suslick 2004).

An essential task is to make the sensors superior to those made by Nature, so that they could sense the compounds with which man had no contact in his historical development and which appeared during the industrial era. These are the poisonous organic compounds and explosive-like vapors.

12.5.2 Electronic Tongue

Taste perception was developed in evolution for the selection and evaluation of food, including avoidance of potentially harmful substances, such as bitter tasting poisons. The principles of the gestation response are still not understood in full but the morphology of the location of specialized nerve cells (taste cells) on the surface of the tongue is well studied. The chemical senses are somehow modified by enzymes present in saliva. Since cell surface receptors for particular taste stimuli have not yet been isolated, a traditional classification into four classes: sweet, salty, sour and bitter is commonly used. Meantime it is known that the cells can respond to more than one type of taste stimulus and this response can be more or less specific. As in odor perception, a pattern of response is formed with an efficient but still unknown mechanism of operation. The output of the sensory system is already not the amount of taste substances present but the taste in all its specific features.

With this very limited amount of information, it is hard to proceed with the functional simulation of the process of perception but an important concept can be borrowed for artificial sensor design (Wiskur et al. 2001). To identify the taste produced by complex mixtures of chemical substances, we need sensor arrays based on a combination of sensors with narrow and broad specificities. The narrow specific sensors (Wiskur et al. 2001) should provide such a fine resolution as the recognition between stereoisomeric forms of the same compounds. For example, most D-amino acids are sweet, while their L counterparts are bitter (Pu 2004). Sensors with a broader specificity are also needed. They should 'cross-talk' and thus participate in the formation of an *integrated response*. The latter do not need to be completely rationally designed to target specific analytes. They simply need to be different (Lavigne and Anslyn 2001). This concept leads us to '*artificial tongue*' devices that should be able to perform sensory evaluation in complex mixtures of chemically unrelated compounds. Such artificial tongues have been developed for the analysis of liquids (Rakow and Suslick 2004).

The major field of application of these devices is, of course, the analysis of drinking water (Martinez-Manez et al. 2005), food (Deisingh et al. 2004) and of beverages such as wines (Di Natale et al. 2004). In these cases, the problem of

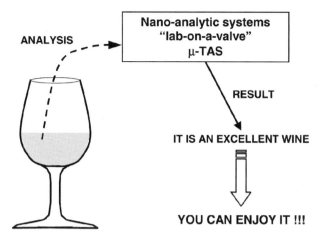

Fig. 12.6 Operational principle for a modern portable laboratory for wine characterization (Simonet and Valcarcel 2006)

correlating sensorial descriptors and chemical parameters should be resolved (Fig. 12.6). In the case of red wine testing, the results demonstrate the capability of such systems to be trained according to the behavior of a practical panel of tasters.

12.5.3   Olfactory and Taste Cells on Chips and Whole-Animal Sensing

In the mammalian olfactory system, the detection of a great number of odorants results from the association of the odorant molecules with olfactory receptors, carried by olfactory sensory neurons. Olfactory and taste sensors can be constructed based on whole-cell sensor technology (see Section 9.5), based on the cultured corresponding nerve cells. Such an approach is very close to the natural recognition of odors and taste. In addition, the methods of molecular biology allow transferring the genes coding these receptors to cells that are convenient for manipulation, for instance, yeast cells (Minic et al. 2005). Together with the receptor, a bioluminescent luciferin-luciferase complex can be incorporated as a reporting element. On this pathway, sensors closely related to biological odor recognition can be constructed and suggested for different uses.

In this respect, it has to be recollected that whole animal biosensors have been used for the purpose of evaluating the quality of the environment and for detecting hazardous compounds for centuries. Fish survival was a common test for the quality of water and canary birds were taken into the mines to indicate the presence of explosive methane gas, which is undetectable by human senses. Different animals can react by their behavior to approaching earthquakes and tsunamis. At the beginning of the eighties Dr. Boris Roytrub (Roitrub and Zlatin 1983) demonstrated an astonishing fact – the detection of 10^{-20} M acetyl choline in a simple test using the contraction of a common medicinal leech, probably a record sensitivity in the days when nobody had heard about single-molecule detection. This is an example to show that the senses of humans and animals are extremely sensitive but still far from being ideal. Modern devices must be strong competitors to them.

12.5.4   Lessons Obtained for Sensing

Electronic noses and electronic tongues have been suggested as technological attempts to mimic the functions of the human chemical senses. Besides this scientifically challenging objective, they have shown potential to be developed into practical instruments for the analysis and characterization of systems of complex composition. However, it is not expected that electronic tongues and noses will be strong competition for precise analytical tools, such as gas chromatograph/mass spectrometer couplings and GC/MS (Gopel 2000). The latter are the established instruments of analytical chemistry and in most applications they should demonstrate a superior performance. They allow characterizing the whole specimen, whereas sensors respond only to certain classes of molecules. The simple, fast and low-cost instrumentation for detecting a relatively small number of analytes is probably the niche for the allocation of electronic noses and tongues. Such sensors may become disposable. Though a one-use nose or tongue looks uncomfortable, this may be the best technical solution.

Since in the mammalian olfactory system a limited number of not very selective cross-reactive receptors are able to generate a *response pattern*, a significant cross-reactivity may also exist in artificial systems. Sensor arrays can be created in such a way that the specificity is distributed across the array's entire reactivity pattern, rather than being contained in a single recognition element. Such patterns can then be incorporated into an artificial neural network for the recognition of mixtures of analytes (Lavigne and Anslyn 2001).

12.6   Sensors Promising to Change Society

What will be the future role of chemical sensors and biosensors? Will their application be limited to special usage like in clinics or they will become mass products available for everyday personal use? During the present phase, we already observe that industrial, community and personal needs combine to provide a strong stimulus for the development of sensors for a broad range of applications. Sensing will come to everyday life but not alone. The merging of sensing and biosensing with digital communication systems and, particularly, with wireless telemetry, will affect society in a very profound manner (Byrne and Diamond 2006).

12.6.1   *Industrial Challenges and Safe Workplaces*

In any automatic system supervising a complex technological process, sensors together with their signal-processing instrumentation have started to play a dominant role. Fluorescence-based sensors successfully compete with sensors based on other response principles in many areas, where their advantages are required. The whole arsenal of industrial sensors may include multichannel optical waveguides for continuous process monitoring, arrays for multifunctional characterization of intermediates or products in complex reaction schemes and microfluidic devices for rapid one-use testing.

Sensors are beginning to be integrated into intelligent automatic systems of decision-making and supervision over a technological process with feedback for implementing corrections in this process. Thus, the suggestion or deployment of appropriate actions in response to the deviation from established process regalement will be made automatically, based on the readings of the sensors that integrated into the powerful decision-support systems. Personnel will be excluded from this role and will need to intervene only in critical situations.

Many industrial areas have already started to use or are ready to use such systems, including the production and quality control of food and beverages and also the production of drugs and plant-protecting chemicals. All areas of the chemical industry may benefit from these possibilities, especially workplaces connected with potential danger to personnel.

Industrial control and measurement instrumentation can be classified into three basic categories: off-line, on-line and in-line. *Off-line* systems require the removal of samples for subsequent analysis by wet-chemical methods or automated laboratory units. Such techniques are time and labor consuming, they are not applicable to any kind of continuous monitoring and do not allow automatic feedback control. *On-line* systems are developed for the continuous sampling of instrumentation with the aim of providing the readout of a result with such a speed that can permit making rapid control decisions. *In-line* systems are those that use sensors that are incorporated directly into the industrial process reactor and report continuously 'from inside', providing all the necessary information for controlling the process.

In many cases, not only the physical parameters are needed for monitoring, such as the temperature, pressure or density in the reaction medium, but also the concentrations of all necessary reactants and reaction products. In addition, it is often required to detect the trace amounts of impurities and, in bioprocess industries, also to provide the control for microbial contamination. All this can presently be done with fluorescence sensors. Some systems have still matured in performance or did not go down in price to open their broad-scale application and this is expected in the near future.

Examples can be presented from the polymer processing industry (Reyes et al. 1999). Very small amounts of dye are added to a polymerized sample to control the industrial process, which allows providing information on the properties of a product on a molecular level.

In different biotechnology processes, the multiparametric sensors have become valuable tools for bioprocess monitoring, with the capacity for continuous improvement. This is important because bioreactors are closed systems in which the microorganisms and different cell lines can be cultivated under defined and controllable conditions and non-invasive control of this process is especially important. Since modern bioprocesses are extremely complex and differ from process to process (e.g., fungal antibiotic production versus mammalian cell cultivation), the appropriate analytical systems must be set up from different basic modules, designed to meet the special demands of each particular process (Ulber et al. 2003). The advantages of obtaining multiparametric information by non-invasive detection will allow putting all the bioprocess technologies under full control.

12.6.2   Biosensor-Based Lifestyle Management

In view of current and prospective achievements of sensing technologies, should we further tolerate the current situation with clinical diagnostics? It is presently still a common practice that even in critical situations the samples for analysis are taken, sent to a specialized laboratory and the doctor with the patient will have to wait for the result. One or two more efforts and this situation will change dramatically and not only in the speed of clinical decisions. An easily accessible self-control on health status will definitely induce self-control on human behavior.

Together with the increase of welfare and the lasting tendency for cultural globalization, more and more of the world population adopt the western-society lifestyle with all its pleasures, many of which are the risk-factors of disease. Additional risk-factors have been brought by industrial pollution. More and more people require medical treatment and even lifetime surveillance. Meantime, for centuries practical medicine has developed along the 'illness treatment' model. This is because the diagnosis and treatment of disease was based on experience acquired by doctors by learning or by personal practice that classified illnesses with reference to an 'average' patient. Despite the efforts of many generations of the best doctors, this system could not, in principle, become personalized. The doctors prescribe the treatments that 'helped other people' and often do that by trial and error. This situation will change only in a sensor-rich environment.

Based on mutiparametric genomic and proteomic personal data, if properly treated and analyzed, the 'wellness support' model will be constructed. It should become the basis of all clinical decisions, realizing the principle of personalized medicine. Moreover, if transformed in a clearly understood way, this information can be actively used by the patient himself. The common people will obtain the tools to monitor their own condition and maintain it properly. The notice 'SMOKING KILLS' does not stop a heavy smoker, because he thinks that smoking will probably kill someone else, not him. But if he observes the deterioration of his own health-related personal indices approaching a critical limit, then the right decision to quit smoking becomes personally motivated.

Such a redistribution of responsibility for health maintenance from doctors to individuals must have a great psychological effect. Moreover, such factors as the distance to the hospital and availability of nursery personnel will begin to play a minor role. Adding the sensing, communication and data-processing capabilities at the point-of-care and commonly at the patient's home, will be one of the most significant contributions to radical changes that must happen in presently centralized health systems.

Post-genomic individual chemical diagnostics are based on an already well-established concept that not only inherited diseases but also common acquired diseases, such as cancer and heart failure are related to the genetics of the individual. Therefore, their risk can be evaluated and minimized. It is expected that in the years to come, DNA screening and expression sensing will become a routine examination not only in diagnostic tests for genetic defects, but also to identify conditions related to metabolic or environment-induced stresses. The great number of targets can serve as medical markers that can predict the onset of malady and monitor its development. They will be established and used in common practice. As we observed in previous chapters of this book, the technological possibilities for this already exist and work in this direction is proceeding.

As a result, strong feedback can be established, allowing knowledge-based intervention into a disease process along the following scheme:

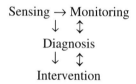

All these diagnostic devices have to become user-friendly and cost-effective. They are expected to provide minimum trouble to even unskilled persons, who will be able to initiate the system and obtain understandable results. There are many suggestions regarding the construction of these devices. For instance, they can be made as wrist-type watches or even *integrated into textile clothes.* We observe an increase in activity for development in this direction. Thus, the method of detecting amines based on a smart dye painting textile and producing both an absorption and fluorescence response have been reported (Staneva et al. 2007).

Another interesting possibility that emerges is the combination of sensing with *sensing-guided drug delivery.* The first steps in this direction have already been made, for minimally invasive feedback-controlled insulin delivery. The miniaturized medical diagnostic and treatment devices are on the way to monitoring the glucose level in diabetic patients and to microinjecting insulin automatically at times of reduction of its level (Hsieh and Zahn 2007).

Time will show which of these interesting research and development products will be selected for practical use.

12.6.3   Living in a Safe Environment and Eating Safe Products

The *pollution of the environment* caused by industrialization, growth of human population and rise in consumption of industrial products is a serious problem that, to a greater and greater extent threatens the existence of men and natural habitats. The environment pollution problem has many sides. The pollutant dosage effects are non-linear; they include both synergistic and antagonistic components as well as the potential effects of unknown, modified or chemically undetected substances. Thus, monitoring the environment by detecting pollutants with chemical sensors is necessary but not sufficient; the role of biological tests, including the application of whole-cell biosensors, becomes increasingly important. Thus, bioelectronics, nanotechnology, miniaturization and especially biotechnology, seem to be growing areas that will have a marked influence on the development of new biosensing strategies in the near future (Rodriguez-Mozaz et al. 2004).

Likewise, the health system, the domains of environment monitoring and control technologies in agriculture currently use advanced and complex techniques that are still based on *in-vitro* test facilities. In most of these procedures, the use of rapid and powerful instrumental techniques for the final separation and detection of the analytes contrasts with the time-consuming and usually manual methods used for sample preparation, which slows down the analysis. Revolutionary changes are expected in these areas (Rasooly and Jacobson 2006). The miniaturization in sample treatment will speed up its preparation (Ramos et al. 2005) and multiplex sensing will substitute the more complicated chromatographic techniques. Among the novel techniques, sensors and sensor arrays based on bioluminescent bacteria are expected to gain popularity. They will be in common use to detect various pollutants. Meantime, with their aid, it is difficult to predict the effects of these pollutants on the human physiology and only cultured mammalian cells offer this possibility (Section 9.5).

The control for food and beverage contaminants also exhibits dramatic changes, with the introduction of sensor arrays (Sapsford et al. 2006) and will continue to do so. Rapid detection is needed for *foodborne contaminants* that come in a variety of sizes ranging from simple chemical compounds to entire bacterial cells and the development and application of correspondent multi-analyte sensor systems are close.

It is expected in the near future that consumers, instead of smelling food products, will start to use sensors indicating their freshness and quality. The characteristic compounds that can be used for sensing the quality of meat are the amines putrescine and cadaverine (Landete et al. 2007; Pircher et al. 2007). The respective fish freshness analyte can be hypoxanthine (Venugopal 2002). Such *'freshness markers'* can be found for every consumer product. Moreover, the simultaneous detection of a number of these markers will reduce the number of false decisions and will make any falsification impossible.

12.6.4 Implantable and Digestible Miniature Sensors Are a Reality

Recent progress in technology allows predicting that very soon all or nearly all in-tube chemical and biochemical testing will be substituted by advanced multiparametric sensor systems. They will be miniaturized and, if necessary, for one-time use. During the next phase, smart implantable or even digestible sensors will become rudimentary in clinical analysis and welfare monitoring. Let us summarize the pre-requisites for this.

Implantable sensors are already in active development, they are mostly related to the highly needed continuous monitoring of glucose levels (Schiller et al. 2007; Yu et al. 2007). As the support media they use specially prepared hydrogels and as a response mechanism – the competition assay between glucose and the competitor bearing either the fluorescent dye or the quencher. They compete for the sensor binding site formed by a fluorescent boronic acid derivative.

The present problem with such implantable sensors is the transmission of the signal to the recording system. This problem, known as *real-time biotelemetry*, has been addressed and its interesting solutions have already been found (Johannessen et al. 2006; Wang et al. 2007). Such telemetry microsystems may include multiple sensors, integrated instrumentation and a wireless interface.

Based on such platforms, miniaturized *digestible sensors* can be devised. Such a sensor has already received the name 'lab-in-a-pill'. One of the first prototypes (Fig. 12.7) can measure the temperature and pH within a gastrointestinal tract with a wireless communication (Johannessen et al. 2006). Moreover, an incorporated magnet allows tracing and guiding the sensor inside the tested body. New developments in this line are expected in the future in the extension of the number of detected parameters.

The diagnostics of cancer in the gastrointestinal tract is presently based on patient investigations with catheter-based video endoscopes, which involves a 1.5–2 cm diameter tube being inserted into the body. The tube contains fiber optic cables carrying visible light and ending with a small camera. The most advanced of these instruments use UV light for excitation and the whole visible range for fluorescence detection Tissue autofluorescence, which is different for normal and bleeding or malignant tissue, is detected. Such a procedure is invasive, producing troubles for the patient. It is also rather costly, due to the need for specialized equipment, operating suites, as well as the presence of skilled personnel. An attractive alternative could be to use ingestible capsule-sized cameras, typically the size of a large vitamin pill, to provide visible illumination and acquire the images in fluorescent light. Prototypes of such *fluorimeters-in-a-pill* have already been described (Kfouri et al. 2008). The construction of one of them is presented in Fig. 12.8.

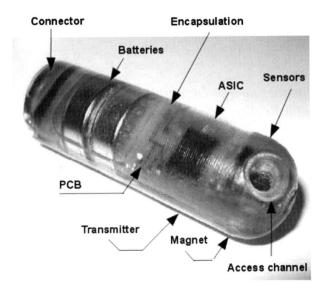

Fig. 12.7 The 'lab-in-a-pill' that can be a digestible sensor of temperature and ph in a gastrointestinal tract. Its functional elements are shown. The whole unit is encapsulated in epoxy resin (Reproduced with permission from Johannessen et al. 2006)

Spectrally resolved detection can be achieved in these systems. Diagnostics is based on the known facts that abnormal tissues display a significant decrease in fluorescence intensity, along with an increase in the ratio of red-to-green fluorescence. The detection of this difference can be improved by the introduction into the patient's body of a fluorescent contrast agent before the examination.

12.6.5   Prospects

It is not a dream but a realistic prediction that sensor-aided early diagnosis as well as a healthy and preventive lifestyle will help in slowing the onset of many health problems and save many millions of lives. In order to achieve this goal, the long-term and if necessary, the continuous monitoring of human vital signs are required. An ordinary person after obtaining access to the most detailed information on his/her health status will be able and willing to change their lifestyle accordingly. This is quite achievable in the near future, with the mass production of inexpensive and user-friendly portable systems.

Monitoring the environment and providing control on its pollution is a necessary step for its improvement. This problem is commonly addressed to governments and community leaders. The correctness of their activities will heavily depend upon the information presented to them and to society as a whole. The primary information will be obtained from sensing.

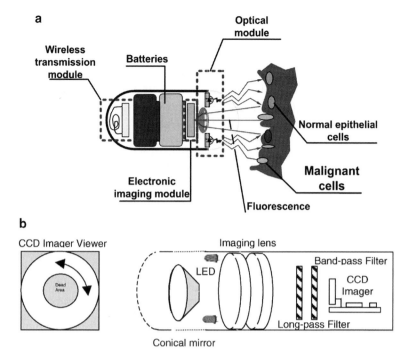

Fig. 12.8 The prototype of a miniature, 'in a pill', wireless fluorescence imaging system for noninvasive clinical diagnosis of the gastrointestinal tract. (a) Conceptual design. The 'pill' contains an optical module for the excitation and collection of fluorescence emission and the wireless transmission module. (b) Schematic showing the optical imaging module (right) and field of view on the CCD imaging sensor (left). UV excitation is achieved via the 360 nm LEDs. Side view fluorescence images are projected to the CCD through the conical mirror and two imaging lenses. A long-wavelength-pass filter is used to block scattered excitation light, while the spectral band selection is achieved using a narrow bandpass filter placed in front of the CCD (Reproduced with permission from Kfouri et al. 2008)

A healthy workplace, healthy food, healthy environment and healthy lifestyle, this is an optimistic prognosis for the development of human society in the XXI century. With this advancement will people become happier? This remains to be seen.

Sensing and Thinking 12: Where Do We Stand and Where Should We Go?

The last decade has seen dramatic transformations in chemical sensing and biosensing technologies. General trends have been marked towards the optimization of sensor receptor and reporter units and their coupling. The novel materials and techniques of their fabrication and new techniques for optical detection contribute to this rapid progress. All this allows approaching the nanometer length and volume scale, applying novel means of signal transduction and fabricating novel devices. The processing of a huge amount of data can be acquired, so new, powerful computerized tools for their analysis are needed and they are being rapidly developed.

In this respect, two important problems are outstanding: how to generate a huge amount of data and how to analyze it. These problems will remain actual for a long time and many of their solutions are expected to come. Multiparametric input can be provided by multi-sensor microarrays formed of sensor elements selected from big libraries and by high-resolution images. Multiparametric outputs have to be dealt with by applying sophisticated algorithms, which still have to be developed. After achieving this, what remains? The presentation of the results of this analysis in a conceivable and user-friendly way. This goal will be achieved under the pressure of a strong demand by users.

Questions and Problems

1. Compare the DNA and the protein array techniques. What factors do not allow developing them within the same scale of methodology? Suggest the most prospective arrays for proteomics, specifying the receptors, reporters and transduction principles.
2. The suggestion to apply antibodies as receptors in proteome analysis is illustrated in Fig. 12.3. Is it possible to apply library selected peptides or DNA aptamers for this purpose?
3. Which one, in your opinion, would be the more promising strategy for developing arrays for proteomics – with positional or optical encoding? Provide pro and contra arguments regarding each of these possibilities.
4. Describe the essence of the combinatorial approach and outline the steps in its application. Select any polyphenolic compound or endocrine disruptor as a target and describe your proposed activities in developing a sensor for its detection.
5. What are disease biomarkers and what are the criteria of their establishment? Explain the sequential steps for their discovery.
6. What are the present and what will be the future trends in the discovery of new drugs? Explain the essence of High-Throughput Screening and evaluate the applicability of different fluorescence sensing technologies for this.
7. What is the need for simulating the human senses? What lessons for the development of fluorescence sensors of a new generation have we learned? Explain the concepts of pattern recognition and integrated response.
8. Analyze different types of instrumentation for industrial control and measurement and, based on the knowledge obtained in Chapter 10, provide a choice between different types of fluorescence-based devices that can be used for this purpose.

9. Explain the essential features that make the 'illness treatment model' and 'wellness support model' very different on the personal and community levels.
10. Suggest the most convenient method of application of the wearable sensor for the analysis of the chemical composition of sweat. Propose one-time use and continuous monitoring versions.
11. Imagine that we reach the goal of developing efficient and applicable large scale sensors for personalized medicine. Explain how these sensors of the future are expected to change your personal life.

References

Apostolidis A, Klimant I, Andrzejewski D, Wolfbeis OS (2004) A combinatorial approach for development of materials for optical sensing of gases. Journal of Combinatorial Chemistry 6:325–331

Aslan K, Geddes CD (2005) Microwave-accelerated metal-enhanced fluorescence: Platform technology for ultrafast and ultrabright assays. Analytical Chemistry 77:8057–8067

Aslan K, Geddes CD (2006) Microwave-accelerated and metal-enhanced fluorescence myoglobin detection on silvered surfaces: potential application to myocardial infarction diagnosis. Plasmonics 1:53–59

Baker GA, Pandey S, Bright FV (2000) Extending the reach of immunoassays to optically dense specimens by using two-photon excited fluorescence polarization. Analytical Chemistry 72:5748–5752

Borrebaeck CA (2000) Antibodies in diagnostics - from immunoassays to protein chips. Immunology Today 21:379–382

Borrebaeck CAK (2006) Antibody microarrays; technology and applications in oncoproteomics. Molecular & Cellular Proteomics 5:S342–S342

Bosch ME, Sanchez AJR, Rojas FS, Ojeda CB (2007) Optical chemical Biosensors for high throughput screening of drugs. Combinatorial Chemistry & High Throughput Screening 10:413–432

Byrne R, Diamond D (2006) Chemo/bio-sensor networks. Nature Materials 5:421–424

Chan FK, Holmes KL (2004) Flow cytometric analysis of fluorescence resonance energy transfer: a tool for high-throughput screening of molecular interactions in living cells. Methods in Molecular Biology 263:281–292

Christodoulides N, Mohanty S, Miller CS, Langub MC, Floriano PN, Dharshan P, Ali MF, Bernard B, Romanovicz D, Anslyn E, Fox PC, McDevitt JT (2005) Application of microchip assay system for the measurement of C-reactive protein in human saliva. Lab on a Chip 5:261–269

Comeau JWD, Costantino S, Wiseman PW (2006) A guide to accurate fluorescence microscopy colocalization measurements. Biophysical Journal 91:4611–4622

Deisingh AK, Stone DC, Thompson M (2004) Applications of electronic noses and tongues in food analysis. International Journal of Food Science and Technology 39:587–604

Devarajan P (2007) Emerging biomarkers of acute kidney injury acute kidney injury. Contributions to Nephrology 156:203–212

Di Natale C, Paolesse R, Burgio M, Martinelli E, Pennazza G, D'Amico A (2004) Application of metalloporphyrins-based gas and liquid sensor arrays to the analysis of red wine. Analytica Chimica Acta 513:49–56

Durick K, Negulescu P (2001) Cellular biosensors for drug discovery. Biosensors & Bioelectronics 16:587–592

Eggeling C, Brand L, Ullmann D, Jager S (2003) Highly sensitive fluorescence detection technology currently available for HTS. Drug Discovery Today 8:632–641

Ercelen S, Klymchenko AS, Mely Y, Demchenko AP (2005) The binding of novel two-color fluorescence probe FA to serum albumins of different species. International Journal of Biological Macromolecules 35:231–242

Finan JE, Zhao RY (2007) From molecular diagnostics to personalized testing. Pharmacogenomics 8:85–99

Freire SL, Wheeler AR (2006) Proteome-on-a-chip: mirage, or on the horizon? Lab on a Chip 6:1415–1423

Frishman D (2007) Protein annotation at genomic scale: the current status. Chemical Reviews 107:3448–3466

Gauglitz G (2000) Optical detection methods for combinatorial libraries. Current Opinion in Chemical Biology 4:351–355

Gomez-Hens A, Aguilar-Caballos MP (2007) Modern analytical approaches to high-throughput drug discovery. Trac-Trends in Analytical Chemistry 26:171–182

Gopel W (2000) From electronic to bioelectronic olfaction, or: from artificial "moses" to real noses. Sensors and Actuators B-Chemical 65:70–72

Harvey MD, Bablekis V, Banks PR, Skinner CD (2001) Utilization of the non-covalent fluorescent dye, NanoOrange, as a potential clinical diagnostic tool - nanomolar human serum albumin quantitation. Journal of Chromatography B 754:345–356

Hsieh YC, Zahn JD (2007) On-chip microdialysis system with flow-through sensing components. Biosensors & Bioelectronics 22:2422–2428

Hu S, Loo JA, Wong DT (2007) Human saliva proteome analysis and disease biomarker discovery. Expert Review of Proteomics 4:531–538

Inuyama Y, Nakamura C, Oka T, Yoneda Y, Obataya I, Santo N, Miyake J (2007) Simple and high-sensitivity detection of dioxin using dioxin-binding pentapeptide. Biosensors & Bioelectronics 22:2093–2099

Jain KK (2007a) Applications of nanobiotechnology in clinical diagnostics. Clinical Chemistry 53:2002–2009

Jain KK (2007b) Cancer biomarkers: current issues and future directions. Current Opinion in Molecular Therapeutics 9:563–571

James D, Scott SM, Ali Z, O'Hare WT (2005) Chemical sensors for electronic nose systems. Microchimica Acta 149:1–17

Johannessen EA, Wang L, Reid SWJ, Cumming DRS, Cooper JM (2006) Implementation of radiotelemetry in a lab-in-a-pill format. Lab on a Chip 6:39–45

Kessler MA, Meinitzer A, Petek W, Wolfbeis OS (1997) Microalbuminuria and borderline-increase albumin excretion determined with a centrifugal analyzer and the Albumin Blue 580 fluorescence assay. Clinical Chemistry 43:996–1002

Kfouri M, Marinov O, Quevedo P, Faramarzpour N, Shirani S, Liu LWC, Fang Q, Deen MJ (2008) Toward a miniaturized wireless fluorescence-based diagnostic imaging system. IEEE Journal of Selected Topics in Quantum Electronics 14:226–234

Kuramitsu Y, Nakamura K (2006) Proteomic analysis of cancer tissues: shedding light on carcinogenesis and possible biomarkers. Proteomics 6:5650–5661

Lafratta CN, Walt DR (2008) Very high density sensing arrays. Chemical Reviews 108:614–637

Landete JM, de las Rivas B, Marcobal A, Munoz R (2007) Molecular methods for the detection of biogenic amine-producing bacteria on foods. International Journal of Food Microbiology 117:258–269

Lavigne JJ, Anslyn EV (2001) Sensing a paradigm shift in the field of molecular recognition: From selective to differential receptors. Angewandte Chemie-International Edition 40:3119–3130

Maggioni G, Manera MG, Spadavecchia J, Tonezzer M, Carturan S, Quaranta A, Fernandez CDJ, Rella R, Siciliano P, Della Mea G, Vasanelli L, Mazzoldi P (2007) Optical response of plasma-deposited zinc phthalocyanine films to volatile organic compounds. Sensors and Actuators B-Chemical 127:150–156

Martinez-Manez R, Soto J, Garcia-Breijo E, Gil L, Ibanez J, Llobet E (2005) An "electronic tongue" design for the qualitative analysis of natural waters. Sensors and Actuators B-Chemical 104:302–307

McShane LM, Altman DG, Sauerbrei W, Taube SE, Gion M, Clark GM (2005) Reporting recommendations for tumor marker prognostic studies. Journal of Clinical Oncology 23:9067–9072

Minic J, Persuy MA, Godel E, Aioun J, Connerton I, Salesse R, Pajot-Augy E (2005) Functional expression of olfactory receptors in yeast and development of a bioassay for odorant screening. FEBS Journal 272:524–537

Mitsubayashi K, Dicks JM, Yokoyama K, Takeuchi T, Tamiya E, Karube I (1995) A Flexible Biosensor for Glucose. Electroanalysis 7:83–87

Moats RK, Sullivan BM (2004) Combinatorial augmentation for a multi-pathogen biosensor: signal analysis and design. Biosensors & Bioelectronics 19:1673–1683

Nakamura C, Inuyama Y, Goto H, Obataya I, Kaneko N, Nakamura N, Santo N, Miyake J (2005) Dioxin-binding pentapeptide for use in a high-sensitivity on-bead detection assay. Analytical Chemistry 77:7750–7757

Pfleger KDG, Eidne KA (2006) Illuminating insights into protein-protein interactions using bioluminescence resonance energy transfer (BRET). Nature Methods 3:165–174

Piehler J (2005) New methodologies for measuring protein interactions in vivo and in vitro. Current Opinion in Structural Biology 15:4–14

Pircher A, Bauer F, Paulsen P (2007) Formation of cadaverine, histamine, putrescine and tyramine by bacteria isolated from meat, fermented sausages and cheeses. European Food Research and Technology 226:225–231

Przybylo M, Borowik T, Langner M (2007) Application of liposome based sensors in high-throughput screening systems. Combinatorial Chemistry & High Throughput Screening 10:441–450

Pu L (2004) Fluorescence of organic molecules in chiral recognition. Chemical Reviews 104:1687–1716

Rakow NA, Suslick KS (2004) Novel materials and applications of electronic noses and tongues (vol 406, pg 710, 2000). MRS Bulletin 29:913–913

Ramos L, Ramos JJ, Brinkman UAT (2005) Miniaturization in sample treatment for environmental analysis. Analytical and Bioanalytical Chemistry 381:119–140

Rasooly A, Jacobson J (2006) Development of biosensors for cancer clinical testing. Biosensors & Bioelectronics 21:1851–1858

Reyes FL, Jeys TH, Newbury NR, Primmerman CA, Rowe GS, Sanchez A (1999) Bio-aerosol fluorescence sensor. Field Analytical Chemistry and Technology 3:240–248

Rodriguez-Diaz RC, Aguilar-Caballos MP, Gomez-Hens A (2003) Simultaneous determination of ciprofloxacin and tetracycline in biological fluids based on dual-lanthanide sensitised luminescence using dry reagent chemical technology. Analytica Chimica Acta 494:55–62

Rodriguez-Mozaz S, Marco MP, de Alda MJL, Barcelo D (2004) Biosensors for environmental applications: Future development trends. Pure and Applied Chemistry 76:723–752

Roitrub BA, Zlatin RS (1983) A method of increasing the sensitivity of the acetylcholine bioassay. Fiziologicheskii Zhurnal 29:237–239

Samaranayake L (2007) Saliva as a diagnostic fluid. International Dental Journal 57:295–299

Sapsford KE, Ngundi MM, Moore MH, Lassman ME, Shriver-Lake LC, Taitt CR, Ligler FS (2006) Rapid detection of foodborne contaminants using an Array Biosensor. Sensors and Actuators B-Chemical 113:599–607

Schiller A, Wessling RA, Singaram B (2007) A fluorescent sensor array for saccharides based on boronic acid appended bipyridinium salts. Angewandte Chemie-International Edition 46:6457–6459

Schneider SE, O'Neil SN, Anslyn EV (2000) Coupling rational design with libraries leads to the production of an ATP selective chemosensor. Journal of the American Chemical Society 122:542–543

Shepard JRE (2006) Polychromatic microarrays: simultaneous multicolor array hybridization of eight samples. Analytical Chemistry 78:2478–2486

Simonet BM, Valcarcel M (2006) Analytical chemistry in modern society: what we can expect. Microchimica Acta 153:1–5

Staneva D, Betcheva R, Chovelon J-M (2007) Optical sensor for aliphatic amines based on the simultaneous colorimetric and fluorescence responses of smart textile. Journal of Applied Polymer Science 106:1950–1956

Storrow AB, Gibler WB (1999) The role of cardiac markers in the emergency department. Clinica Chimica Acta 284:187–196

Szurdoki F, Ren DH, Walt DR (2000) A combinatorial approach to discover new chelators for optical metal ion sensing. Analytical Chemistry 72:5250–5257

Taussig MJ, Stoevesandt O, Borrebaeck CAK, Bradbury AR, Cahill D, Cambillau C, de Daruvar A, Dubel S, Eichler J, Frank R, Gibson TJ, Gloriam D, Gold L, Herberg FW, Hermjakob H, Hoheisel JD, Joos TO, Kallioniemi O, Koegl M, Konthur Z, Korn B, Kremmer E, Krobitsch S, Landegren U, van der Maarel S, McCafferty J, Muyldermans S, Nygren PA, Palcy S, Pluckthun A, Polic B, Przybylski M, Saviranta P, Sawyer A, Sherman DJ, Skerra A, Templin M, Ueffing M, Uhlen M (2007) ProteomeBinders: planning a European resource of affinity reagents for analysis of the human proteome (vol 4, pg 13, 2007). Nature Methods 4:126–126

Tian H, Ip L, Luo H, Chang DC, Luo KQ (2007) A high throughput drug screen based on fluorescence resonance energy transfer (FRET) for anticancer activity of compounds from herbal medicine. British Journal of Pharmacology 150:321–334

Ulber R, Frerichs JG, Beutel S (2003) Optical sensor systems for bioprocess monitoring. Analytical and Bioanalytical Chemistry 376:342–348

Uttamchandani M, Wang J, Yao SQ (2006) Protein and small molecule microarrays: powerful tools for high-throughput proteomics. Molecular Biosystems 2:58–68

Venugopal V (2002) Biosensors in fish production and quality control. Biosensors & Bioelectronics 17:147–157

Von Lode P, Rainaho J, Pettersson K (2004) Quantitative, wide-range, 5-minute point-of-care immunoassay for total human chorionic gonadotropin in whole blood. Clinical Chemistry 50:1026–1035

Wang J (2000) From DNA biosensors to gene chips. Nucleic Acids Research 28:3011–3016

Wang L, Drysdale TD, Cumming DRS (2007) In situ characterization of two wireless transmission schemes for ingestible capsules. IEEE Transactions on Biomedical Engineering 54:2020–2027

Wiskur SL, Ait-Haddou H, Lavigne JJ, Anslyn EV (2001) Teaching old indicators new tricks. Accounts of Chemical Research 34:963–972

Yang Z, Zhou DM (2006) Cardiac markers and their point-of-care testing for diagnosis of acute myocardial infarction. Clinical Biochemistry 39:771–780

Yapijakis C, Panis V, Koufaliotis N, Yfanti G, Karachalios S, Roumeliotou A, Mantzavinos Z (2006) Immunological and molecular detection of human immunodeficiency virus in saliva, and comparison with blood testing. European Journal of Oral Science 114:175–179

Yu BZ, Ju YM, West L, Moussy Y, Moussy F (2007) An investigation of long-term performance of minimally invasive glucose biosensors. Diabetes Technology & Therapeutics 9:265–275

Zhu H, Snyder M (2003) Protein chip technology. Current Opinion in Chemical Biology 7:55–63

Epilogue

The community of common citizens should prepare itself for a great revolution. This revolution should produce changes in human lifestyles greater than mobile phones and the Internet. At the beginning of the XXI century, the most developed countries have reached the level of computerized societies. In the decade ahead, we will enter into a *sensorized society*. This means that sensors will become part of everyday life and the results of their application will become strongly involved in human decisions both on the personal and community levels.

To say that fluorescence sensing technologies develop rapidly is to say nothing. In fact, the development is explosive. The reader may compare the present usage of analytical methods with that achieved in advanced laboratories. Until now, in many industrial laboratories the measurements are made by taking a sample and analyzing it for a single analyte. When multiple analytes have to be measured from a single sample, the sample is conventionally divided into appropriate aliquots and each aliquot is analyzed separately, often according to different protocols. This approach is exemplified by many of the clinical analyzers employed in today's hospital laboratories. The patient may need to wait hours or days for the results of the analysis. Now, there are novel sensor technologies that bring about revolutionary changes. Analysis becomes possible for hundreds or thousands of analytes simultaneously, its results can be obtained in minutes or even seconds, they can be automatically communicated to a central unit for on-line analysis and decision making. Very soon, monitoring the technological process will be made possible from a mobile phone and the patient that needs to monitor a complex of health-related parameters will go and buy a smart, pocket-sized instrument in the supermarket.

Such a rapid technological development is heated up by the fact that in every aspect of sensor design and application there is strong competition from different technologies based on different principles. This can be seen starting from the first pages of this book. Will direct sensing beat the sandwich assays in performance and what will be the role of competitive assays? (Chapter 1). How does the theoretically imposed limit on the dynamic range of target detection have to be overcome? (Chapter 2). What is the best choice of fluorescence detection method among a number of powerful techniques offered by optical instrumentation? (Chapter 3). What are the best fluorescence reporters of organic (e.g., dyes) and inorganic (e.g., Quantum Dots) or biosynthetic (e.g., green fluorescent protein) origin? (Chapter 4). How does one select optimal recognition units that originate from chemical and biological worlds? Should it be a rational design or a process of selection involving huge libraries? (Chapter 5). How does one combine the recognition and reporting units, providing proper signal transduction? This can be done by different physical mechanisms that may involve the effects of proximity and conformational variables (Chapter 6). How do we provide the assembly of smart supramolecular sensors on different surfaces or their nano-scale self-assembly in solution? (Chapter 7). Exciting non-conventional methods exist for excitation and monitoring emission, how do we use them optimally? (Chapter 8). Using all this immense information, many possibilities exist for constructing the instrumentation – which one is the best for particular needs? (Chapter 9). Are we presently able to respond to the challenges of detecting the most important targets of the physical, chemical and biological worlds? (Chapter 10). Can we transfer these sensing technologies to observations inside the living cells? What new opportunities in clinical diagnostics and treatment will that open? (Chapter 11). Finally, what are the prospects of sensing technologies and what opportunities and limitations will we observe beyond the horizon? (Chapter 12).

Throughout this book we did not attempt to respond to all these questions because of the immense number of existing possibilities, both realized and yet unrealized. Only one final remark should be made. In the present days, we often observe excellent technological solutions based on seemingly outdated concepts and, at the same time, ideas offering revolutionary breakthroughs but presented only on the level of 'proof of concept' and based on a rather primitive form of technology. This is indispensable in explosive development. Usable products based on frontier science combined with frontier technology – this is the challenge for future generations of researchers.

The author must confess that this frontier of science and technology cannot be recognized easily because sensor research is the crossing point of many disciplines. It is the field of mutual penetration of science and technology of inorganic and organic man-made materials with biotechnology that is derived from studies of living matter. The new technical solutions are the result of combining the highly specific recognition ability of biomolecules with the unique structural character of inorganic nanomaterials, such as nanocrystals, nanotubes and nanowires. Endowing these materials with reporting functions is the crossing point of these disciplines with photophysics and photochemistry, plus optics and photonics, microelectronics and many more. Various kinds of researchers are required to cooperate in order to achieve a valuable result, stimulating research and bringing out a useful product.

The potentially existing multitude of solutions will be finally selected by the end users. To summarize, we can outline only the general tendencies that are becoming the criteria for this selection. They are *miniaturization, simplification* and *multifunctionality*. Sensors with such advanced properties are expected to be mass produced and be as useful to everyone as mobile phones.

This field of sensor development, bringing together research and technology, should be extremely attractive for fresh bright minds that will make a selection among these possibilities for the exploration of their personal skills. New skills and, moreover, a new way of thinking, is also needed from the side of users. The ways we make measurements, process the data from these measurements and use the information that can be extracted from these data – all that should change profoundly.

Every day brings new information in the form of presentations, publications and patents in this hot field. Navigating in this ocean of information becomes difficult without guidance. The author will be happy if this book is useful to the reader by serving as such a guide

Appendix
Glossary of Terms Used in Fluorescence/ Luminescence Sensing

Anisotropy is a parameter used in the analysis of polarized fluorescence emission. It characterizes fluorescence probe rotation during the excited state lifetime. Polarized excitation selects dye molecules with particular orientation in space and their emission will be also polarized in the case of absence of their rotation. Rotation during fluorescence lifetime will depolarize its emission, resulting in a mechanism with which to measure the rigidity (viscosity) of the environment containing the dye or the mass of the rotating unit to which it is attached. Fluorescence anisotropy is defined as the ratio of the difference between the emission intensity parallel to the polarized electric vector of the exciting light and the intensity perpendicular to the vector, divided by the total intensity. Possible and useful are measurements of anisotropy in time-resolved format.

Antibodies (Ab) are the proteins synthesized by the immune system to recognize, bind and produce the functional response to molecules, particles, viruses and cells foreign to this organism (antigens). Two types of antibodies are used as recognition elements in sensing: monoclonal, produced in cell cultures called hybridomas and gene-recombinant antibodies. The latter can be of reduced size (Ab fragments).

Aptamers are single-strand DNA or RNA molecules that have been selected from random pools based on their ability to bind other molecules. Using repetitive binding-enrichment procedure, aptamers have been obtained that bind nucleic acids, proteins, small organic compounds, and even entire organisms with high affinity. The basis of their target recognition is the tertiary structure formed by the single-stranded oligonucleotides that allows target-induced conformational changes.

Argon ion laser is the laser that is the most commonly used in microscopy and flow cytometry. Its principal emission lines are at 488.0 and 514.5 nm.

Array technology is any high-throughput methodology permitting analysis of many (hundreds, thousands) binding events of different specificity in parallel, usually using spotting of many sensors on the surface of the same unit (spotted arrays) or dispersing them (suspension arrays).

Avidin is a protein that binds a small molecule biotin with high specificity and affinity. Avidin or its bacterial analog streptavidin are the most frequently used to form self-assembled macromolecular structures.

Bar code is the identification system that is commonly used in supermarkets. It employs a series of machine-readable lines of varying widths of black and white and is read with a laser scanner. In sensing technologies it is the system for identification of sensor particles in multiplex detection arrays.

Binder is the sensor recognition unit that interacts with the target with good affinity and specificity. It has to be obtained in response to the requirement for the target structure that can be as small as an ion chelating group and as large as a sizeable macromolecule, such as protein or DNA or even an imprinted polymer structure.

Binder-reporter coupling is the coupling between sensing elements (binders) with fluorescent elements (reporters) that provides transduction of the signal on target binding into a detectable signal.

Biochip is a broad term for various types of highly parallel functioning miniaturized detection devices that are deposited on solid supports. They include various types of DNA, protein chips and microarrays.

Bioluminescence is a biochemical oxidative process that results in the release of energy as emitted light. Firefly luminescence, which requires the enzyme luciferase to catalyze a reaction between the substrate luciferin and molecular oxygen (in the presence of adenosine triphosphate), is a commonly employed example of bioluminescence. This phenomenon occurs in a wide variety of marine organisms and insects. The luciferin-luciferase complex can be genetically engineered within living cells, allowing many possibilities for sensing.

Biosensor is a device that allows the detection and quantification of biomolecules, commonly by biospecific interaction between macromolecules (molecular recognition). Usually, this occurs in a single step. Molecular recognition, which is the formation of a complex between the biomolecule of interest (target) and its recognition element (binder or receptor), generates a recordable signal.

Brightness is an important characteristic of a fluorescence reporter, which determines the absolute sensitivity of detection. It is often defined as the product of molar absorbance at the wavelength used for fluorescence excitation and the fluorescence quantum yield. A more precise definition involves the account of maximum rate at which the fluorophore is able to emit photons and the number of excitation-emission cycles before photobleaching.

Calibration in fluorescence sensing is an operation, as a result of which at every sensing element (molecule, nanoparticle, etc.) or at every site of the image the fluorescence signal becomes independent of any other factor, except the concentration of bound target.

Charge coupled device (CCD) is a monolithic semiconductor device used for collecting images. It is arranged as an array of elements such that the output from one element provides the stimulus for the next. A response in the form of an image in artificial colors can be created in these devices.

Chemical sensor is a designed molecule or miniaturized analytical device that delivers real-time and online information on the presence of specific compounds in complex samples. According to the definition recommended by IUPAC, "a chemical sensor is a device that transforms chemical information, ranging from the concentration of a specific sample component to total composition analysis, into an analytically useful signal".

Combinatorial chemistry is a branch of chemistry that is concerned with the mass synthesis (creating a 'library') of compounds and their systematic evaluation (screening) for desired properties or functions.

Confocal laser scanning microscopy is a technique that allows high resolution within the focal plane by using a so-called pinhole that discriminates the out-of-plane emission. This enables a sharp image at different depths in the specimen. To obtain the image, a focused laser beam is scanned laterally along the **x** and **y** axes of a specimen in a raster pattern, to generate fluorescence emission. The latter is sensed by a photomultiplier tube and displayed in pixels on a computer monitor. The pixel display dimensions are determined by the sampling rate of the electronics and the dimensions of the raster. This technique enables the specimen to be optically sectioned along the **z** axis. Multiple images taken at different depths can be composed into a three-dimensional object representation.

Dextran is a polysaccharide with a high molecular weight of 3,000 to 70,000, which can be used in sensing technologies for different purposes. It can serve as a support for supramolecular structures and as a competitor in glucose sensing. It can be labeled with a fluorescent dye and constitute the shell of a nanoparticle.

Diffraction limit is the limit of resolution that cannot be overcome in standard optical methods because of the diffraction property of light. This limit is about 200–400 nm, a half wavelength of illumination source.

Direct sensors are sensors in which the transduction of the signal produced on complex formation with analytes does not need any additional steps or third interacting partners. Direct sensors demonstrate the highest speed of response that is limited only by the rate of target binding.

Dynamic range is the range between the maximum to minimum measurable values of the parameter used for sensing response. In imaging, it is the difference in intensity between the lightest and darkest points of the image.

Electronic State is the well specified overall configuration of electrons in an atom or molecule. Any given molecule can exist in one of several electronic states (ground or multiple excited), depending upon the total electron energy and on the symmetry of the electron spins in the orbitals. Under normal conditions, the majority of molecules exist in the electronic state with the least energy (the ground state). When the molecule absorbs a photon, electrons are excited to higher energy states (excited states).

Emission spectrum is the wavelength or the energy (wavenumber) function of the intensity of light emitted by a fluorophore after it has been excited by absorbing a light quantum. Usually, the emission spectrum is located at lower energies than the absorption spectrum (Stokes shift) and is independent of excitation wavelengths, even if the fluorophore is excited to different high-energy electronic states (Kasha rule).

Epi-fluorescence microscopy (EFM) is a microscopic technique in which the illumination source (often termed a vertical illuminator) is placed on the same side of the specimen as the objective, which serves a dual role as both condenser and imaging lens system. Monochromatic excitation light is used to excite the fluorophores in the specimen via the microscope objective. The fluorescence light emitted from the specimen is again captured by the objective lens and filtered to the appropriate detection wavelength window by passage through a dichroic mirror in the microscope. This part of the emitted light is detected by an electronic camera or can be observed at the microscope oculars.

Excited state intramolecular proton transfer (ESIPT) is a reaction occurring between two groups of atoms located closely within the same molecule, the donor and the acceptor. ESIPT results in reorganization of the electronic system, leading to dramatic transformations in fluorescence spectra.

Excitation spectrum is the wavelength or the energy (wavenumber) function of the intensity of light excited by a fluorophore and detected at a fixed emission wavelength. Usually, the excitation spectrum matches the absorption spectrum.

Enzyme linked immunosorbent assay (ELISA) is a sensor technique based on sandwich methodology. It involves antibody or antigen sensing by complementary antigens or antibodies immobilized on the surface. The binding is detected by additional antibodies coupled with an enzyme that produces a colored enzyme reaction. The method requires the addition of auxiliary reagents and/or the separation between free and bound interacting partners.

Energy of light quanta. Matter absorbs light in discrete quanta (photons). The energy of a light quantum is proportional to the position of the corresponding band in the spectrum on a wavenumber scale and is inversely proportional to its position on a wavelength scale. The relation is: $E = h\nu = 10^7 \, hc/\lambda$, where ν is the wavenumber in cm^{-1} and λ is the wavelength in nm, h is the Planck's constant and c the velocity of light.

Excimers and **exciplexes** are dimers formed in the excited state. Their formation requires the proximity and proper orientation of interacting partners. Often, they are non-fluorescent. In some cases, however (such as for pyrene exciplexes), a bright structureless and long-wavelength shifted fluorescence spectrum can be detected.

Excitation is the process of absorption of light quanta that raises a molecule from its ground energy state to an excited state of higher energy.

Excitons are imaginary particles (quasi-particles) that allow describing the propagation of electronic excitation in the solid. Excitons are composed of negative and positive charges (electron + hole) and are electrostatically neutral. In sensor technologies, they are used for describing the properties of semiconductor nanoparticles and conjugated polymers.

Fiber optics represents a flexible bundle of glass, plastic or quartz fibers, with a cladding of lower refractive index, designed for transmission of light without loss at long distances. Light is transmitted along individual fibers by total internal reflection. Fiber optic elements are broadly used for illumination and imaging.

Filter (optical) is a device that reduces the radiant power or spectral range of transmitted light. Bandpass filters isolate a particular wavelength range, whereas cut-off filters only permit the transmission of radiation longer or shorter than the specified wavelength.

Flow cytometry is a technique allowing the cells or particles to be counted and analyzed during their flow through a very thin capillary, at a rate of thousands per second. The analysis may involve several channels of fluorescence detection and therefore several markers can be used simultaneously and together with light-scattering detection.

Flow cytometric analysis is usually displayed on a computer window as a two-dimensional image with dots representing counted objects (cells or particles) as a function of forward *versus* side-scatter, or forward scatter *versus* fluorescence intensity, or two fluorescence intensities at different wavelengths.

Fluorescent label is a fluorescent reporter group that is bound covalently to a natural or synthetic macromolecule, nanoparticle or at an interface. In proteins the easiest modification is achieved upon labeling at their $-SH$ and $-NH_2$ groups.

Fluorescent latex particles are mono-dispersed polymeric, commonly polystyrene, beads (microspheres) having a diameter between 0.01 and 15 µm, with an incorporated fluorescent dye. They allow high brightness and are used for observing the transport in axons, research into phagocytosis and measurement of blood flow or as the standard for determining resolution and microscopy performance in fluorescence microscopy.

Fluorescence biosensors are biosensor devices for which the transduction mechanism transforming the act of biospecific recognition into response is based on fluorescence.

Fluorescence correlation spectroscopy is a technique that uses measurement of fluorescence intensity fluctuations in a microscopic volume. This volume is formed by a tightly focused laser beam. The response is sensitive to diffusion rate in and out of this volume, which allows the detection of intermolecular interactions.

Fluorescence lifetime (or, more generally, the excited-state lifetime) is the characteristic time during which a molecule remains in an excited state prior to returning to the ground state. Usually, it is an averaged parameter that does not depend upon the dye concentration. In a homogeneous system without excited-state reactions, it follows the law of exponential decay. In this case, the fluorescence lifetime can be defined as the time in which the initial fluorescence intensity of a fluorophore decays to 1/e (approximately 37%) of the initial intensity. This quantity is the inverse of the rate constant of the fluorescence decay from the excited state to the ground state.

Fluorescence lifetime imaging microscopy (FLIM) is a sophisticated technique that enables simultaneous recording of both the fluorescence lifetime and the spatial location of fluorophores throughout every site in the image. FLIM results are independent of localized fluorophore concentration, photobleaching artifacts and path length (specimen thickness), but are sensitive to excited state reactions such as resonance energy transfer or dynamic quenching. Long-lifetime luminophores are sensitive to quenching by diffusing oxygen. The methodology provides a mechanism to investigate different intracellular processes in spatial and temporal dimensions and intracellular sensing of various parameters, such as pH, ion concentration, solvent polarity and oxygen tension.

Fluorescence recovery after photobleaching (FRAP) is a technique developed for studying the translational mobility (lateral diffusion coefficients) of fluorescently labeled macromolecules, fluorescent nanoparticles and small fluorophores. In FRAP, a very small selected volume is subjected to intense laser illumination to produce complete photobleaching of the fluorophores. After the photobleaching pulse, the rate and extent of fluorescence intensity recovery in the bleached region is monitored as a function of time, to generate information about repopulation by fluorophores and the kinetics of recovery.

Förster resonance energy transfer (FRET) is a physical process in which the primary excited fluorescent molecule (donor), instead of emission transfers its excitation energy to another molecule (acceptor), so that the latter can be either emissive or quenched. Efficiency of transfer depends on donor-acceptor distance and, in the case of weak electronic coupling between them, the distance dependence is observed in the range of 1–10 nm and is very steep. This forms a background for the design of molecular sensors, in which this distance changes in the binding event. Efficiency of this transfer depends on the properties of donor and acceptor (overlap of fluorescence and absorption spectrum between them), which allows efficient selection of donor-acceptor pairs. The acronym FRET is often presented as 'Fluorescence Resonance Energy Transfer', which is incorrect, since this mechanism does not necessarily involve light emission in the form of fluorescence.

Fluorophore is a molecule or particle and also their structural domain or specific region, which is capable of emitting fluorescence.

Fluorescence probe is a fluorescent molecule designed to localize within a specific region of a structurally heterogeneous specimen and/or to respond to a specific stimulus by change of its parameters of fluorescence.

Fluorescence reporter is a functional unit that provides a response to target binding by changing one or several fluorescence parameters. In most cases, it is an organic dye that contains an extended π-electronic system with excitation and emission in the convenient visible range of the spectrum. Coordinated transition metal ions that produce luminescence emission with extended lifetimes can be also used. Systems that use green fluorescent protein (GFP) and its analogs are applied in cellular research. Some metal and metal oxide nanoparticles known as Quantum Dots can function as reporters.

FRET microscopy is an adaptation of the resonance energy transfer phenomenon to fluorescence microscopy in order to obtain quantitative temporal and spatial information about the binding and interaction of proteins, lipids, enzymes and nucleic acids in living cells. FRET microscopy is performed using either steady-state or time-resolved techniques however, time-resolved FRET imaging has the advantage of more accurately mapping the transfer efficiency and therefore, donor-acceptor distance.

Green fluorescent protein (GFP) is a protein that was originally isolated from jellyfish and emits green fluorescence under blue-light excitation. The emission originates from spontaneously formed fluorophores by the reaction of three amino acids in its sequence during protein folding inside the living cells. The GFP gene has been isolated and improved to obtain different color variants. By gene fusion and expression in living cells, different proteins and structures formed by them can be visualized, which allows one to determine the location, concentration, interactions and dynamics. Visible fluorescent proteins of different colors can be isolated from different species or obtained by mutations.

Image analysis is a methodology used to extract data from images rather than counting or measuring directly. The advantages are that the captured images are permanent, allowing more time for study and they can be subjected to enhancement and automated analysis techniques to improve data quality. Image analysis begins with a digitized image, which means that the image is composed of a rectangular array of individual points of color, called pixels. The pixel value represents a color that can be displayed on a computer monitor or some other display device. Using computers, the image pixels can be manipulated mathematically to change the appearance of the image or to determine some quality of the image.

Immunoassay is a general term for any technique or device that relies on recognition between an analyte called an antigen and specific protein, the antibody (Ab). Various detection techniques use specific antigen-antibody complexation for analytical purposes. Radioimmunoassays, enzyme immunoassays and fluorescence immunoassays are well-established in clinical diagnostics.

Immunosensors are important analytical tools for monitoring antibody-antigen reactions in real time. Recent developments in immunosensors have produced systems that allow rapid and continuous analysis of the binding event without the requirement for added reagents or separation/washing steps.

Intercalators are small planar molecules that insert specifically between adjacent base pairs in double-stranded DNA.

Intramolecular charge transfer (ICT) is the process that changes the overall charge distribution in a molecule. It is a frequently observed process in the excited states of organic dyes.

Lab-on-a-chip is a highly miniaturized device that is able to carry out instant analysis, often in combination with sample preparation and separation of its components.

Light emitting diode (LED) is a small, efficient, low-voltage, low-power, cold light source that can be supplied at any visible wavelength and produced narrow-spectrum light emission.

Luminescence is the emission of light from any substance (usually a molecule or atom) that occurs from an electronically excited state generated either by a physical (light absorption), mechanical or chemical mechanism. Luminescence is formally divided into fluorescence and phosphorescence. The generation of luminescence can occur through excitation by ultraviolet or visible light photons (photoluminescence), an electron beam (electrochemiluminescence), application of heat (thermoluminescence), chemical energy (chemiluminescence) or a biochemical enzyme-driven reaction (bioluminescence).

Microarrays are high-throughput versions of the array technology that allow spotting sensor molecules with a high density.

Molecular library is a body of entities designed to handle a large numbers of molecules, usually with similar properties, at the same time.

Molar absorbance is a direct measure of the ability of a molecule to absorb light. It is also an important molecular parameter used for converting units of absorbance into units of molar concentration for a variety of chemical substances. It is determined by measuring the absorbance at a given wavelength characteristic of the absorbing molecule (usually, the band maximum) for a 1 molar (M) concentration (1 mol/l) of the target chemical in a cuvette having a 1 cm path length.

Multiplex analysis is simultaneous analysis of many targets, usually in the same sample and using the same assay conditions.

Membrane potential is an important parameter characterizing electric fields in and across biological membranes. In a living cell, three types of membrane potential are distinguished with the use of fluorescence probes. It is the surface potential produced by charged proteins and lipids on the membrane surface and partially screened by counter-ions. This potential becomes strongly negative on apoptosis, due to the externalization of anionic lipids. The cells usually possess strong (200–500 V/cm) dipole potential, produced mostly by lipid carbonyl groups and also transmembrane potential. The latter is of about -70 mV (negative inside the cells). Mitochondria possess a stronger transmembrane potential, of about -150 mV. A family of fluorescence membrane potential sensitive probes has been developed.

Molecular biosensors are devices of molecular size that are designed for sensing different analytes on the basis of biospecific recognition. They should provide two coupled functions – the recognition (specific binding) of the target and the transduction of information about the recognition event into a measurable signal.

Molecular recognition is a strong and highly specific interaction between biological macromolecules, based on multipoint noncovalent interaction. Often, it requires steric complementarity in contact areas. The contact areas of interacting molecules may be flexible and local structuring occurs in the course of interaction in these cases. Also, a conformational change may occur in one of the components.

Monochromatic light is a light beam that is composed of a single wavelength. In a sense, this is an idealization, since even a laser emission or the emission line of a mercury lamp has a measurable bandwidth.

Near-field microscopy is the principle of increasing the resolution of optical images to sub-wavelength scale. In its realization in scanning near-field optical microscopy (SNOM), the sample is illuminated through a small aperture held less than 10 nm from the sample surface.

Nanosensors are nano-scale hybrid systems that integrate other components, together with recognition and reporting functionalities and their support matrix, allowing a synergistic approach in modulating the target binding and improving or enhancing the response.

Optical sensor is an analytical device that is based on measurement of either intrinsic optical properties of the target molecules (supramolecular structures, cells), or of optical properties of indicator dyes, nanoparticles or labeled biomolecules in solutions or attached to solid supports.

Optodes are integrated optical sensors, the sensing principle of which is based on a chemical transduction of the signal on the analyte binding event on the surface of the polymeric membrane. The coupled fluorescence dye can respond to the produced change of pH. The polymeric membrane can be deposited onto the tip of optical fiber. Such optodes are actively used for detection of Na^+, K^+, Ca^{++} and Cl^- ions in clinical samples.

Photobleaching is the irreversible loss of fluorescence intensity over a period of time due to chemical destruction, leading to nonfluorescent products (color fading). High photostability is a very desirable property for various applications, especially for those in which intense illumination is needed (e.g., in microscopy). The rate of photobleaching can be dramatically reduced by lowering the excitation light flux or by limiting the oxygen concentration in the medium. Quantum Dots exhibit much higher photostability than organic fluorophores.

Phosphorescence is a long-lifetime (measured in milliseconds to seconds) emission from the fluorophore triplet state. This state can be achieved from the initially excited singlet state in a process called intersystem crossing.

Photochemical reactions are the reactions of chemical transformations that need photoexcitation and occur in the excited states. Many reactions of electron and proton transfer as well as isomerizations cannot occur in the ground state because of the high energy barrier separating reactant and product states. This barrier may become negligible in the excited state and therefore, many photochemical reactions are very fast. These reactions are actively used for providing fluorescent responses in sensing. However, some of them produce photobleaching and are not desirable.

Photoinduced electron transfer (PET) is the electron transfer occurring in the excited state. It is a short-distance reaction that can be both intramolecular and intermolecular (with the participation of other molecules and the solvent). PET commonly leads to quenching and is in the background for the operation of many sensors responding to the change of fluorescence intensity.

Photomultiplier tube (PMT) is an electrical device designed to collect and amplify photon signals. Incoming photons strike a target in the face of a photomultiplier to liberate free electrons, which are accelerated onto a dynode that in turn liberates an amplified stream of electrons. Several dynodes are arranged in series to produce a tremendous degree of amplification from each original photon and then transmit the signal to a processing circuit. Unlike area-array detectors such as charge-coupled devices (CCDs), photomultipliers do not form an image.

Pinhole is a device used in confocal microscopy for obtaining a sharp image in a focal plane. It allows one to produce thin optical sections of focal planes in the specimen.

Polarization of fluorescence. When fluorescence is excited by polarized light it is observed as partially depolarized and the extent of depolarization depends on rotation of fluorescent dye (together with rigidly attached molecules or their segments) during the fluorescence lifetime. The measurements of polarization are used in fluorescence polarization assays. Anisotropy of fluorescence is a parameter similar to polarization.

Quantum Dots are semiconductor nanoparticles that are characterized by broad excitation and a very sharp emission spectrum, high brightness and extremely high photostability. The narrow symmetric emission spectrum is tunable by variation of the particle size (within 1–10 nm), which makes them good tracers for cellular research and donors or acceptors for FRET.

Quantum yield of fluorescence is the ratio of the number of photons emitted as fluorescence to the total number of absorbed photons. In other words, the quantum yield represents the probability that a given excited fluorophore will produce an emitted (fluorescence) photon. It is a measure of efficiency of fluorescence compared to other processes occurring during excited-state lifetime: static and collisional quenching, transition to triplet state, etc. Commonly, the quantum yield is determined by integration over the whole fluorescence spectrum and comparison with a well-characterized standard. In some dyes (e.g., rodamins), it may approach 100%.

Quenching is the reduction or complete disappearance of fluorescence emission. Different intermolecular interactions either induce quenching or protect from quenching produced by other interactions. There are two types of quenching: static and dynamic and they influence the parameters of emission differently. Fluorescence quenching is frequently used in different sensing technologies.

Quencher is a molecule or a group of atoms that upon interaction with fluorophores induces a transition from excited to ground state without emission. A substituent may provide this effect by electron transfer or by hydrogen bond formation. External molecules may provide quenching by forming a complex with the fluorescent dye (static quenching) or by a collisional approach to it (dynamic quenching). Dynamic quenching reduces fluorescence lifetime, whereas static quenching does not.

Ratiometric dyes are dyes that respond to different environmental perturbations in such a way that allows convenient detection of these changes by evaluation of the ratio of two independently measured parameters, usually of intensities at selected wavelengths. Ratiometric effects can be produced either by spectral shift or by variation of relative intensities of two or more bands. In the case of spectral shifts, the ratio of intensities at the two wings of the spectrum and in the case of two bands at two band maxima, are recorded.

Recognition elements (receptors or binders) are the functional sensor components that are responsible for strong and selective target binding. They can be chemical constructs such as ion chelators but can be also natural and designed biopolymers (proteins and nucleic acids).

Relaxation is the process of attaining thermal equilibrium of an excited or otherwise perturbed system with the environment. Orientational relaxation of dipoles in the probe environment, leading to long-wavelength shift of their spectra, is the key mechanism of response to variation of polarity of many fluorescence probes.

Response function of the sensor is the dependence of the recorded parameter of fluorescence on the target concentration in the sensed medium. This function is determined in sensor calibration.

Selectivity (or **specificity**) is the capability of a sensor to establish an especially strong interaction with the target analyte, thus recognizing it in the presence of other competing and less gifted targets. Its quantitative estimate is usually made by comparison of the binding constants for specific and nonspecific targets.

Signal-to-noise ratio (S/N) is the standard ratio of the reporter optical signal from a specimen to the unwanted optical signals of the surrounding background. Noise is defined as the square root of the sum of the variances of contributing noise components. It is different from the background emissions that can be produced by a steady contribution from the impurities. In fluorescence sensing, both the noise and the background emission can be suppressed by using brighter reporting emitters and increasing their concentration.

Single molecule detection represents a family of techniques that detect intermolecular interactions on the level of single molecules. They allow obtaining responses from single molecules in conditions of their high dilution and a dramatic decrease of illumination/detection volume, down to femtoliters. This technique is extremely useful in detecting rare events and in studies of heterogeneous systems, where the interesting effects can be hidden within ensemble-averaged signals obtained in common conditions.

Steady-state fluorescence encompasses all spectroscopic and imaging measurements performed with constant illumination and observation.

Stokes shift is the shift in energy between absorbed and emitted quanta. It is measured as the difference in positions of absorption and emission band maxima on an energy (wavenumber) scale (in cm^{-1}). The Stokes shift is commonly observed in the direction of low energies and long wavelengths. Dyes exhibiting a stronger Stokes shift are preferable for fluorescence sensing and imaging because this enables more efficient isolation of excitation and emission light by optical filters. For these dyes, the effects of light-scattering, re-absorption of emitted light and quenching at high concentrations are smaller.

Stroboscopic techniques are used in illumination and in detection of fluorescence in the time domain. For excitation, they use short-duration intense pulses (strobes) of light, often repeated at regular intervals. In time-resolved spectroscopy, the strobes are synchronized with pulsed excitation repeated cycles of opening-closing the high voltage on the photodetector.

Suspension arrays are arrays of sensor elements that are not fixed in position but float freely and are available for target binding in solution. The size of these elements can be $\sim5\,\mu m$ or less. For their identification they should possess optical or another type of 'barcodes' and be analyzed one by one by flow cytometry or its specialized version.

Transducer is a functional element that converts a change in properties such as molecular mass, refraction index or formation/disruption of noncovalent bonds, upon complex formation into a recordable analytical signal. In fluorescence sensing, the most important mechanisms of transduction are the influence of target binding on electronic polarization, conformation, freedom of rotation and ionization of fluorescence reporters. Intermolecular interactions resulting in quenching, formation of excimers and exciplexes and also distance-dependent FRET can also be used. In contrast with classical bioassays, such as ELISA the detection of an interaction between the recognition unit and the target with the aid of transducer, is direct and rapid.

Two-photon excitation is electronic excitation by simultaneous absorption of two light quanta. The sum of their energies should fit the energy of common single-photon excitation. Since simultaneous absorption of two photons is a low-probable event, two-photon excitation has to be achieved by properly focused pulsed lasers that provide extremely high instantaneous intensity. In this case, the rate of excitation is not a linear but a square function of light intensity. This allows achieving a very high contrast and sharpness of images. In two-photon fluorescence microscopy, the problems of photobleaching and autofluorescence are much less important than in confocal microscopy.

Visible light is the part of the spectrum (400–760 nm) that can be detected by the naked eye. This part of the light spectrum is used in most fluorescence sensor applications because it is less damaging and has better transmission in many media compared to ultraviolet light. Inexpensive optical elements made of plastic and glass can be used with visible light.

Index

A

Acrylodan 104, 105, 131, 151, 216, 217
Affibodies 218
Affinity 2, 8–10, 15, 17, 19, 20, 26, 33,
    39–43, 45, 46, 48–52, 61, 62, 76, 81,
    128, 143, 149, 158, 167, 168, 197,
    201–203, 206, 209, 210, 212, 213,
    215, 216, 219, 221–226, 231, 232, 236,
    238–240, 280, 301, 303, 304, 306, 307,
    309, 315, 321, 328, 358, 385, 421, 428,
    430, 443, 447, 476, 477, 482, 484, 492,
    493, 496, 498, 512, 514, 518, 519, 527,
    549, 550
Affinity coupling 303, 304, 307, 315, 321,
    328, 385
Affinity-based limit 40
Ag+ ion 89
Alexa dyes 125, 270
Alkyl silane 317
Alkylsiloxane 317
Allosteric effector 25, 27
Allosteric sites 24
Amino groups modifications 312
Amphiphilic molecules 321, 421
Amplification effect 93, 110
Amplification of reporting signal 24, 136
Amplified stimulated emission 35–352, 367
Amyloid fibrils 302
Angle-ratiometric fluorescence sensing 360
Angle-resolved emission 359
Anisotropy-based sensing 76–81
Antenna effects 146, 149, 162, 264, 265, 291,
    436, 437
Antibiotics 87, 238
Antibodies 5, 9, 10, 14, 18, 21–24, 30, 33, 43,
    60, 67, 71, 80, 81, 105, 149, 151, 161,
    166–170, 175, 184, 198, 204, 209–214,
    217, 218, 223–226, 231–233, 236, 239,
    268, 270, 271, 283, 303, 304, 307, 312,
    313, 316, 337, 338, 340, 357, 382, 384,
    387–389, 398, 438, 440, 442–445, 447,
    457, 484, 497, 511–513, 517, 518, 522,
    540, 549, 552, 555, 566
Anticalins 217
Antigen (Ag) binding sites 5, 30, 105, 209, 211
Antigenic determinants 21, 209, 223, 239, 303
Antimicrobial peptides 224, 443, 447
Apoptosis 14, 141, 486, 487, 490, 497, 500,
    528, 556
Aptamer microarrays 232
Aptamers 18, 26, 82, 90, 168, 225–240, 274,
    276, 282, 284, 287, 313, 315, 387, 438,
    440, 441, 444, 447, 513, 518, 540, 541
Aptamers structure-switching 228
Aptazymes 232
Association between Quantum Dots 284
Association of nanoparticles 252, 282
Association-dissociation induced PET 252
Autofluorescence 76, 121, 344, 426, 456, 483,
    485, 537, 560
Avalanche photodiodes 377, 470
Azacrown ether derivatives 141, 198

B

Bacterial periplasmic binding proteins 105,
    215, 494
Bacterial spores 444
Barcode 169, 263, 389, 390, 392, 394, 520,
    521, 559
Barcodes for microsphere arrays 389
Binding constant 31, 32, 41–52, 70, 285, 385,
    475, 558
Binding stoichiometry 26
Biochips 315, 383, 432, 550
Bioluminescence 65, 91, 335, 342–345, 366,
    400, 485, 527, 550, 555
Biomembrane lipid asymmetry 486

Biomembranes 43, 123, 130, 132, 134,
        135, 140, 239, 320, 323, 397, 421,
        457, 467, 488
Biosensor 4, 5, 7, 9, 11, 18, 24–28, 31, 73, 82,
        99, 102, 136, 141, 143, 153, 168, 185,
        197, 212, 213, 215, 216, 219, 224, 235,
        270, 280, 318, 325, 344, 345, 357, 396,
        398, 400, 417, 422, 438, 442, 443, 478,
        480, 489, 516, 518, 524, 526–528, 531,
        532, 534, 536, 550, 553, 556
Biotelemetry, real-time 537
Biotin 17, 43, 154, 157, 168, 174, 218, 232,
        268, 270, 303, 304, 307, 315, 328, 357,
        385, 549
Biotinilated hydrogel 315
Bivalent and polyvalent binding 49
Blood sensing 523
Blood, trace amounts 337
Bobrovnik method 54
BODIPY dyes 125, 493
Boronic acid 32, 104, 198, 425, 526, 537
Bottom-up approach 299, 201, 304
B-phycoerythrin 184
Brownian rotational diffusion 80

C
Ca⁺⁺ ions 74, 478
Ca²⁺-binding proteins 104
Calibration 14, 38, 57, 71, 72, 74, 75, 83, 91,
        111–114, 228, 385, 386, 408, 424, 483,
        519, 550, 558
Calixarenes 146, 203, 205, 206, 208, 239, 261
Calmodulin 272, 273, 280, 479, 491
Calorimetry 6, 7
cAMP 97, 149, 493
Cancer 13, 82, 168, 170, 224, 231, 232, 409,
        432, 437, 487, 489, 491, 497, 498, 500,
        514, 524, 525, 528, 535, 537
Cancer cell marker 168, 170, 497, 498
Cancer-associated proteins 231
Carbon nanotubes 288, 299, 300, 328
Caspase 3 497, 528
Catalytic sensors 10, 25
Cathodic luminescence 339
Cell arrays 400
Cell biosensors 102, 185, 344, 345, 400, 489,
        527, 536
Cell calcium 40, 493
Cell cAMP 97, 493
Cell cytosolic maltose 494
Cell labeling 482
Cell sodium and potassium 492
Cells in microfluidic devices 398

Cells living and fixed 397
Cellular signaling 492
Changes of protein conformation 30
Charge-coupled device (CCD) 377, 390, 458,
        459, 515, 539, 550, 557
Chemical cross-linking 314, 326
Chemiluminescence 65, 91, 335–342, 363, 366
Cholera toxin 325
Cholesterol 200, 239, 323–325, 329, 427, 524
Cl⁻ ion 73, 87, 556
Cl⁻ sensor, ratiometric 73
Clinical diagnosics, noninvasive 347
Collective effect in quenching 181
Collisional quencher of phosphorescence 87
Combinatorial chemistry 516, 551
Combinatorial discovery of fluorescent
        dyes 143
Combinatorial library selection 239
Compact disc format 393
Competition involving two binding sites 19
Competitive immunoassays 81
Competitor displacement 15–19, 33, 79, 81,
        97, 168, 178, 202, 238, 270, 281
Complex formation with cyclodextrins 206
Computer design of fluorescent dyes 49
Concentration-dependent quenching 155
Confined plasmon resonance 172
Confocal microscopy 124, 456, 460–462, 466,
        467, 487, 557, 560
Conformational flexibility 69
Conjugated polymer 93, 119, 160, 175–188,
        198, 205, 208, 234, 235, 238, 240,
        252, 262–265, 267, 269, 272, 274,
        276, 282, 284, 287, 319–321, 347,
        352, 362, 387, 410, 411, 436, 438,
        441, 447, 471, 529, 551
Contact sensing 29, 30, 211
Continuous monitoring 4, 15, 17, 32, 37, 315,
        381, 401, 524, 533, 537, 538, 541
Copper nanoparticles 173
Core-shell compositions 300, 301, 328
Core-shell liposomes 325
Co-synthetic incorporation 313, 481
Coupled plasmon absorbance 282
Covalent labeling 14, 29, 127, 131, 141, 231,
        276, 314
C-reactive protein 524
Critical micelle concentration 308
Crown ethers 174, 198, 256, 492, 493
Cryptands 198, 199
Cu²⁺ ions 19, 161, 309, 425
Cuvette-type fluorescence detection 373
Cyanine dyes 13, 123, 126–129, 181, 205,
        229, 278, 347, 413, 428

Cyclodextrins 89, 200–202, 206–208, 239, 392, 427, 438
Cytokines 14, 231, 492, 524

D

Defensin A 218
Delayed fluorescence 141, 142
Dendrimers 154–161, 173, 175, 207, 208, 239, 263, 308, 320, 347, 442, 471
Dendritic conjugated polymers 181
Deoxyribozymes 25, 26, 278, 287
Depolarization of emitted light 76
Detection of pathogens 437, 442, 443, 524
Detergent micelles 150, 308, 311
Dielectric relaxation 102, 135, 154, 257, 416, 417, 421
Differential expression pattern 510
Diffraction limit 455, 456, 464, 465, 468, 500, 551
Diffusion-controlled chemical inputs 289
Digestible sensors 537, 538
Dioxin 519
Diphtheria toxin 168
Dipole moment 93, 102, 110, 130, 131, 134, 135, 138, 144, 254, 256, 257, 261, 291, 362, 413
Dipole moment increase/decrease 413
Direct molecular immunosensor 104
Direct recording of digital signal 284
Direct sensors 2, 3, 10, 20, 28, 29, 31–33, 202, 211, 212, 357, 551
Disease-related gene mutations 524
Disorder, static and dynamic 418, 420
Dissociation constant 15, 41–43, 73, 197, 207, 226, 257, 275, 304, 385, 476, 477, 527
Dissociation-Enhanced Lanthanide Fluoroimmunoassay (DELFIA) 149
DNA detection, dye dimmers 156, 253, 428
DNA detection, intercalators 17, 49, 128, 428, 435, 555
DNA flat microarrays 432
DNA hybridization 13, 18, 23, 33, 56, 93, 161, 1698, 171, 235, 274, 301, 315, 374, 365, 385, 387, 434, 511, 522
DNA or RNA hybridization assays 13–15, 168, 270, 388
DNA sensing, conjugated polymer 436
DNA, double-strand (dsDNA) 235, 240, 300, 427, 437, 447
DNA, molecular beacons 26, 30, 98, 173, 174, 253, 276–279, 282, 291, 357, 385, 387, 433, 435

DNA, sandwich assays 20, 21, 23, 24, 27, 33, 187, 339, 391, 401, 433, 434, 545
DNA, sequence-specific recognition 429
DNA, single-stranded 173, 225, 227, 234, 240, 274, 300, 301, 429, 431, 432, 435, 447
Domain antibodies 210
Drug discovery 17, 82, 223, 268, 384, 388, 439, 515–517, 526–528
Dye isomerization dynamics 420
Dyes replacing DNA bases 231
Dynamic range of the assay 3, 79

E

Efficiency of energy transfer 92
Electric field 102, 134, 135, 139, 140, 172, 174, 187, 254, 323, 364, 374, 376, 487, 448, 556
Electric field sensing 140
Electrochemical immunoassays 10
Electrochemiluminescence 65, 335–341, 366, 520, 555
Electrochromism 134, 139
Electrogenerated chemiluminescence 338, 339, 341, 342, 500
Electroluminescent single molecules 288, 341
Electronic conjugation 58, 176, 198, 206, 254, 257, 282
Electronic excitation 65, 83, 91, 135, 176, 177, 250, 413, 552
Electronic nose 411, 529, 532
Electronic polarizability 140
Electronic tongue 530, 532
Electrostatic interactions 42, 166, 198, 270, 272, 274, 302, 305, 317, 428, 429
Emission anisotropy 30, 65, 66, 120, 155
Emission of light quanta 65
Enantioselective detection 208
Enantioselective molecular sensing 205
Endocrine disruptors 202, 519, 540
Engineered Cys locations 312
Enhancement near metal nanoparticles 359
Enzyme activity 9, 24, 96, 345
Enzyme-based dye transfer 478
Enzyme-linked immunosorbent assay (ELISA) 21, 22, 26, 31, 33, 211, 231, 235, 337, 440, 445, 552, 559
Epi-fluorescence microscopy 457, 458, 469, 551
Equilibrium in ESIPT reaction 109
Ethidium bromide 17, 49, 93, 126, 264, 287, 428
Evanescent wave fluorescence 8, 354, 359

Excimer of pyrene 88
Exciplexes 88–91, 155, 265, 409, 552, 559
Excited state 30, 65, 68–71, 84, 85, 87–89,
        91, 95, 98, 100–102, 105–112, 120–
        122, 131, 134–138, 140, 141, 143, 144,
        146, 147, 150, 156, 163, 164, 177, 178,
        198, 201, 205, 206, 235, 249–251,
        253–271, 335, 336, 338, 339, 342, 344,
        345, 347, 349–352, 354, 357, 360, 363,
        366, 409, 416, 420, 423, 461, 466, 549,
        551–553, 55, 57, 558
Excited-state intramolecular proton transfer
        (ESIPT) 109, 137, 259, 409, 552
Excited-state proton transfer 201, 259
Excited-state reaction 30, 85, 106, 107, 111,
        112, 141, 250, 258, 409, 553
Excitons 128, 163, 170, 171, 177, 178, 186,
        320, 352, 464, 552
Explosives 7, 18, 168, 352, 411

F
Far-field optical nanoscopy 457
Fatty acid 104, 126, 167, 319
Fiber optics technology 380
Flow cytometers 390, 475, 520
Fluorescein 60, 66, 71, 107, 12–124, 139,
        142, 229, 235, 264, 266, 288, 467, 469,
        477, 486, 491, 492
Fluorescein arsenical helix binder
        (FLASH) 477
Fluorescence correlation microscopy 474, 475
Fluorescence correlation spectroscopy 285, 474
Fluorescence decay rate 84, 113
Fluorescence emission spectrum 65, 67, 90,
        92, 125, 207, 258, 281, 414
Fluorescence excitation spectrum 65
Fluorescence lifetime 65, 66, 70, 76, 78,
        80, 82–84, 104, 110, 122, 135, 136,
        229, 260, 416, 463, 474, 483, 549,
        553, 557, 558
Fluorescence polarization immunoassays
        81, 211
Fluorescence quantum yield 68, 120, 121, 124,
        128, 158, 178, 183, 205, 227, 231, 238
Fluorescence quenching 7, 69, 75, 109, 131,
        132, 155, 173, 177, 178, 181, 198, 204,
        205, 230, 290, 300, 310, 357, 386, 411,
        416, 420, 424, 463, 488, 558
Fluorescence quenching, dynamic 69, 420
Fluorescence quenching, static 155
Fluorescence recovery after
        photobleaching 467, 553
Fluorescent angiography 496

Fluorescent in situ hybridization (FISH) 494
Fluorescent polyelectrolites 180
Fluorescent proteins 94, 98, 119, 182–187,
        267, 269, 344, 345, 397, 399, 463, 473,
        477–480, 482, 489, 491, 493–495, 511,
        514, 546, 554
Fluorimeter-in-a-pill 537
Fluorogenic substrates 24, 27
Fluorophore 10, 12, 20, 26, 65, 70, 71, 76,
        78–80, 84, 88, 90, 92, 96, 99, 103, 105,
        106, 110, 120, 125, 126, 130, 134, 137,
        140, 143, 155, 158, 161, 168, 182, 183,
        185, 187, 199, 204, 205, 215, 217, 224,
        225, 228, 230, 251, 253, 257, 258, 262,
        277, 287, 288, 341, 344, 360–362, 365,
        366, 424, 429, 431, 456, 459, 461, 464,
        467, 481, 482, 485, 487, 488, 492, 499,
        517, 522, 550–554, 557, 558
Fluoroprobe 131, 412
Focal plane and volume 461
Foodborne contaminants 536
Förster resonance energy transfer (FRET) 18,
        91–113, 122, 124, 129, 141, 142, 153,
        155, 156, 161, 162, 168, 169, 173, 181,
        184, 187, 198, 202, 205, 211, 227, 228,
        230, 231, 235, 237, 251, 253, 261–272,
        275, 278, 281, 284, 287, 290, 291, 316,
        328, 337, 344, 349, 350, 354, 357, 367,
        386, 389, 401, 413, 425, 433–438, 447,
        463, 464, 467, 468, 470, 471, 474,
        477–479, 481, 485, 491–494, 514, 515,
        528, 554, 557, 559
Frequency-modulation technique 374
Freshness markers 536
FRET conformational change 97–99
FRET imaging 98, 554
FRET immunoassays 211
FRET quenching 98, 181
FRET to non-fluorescent acceptor 98, 113, 278
FRET with competitor 18, 211
FRET, cascade, 262
FRET, directed 262
FRET-gating 263

G
Gas sensors 410, 411, 529
Gene expression profile 509
Genome 62, 235, 385, 432, 437, 490,
        507–510
Glucose 9, 18, 32, 38, 73, 105, 154, 153,
        198, 200, 215, 216, 425, 479, 480,
        524, 535, 537
Glucose binding protein 32, 73, 105, 216

Glucose biosensor competitive 18
Glucose sensing 32, 283, 425, 426, 551
Glucose, noninvasive monitoring 426
Glycan sensors 442
Glycolipids and glycoproteins 441
Gold and silver surfaces 316, 317, 321
Gold nanoparticles 171, 173, 174, 176, 181,
        187, 266, 269, 270, 283, 284, 320, 360
Green fluorescent protein (GFP) 182–185,
        187, 344, 345, 397, 399, 423, 463, 478,
        480, 482, 489, 491, 497, 546, 554

H

Heart disease biomarkers 525
Heavy metal ions 283, 425, 447
Heterocyclic hydrocarbons 310
Hetero-FRET 155, 156, 263
Heterogeneous formats 2, 5, 6, 149
Hetero-transfer 93, 162
High throughput screening (HTS) 82, 169,
        349, 350, 364, 435, 515, 519, 526–528,
        540
Hinge-bending motions 215, 279, 479
Histidine tag 477
Homo-FRET 122, 155, 156, 161, 162, 205,
        262, 265, 291
Homogeneous assays 2, 4, 10, 11, 16, 81,
        161, 168, 228, 229, 253, 263, 309,
        340, 345, 371
Homo-transfer 93
Horseradish peroxidase 337
Hot electrons injection 340, 341
Human immunodeficiency virus (HIV) 224,
        397, 429, 445, 476, 484, 523
Human serum albumin 104, 105, 526
Hydrogel layers 315, 329
Hydrogen bond 28, 50, 69, 105, 109, 110,
        130, 132, 134, 137, 230, 276, 290, 303,
        305, 306, 323, 329, 410, 428–430, 558
Hydrogen-bonding complementarity 306
Hydrophobic interactions 240, 302, 305
Hydroxychromone derivatives 110, 138, 488
Hydroxychromone dyes 137, 310, 323, 413,
        414, 418, 423, 427, 500

I

Illness treatment model 534, 541
Imaging cancer tissues 497
Immunoglobulin E detection 82
Immunosensor techniques 21
Implantable sensors 537
Imprinted polymers 82, 236–238, 240, 550

Inclusion complexes 200, 201, 208
Indicator 3, 20–24, 26, 27, 33, 98, 382,
        422–424, 433–436, 490, 492, 493,
        525, 556
Individual event analysis 471
Indocyanine green 496
Induced and assisted folding 274, 275
Induced fitting 223
Industrial sensor applications 409
Infectious diseases 442, 523, 524
Intensity-weighted format 59
Interactions in a large volume 54
Interactions in a small volume 54
Interactome 344, 388, 478, 490, 499, 508,
        509, 513, 515
Intercalation 17, 126, 128, 428, 429
Intercalator dyes 17, 49
Internal calibration 71, 111–114, 385, 386
Intracellular imaging 166, 490
Intramolecular charge transfer (ICT)
        104–108, 131, 132, 198, 201, 204, 205,
        254–261, 287, 290, 291, 411–413, 425,
        492, 555
Intramolecular dynamics of protein 280
Intramolecular electron transfer
        (IET) 69, 253
Ionic liquids 417, 418
Isoemissive point 58, 103, 104
Isomerization 56, 109, 271, 282, 290, 416, 557

J

J-aggregates 128, 129

K

Kasha heavy atom effect 142
Kasha rule 69, 551
Ketocyanine dyes 50, 131–134
Kidney-related diseases 525
Kinetics of target binding 55, 63, 390
Klotz plot 45, 46

L

Label-based techniques 10, 33
Labeled pool of potential targets 12, 13, 33
Label-free approach 7–10, 12
Lab-on-a-chip 371, 391, 393, 394, 396, 398,
        402, 555
Langmuir-Blodgett films 317–321, 329
Lanthanide-chelator complexes 146
Lanthanides 144–151, 153, 187, 264, 267–269,
        341, 346, 349, 375, 423, 463, 477

Lasers 68, 85, 119, 123, 124, 126, 165, 285,
    346, 347, 349–353, 363, 374–377, 390,
    391, 393, 394, 459, 461–463, 465–467,
    470, 471, 475, 549–551, 553, 556, 559
Lasing threshold 35–352
Layer-by-layer technique 317, 320, 321
Lead 205
Lectins 32. 273, 304, 425, 442
LE-ICT switching 254
Leucine-rich repeat proteins 218, 219
Ligand-binding proteins 30, 31, 167, 168,
    213–219, 225, 236, 239, 270, 279, 280,
    282, 440
Ligands with different affinities 19, 46, 50
Light emitting diodes 341, 375, 376, 394,
    539, 555
Light quenching 71
Light scattering 39, 67, 74–76, 83, 85–87,
    103, 122, 142, 144, 147, 154, 165, 211,
    271, 297, 300, 344, 349, 360, 362, 374,
    390, 391, 485, 495, 496, 523, 552, 559
Light sources 68, 72, 75, 85, 106, 107, 121,
    165, 263, 335, 339, 343, 344, 366,
    372–378, 380, 383–391, 458, 460, 464,
    468, 485, 500, 555
Light-addressable potentiometric sensors 32,
    352, 353, 367
Light-harvesting 160, 264, 290
Light-scattering sensitivity 83
Limit of detection (LOD) 38, 39, 166
Limit of quantitation (LOC) 39
Linearly polarized light 76, 79, 170
Lipid bilayer 305, 322–327, 329, 427, 487, 488
Lipid bilayers, stabilized 325
Lipid vesicles, liposomes 155, 322
Lipobeads 325, 326
Lipocalins 217, 218
Liquid-liquid interfaces 421
Logic gates, molecular 287, 288
Logic operations 286
Low-power microwave heating 364, 522
Luciferin-luciferase system 343, 366
Luminescence 65, 81, 85, 145, 147–149,
    151, 153, 160, 167–172, 230, 249, 260,
    264, 265, 268, 289, 335–339, 341, 342,
    350, 352, 366, 374, 382, 392, 407–411,
    423, 445, 446, 523, 527, 530, 549, 550,
    554, 555
Luminol 336, 337

M
Maltose binding protein 49–97, 104, 202,
    215, 216, 239, 280, 291, 479, 494
Membrane dipole potential 141

Membrane potentials in mitochondria 125, 128
Mercury 205, 271, 425, 556
Metabolome 499, 508, 509
Metal nanoclusters 174
Metal plasmonic effects 357
Metal-ligand charge transfer (MLCT) 151
Microarray fabrication 384
Microarray immunosensors 14
Microarrays, spotted 383, 385, 401, 520
Microbes 21, 224, 442–444, 508, 510
Microcantilever 8, 32
Microfluidic capillary channels 394
Microfluidic devices 318, 325, 372, 378, 382,
    391, 393–396, 398–400, 402, 473, 533
Microfluidic immunosensor 394
Microfluidic sample preparation 393, 395
Microfluidic single-molecule detection 475
Micropatterning 307
Microplate readers 374, 520
Microsphere suspension arrays 389, 390
Micro-total-analytical-systems 394
Microwave acceleration 364, 522
Mini-emulsion polymerization 156
Minimal protein scaffolds 218
Mix-and-read sensing 28
Molar absorbance 58, 61, 75, 120, 121, 125,
    126, 128, 141, 145, 151, 153, 165, 177,
    183, 184, 187, 265, 347, 550, 555
Molecular beacons 26, 30, 98, 173, 174,
    253, 276–279, 282, 291, 357, 387,
    433, 435
Molecular computing 288
Molecular dynamics 413, 417, 420
Molecular recognition 15, 23, 31, 171,
    197, 202, 206, 217, 221–223, 232,
    236, 239, 249, 274, 299, 302, 304,
    365, 387, 430, 480, 486, 490–493,
    511, 518, 522, 550, 556
Molecular rotors 416, 417, 420, 470
Molecular wires 177, 288, 300
Monoclonal antibodies 209, 226
Multiplexed system 383
Multivalent interactions 306, 441

N
Nanoarrays 388, 510, 515
Nanopatterns generation 384
Near-field fluorescence microscopy 456
Near-IR dyes 279, 426, 486, 496
Neutral molecules sensing 252
Noble metal ions 142, 144
Non-fluorescent dimers 71, 156
Nonlinear and multiparametric operation 287
Non-linear effects in response 57

Non-natural amino acids 314
Nonradiative rate constant 360

O

Olfactory system 439, 529, 531, 532
Optical encoding 483, 520, 521, 532
Optical fibers 357, 372, 378, 380–383, 401,
    417, 556
Optical waveguides 75, 325, 355, 356, 372,
    374, 378, 380, 381, 383, 394, 398, 402,
    464, 533
Optode 321, 380–382, 423, 556
Optofluidics 393
Organic peroxides 336
Overlap integral 92, 93, 95, 141, 155, 198
Oxygen 74, 84, 87, 88, 122, 123, 138, 142,
    144, 146, 151, 153, 183, 260, 336, 337,
    343, 363, 382, 410, 411, 423, 424, 446,
    447, 467, 500, 553, 557
Oxygen sensing 142, 382, 423, 424, 447

P

Pattern recognition 62, 529, 540
Peptide beacons 224, 279, 445
Peptide nucleic acid (PNA) 233–235, 240,
    278, 279, 287, 316, 347, 384, 387, 436
Peptide scaffolds 222, 301
Peptide synthesis 220, 221, 302, 429
pH sensing 73, 261, 382, 422
Phage display 210, 212, 213, 223, 225, 512
Pharmacokinetics 527
Phase-modulation technique 85
Phosphate binding protein 216
Phospholipid bilayers on planar surfaces 325
Phospholipid monolayers 318
Phosphorescence 65, 87, 91, 141, 142, 144,
    151, 153, 230, 237, 267, 276, 363, 408,
    424, 555, 557
Photobleaching 72, 76, 78, 103, 122, 158,
    161, 171, 172, 187, 269, 300, 363, 424,
    461, 462, 466, 467, 469, 490, 496, 500,
    553, 557, 560
Photocell sensors 353
Photochromes 287
Photochromic FRET 95
Photodetectors 121, 366, 367, 373, 376,
    379, 559
Photodynamic therapy 497
Photoinduced electron transfer (PET) 150,
    155, 167, 173, 181, 198, 228, 250–254,
    260, 261, 275, 278, 281, 287, 290, 291,
    316, 353, 357, 359, 425, 433, 471, 484,
    492, 493, 557

Intermolecular 557
Intramolecular 557
Phototoxicity 120, 461, 462, 466, 467
pH-sensitive dyes 87
Phycobiliproteins 184
Planar waveguides 377, 382, 383, 394, 444
Plasmonic interactions 335
Plastic scintillator 376
Platelet-derived growth factor 203, 229
Polarity 101, 102, 105, 108, 112, 123, 125,
    130–132, 134, 135, 137, 139, 140, 155,
    158, 187, 201, 223, 231, 254, 257, 291,
    308, 310, 311, 319, 323, 326, 411–418,
    420–422, 429, 430, 446, 447, 487–489,
    553, 558
Polarity probing 412
Polarity, empirical scales 412
Polarity-sensitive dyes 101, 155, 319, 416
Polarization assays 59, 76–82, 122, 170, 367,
    527, 557
Pollution of the environment 536
Polyclonal antibodies 209
Polymerase chain reaction (PCR) 226, 363,
    394, 430, 432, 434, 437
Polymers, highly elastic state 420
Polymers, structure and dynamics 420
Polymersomes 326
Polynucleotide detection 283
Polynucleotide scaffolds 301
Polysaccharides 441, 551
Polystyrene latexes 157, 169
Porous silicon 169, 170, 378, 529
Porphyrins 121, 142, 152, 153, 159, 206–208,
    239, 301, 328, 344, 410, 495
Porphyrins, zinc binding 206
Positional encoding 520
Pressure sensors 446
Prodan 130, 132, 134, 412
Programmable array microscope 461
Prostate-specific antigen 161, 525
Protecting the information 288
Protein arrays 439–441, 497, 511, 540
Protein disease biomarkers 525
Protein folding code 220
Protein function arrays 14
Protein labeling coupled with synthesis 481
Protein scaffolds 215, 218, 518
Protein, specific recognition 13, 219, 238,
    275, 299, 417, 439, 477, 486, 546, 553
Protein, total content 438
Protein-protein interactions 14, 30, 184, 221,
    232, 345, 387, 388, 440, 484, 508
Proteome 62, 213, 388, 438, 490, 499,
    507–512, 515, 525, 540
Proteome profiling 510

Protonation-deprotonation of dyes 261
Push-pull structures 101

Q

Quantum confinement 163, 170, 282
Quantum Dot 18, 23, 94, 96, 97, 119, 161–172,
    174, 175, 181, 186, 187, 205, 252, 253,
    265, 267–270, 281, 284, 301, 318, 325,
    337, 341, 344, 354, 359, 362, 389, 392,
    397, 425, 426, 438, 444, 467, 483–486,
    489, 494, 496, 498, 510, 546, 554, 557
Quantum Dot, self-illuminating 344, 485
Quantum yield of photodestruction 87, 467
Quenching by spin labels 253

R

Radiative and non-radiative decay rate 360,
    361, 366
Radiative rate constant 68, 359
Raman scattering bands 67, 68, 358
Ratio of band intensities 110, 138
Rational design 30, 143, 183, 220, 224, 239,
    516, 546
Read-out and data analysis 386
Reagent-independent sensors 2, 28–31, 236
Receptor 1, 3, 4, 6–9, 12–18, 20–25, 31, 33,
    37, 39–50, 52–63, 82, 89, 97, 126, 167,
    168, 197, 203, 204, 207, 211, 212, 215,
    217, 218, 232, 239, 240, 249, 251, 252,
    269, 270, 274, 275, 281, 307, 309, 311,
    314–318, 321, 325–328, 337, 341, 342,
    345, 354, 356, 364, 382, 383, 392, 383,
    392, 396, 425, 426, 431, 438, 439, 443,
    44, 447, 474, 485, 489, 490, 493,
    511–513, 516–518, 527, 529–532, 539,
    540, 550, 558
Recognition unit 1, 27, 29, 154, 157, 161,
    166, 167, 197–240, 249, 256, 261, 275,
    288, 311, 325, 327, 328, 422, 425, 437,
    438, 440, 443, 447, 493, 495, 511, 546,
    550, 559
Recombinant antibody fragments 210, 224, 440
Recombinant selection in vitro 512
Reference emitters 148, 151
Reference reporter 73
Refraction of light 354
Remote sensing 30, 97, 211
Reporter 1, 3–7, 9, 10, 15–17, 26–31, 33, 51,
    57, 58, 61, 68–70, 73–76, 78, 81, 87–
    91, 93, 98–100, 104, 106–108, 112,
    113, 119–187, 197, 204, 205, 210–212,
    214–216, 219, 220, 223, 226–228, 237,

239, 240, 249, 251, 255, 257, 258, 261,
    265, 267, 270–272, 275, 277, 278, 280,
    281, 287, 288, 299, 300, 307, 309, 310,
    313, 314, 318, 325, 327, 328, 338,
    340–342, 350, 352, 354, 358, 367, 374,
    382, 386, 390, 396, 397, 399, 401, 410,
    412, 427, 435, 445, 447, 456, 473, 470,
    482, 483, 496, 499, 517, 520, 531, 523,
    539, 540, 545, 550, 553, 554, 558, 559
Response function, linear 59
Response function, non-linear 59
Reversible target-receptor binding 37
Rhodamine 67, 71, 122–125
Ribosome display 223
Ribozymes 24–26, 278
RNA 13–15, 26, 41, 89, 90, 128, 165, 168,
    180, 225, 227, 231, 233, 234, 270, 277,
    278, 301, 357, 363, 365, 388, 427, 429,
    430, 432, 435, 440, 443, 444, 473, 508,
    509, 549
Rotamers 272
Rotational correlation time 78, 80, 151, 413
Rotational diffusion 37, 76, 79, 80, 413, 420
Rotational mobility 29, 30, 77, 79, 81, 249, 553
Ruthenium 144, 151, 52, 167, 268, 269, 281,
    338, 410, 424

S

Saliva 426, 522–524, 526, 530
Sandwich immunoassay 211, 394
Sandwich-type fluorescence assay 20, 383
Saturation effects 40
Scale of concentrations 38, 43, 492
Scatchard plot 46
Scattered light 83, 85, 88, 283, 347, 360,
    375, 390
Second harmonic generation 135
Selection-amplification techniques 225
SELEX 226, 227, 444, 517
Self-assembled monolayers 316–318
Self-assembled tubular structures 302
Self-assembly 89, 146, 232, 299, 301–307,
    311, 314, 320–322, 325, 326, 328, 329,
    484, 546
Self-illuminating Quantum Dots 344, 485
Self-quenching 71, 125, 155, 156, 162,
    184, 263
Sensing of ions 19, 251
Sensing-guided drug delivery 535
Sensor absolute sensitivity 267
Sensor dynamic range 62
Sensor response function 48, 286
Sensor selectivity 40

Sensors integrated into textile clothes 535
Sensors, assembly on the surface 314, 318
SH-groups modifications 312
Si nanoparticles 170
Signal-to-noise ratio 39, 102, 341, 376, 379, 456, 460, 558
Silica nanoparticles 23, 119, 154, 158, 161, 169, 268, 301, 383, 443
Silver nanoparticles 171, 173–176, 283, 301, 360, 362, 364, 365
Single molecule detection 469, 471, 472, 475, 499, 532, 559
Single nucleotide polymorphism (SNP) 161, 235, 432, 437
Single-photon counting 85, 376, 377, 463
Singlet oxygen 122, 363, 467
S-layer protein 306, 307, 326, 327, 329
Sol-gel matrices 158
Sol-gel process 316
Solid-liquid interfaces 421
Solid-solid interfaces 421
Solid-state electroluminescence 341
Spectral shift 32, 100–105, 112, 113, 130–132, 134–136, 140, 187, 228, 254, 155, 261, 290, 413, 425, 492, 558
Split proteins 514
Spotted arrays 55, 374, 388, 521, 549
Stark effect 134, 140
Stern-Volmer relationship 70, 84
Stimulated emission depletion microscopy 465, 466, 468
Stokes shift 67, 68, 71, 75, 83, 87, 93, 103, 120, 122, 125, 127, 141, 142, 146, 147, 151, 153, 155, 157, 162, 164, 175, 179, 184, 254, 420, 428, 457, 551, 559
Stokes-Einstein relation 80
Streptavidin 167, 174, 232, 303, 304, 306, 307, 315, 328, 385, 549
Stroboscopic technique 85, 463, 559
Styryl dyes 135, 347
Supercritical fluids 418
Superquenching 177–180, 182, 187, 235, 263, 272, 320, 436, 437, 441
Supporting surfaces 357
Supramolecular structures 24, 31, 154, 249, 299–329, 471, 512, 551, 556
Surface plasmon field enhancement 358
Surface plasmon resonance 6–9, 11, 172, 315, 358, 359

113, 123, 130, 136, 149–151, 153, 158, 160, 165–169, 171, 178, 180, 182, 184, 186, 187, 197, 198, 202–240, 249–253, 255, 257–259, 261–163, 270–272, 274, 276–286, 289, 290, 299–301, 309–316, 318, 321, 325, 327–329, 335, 337, 381–391, 394, 396, 397, 399, 401, 402, 407–447, 463, 469, 473–476, 478, 480–485, 489, 490, 494, 496–499, 507, 510–512, 514, 516–519, 521, 522, 524, 526–528, 530, 535, 540, 545, 546, 549–551, 554, 556–559
Target-induced folding 274
Target-receptor affinity 39, 61
Taste perception 530
Temperature sensors 408, 409
Template-assisted assembly 307
Template-directed polymerization 158, 236
Theophylline 26
Thermal quenching of luminescence 408
Thermodynamic equilibrium 15, 29, 56
Thermosensitive polymer 89
Time-resolved anisotropy 85, 261
Time-resolved decay 340
Time-resolved fluorimeter 85
Time-resolved imaging 463, 468, 499
Time-resolved immunoassays 211
Tissues, transparency window 300, 495
Titration experiment 44, 50, 430
Topological saturation 303
Total internal reflection 355, 356, 378, 380, 421, 443, 471, 552
Total internal reflection microscopy 456, 458–460, 468, 470, 514
Toxicity testing 399, 496
Transcriptome 385, 508–510, 515
Transcriptome profiling 432, 509
Transducer 1, 4, 6, 160, 249, 261, 274, 288, 299, 316, 323, 328, 354, 382, 559
Transmembrane potential 487, 556
Trinitrotoluene 18, 168, 169, 238, 352, 411
Tryptophan 12, 130
Tubular structures 299, 300, 302
Two-dimensional crystals 306
Two-photon excitation 128, 165, 171, 269, 345–351, 363, 375, 461, 463, 468, 488, 496
Two-photon microscopy 375, 396, 456, 462, 465, 470, 484, 488, 496

T
Target 1–33, 37–43, 48–57, 59–63, 65, 68–70, 72–84, 86, 88–91, 97, 102, 107, 109,

U
Ultra-short laser pulses 347, 363
Up-conversion emitters as FRET donors 349

Up-conversion nanocrystals 348
Upconverting luminophores 523

V
Viral antigen 445
Viruses 11, 21, 43, 224, 273, 304, 357, 365,
        407, 442, 444, 445, 457, 477, 510, 549

W
Waveguides, planar 377, 382, 383, 394, 444
Wavelength-shifting probes 100, 492

Wavenumber scale 67, 100, 112, 122,
        552, 559
Wellness support model 534, 541
Whole body sensing 271
Whole-animal sensing 531
Whole-blood sensing 523

Z
Zinc finger 98, 104, 275, 276, 429, 492
Zn^{+2} ion 257
Zn^{2+} 82, 98, 207, 411, 491, 492
π-electrons 89, 101, 177, 206

Color Plates

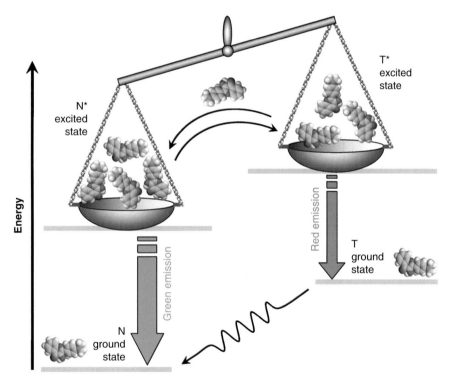

Color Plate 1 The scheme illustrating the principle of signal amplification in two-band ratiometric fluorescence response. The two excited species (N* and T*) are in dynamic equilibrium. For each of them the change of intermolecular interactions results in the change of energy separation between ground (N or T) and excited (N* or T*) states. The equilibrium in relative populations of two forms shifts in the direction of the form possessing lower energy, which gives higher fluorescence intensity from a more populated state. According to Boltzmann law, the difference in state populations is an exponential function of the difference in state energies. Therefore, a strongly amplified response is observed in band intensity ratios in comparison with spectral shifts (Demchenko 2006)

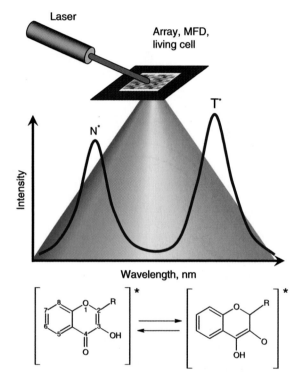

Color Plate 2 The principle of two-band ratiometric fluorescence sensing with functional 3HC dyes. On excitation of any labeled molecular or supramolecular structure (sensor array, microfluidic device, MFD, or living cell) the fluorescence intensity is distributed between two bands, N* and T*, centered in green and orange-red regions of the spectrum (above) due to establishment of dynamic equilibrium between two excited-state forms (below). The sensing signal is produced by the change of relative intensities of these two bands (Demchenko 2006)

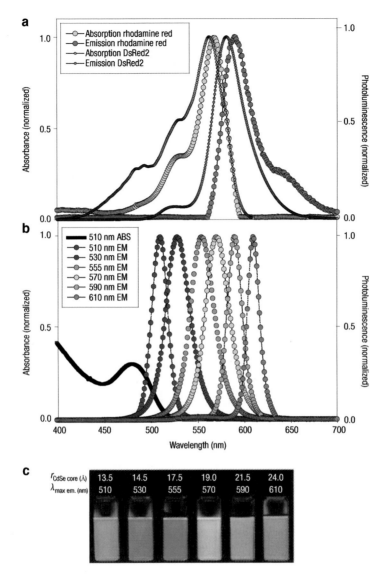

Color Plate 3 Comparison of spectral properties to those of QDs highlighting how multiple narrow, symmetric QD emissions can be used in the same spectral window as that of an organic or genetically encoded dye. (**a**) Absorption and emission spectra of rhodamine red, a common organic dye and genetically encoded DsRed2 protein. (**b**) Absorption and emission of six different QD dispersions. The black line shows the absorption of the 510-nm emitting QDs. Notably, at the wavelength of lowest absorption for the 510-nm QD, ~450 nm, the molar absorbance is greater than that of rhodamine red at its absorption maxima (~150,000 versus 129,000 M^{-1} cm^{-1}). (**c**) Photo demonstrating the size-tunable fluorescence properties and spectral range of the six QD dispersions plotted in (b) versus CdSe core size. All samples were excited at 365 nm with a UV source. For the 610-nm-emitting QDs, this translates into a Stokes shift of ~250 nm (Reproduced with permission from Medintz et al. 2005)

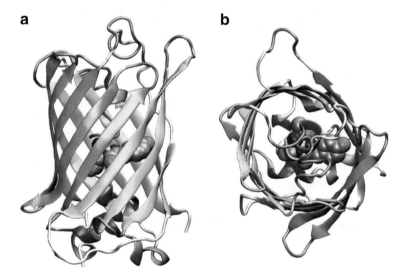

Color Plate 4 The three-dimensional structure of green fluorescent protein from jellyfish *Aequorea Victoria* in two projections. The sequence of 238 amino acid residues (26.9 kDa) is assembled into 11 β strands (shown as yellow ribbons) forming the so-called β-can. The fluorophore, located inside the beta-barrel, is shown is space-filling representation. It is formed inside this structure as a part of the central helix and is highly protected from interaction with the solvent

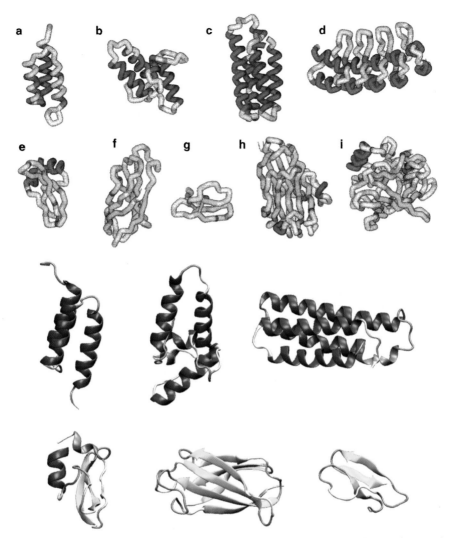

Color Plate 5 Representative protein display scaffolds that can be selected for grafting the functional recognition units and library construction of specific molecular recognition binders. Scaffold proteins in a–d consist of α-coils, the depicted in e small kunitz domain inhibitor shows an irregular α-coil and β-sheet architecture, whereas f–i show scaffolds predominantly consisting of β-sheet frameworks. α-Helices are depicted in red; β-sheets, in blue; disulfide bonds, in orange; and positions subjected to random or restricted substitutions, in yellow. The PDB IDs used to generate this figure are given in parentheses: (**a**) Affibody: Z-domain of protein A (1Q2N), (**b**) immunity protein: ImmE7 (1CEI), (**c**) cytochrome b562 (1M6T), (**d**) repeat-motif protein: ankyrin repeat protein (1SVX), (**e**) kunitz-domain inhibitor: Alzheimer's amyloid b-protein precursor inhibitor (1AAP), (**f**) 10th fibronectin type III domain (1FNA), (**g**) knottin: cellulose binding domain from cellobiohydrolase Cel7A (1CBH), (**h**) carbohydrate binding module CBM4-2 (1K45); and (**i**) anticalin FluA: bilin-binding protein (1T0V) with cavity randomization for fluorescein binding (Reproduced with permission from Hosse et al. 2006)

Below: Presentation of 'minimal' protein scaffolds in ribbon-based graphics. The α-helices are purple and β-sheets are yellow

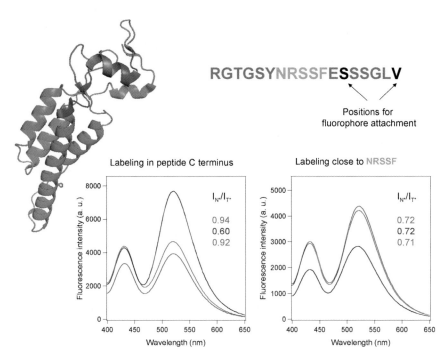

Color Plate 6 The interaction of peptide representing the segment 134–151 of tobacco mosaic virus protein with the specific antibody fragment Fab57P as a model target. The segment of protein structure and the corresponding peptide sequence are shown in the same color code. The pink part of the sequence corresponds to the antigenic binding site for Fab57P. Below are the fluorescence spectra of the dye FC covalently attached to indicated sites in the absence (red) and at presence (blue) of saturating amounts of specific antibody fragment Fab57P. Left: the mutant V151C-FC. The spectra change dramatically on target binding with the decrease of fluorescence ratio between the two bands (I_{N^*}/I_{T^*}) from 0.94 to 0.60. Right: the mutant S146C-FC. The ratio of fluorescence intensities does not change. The F(ab')2 fragment with unrelated specificity (green) was used as a negative control (Enander et al. 2008)

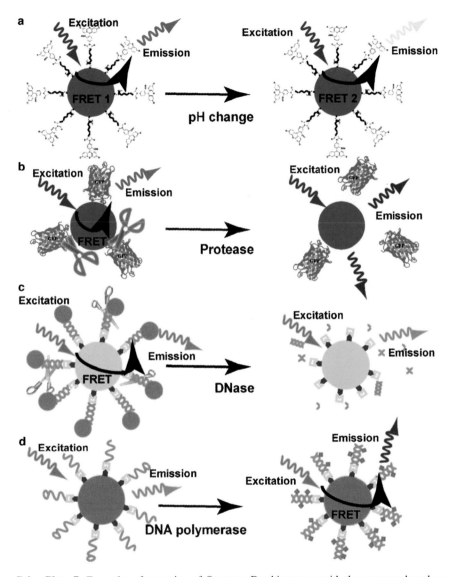

Color Plate 7 Examples of operation of Quantum Dot biosensors with the response based on FRET mechanism. (**a**) pH change via pH-sensing dyes attached to a QD; (**b**) cleavage of a GFP variant with an inserted sequence recognized by a protease (e.g., trypsin) to release GFP from the QD surface; (**c**) digestion by DNase of dsDNA (labeled with fluorescent dUTP) bound to a QD; (**d**) incorporation of fluorescently labeled dUTPs into ssDNA on a QD by extension with DNA polymerase (Reproduced with permission from Suzuki et al. 2008)

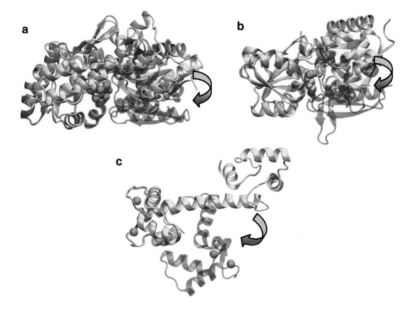

Color Plate 8 Conformational changes in hinge-bending proteins induced by the ligands. (**a**) Maltodextrin-binding protein (PDB codes 1ANF, 1OMP). (**b**) Lysine-, arginine-, ornitine-binding protein (PDB codes 2LAO, 1LST). (**c**) Calmodulin (PDB codes 1CLL, 1CDL). Open ligand-free conformations of the proteins are yellow, closed ligand-bound conformations are semitransparent blue. The ligands (maltose, lysine and Ca2+ ions respectively) are shown in space-fill representation. Closed and open forms of the proteins are aligned by the first domain. The motion of the second domain induced by the ligand binding is clearly visible and indicated by the arrows

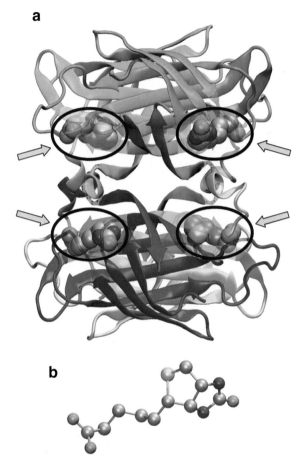

Color Plate 9 Streptavidin tetramer (PDB code 1SWE) in complex with biotin. The protein is shown in cartoon representation. The biotin molecules are in space fill representation. Streptavidin monomers are colored red, green, blue and yellow respectively. The parts of the protein, which are in front of the biotin molecules are made semi-transparent for clarity. Biotin molecules are circled and marked by the arrows. Below: Structure of single biotin molecule from the complex shown above. The carbon is cyan, oxygen is red, nitrogen is blue, and sulfur is yellow. The hydrogen atoms are not shown

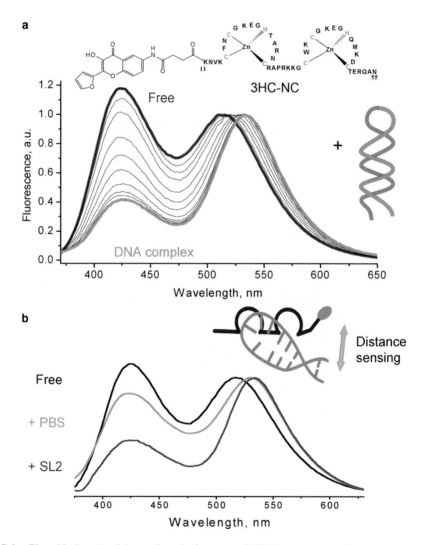

Color Plate 10 Results of the studies of a fragment of NCP7 protein with different oligonucle-otides. (**a**) The structure NCp7 peptide fragment labeled with 3-HC reporter dye and evolution of its fluorescence spectrum on titration of the peptide with oligonucleotide dTAR (DNA). (**b**) The response of 3-HC reporter discriminates the complexes formed with different oligonucleotides: PBS and SL2. In SL2-NCp7 complex the N-terminus of the peptide is closer to the oligonucle-otide than in PBS-NCp7 complex (according to the NMR data). In fluorescence spectra this fact is observed as lower relative intensity of the short-wavelength band. The concentration of labeled peptide was 100 nM (Courtesy of Dr. Y. Mély, Université Louis Pasteur, Strasbourg, France)

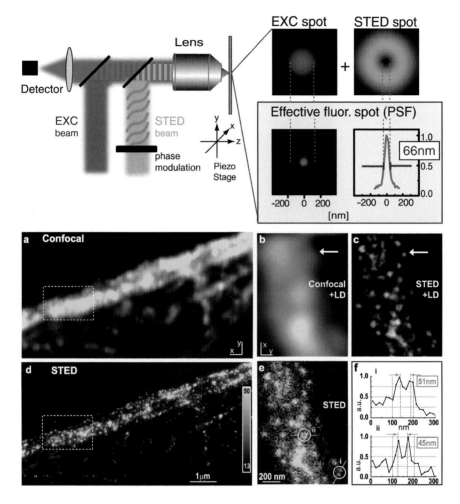

Color Plate 11 The basic principle of STED microscopy (above). The excitation light beam (EXC beam, in blue) is steered by a mirror through the objective lens, and due to diffraction is focused to a spot ca. 200 nm in diameter on the sample. By scanning this blue excitation spot over the sample (the cell) and recording the fluorescent light emitted at longer wavelengths from the labeled molecules with a computer, one can form an image of the sample. The smaller the excitation spot is, the higher the resolution of the microscope. Due to diffraction, the excitation spot cannot be made smaller than ~200 nm by focusing with a lens, which limits the resolution of common microscope. To overcome this limit, a second beam is used (STED beam, in orange) to quench the fluorescent markers before they fluoresce. Because the STED beam is doughnut-shaped and centered over the excitation spot, one is able to preferentially quench the markers at the outer edge of the excitation spot and not those in the center. The result is a much smaller effective fluorescence spot (green), here reduced to a diameter of tens of nanometers. Imaging neurofilaments in human neuroblastoma (Donnert et al. 2006) (below). Low-magnification confocal image indicates the site of recording. (**a–d**) Contrary to the confocal recording (**a**), the STED recording (**b**) displays details 30 nm, as also highlighted by the comparison of image subregions shown in **c** and **d** bordered by dashed lines in **a** and **b**, respectively. (**c** and **e**) Subregion after linear deconvolution (LD). Note that the deconvolved confocal image does not yield a substantial gain in information. (**f**) Profiles of raw data demonstrate the ability of (undeconvolved) STED data to reveal object structures that are far below the wavelength of light (Courtesy of Dr. Stefan W. Hell, Gottingen, Germany)

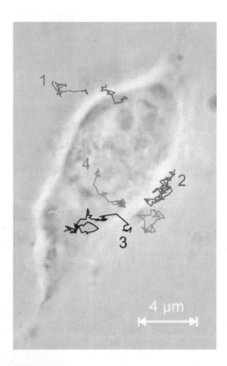

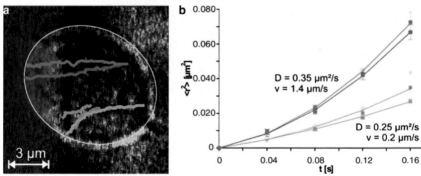

Color Plate 12 Interaction of adeno-associated virus (AAV) with a living He-La cell. Above: Trajectories of single AAV particles labeled with dye Cy5 indicating infectious entry pathways of AAVs into a living HeLa cell. The traces describe various stages of AAV infection (e.g., diffusion in solution (1 and 2), touching at the cell membrane (2), penetration of the cell membrane (3), diffusion in the cytoplasm (3 and 4), penetration of the nuclear envelope (4), and diffusion in the nucleoplasm). Below: Transport of AAV-Cy5 within the nuclear area. (**a**) The visualization of five trajectories projected onto the white light image of the cell nucleus. In this case, all trajectories showed uni-directional motion from the left to the right side. (**c**) The mean square displacement plotted with time. (Reprinted with permission from Seisenberger et al. 2001)

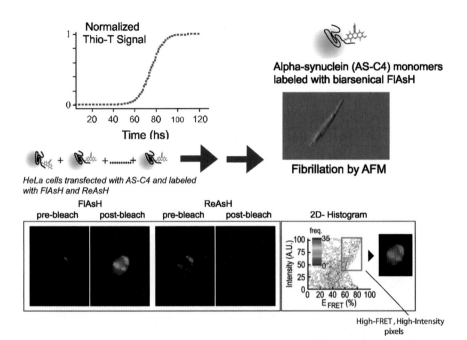

Color Plate 13 Observation of α-synuclein aggregates with site-specific labeling with biarsenical dyes (Roberti et al. 2007). Aggregation kinetics can be observed with Thio-T dye (showing latent phase, aggregation phase and saturation), and the formed fibrils are visualized with atomic force microscope (AFM). Biarsenical dyes FlAsH and ReAsH can be attached to terminal 12-mer peptide sequence of α-synuclein. In an aggregate, both FRET between FlAsH and ReAsH and the reduction of FRET by bleaching the acceptor (ReAsH) can be observed as the enhancement of donor (FlAsH) emission in a microscope image (Courtesy of Dr. E.A. Jares-Erijman, Universidad de Buenos Aires, Argentina)

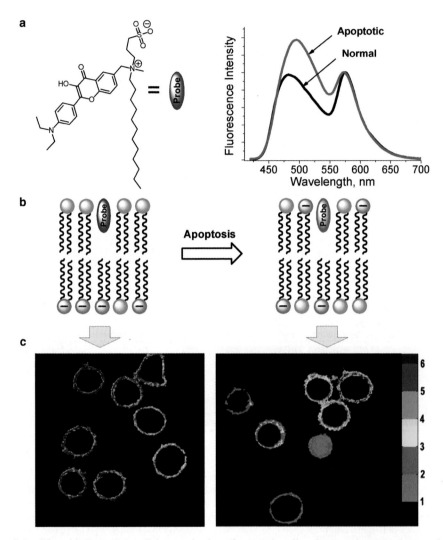

Color Plate 14 Detection and characterization of apoptotic cells using membrane-incorporated dye (Shynkar et al. 2007). (**a**) Structure of 3-hydroxychromone wavelength-ratiometric dye F2N12S (probe) and its fluorescence spectra in normal and apoptotic cells. (**b**) Cartoon illustrating the mechanism of fluorescence response. Normal cells exhibit asymmetry in lipid composition between inner and outer leaflet of their membranes: anionic lipids are oriented inside the cell. In apoptotic cells they appear in the outer leaflet. The dye incorporates spontaneously into outer leaflet and reports on the appearance of anionic lipids. Its response is based on the ability to sense the change of surface potential and of hydration in this leaflet. (**c**) Confocal fluorescence ratiometric images of human lymphoid CEM T cells and the same cells with induced apoptosis by treatment with actinomycin D. The ratios of intensities of the T* band to those of the N* band are displayed in pseudocolor by using the color code on the right scale

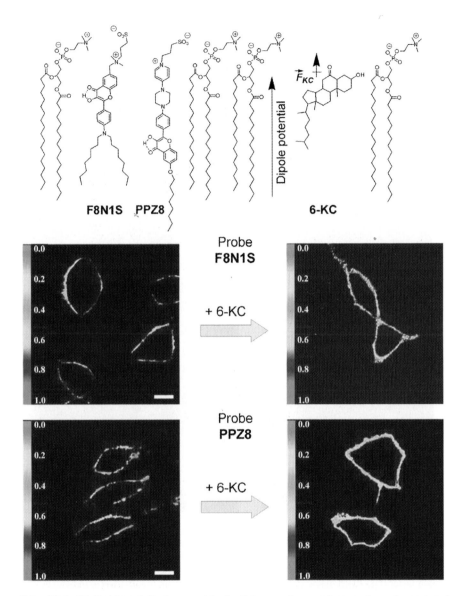

Color Plate 15 Probing of dipole potential of cellular membrane using membrane-incorporated dyes (Shynkar et al. 2005). Schematic representation of lipid monolayer with four phospholipids (above) shows the location in opposite orientations of two wavelength-ratiometric 3-hydroxyflavone dyes F8N1S and PPZ8. Being electrochromic, they respond to the change of membrane electric field by spectroscopic changes in different directions. Also depicted is the 6-ketocholestanol (6-KC) molecule that being a strong dipole can incorporate spontaneously into the membrane and increase its dipole potential. In confocal microscopic images it is clearly seen that treatment of cells with 6-KC produces opposite effect in changing the color of fluorescence emission of two dyes

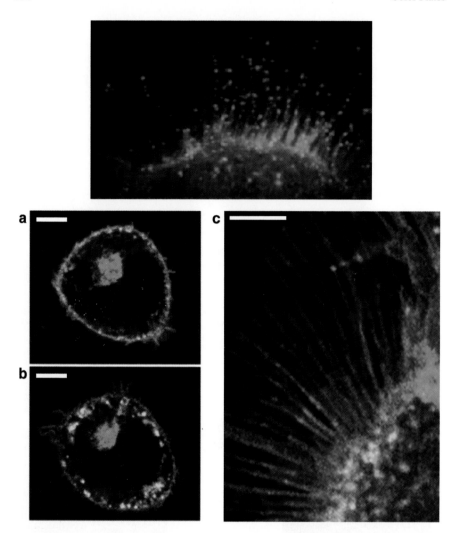

Color Plate 16 Visualization of epidermal growth factor receptors on the surface of carcinoma cells (Lidke et al. 2005). Top: A Programmable Array Microscope (PAM) image showing a maximum intensity projection of 20 optical sections of a living human carcinoma cell expressing a GFP-fusion protein of epidermal growth factor receptor (green) binding Quantum Dot-epidermal growth factor (QD-EGF) ligand (seen as red spots). The binding of QD-EGF to individual EGF receptors on the filopodia and the cell membrane is observed. Bottom: Single confocal sections of living cells showing QD-EGF ligand binding to the surface expressed membrane receptor and subsequent activation. The cells express a green fluorescent protein-fused EGF receptor. (**a**) Red QD-EGF and green GFP-EGF receptor co-localize on the cell membrane 3 min after addition of ligand and (**b**) are rapidly internalized together into endosomes in the cytoplasm. (**c**) Individual receptors labeled with QD-EGF on long sensory filaments called filopodia can be visualized and seen to transport to clathrin coated pits for internalization at the cell body (Courtesy of Dr. D. Amdt-Jovin, Gottingen, Germany)